CLOSE BINARIES IN THE 21ST CENTURY: NEW OPPORTUNITIES AND CHALLENGES

Edited by:

ÀLVARO GIMÉNEZ
ESA-ESTEC, Noordwjk, The Netherlands

EDWARD GUINAN
Villanova University, USA

PANAGIOTIS NIARCHOS
University of Athens, Greece

SLAVEK RUCINSKI
University of Toronto, Canada

Reprinted from *Astrophysics and Space Science*
Volume 304, Nos. 1–4, 2006

Library of Congress Cataloging-in-Publication Data is available

ISBN 1-4020-5026-8 (hardbook)
ISBN 1-4020-5027-5 (eBook)
ISBN 978-1-4020-5026-8 (hardbook)
ISBN 978-1-4020-5027-5 (eBook)

Published by Springer,
P.O. Box 17, 3300 AA Dordrecht, The Netherlands.

Cover figure: A view of the town of Ermoupolis, Syros, Greece

Printed on acid-free paper

All Rights Reserved
© Springer 2006
No part of the material protected by this copyright notice may be reproduced or utilized in any form or by any means, electronic or mechanical, including photocopying, recording or by any information storage and retrieval system, without written permission from the copyright owner.

Printed in the Netherlands

TABLE OF CONTENTS

Àlvaro Giménez, Edward Guinan, Panagiotis Niarchos and Slavek Rucinski / Preface	1–2
Edward F. Guinan and Scott G. Engle / The Brave New World of Binary Star Studies	3–9
E. Budding, C. Bembrick, B.D. Carter, N. Erkan, M. Jardine, S.C. Marsden, R. Osten, P. Petit, M. Semel, O.B. Slee and I. Waite / Multisite, Multiwavelength Studies of the Active Cool Binary CC Eri	11–14
Antonio Frasca, Ettore Marilli, Patrick Guillout, Rubens Freire Ferrero, Elvira Covino and Juan M. Alcalá / Late-type X-ray Emitting Binaries in the Solar Neighborhood and in Star Forming Regions	15–18
Àlvaro Giménez / Analysis of Extra-Solar Planetary Transits Using Tools from Eclipsing Binaries	19–22
Qian Shengbang, Yang Yuangui, Zhu Liying, He Jiajia and Yuan Jingzhao / Photometric Studies of Twelve Deep, Low-mass Ratio Overcontact Binary Systems	23–26
Zeynep Bozkurt and Ömer Lütfi Değirmenci / Updated UBV Light-Curve and Period Analysis of Eclipsing Binary HS Herculis	27–31
H.V. Şenavcı, B. Albayrak, S.O. Selam and T. Ak / The Period Variation of V839 Oph	33–35
Pavel Mayer, Marek Wolf, P.G. Niarchos, K.D. Gazeas, V.N. Manimanis and Drahomír Chochol / Investigation of Times of Minima of Selected Early-Type Eclipsing Binaries	37–39
R.R. Shobbrook and S. Zola / Photometric Studies of Bright Southern Binary Systems: ϵ Cra and ψ Ori	41–43
S. Nesslinger, H. Drechsel, R. Lorenz, P. Harmanec, P. Mayer and M. Wolf / Photometric Solution of the O-type Eclipsing Binary V1007 Sco	45–47
T. Hegedüs, J. Nuspl, N. Markova, H. Markov, H. Rovithis-Livaniou, I. Vince and J. Vinkó / First Results of the Central-East-South European Binary Star Study Group (CESEB)	49–51
Zsolt Kővári, Katalin Oláh, János Bartus, Klaus G. Strassmeier and Thomas Granzer / Spot Modelling of ζ Andromedae	53–55
M.C. Gálvez, D. Montes, M.J. Fernández-Figueroa and J. López-Santiago / Chromospheric Activity and Orbital Solution of Six New Late-type Spectroscopic Binary Systems	57–59
V.N. Manimanis and P.G. Niarchos / The Near-Contact System V387 Cygni: BVRI Photometry and Modelling	61–63
K. Yüce, S.O. Selam, B. Albayrak and T. Ak / Monitoring Secular Orbital Period Variations of Some Eclipsing Binaries at the Ankara University Observatory	65–67
J.M. Kreiner, S. Zola, W. Ogloza, B. Pokrzywka, M. Drozdz, G. Stachowski, B. Zakrzewski, P.G. Niarchos, K. Gazeas, S.M. Rucinski, M. Siwak, D. Koziel, D. Kjurkchieva and D. Marchev / The W UMa-type Stars Program: First Results, Current Status and Perspectives	69–71
Peter P. Eggleton and Ludmila Kisseleva-Eggleton / A Mechanism for Producing Short-Period Binaries	73–77
Kazimierz Stępień / Evolutionary Status of Late-type Contact Binaries: Case of a $1.2 + 1\ M_\odot$ Binary	79–82
C. Foellmi, A.F.J. Moffat and S.V. Marchenko / Wind Ionization Structure of the Short-Period Eclipsing LMC Wolf-Rayet Binary BAT99-129: Preliminary Results	83–86
I. Ribas / Masses and Radii of Low-Mass Stars: Theory Versus Observations	87–90
D. Chochol, T. Pribulla, M. Vaňko, P. Mayer, M. Wolf, P.G. Niarchos, K.D. Gazeas, V.N. Manimanis, L. Brát and M. Zejda / Light-Time Effect in the Eclipsing Binaries GO Cyg, GW Cep, AR Aur and V505 Sgr	91–94

Stephen Skinner, Manuel Güdel, Werner Schmutz and Svetozar Zhekov / X-ray Observations of Binary and Single Wolf-Rayet Stars with XMM-Newton and Chandra — 95–97

Roman Rodica and Mioc Vasile / Libration Points in Schwarzschild's Circular Restricted Three-Body Problem — 99–101

Laurits Leedjärv / Symbiotic Stars in the Context of Binary Evolution and Metallicity — 103–105

S. Zola, K. Gazeas, J.M. Kreiner and B. Zakrzewski / Properties of Components in Contact Systems — 107–109

A. Kalimeris and H. Livaniou-Rovithis / On ($\dot{m} - \dot{P}$) and ($\dot{J} - \dot{P}$)-Type Relations for Close Binaries — 111–113

S. Tsantilas, H. Rovithis-Livaniou and G. Djurasevic / Radiation Pressure and Surface Gravity of Close Binaries: First Results from Observational Data Analysis — 115–117

Waldemar Ogloza and Bartlomiej Zakrzewski / The Spatial Distribution of W UMa-Type Stars — 119–121

K.D. Gazeas, P.G. Niarchos and G.-P. Gradoula / Modeling the 2004.5 Brightening of the Contact Binary V839 Oph — 123–125

E. Antonopoulou, N. Nanouris, G. Kontogiannis, G. Giannikakis and O. Dionatos / First Photometric Observations and Preliminary Results of Binaries FO Aurigae and V1025 Herculis — 127–128

B. Albayrak, S.O. Selam, K. Yüce, M. Helvacı and T. Ak / Monitoring Possible Light-Time Effect in the Orbital Period of Some Eclipsing Binaries at the Ankara University Observatory — 129–131

M. Vaňko, J. Tremko, T. Pribulla, D. Chochol, Š. Parimucha and J.M. Kreiner / Distributions of Geometrical and Physical Parameters of Contact Binaries — 133–135

Christiaan Sterken / Some Thoughts on Pulsation and Binarity — 137–141

Katalin Oláh / Active Longitudes in Close Binaries — 143–146

Gösta Gahm / Enhanced Activity in Close T Tauri Binaries — 147–150

S.V. Jeffers / Modelling Eclipsing Binaries with Dense Spot Coverage — 151–154

D.M.Z. Jassur, M.H. Kermani and G. Gozaliasl / Orbital Period Changes and Long Term Luminosity Variation in Active Binary CG Cyg — 155–158

Miloslav Zejda, Zdeněk Mikulášek, Marek Wolf and Ondřej Pejcha / Short Time-Scale Variability in the Light Curve of TW Draconis — 159–161

Stephen Skinner, Manuel Güdel, Kevin Briggs, Stanislav Melnikov and Marc Audard / X-ray Emission from the Pre-Main Sequence Systems FU Orionis and T Tauri — 163–165

D. Mkrtichian, S.-L. Kim, A.V. Kusakin, E. Rovithis-Livaniou, P. Rovithis, P. Lampens, P. van Cauteren, R.R. Shobbrook, E. Rodriguez, A. Gamarova, E.C. Olson and Y.W. Kang / A Search for Pulsating, Mass-Accreting Components in Algol-Type Eclipsing Binaries — 167–169

Pedro J. Amado, Susana Martín-Ruíz, Juan Carlos Suárez, Armando Arellano Ferro, Andrés Moya, Ignasi Ribas and Ennio Poretti / HD 172189, a Cluster Member Binary System with a δ Scuti Component in the Field of View of COROT — 171–173

Petr Zasche, Miloslav Zejda and Luboš Brát / Eclipsing Binaries with Possible Light-Time Effect — 175–177

Marek Wolf, P.G. Niarchos, K.D. Gazeas, V.N. Manimanis, Lenka Kotková, Anton Paschke and Miloslav Zejda / Eccentric Eclipsing Binary YY Sagittarii — 179–181

Thierry Morel, Giuseppina Micela and Fabio Favata / Photospheric Abundance Peculiarities in RS CVn Binaries — 183–185

A. Elmaslı, S.O. Selam, B. Albayrak and D. Özuyar / Possible Light-Time Effect in the Eclipsing Binary System TY Boo — 187–188

Sergio Messina and Edward F. Guinan / Preliminary Photometric Evidence of Starspot Umbrae and Penumbrae on Late-Type Active Stars	189–191
T.H. Dall, K.G. Strassmeier and H. Bruntt / Late-Type Active Stars: Rotation & Companions	193–195
John Southworth and Jens Viggo Clausen / Eclipsing Binaries in Open Clusters	197–200
R.W. Hilditch / Eclipsing Binaries in Local Group Galaxies	201–204
A.Z. Bonanos, K.Z. Stanek, R.P. Kudritzki, L. Macri, D.D. Sasselov, J. Kaluzny, D. Bersier, F. Bresolin, T. Matheson, B.J. Mochejska, N. Przybilla, A.H. Szentgyorgyi, J. Tonry and G. Torres / The First DIRECT Distance to a Detached Eclipsing Binary in M33	205–207
S. Nesslinger and H. Drechsel / Absolute Parameters of Early-Type Close Binaries in the LMC	209–212
Russell Robb, John Vincent and Joanne Rosvick / Searching for Planetary Eclipses in Open Clusters: NGC 7086	213–215
Pierre North / Binaries with Total Eclipses in the LMC: Potential Targets for Spectroscopy	217–219
Ian Todd, Don Pollacco, Ian Skillen, D.M. Bramich, Steve Bell and Thomas Augusteijn / Eclipsing Binary Stars in Nearby Galaxies	221–223
J. Pritchard, J.B. Marquette, Patrick Tisserand, J.P. Beaulieu, E. Lesquoy and A. Milsztajn / EROS-II Variable Stars: Eclipsing Binaries in the EROS Microlensing Surveys	225–227
I. Ribas, J.C. Morales, C. Allende Prieto, C. Jordi, D.H. Bradstreet and S.J. Sanders / First Results from ROTES: The ROtse Telescope Eclipsing-binary Survey	229–231
Frédéric Pont and François Bouchy / Planets and Planet-Sized Binaries from the OGLE Transit Survey	233–236
Genya Takeda and Frederic A. Rasio / Eccentricities of Planets in Binary Systems	237–240
Keith Hsuan and Rosemary A. Mardling / A Three Body Solution for the DI Her System	241–244
Valentin D. Ivanov, G. Chauvin, C. Foellmi, M. Hartung, N. Huélamo, C. Melo, D. Nürnberger and M. Sterzik / Common Proper Motion Search for Faint Companions Around Early-Type Field Stars – Progress Report	245–247
R.U. Claudi, M. Cancian, M. Barbieri, R. Gratton, S. Desidera, M. Montalto, G.P. Piotto and S. Scuderi / The Asiago Extrasolar Planet Transit Search	249
D. Pollacco, I. Skillen, A. Cameron, D. Christian, J. Irwin, T. Lister, R. Street, R. West, W. Clarkson, N. Evans, A. Fitzsimmons, C. Haswell, C. Hellier, S. Hodgkin, K. Horne, B. Jones, S. Kane, F. Keenan, A. Norton, J. Osborne, R. Ryans and P. Wheatley / The WASP Project and SuperWASP Camera	251–253
Vassiliki Kalogera / X-Ray Binaries in Nearby Galaxies	255–259
A.M. Cherepashchuk / Synthesis of Line Profiles and Radial Velocity Curves for X-Ray Binary Systems	261–264
Vladislav Pustynski and Izold Pustylnik / Hot Subdwarfs in Precataclysmic Binaries	266–268
Linda Schmidtobreick and Katherine Blundell / On the Binary Nature of SS 433	269–272
D.V. Bisikalo, P.V. Kaygorodov, A.A. Boyarchuk and O.A. Kuznetsov / On Possible Nature of Pre-eclipse Dips in Light Curves of Semi-detached Systems with Steady-state Disks	273–276
L.M. Dray, J.E. Dale, M.E. Beer, A.R. King and R. Napiwotzki / After the Supernova: Runaway Stars and Massive X-ray Binary Populations with Metallicity	277–280
Gerardo J.M. Luna, J.L. Sokoloski and Roberto D.D. Costa / X-rays from the Symbiotic Star RX Pup	281–283
Irina Voloshina and Valerian Sementsov / Results of Optical Monitoring of Cataclysmic Variable GK Per in 2004	285–287

Fabíola Mariana Aguiar Ribeiro and Marcos Diaz / Tomographic Simulations of Accretion Disks in Cataclysmic Variables – Flickering and Wind — 289–291

Thierry Morel and Yves Grosdidier / X1908+075: A Late O-Type Supergiant with a Neutron Star Companion — 293–295

Claus Tappert, Boris T. Gänsicke, Ronald E. Mennickent and Linda Schmidtobreick / Spectroscopy of the Candidate Pre-CV LTT 560 — 297–299

Linda Schmidtobreick, Claus Tappert, Alessandro Ederoclite, Ivo Saviane and Elena Mason / The Enigmatic Behaviour of the Old Nova RR Pic — 301–303

L. Hric, R. Gális, P. Niarchos and K. Gazeas / Symbiotic Binary YY Her – Eclipses, Flares and Outbursts — 305–307

Dubravka Kotnik-Karuza, Michael Friedjung and Patricia Whitelock / Analysis of Near Infrared Observations of the Symbiotic Mira RR Tel — 309–311

Christos Papadimitriou, Emilios Harlaftis and Danny Steeghs / The Structure and the Partial Opacity of the Spiral Shocks on the Novalike Cataclysmic Variables — 313–314

Michael Friedjung, Andrej Dobrotka, Alon Retter, Ladislav Hric and Rudolf Novák / Deductions from the Indication of Eclipses of Nova V1493 Aql — 315–317

O. Giannakis, E.T. Harlaftis, P.G. Niarchos, S. Kitsionas, H. Barwig, M. Still and R.G.M. Rutten / Mapping of the Disc Structure of the Neutron Star X-ray Binary X1822-371 — 319–320

Slavek M. Rucinski / The DDO Short-Period Binary RV Program — 321–325

K. Pavlovski, D.E. Holmgren, P. Koubský, J. Southworth and S. Yang / Abundances from Disentangled Component Spectra of Close Binary Stars: An Observational Test of an Early Mixing in High-Mass Stars — 327–330

Klaus G. Strassmeier / Doppler Imaging of Rapidly-Rotating M Stars — 331–334

Petr Hadrava / Disentangling of the Spectra of Binary Stars – Principles, Results and Future Development — 335–339

Tsevi Mazeh, Omer Tamuz and Pierre North / Automated Analysis of Light Curves of OGLE LMC Binaries: The Period distribution — 341–344

A. Prša and T. Zwitter / Disentangling Effective Temperatures of Individual Eclipsing Binary Components by Means of Color-Index Constraining — 345–348

J. Devor and D. Charbonneau / A Method For Eclipsing Component Identification In Large Photometric Datasets — 349–352

Szilárd Csizmadia, Zsolt Kővári and Péter Klagyivik / Hα Photometry of Two Contact Binaries — 353–355

T. Borkovits, Sz. Csizmadia and L. Patkós / A Library of Eclipsing Binary Light-Curves: Fast Initial Parameters Estimation, and on the Uniqueness of the Inverse Problem — 357–359

Zdeněk Mikulášek, Marek Wolf, Miloslav Zejda and Petra Pecharová / On Methods for the Light Curves Extrema Determination — 361–363

D. Montes, M.C. Gálvez, M.J. Fernández-Figueroa and I. Crespo-Chacón / LU Vel (GJ 375): A M3.5Ve Flare and Double-Lined Spectroscopic Binary — 365–367

S.V. Jeffers, A. Collier Cameron, J.R. Barnes and J.P. Aufdenberg / Dense Spot Coverage and Polar Caps on SV Cam — 369–371

Zsolt Kővári, Katalin Oláh, János Bartus, Klaus G. Strassmeier, Michael Weber, Albert Washuettl, John B. Rice and Szilárd Csizmadia / Doppler Images of ζ Andromedae — 373–375

Margarita Karovska / Future Prospects for Ultra-High Resolution Imaging of Binary Systems at UV and X-ray Wavelengths — 377–380

Carla Maceroni and Ignasi Ribas / The Impact of CoRoT on Close Binary Research — 381–384

P.G. Niarchos / On the Gaia Expected Harvest on Eclipsing Binaries — 385–388

D. Koch, W. Borucki, G. Basri, T. Brown, D. Caldwell, J. Christensen-Dalsgaard, W. Cochran, E. Dunham, T. N. Gautier, J. Geary, R. Gilliland, J. Jenkins, Y. Kondo, D. Latham, J. Lissauer and D. Monet / The *Kepler Mission*: Astrophysics and Eclipsing Binaries — 389–393

Klaus G. Strassmeier and The AIP STELLA team / STELLA: Two New Robotic Telescopes for Binary-Star Research — 395–398

Guy S. Stringfellow and Frederich M. Walter / Getting SMARTS on Novae: Highlights of the Early Evolution of Nova V475 Sct — 399–401

- Formation and Evolution of Close Binaries: Star + Star Systems and Planet + Star Systems
- Binary Stars as Astrophysics Laboratories
- Binaries in Clusters and Nearby Galaxies
- Binary Star-Planet Systems: New Developments and New Things learned
- Recent Developments for Close Binaries with Degenerate Components
- New and Improved Tools for Analyzing Light and Radial Velocity Observations
- Future expectations: Impacts of New Instrumentation and Technologies on Close Binary Star/Exoplanet Research

The conference was organized under the auspices of Commission 42 of the IAU, the European Space Agency (ESA), the National and Kapodistrian University of Athens, Greece, the University of the Aegean, Greece, and the Hellenic Astronomical Society. This meeting could not be realized without the financial support of the following sponsors: National and Kapodistrian University of Athens, Greece; University of the Aegean, Greece; Hellenic Astronomical Society; Municipality of Ermoupolis, Syros; Greek Ministry of National Education and Religious Affairs; Alpha-Bank, and Plaisio Computers AEBE. Their generous financial support is deeply acknowledged.

We appreciate very much the hospitality of the town of Ermoupolis, which offered its splendid cultural centre as a really exceptional meeting site. The success of the meeting must also be ascribed to the very pleasant surroundings and the warm hospitality offered by the town of Ermoupolis, capital of Syros, its official representatives, and its citizens. We also like to thank the Union Local Authorities Cyclades for a marvellous reception held in the premises of the cultural centre. Thanks are also due to Dr. M. Stefanakis of the University of Aegean for his interesting lecture on the ancient culture of the Aegean islands, delivered before the reception. Special thanks are due to Mr. Constantinos Stathoulis who prepared and maintained the web site of the conference. What remains for us is to acknowledge the precious help and advice provided by our colleagues in the Scientific and Local Organizing Committees, and the Secretariat (Anastasia Niarchou, Chloe Vamvatira-Nakou and Georgia-Peristera Gradoula), who contributed to the success of such an unforgettable meeting.

The Brave New World of Binary Star Studies

Edward F. Guinan · Scott G. Engle

Received: 23 February 2006 / Accepted: 6 June 2006
© Springer Science + Business Media B.V. 2006

Abstract In this paper we discuss some of the new and exciting developments in the study of binary stars. Recent technological advances (such as CCDs) now make it possible (even easy) to study faint, astrophysically important binaries that in the past could only be done with large 4 + meter class telescopes. Also, the panoramic nature of CCDs (and the use of mosaics), permit large numbers of stars to be imaged and studied. At this conference, most of the observational material discussed was secured typically with smaller aperture 0.5 – 2 m telescopes. Excellent examples are the discovery of over 10^4 new ∼13 – 20 mag eclipsing (and interacting) binaries now found in nearby galaxies from the EROS, OGLE, MACHO and DIRECT programs. As briefly discussed here, and in more detail in several papers in this volume, a small fraction of these extragalactic eclipsing binaries are now serving as "standard candles" to secure accurate distances to the Magellanic Clouds, as well as to M31 and M33. Moreover, the discovery of increasingly larger numbers of eclipsing binaries has stimulated the development of automatic methods for reducing and analyzing the light curves of thousands of systems. In the near future, hundreds of thousands (possibly millions) of additional systems are expected to be discovered by Pan-STARRS, the Large Synoptic Survey Telescopes (LSST), and later by GAIA. Over the last decade, new classes of binary systems have also been found which contain Jupiter-size planets and binaries containing pulsating stars. Some examples of these important binaries are discussed. Also discussed are the increasing numbers (now eight) of eclipsing binary planet–star systems that have been found from high precision photometry. These systems are very important since the radii and masses of the hosted planets can be directly measured. Moreover, from the upcoming COROT and KEPLER missions hundreds of additional transiting planet-star systems are expected to be found. All in all, we hope in this paper to highlight some of the current developments and new directions in the "Brave New World" of binary star studies.

Keywords Eclipsing Binaries

1. Overview of binary and multiple star systems

Surveys and census of stars in the solar neighborhood show that ∼60–70% of all stars belong to binary or multiple star systems (e.g. Abt, 1983). This means that understanding the dynamics and physical properties of stellar systems and possibly developing laws to govern them can have a broad impact on our knowledge of the majority of stars in the universe. The study of binary stars is greatly aided by the presence of eclipses. Obtaining the light curve (when combined with a radial velocity data) of a binary system can yield a number of fundamental properties of both stars in the system and its orbit. Previous studies have found that binary systems are an extremely diverse class of objects. Their periods can be as small as those of the AM CVn binaries – 18 minutes (AM CVn) and 46 minutes (GP Com) – or the double degenerate systems – 321 seconds (RX J0806.3 + 1527) – or as large as (or larger than) the F0 Ia star & disk system ϵ Aur – 27.2 years. The size of their orbits can range from a < 0.05 $R_\odot$ out to ∼28 AU and beyond, and they can include every type of star imaginable, dead or alive, young and old, from PMS to neutron stars & black holes, as well as those whose components consist of planets and brown dwarfs. Also, certain binary stars having accretion discs (such as CVs and X-ray binaries) serve

E. F. Guinan (✉) · S. G. Engle
Villanova University – Department of Astronomy & Astrophysics – 800 E. Lancaster Ave. – Villanova, PA 19085, USA

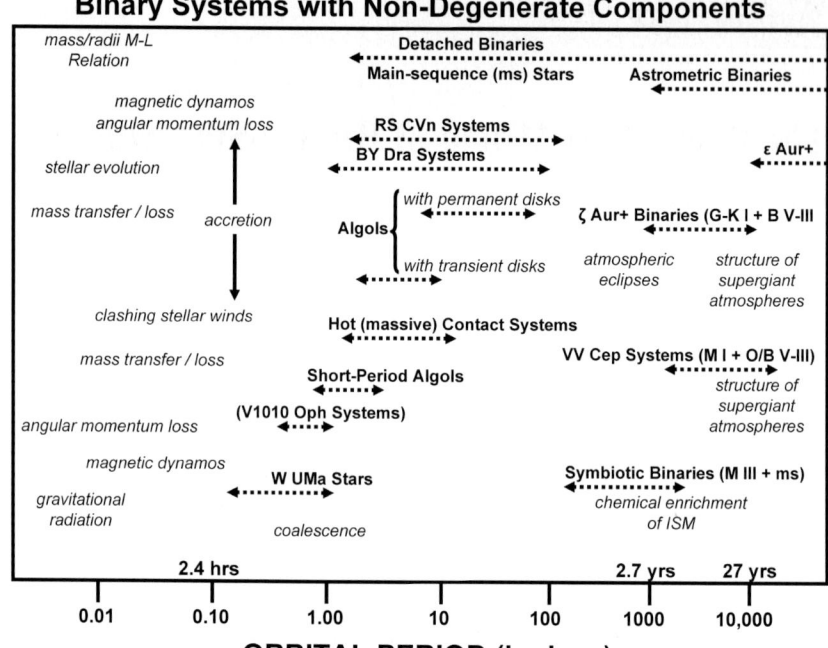

Fig. 1 The period range of all binary systems with non-degenerate components

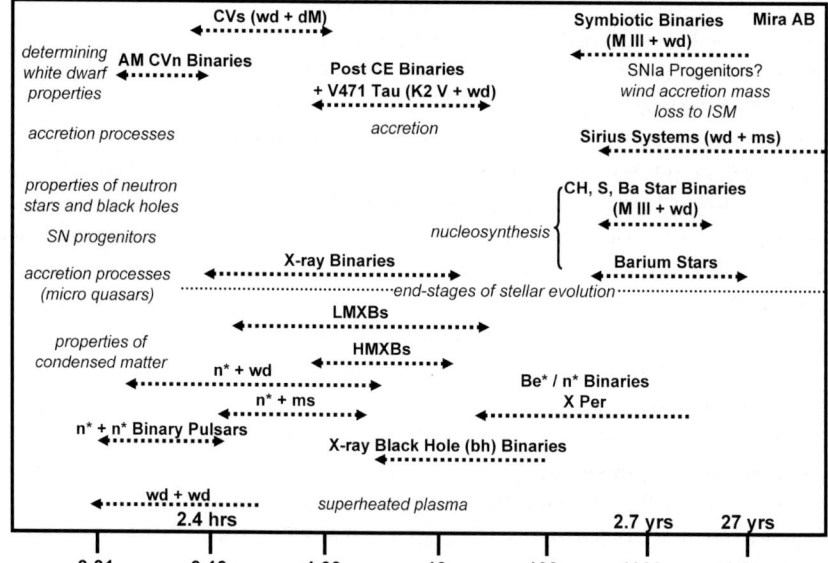

Fig. 2 The period range of all binary systems with degenerate components

as miniature proxies (*micro-quasars*) for studying accretion processes in Seyfert galaxies and quasars.

A broad range of binary systems are known and most types are shown in Figs. 1, 2, and 3, along with their respective impacts on astrophysics. Fig. 1 includes most known types of binaries with non-degenerate components, while Fig. 2 shows binaries with one or two degenerate components (white dwarfs, neutron stars, or black holes). Fig. 3 shows the relatively recent classes of binary star–planet, binary star–brown dwarf and brown dwarf–brown dwarf binary systems. In the figures, the binary systems are plotted according to orbital period (or separation). The shortest period systems are plotted to the left while the binaries with the longest orbital periods are located on the right portion of the figures.

Fig. 3 The period range of all binary systems with planetary or brown dwarf components

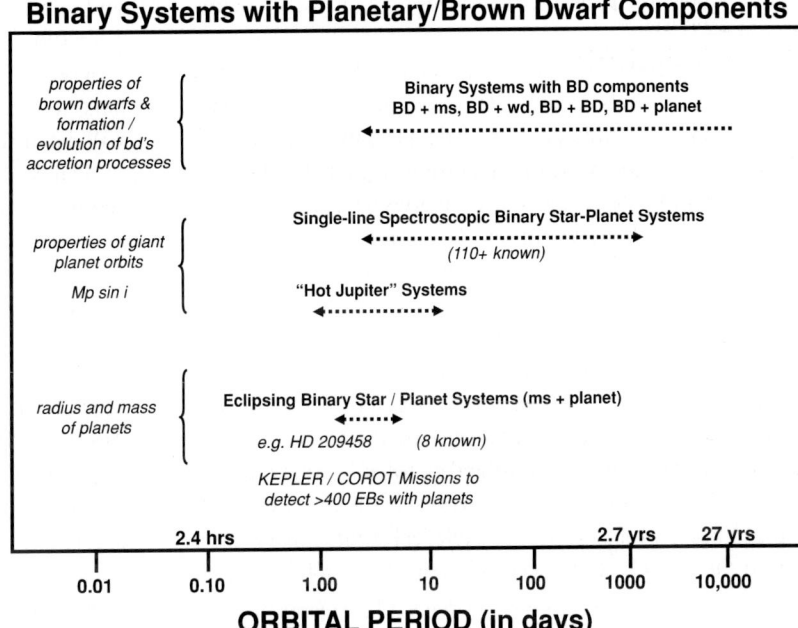

2. The importance of eclipsing binaries to astrophysics

A small but extremely valuable fraction (<0.3%) of all binary stars are favorably aligned in space to produce mutual eclipses of their components. These eclipsing binary stars (EBs) produce characteristic light curves which can be solved; i.e., analyzed by specialized software to determine the parameter values of a realistic model that best fits, in the sense of least-squares, the observed light curve data. Nearly every type of main sequence stars, subgiants, giants and supergiants with the entire range of spectral types and masses represented. Subdwarfs, white dwarfs, neutron stars, black holes, brown dwarfs (L and T dwarfs) and more recently giant planets, e.g., HD 209458b and at this time several others, have been found to be members of eclipsing systems. Further, nearly every type of variable star is a member of an eclipsing binary system. As discussed later, Cepheid variables have been identified as members of eclipsing binaries (EBs) in the Magellanic Clouds (Alcock et al., 2002;). Membership of a pulsating star as a component of an eclipsing binary permits its important physical properties to be determined which, if alone in space, could not otherwise be found.

The information provided by the study of EBs, however, transcends even the fundamental data provided about stellar masses, radii, and luminosities – important as these quantities are to stellar astrophysics. As discussed in several papers in this volume certain types of EBs and systems located in special environments, such as card-carrying members of star clusters and exterior galaxies, have characteristics (or locations) that make them well suited as "astrophysical laboratories" for the study of many diverse and important problems in astrophysics, physics and cosmology (see Guinan et al., 2004; Ribas et al., 2005 for a more thorough discussions). Some of the more easily recognized uses of EBs, providing vital astrophysical information, are in the fields of stellar physics and astrophysics. For example, EBs provide vital information (that would not be known otherwise) about stellar atmospheres (limb darkening, gravity brightening, atmospheric eclipse studies), stellar interiors, structure and convective core over-shooting (through apsidal motion studies of binaries with eccentric orbits), stellar magnetic dynamos and magnetic activity (from X-ray, UV, optical and radio eclipse mapping methods of stellar coronae and chromospheres), and on plasma physics (binaries with accretion disks). Also, the membership of black hole candidates in close binaries lends strong confirmation that these rare objects exist by yielding information about their masses, ages, and X-ray characteristics.

The modeling of stars is the foundation of astrophysics; it is where computations first made a qualitative and quantitative difference to the field. Numerical modeling of single stars is a relatively mature (but by no means complete) sub-discipline in astrophysics today (e.g. see Pols et al., 1997, for a "survey" of models). Two out of three stars, however, occur in binary or multiple star systems (Abt, 1983), and the modeling of these more complex star systems is rife with unsolved problems (e.g. Kisseleva–Eggleton, 2001). Additionally, binary stars play important roles in many current and future areas of astrophysics, including:

- **Providing fundamental stellar astrophysical quantities** (masses, radii, ...) – see Guinan, 1997; and applied to low mass stars – see Ribas and a number of other papers in this volume.

- **Providing tests of binary and single stellar evolution theory** (e.g. Pols et al., 1997; Kisseleva–Eggleton and Eggleton in this volume). Of particular importance are binaries that are members of star clusters (e.g., in this volume, see papers by Kaluzny (close binaries in Globular clusters)) and by Southworth and Clausen (close binaries in open clusters). Also important are binaries with pulsating star components (in this volume – see papers by Sterken, Mkritchian, and Nuspl) and spotted stars (in this volume – e.g. papers by Olah, Gahm, Jeffers and others). Also see Section 3 of this paper where two very rare eclipsing binaries with Cepheid components are discussed.

- **Eclipsing binaries as "standard candles"** – Accurate distance determinations within and outside the Galaxy, which are possible for selected suitable eclipsing binary stars (see Fitzpatrick et al., 2002; Guinan et al., 2004; Harries et al., 2003). In this volume see papers by Bonanos et al.; Nesslinger; North.; Todd et al., Pritchard et al., Kalogera et al. Over the last decade more than ten thousand eclipsing binaries have been discovered mostly in the Magellanic Clouds, the Andromeda Galaxy (M31) and in the Triangulum Galaxy (M33) – see Guinan et al., 2004; Bonanos et al., 2003; Hilditch in this volume. Although thousands of EBs are known and have light curves, only a few dozen systems have the spectroscopic radial velocity curves needed to obtain their fundamental properties and distances. As discussed by Hilditch (this volume), the systems that have double-line spectroscopic radial velocity curves and light curves can be used as first class standard candles to determine accurate distances to nearby galaxies and to calibrate the zero point of the cosmic distance scale. Extremely interesting Local Group galaxies for study include the very metal deficient dwarf galaxies Ursa Minor, Leo I, Draco, and the Sagittarius dwarf spheroidal system (see Grebel, 1997; Guinan et al., 2004).

- **Discovery of eclipsing binary star–planet systems**, e.g., so far eight eclipsing transiting planet–star have been discovered (e.g. see in this volume – papers by Gimenez; Bouchy and Pont; Robb et al.,). However, hundreds of eclipsing star–planet systems are expected to be discovered by the upcoming KEPLER Mission. (http://kepler.nasa.gov/). KEPLER is even capable of discovering the transits of Earth-size planets – in this volume – see Koch; a discussion of the impact on close binaries that will be made by COROT (http://smsc.cnes.fr/COROT/) is discussed by Maceroni in this volume. A more complete discussion is given in Section 3.

- **Exploiting the large numbers of eclipsing binaries being discovered** – New, valuable astrophysical information can now be gleamed from the thousands of new eclipsing binaries so far discovered from wide-field panoramic surveys such as the OGLE program (http://bulge.princeton.edu/~ogle/) and the MACHO program (www.macho.mcmaster.ca/). Also, during the next decade, over one hundred new eclipsing binaries will be discovered from other wide-field surveys such as Pan-STARRS and the Large Synoptic Survey Telescope (www.lsst.org/). In particular, as discussed here by Niarchos, the enormously important future space mission GAIA (www.rssd.esa.int/gaia/) will deliver $\sim 8 \times 10^6$ EBs of which $\sim 10^5$ will have radial velocities, calibrated colors, and distances (Zwitter et al., 2003). New approaches for mining these rich datasets are discussed in this volume by Mazeh et al., Prsa and Zwitter, and Devour & Charbonneau.

- **The rare finds of extragalactic surveys** – The increasingly large samples of extragalactic binaries (now in the thousands) found from photometric surveys are providing unprecedented opportunities to have sufficient samples of binary stars to test theories of binary system formation and evolution. Currently rare systems of interest include: (i) EBs that are much wider than the radii of both components, and so eclipse for less than 1% of an orbit; (ii) EBs of two red giants, that eclipse only every few years; (iii) EBs in which a star pulsates; and (iv) triple systems sufficiently well inclined that eclipses occur in both orbits. There are now only a handful of each of the first three categories, and only one (arguably) of the fourth. But out of 10^7 EBs we can expect perhaps 10^3–10^4 of each category, and can improve enormously the statistical distributions of periods, mass ratios and triple-star parameters.

- **Gravitational-wave astronomy** – gravitational radiation from close binary stars will dominate the signal recorded by the upcoming Laser Interferometer Space Antenna (LISA) mission (lisa.jpl.nasa.gov/) and the Laser Interferometer Gravitational-Wave Observatory (LIGO) project (www.ligo.caltech.edu). This is especially true for ultrashort period double degenerate binaries which should be relatively strong gravity wave sources (e.g. see Corey et al., – this volume).

3. Examples of eclipsing binaries as important astrophysical laboratories

3.1. Eclipsing binary star–planet systems

Over a decade ago, with the discovery of a Jupiter–size planet in a tight ~ 3.524 day orbit around the nearby G2V star – 51 Pegasi, a new class of single–line spectroscopic planet–star binary systems was born (see Mayor and Queloz, 1995; Marcy and Butler, 1995). Using high precision (± 5 m/s and better) spectroscopy, 175 + extrasolar planets have been discovered so far. Most of these have been found by the small Doppler reflex orbital motions of their mostly main-sequence

Fig. 4 Eclipsing binary star–planet systems

Eclipsing Binary Star-Planet Systems:

Eight candidates are now known but 100s are expected in the near future. Photometry, combined with radial velocities, will yield radii and masses for these extrasolar planets.

- HD 209458 (dG0, P=3.524d, M=0.69M_J)
  (Brown et al. 2001 ApJ, 552, 699)
- OGLE-TR-10b (~dG2, P=3.101d, M=0.57M_J)
  (Konacki et al. 2005 ApJ, 624, 372)
- OGLE-TR-56b (~dG2, P=1.212d, M=1.45M_J)
  (Torres et al. 2004 ApJ, 609, 1071)
- OGLE-TR-111b (~dK1, P=4.0d, M=0.53M_J)
  (Pont et al. 2004 A&A, 426, L15)
- OGLE-TR-113b (~dK2, P=1.43d, M=1.35M_J)
  (Bouchy et al. 2004, A&A, 421, L13)
- OGLE-TR-132b (~dF8, P=1.69d, M=1.19M_J)
  (Moutou et al. 2004 A&A, 424, L31)
- TrES-1 (K0 V, P=3.030d, M=0.75M_J)
  (Alonso et al. 2004 ApJ, 613, 153)
- HD 189733 (dK0, P=2.219d, M=1.15M_J)
  (Bouchy et al. 2004 A&A, 444, 15)

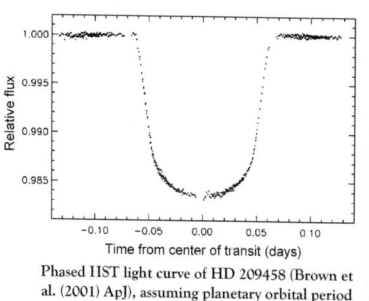

Phased HST light curve of HD 209458 (Brown et al. (2001) ApJ), assuming planetary orbital period of 3.52474 days.

- Kepler/COROT Missions will detect 100s more transiting exoplanets and Kepler, with its μ-mag photometric precision, is even capable of detecting Earth-size planets.
- High-Precision Spectroscopy Gives Atmosphere/Albedo of Planet, and High-Precision Photometry can give R_p / R_* = Reflection Effect, possible Detection of Large Moons

G–K–M host stars caused by the pull of orbiting planets. In terms of planet properties – like mass and orbit size, the majority of these planets typically have masses between ~0.5 – 5 M_J. The shorter period, close-in planets (like 51 Peg b and HD 209458 b) are known as "Hot Jupiters" because of their proximity to the host stars and the expected heated and bloated atmospheres (~500–1500K). Because of the faintness of the planets, these giant planet–star systems are single-line spectroscopic binaries or multiple planet systems. As in the case of their stellar binary big cousins, the spectroscopic studies of these single-line planet–star systems yield the orbital and physical properties such as orbital period, eccentricity, and the product of mass and the sine of the orbital inclination – $M_P \sin^3 i$, as well as the semi-major axis of their orbits. Summaries of the orbital and physical properties of systems with exoplanets are given by the Marcy and Mayor websites. (http://exoplanets.org/).

A major breakthrough in the study of extrasolar planets came with the discovery of shallow (eclipse depth ~0.013 mag) planetary transit eclipses of the single-line planet–star binary – HD209458 (Charbonneau et al., 2000; Henry et al., 2000). Since that time, seven additional eclipsing planet–star systems have been discovered and information about these systems are given in Fig. 4. As in the case of the stellar eclipsing binaries, the occurance of eclipses permits the diameters of the orbiting planets to be determined (relative to the host star). Also, with the orbital inclination known from the light curve, the degenercy in $M_P \sin^3 i$ is broken and the planetary masses are determined (relative to their host stars).

The eight known eclipsing binary planet–star systems are listed in Fig. 4 along with the HST light curve of the transit eclipse of HD 209458 from Brown et al., 2001. Also briefly noted are expected numbers of new systems anticipated from the upcoming COROT and KEPLER space missions. As noted in the figure, these missions will carry out high precision photometry and are expected to find hundreds of additional transiting systems. The KEPLER mission is designed to be able to detect transit eclipses of Earth-size planets. KEPLER is expected to detect up to 50 such systems. Also briefly noted in Fig. 4 are the expectations of high signal-to-noise spectroscopic observations. For example, the stellar spectrum observed in and out of transit can be searched for features imposed by transmission of starlight through the outer parts of the planet's atmosphere (Seager and Sasselov, 2000; Brown, 2001). As indicated in Figs. 3 and 4, eclipsing (or near-eclipsing) systems are important targets to search for secondary eclipses of the heated planet and reflection effects (gives the planet's albedo) using very high precision photometry. More recently Charbonneau et al., (2005) have reported the detection of thermal emission from the "Hot Jupiter" planet orbiting TrES-1 by observing the system with the Spitzer Space Telescope outside and inside the total occultation eclipse of the planet.

3.2. Eclipsing binary systems with Cepheid components

Eclipsing binary systems that contain pulsating star components provide excellent examples of the importance of eclipsing binaries as "astrophysical laboratories." Several papers in this volume focus on these opportunities (see Sterken – this volume – for a review). When these eclipsing binaries are double-line systems, the masses, radii, and luminosities of the pulsating members (as well as the constant members) can be directly determined from the analyses of their light and radial velocity curves. Moreover, the study of pulsating

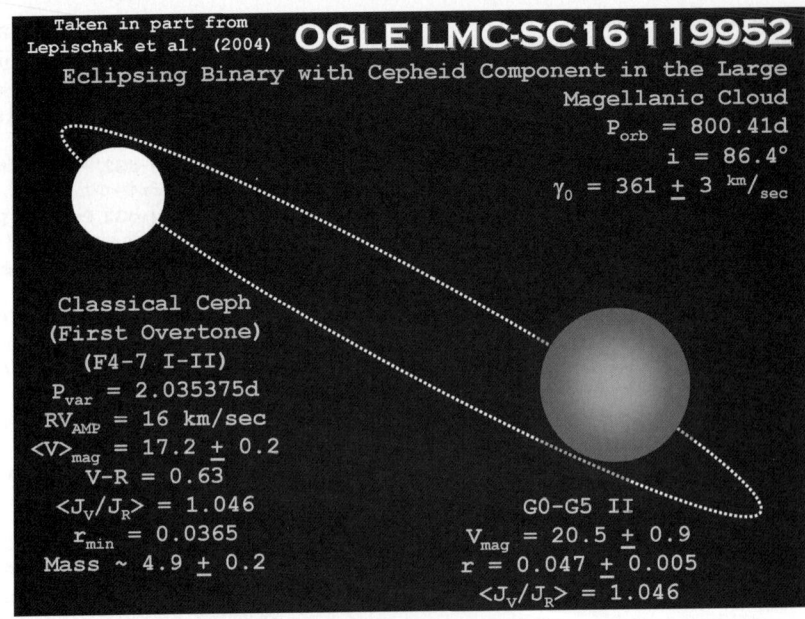

Fig. 5 A model for the LMC EB SC16 which contains a first overtone Cepheid

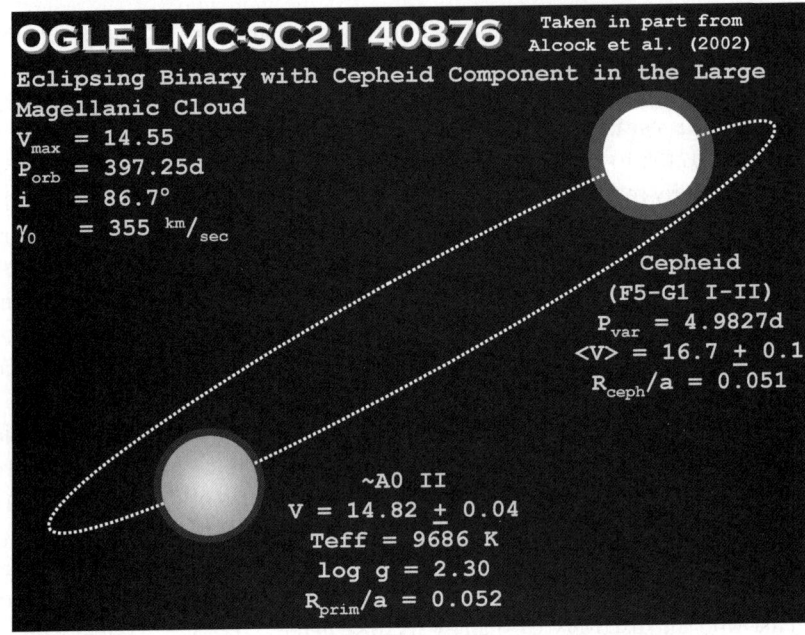

Fig. 6 A model for the LMC EB SC21 which contains a (possible Type II) Cepheid

stars in eclipsing binaries offers a splendid opportunity to investigate the structure and evolution of a wide variety of pulsating variables. The study of these systems provide tests of pulsational theories for the different types of pulsating stars now found in close binaries. Below, a brief discussion is given of two recently discovered eclipsing binaries in the Large Magellanic Cloud that contain Cepheid components.

The OGLE and MACHO microlensing photometry programs of the Magellanic Clouds have serendipitously discovered three eclipsing binaries that have Cepheid components (Udalski et al., 1999; Alcock et al., 2002. The astrophysical and cosmological significance of finding a Cepheid (in particular, a Classical Cepheid) as a member of an eclipsing binary are enormous. Classical Cepheids form the corner stone of the extragalactic distance scale and the determination of the Hubble constant (H_0). Moreover, the study of Cepheids in eclipsing binaries offers important opportunities to investigate the structure and evolution of this important class of pulsating stars since these systems will yield (among other quantities) for the first time the absolute radii (and changes in the radius), masses, luminosities and accurate distances of these important pulsating stars.

Ultraviolet to optical spectrophotometry of the two best-suited systems has been carried out by us (with colleagues E.L. Fitzpatrick, I. Ribas, and F.P. Maloney) with HST/STIS to secure radial velocity measures as well as determinations of accurate values of the temperatures, log g, [Fe/H], and ISM absorptions (see Guinan et al., 2005). These systems are: the 14.5 mag, 397 day LMC EB (SC21 40876 = MACHO 6.6454.5), which contains a Cepheid (Type II ?) with a pulsation period of 4.97 day and the 17th mag, 801 day LMC EB (SC16 119952 = MACHO 81.8997.87) which contains a first overtone Classical Cepheid with a 2.035 day period (cf. Lepischak et al., 2004). The preliminary results of the analysis of HST/STIS spectrophotometry as well as the analysis of the light curves and radial velocity observations are shown in Figs. 5 and 6. Additional radial velocity observations of these systems are planned during 2006/7. These observations will provide an unprecedented opportunity to probe the distance to the LMC, to "self-calibrate" the Period-Luminosity Law of Cepheids and to test the Baade–Wesselink pulsational parallax method. Additionally, the fundamental physical properties and ages of the Cepheids components will also be reliably determined for the first time.

4. Conclusions: The impacts of new or planned missions on the study of close binaries

The exciting recent developments in binary (and multiple) star studies are expected to continue. New opportunities for binary star research are possible (now and in the near future) with high precision photometry, spectroscopy & interferometry. Also, new and vital astrophysical information can be gleamed from the discovery of hundreds of thousands of new eclipsing binaries from wide-field surveys such as the Large Synoptic Survey Telescope (LSST) and Pan-STARR, and later from the GAIA Mission. Also, in the near future ultra-high precision (sub-millimag and micromag) photometry of binary stars will be possible with the COROT and KEPLER missions (as discussed at this meeting), that will lead to new discoveries and new research opportunities. Moreover, binaries of all types (including CVs, X-ray binaries, binary pulsars, chromospherically-active systems, symbiotics etc.) can now be studied across the entire electromagnetic spectrum from γ–rays to the optical with INTEGRAL, XMM, CHANDRA, FUSE, HST, GALEX, and in the infrared with the Spitzer Space Telescope (SST), as well as at radio wavelengths with the VLA and VLBI. Soon, this baseline will be extended into gravity waves with LISA & LIGO.

In the near future the observations of close binaries with more powerful interferometers (both from the ground and from space (e.g. Space Interferometry Mission (SIM) and TPF/Darwin) and with Adaptive Optics Telescopes will permit many close binaries to be resolved, eventually permitting the imaging of the individual stellar surfaces as well as providing images of circumstellar accretion disks (e.g. The Stellar Imager Mission – see Karovska). The availability (at affordable costs) of powerful computers (and networks), and easy access to the World Wide Web are making it possible to "mine" Terabyte astronomical data sets, as well as facilitating the analysis and modeling of the these data. These advances, new technologies and new instruments and the establishment of an International Virtual Observatory, are revolutionizing the study of close binaries and leading to a renaissance and a "Brave New World" of binary star studies. These are, indeed, exciting times to live and to study binary stars.

Acknowledgements We thank Dr. Panos Niarchos for taking the lead in organizing this conference on the beautiful island of Syros, and also for arranging good weather. We wish to also thank the local organizing committee for their enthusiastic and tireless efforts in making this meeting a success, and enjoyable for everyone. This research is supported by NASA HST Grant GO–09176 and NSF/RUI Grant AST05–07542 which we very gratefully acknowledge.

References

Abt, H.A.: ARA& A **21**, 343 (1983)
Alcock, C., et al.: ApJ **573**, 338 (2002)
Bonanos, et al.: AJ **126**, 175 (2003)
Brown, T.M., et al.: ApJ **552**, 699 (2001)
Brown, T.M.: ApJ **553**, 1006 (2001)
Charbonneau, D., et al.: ApJ **626**, 523 (2005)
Charbonneau, D., et al.: ApJ **529**, 45 (2000)
Fitzpatrick, E.L., et al.: ApJ **564**, 260 (2002)
Grebel, E.K.: RvMA **10**, 29 (1997)
Guinan, E.F., et al.: BAAS 37 (in press)
Guinan, E.F., Ribas, I., Fitzpatrick, E.L.: ASPC **310**, 363 (2004)
Guinan, E.F.: ASPC **38**, 1 (1993)
Harries, T.J, Hilditch, R.W., Howarth, I.D.: MNRAS **339**, 157 (2003)
Henry, G.W., et al.: ApJ **529**, 41 (2000)
Kiseleva-Eggleton, L., Eggletong, P.P.: ASPC **228**, 488 (2001)
Lepischak, D., Welch, D.L., van Kooten, P.B.M.: ApJ **611**, 1100 (2004)
Marcy, G., Butler, P.: BAAS **27**, 1379 (1995)
Mayor, M., Queloz, D.: Nature **378**, 355 (1995)
Pols, O.R., et al.: MNRAS **289**, 869 (1997)
Ribas, I., et al.: ApJ **635**, 37 (2005)
Henry, G.W., et al.: ApJ 537, 916 (2000)
Zwitter, T.: ASPC **298**, 329 (2003)

Multisite, Multiwavelength Studies of the Active Cool Binary CC Eri

E. Budding · C. Bembrick · B. D. Carter · N. Erkan ·
M. Jardine · S. C. Marsden · R. Osten · P. Petit ·
M. Semel · O. B. Slee · I. Waite

Received: 7 October 2005 / Accepted: 20 April 2006
© Springer Science + Business Media B.V. 2006

Abstract New data acquired on the active, cool binary CC Eri ranged across the spectrum from Chandra X-ray to broadband photometry and microwave observations using the VLA and ATCA. Also, high-dispersion spectropolarimetry using the AAT enabled Zeeman-Doppler imaging to be performed. Our interpretations infer strong localised concentrations of the stellar magnetic field, manifested by surface activity and related large coronal plasma structures. Comprehensive matching of the modelling parameters awaits more detailed investigation.

This brief interim review includes consideration of the ATCA data. Microwave radio emission is usually low level ('quiescent'), but occasionally flares of several mJy peak intensity are observed. We associate the emission, generally, with wave-like mechanisms, expanding through the outer atmosphere. Related characteristics of this emission are discussed.

Keywords Stars: activity · Stars: coronae · Stars: individual: CC Eri · Radio continuum: stars · Techniques: miscellaneous

1. Summary of multiwavelength observations

A full combining of data to build a single general model was not attempted in this presentation. We should emphasize that this is a preliminary report, given the large body of data involved in the campaign. More detailed accounts will be given later elsewhere. The scheduling allowed overlap with ground-based optical and Chandra X-ray observations. Optical data included ZDI observations at the AAT, made with the University College London échelle spectrograph. A brief review of the evolution of spectropolarimeters applied to ZDI was given by Semel and Lopez Ariste (2001), and the current ZDI instrumental framework can be found from

E. Budding (✉)
Carter Observatory, P.O. Box 2909, Wellington, New Zealand
e-mail: ebudding@comu.edu.tr

E. Budding · N. Erkan
Physics Dept., 18th March University of Canakkale, Turkey

C. Bembrick
Mt Tarana Observatory, PO Box 1537, Bathurst, NSW 2795, Australia

B. D. Carter · I. Waite
Centre for Astronomy, Solar Radiation and Climate, University of Southern Queensland, Baker St., Toowoomba, Qld., Australia

M. Jardine
School of Physics and Astronomy, University of St Andrews, North Haugh, St Andrews, Fife, Scotland

S. C. Marsden
Institute of Astronomy, ETHZ, 8092, Zurich, Switzerland

R. Osten
National Radio Astronomy Observatory, Ivy Rd., Charlottesville, Virginia, USA

P. Petit
Max Planck Institut für Sonnensystemforschung, Max-Planck-Str. 2 37191 Katlenburg-Lindau, Germany

M. Semel
LESIA, Observatoire de Meudon. 5, place Jules Janssen 92195, Paris, France

O. B. Slee
Australia Telescope National Facility, Vimiera & Pembroke Rds., Marsfield 2121, NSW, Australia

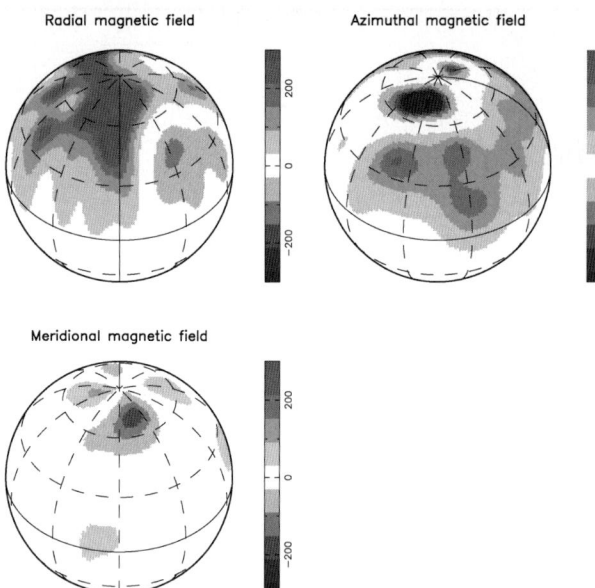

Fig. 1 Maps of the global field distribution over the primary of CC Eri derived from AAT observations using the UCL échelle spectrograph and reduced with ZDI processing software as discussed by Donati et al. (2003)

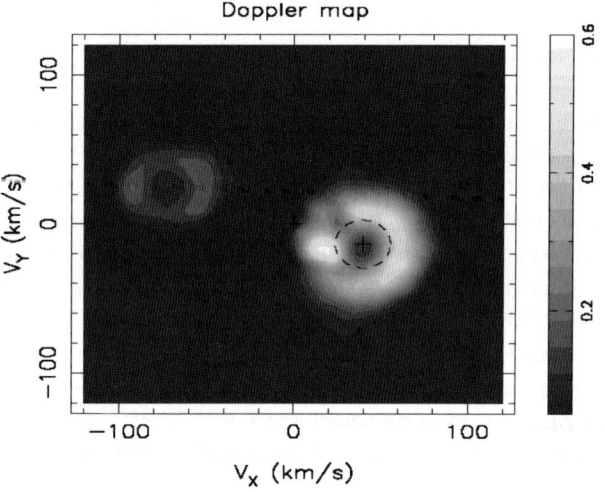

Fig. 2 H_α tomography of the CC Eri system

Donati et al. (2003). Interim results on the radial, meridional and azimuthal field components are shown in Fig. 1. The spectrophotometric work also included prominence mapping done from H_α time-series data. A preliminary map of the binary in velocity space is shown in Fig. 2.

Broadband B optical photometry to monitor the star spot phases and detect impulsive flares came from Mt Tarana Observatory, central western NSW. There was good overall coverage of the 1.56 day phase cycle. A light curve and its heuristic interpretation as a 'starspot' (concentration of maculation) are shown in Fig. 3.

The Chandra observation utilized the High Energy Transmission Grating Spectrometer (HETGS) with the ACIS-S detectors; the HETGS observes from 1.8–30 Angstroms with maximum spectral resolution ∼1000 (Chandra, 2004). The observation lasted for 135ks, but was split into two segments. At the present time we are awaiting the latest software for reducing Chandra data to become available, but preliminary results are shown in Fig. 4. This shows integrations from the medium energy grating (MEG) of Chandra's X-ray transmission system. A comparable Chandra large flare observation on σ^2 CrB with the HETGS was presented by Osten et al. (2003). Dramatic changes clearly occur during an X-ray flare: the continuum enchances noticeably, along with a temporary increase in coronal abundances. An initial model of the X-ray emitting region of the primary star in CC Eri is shown in Fig. 5.

As a result of checking the binary's geometry, a new ephemeris was produced that uses Amado et al.'s (2000) more recent, well-observed conjunction and marginally shorter period. This subtracts about 0.31 from the used CABS phases. There is thus the possibility of an eclipse producing the dip in the flare of Oct 1.35, but a more circumspect interpretation involves the Neupert effect (cf. e.g. McKenzie and Hudson, 1999). A theoretical model, deduced from combining the emission measure with the field distribution, shows the X-ray peak emission regions much closer to the primary photosphere than would correspond to the $\sim 7R_1$ needed to enable any eclipse effect to be observed.

VLA observations were obtained during the same late Sep. to early Oct. period of the campaign. Regarding the Ku band (15 GHz) observations, on 30 Sep. no source was detected, with a 3σ upper limit of 0.45 mJy. A large radio flare occurred on 1 Oct., but the source was undetected on 2 Oct with a 3σ upper limit of 0.39 mJy. The L band observations, do not appear to show a source at the present time, but this requires further confirmation. Ku-band coverage, contemporaneous with the Chandra observations, confirms the flare event in the early part of Oct 1 (inset to Fig. 3).

The ATCA observed a similar event later on the same day (Fig. 6). The 3 nights of ATCA observations were made up of 25-min integrations, separated by 2-min observations of a secondary calibrator, the 4.80 and 8.64 GHz data being recorded simultaneously. The typical error of a flux determination was 0.15 mJy at both frequencies.

The main features of the ACTA data are the high peaks at the end of the second plot and lesser peaks around phases 0.85 and 0.15 (Sep 30.7, Oct 2.7). We interpret these as flares. The rise on Oct 2.7 is seen also in the Chandra observations. The average flux densities increased steadily over three days with the spectral index α (of an assumed power law for the flux at frequency ν, i.e. $S_\nu \sim \nu^\alpha$) being definitely negative on two days, but less so on Oct 1, when a strong, optically thick flare influenced the average.

Typical microwave radio emission of around 1mJy flux density can be explained by a standard gyrosynchrotron

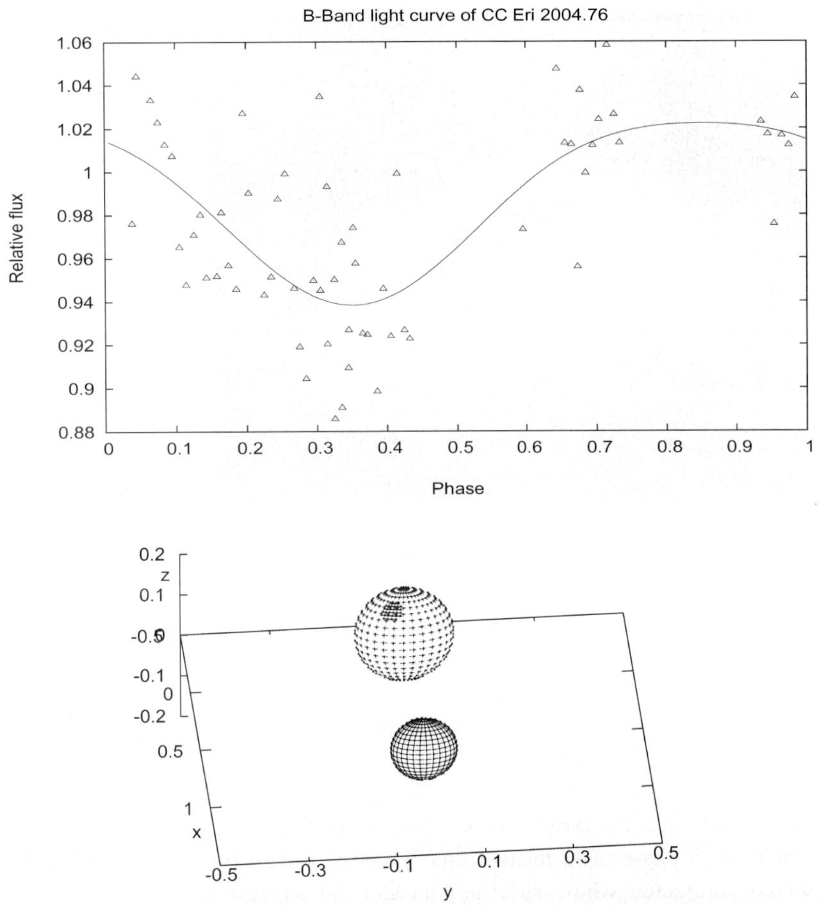

Fig. 3 Optical B photometry of CC Eri from 5 nights data from Mt Tarana Observatory. The diagram shows, as well as the observations, a maculation wave and (GNUPLOT) 3d model of the system, using the results of Amado et al. (2000). Note that the x-scale, along the line of centres of the stars, is contracted in this diagram by ∼4 on the z (polar direction) and y scales. This model shows the maculation at phase 0.047, given by a newly derived ephemeris. Phases, as shown in the light curve, would be advanced on this by 0.31 using the CABS ephemeris (JD 2430001.2905 + 1.56145E a). The new (spectroscopically derived) ephemeris is: JD 2447834.9528 + 1.561492E d.)

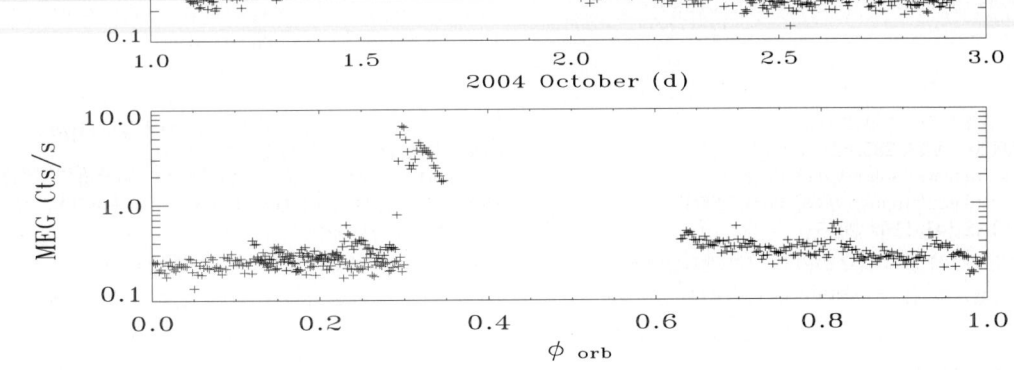

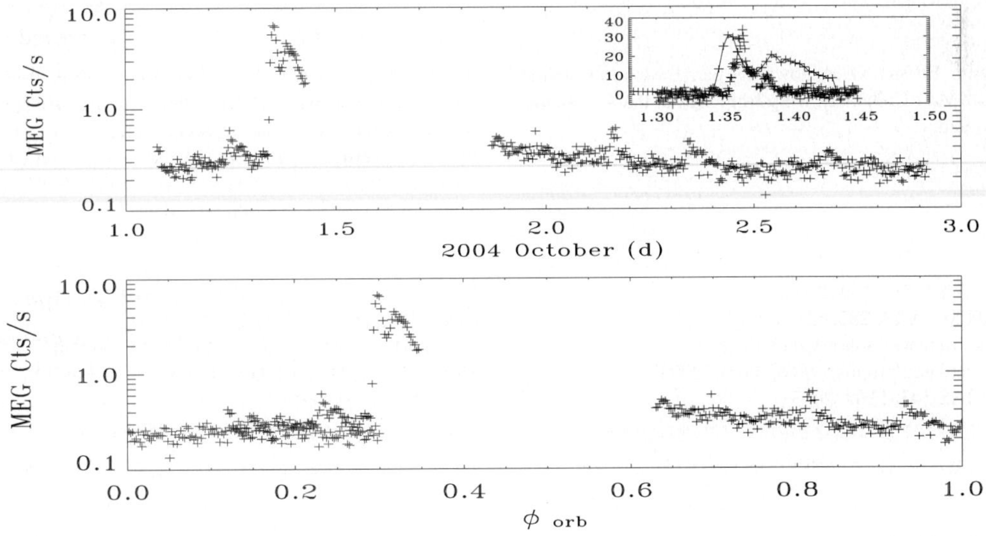

Fig. 4 X-ray light curve integrated from the medium energy grating (MEG) of Chandra's X-ray transmission system. The inset shows contemporaneous VLA coverage during the flare event in the early part of Oct 1 at 15 GHz

model involving coronal magnetic fields, source radii and ambient electron densities in the order of 0.01T, 10^8m and 10^{16} m^{-3}, respectively. There are indications that such sources are expanding. However, there are differences during strong flares. The spectral index then becomes positive and steepens dramatically, and although the known correlation between C and X band flux variations becomes enhanced, the lag effect (C behind X) is reduced. The impression is then of a more compact, denser source, nearer to the stellar surface.

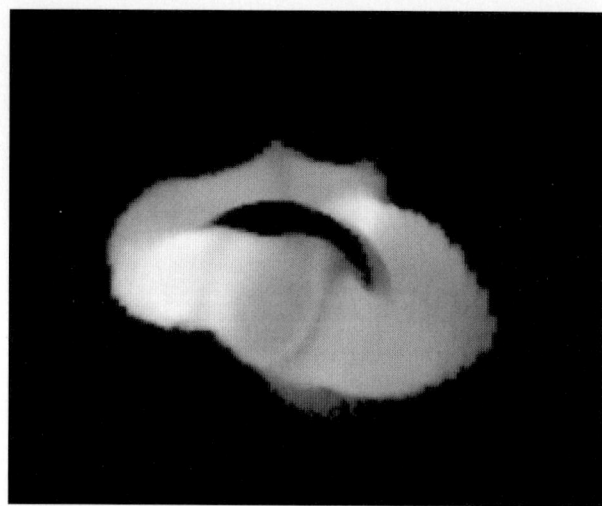

Fig. 5 3-D model of the X-ray emitting region determined from combining flux levels and a calculated field distribution

Derived radio proper motions indicate no significant gravitationally interacting third component to the system (not infrequently found for close binaries)

The general preliminary indications from the Sep-Oct 2004 multi-wavelength survey of CC Eri are thus of an active centre around the phase 0.25 (using the CABS ephemeris), with the possibility of a second concentration around 0.70 phase. The relative flux levels of a flare observed across the whole spectrum on Oct 2.4 support a Benz and Güdel (1994) spectrum, with a power-law index of 0.75.

Acknowledgements EB and NE acknowledge partial support from the Canakkale Onsekiz Mart University. Help from staff at the ATCA was also much appreciated.

References

Amado, P.J. et al.: A&A **359**, 159 (2000)
Benz, A.O., Güdel, M.: A&A **285**, 621 (1994)
Chandra, Chandra Proposers' Observatory Guide,
  http://asc.harvard.edu/proposer /POG/html. (2004)
Donati et al.: MNRAS **345**, 1145 (2003)

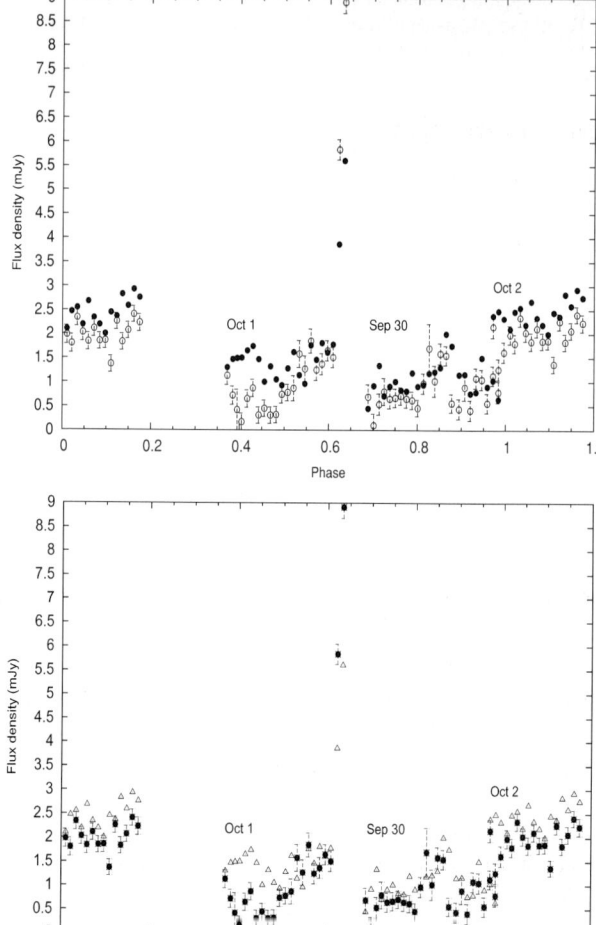

Fig. 6 The flux densities from 25 min integrations. Filled circles denote 4.80 GHz and open circles 8.64 GHz. The start of a significant flare can be seen at the end of the observing interval on Oct 1. Lesser events may be discerned around Sep 30.7 ($\phi \sim 0.85$) and Oct 2.7 ($\phi \sim 0.15$). To avoid confusion, error bars are added only to the 8.64 GHz values. They are similar for the 4.80 GHz data

McKenzie, D.E., Hudson, H.S.: ApJ **519**, L93 (1999)
Osten, R.A. et al.: ApJ **582**, 1073 (2003)
Semel, M., López Ariste, A.: ASP Conf. Ser. **248**, 575. (2001)
Strassmeier, K.G., Hall, D.S., Fekel, F.C., Scheck, M.: (CABS), A&AS **100**, 173 (1993)

Late-type X-ray Emitting Binaries in the Solar Neighborhood and in Star Forming Regions

Antonio Frasca · Ettore Marilli · Patrick Guillout ·
Rubens Freire Ferrero · Elvira Covino · Juan M. Alcalá

Received: 1 November 2005 / Accepted: 1 February 2006
© Springer Science + Business Media B.V. 2006

Abstract We present a spectroscopic and photometric follow-up of binary stars, discovered in a sample of X-ray sources, aimed at a deep characterization of the stellar X-ray population in the solar neighborhood and in Star Forming Regions (SFRs). The sources have been selected from the RasTyc sample, obtained by the cross-correlation between the ROSAT all-sky survey and Tycho catalogues (Guillout et al., 1999). Thanks to the high resolution spectroscopy, we have obtained good radial velocity curves, whose solutions provided us with the mass ratios and minimum masses of the components. We have also obtained an accurate spectral classification with codes specifically developed by us. In addition, we could obtain information on the age of the sources through the LiI-6708 line and on the chromospheric activity level through the Hα line.

We show also some results on very young pre-main sequence (PMS) binaries discovered as optical counterparts of X-ray sources in SFRs. The spectroscopic and photometric monitoring has allowed us to determine the orbital and physical parameters and the rotation periods, that are of great importance for testing the models of PMS evolution.

Keywords Stars: binaries: spectroscopic · stars: X-ray · Stars: pre-main sequence

A. Frasca (✉) · E. Marilli
INAF-Osservatorio Astrofisico di Catania, Italy

P. Guillout · R. F. Ferrero
Observatoire Astronomique de Strasbourg, France

E. Covino · J. M. Alcalá
INAF-Osservatorio Astronomico di Capodimonte (NA), Italy

1. Introduction: Young and active X-ray emitting stars

The study of young and active stars has been traditionally devoted to the members of young open clusters and associations. However, it is expected to find young stars also in the field, since gravitationally-bound open clusters can account only for a few percent of the total galactic star formation rate (cf. Wielen, 1971). This implies that either most clusters and associations disperse very quickly after star formation has started or that most stars are born isolated. Indeed, apart from the typical star formation sites, closely associated with HII regions, OB associations and dark molecular clouds, there is a clear evidence of a huge star forming complex in the solar neighborhood known as the Gould Belt (e.g., Pöppel, 1997; Guillout et al., 1998). Therefore, the study of the sparse population of young stars around the Sun and of its statistical properties is of great importance to have a more coherent picture of the star formation process in our Galaxy.

From the perspective of optical observations, the selection of young solar-type and low-mass stars in the field is a very difficult task, but it becomes much easier to detect them on the basis of the X-ray emission coming from their active coronae. The most comprehensive sample of stellar X-ray sources constructed so far arises from the cross-correlation between the ROSAT all-sky survey (RASS) and the TYCHO mission catalogues that has led to the selection of about 14 000 stellar X-ray sources (RasTyc sample, Guillout et al., 1999). We have surveyed, using high-resolution spectroscopy, a sub-sample of the RasTyc catalogue composed of about 400 late-type northern stars by using the 193-cm and 152-cm telescopes of the *Observatoire de l'Haute Provence* (OHP). A subsequent spectroscopic and photometric monitoring of the more interesting sources has been performed with the 91-cm telescope of the *Osservatorio Astrofisico di Catania* (OAC).

2. The equations

The light curve of a planetary transit can be reduced to the relative loss of light, α, due to the occultation of part of the stellar surface by the planet. The total luminosity of the system, l, at any orbital phase θ, can be expressed as $l(\theta) = 1 - \alpha(\theta)$, where circular discs have been adopted as well as normalized luminosities for the components and no contribution of the planet relative to the star. Moreover,

$$\alpha = \int_S I \cos \gamma \, d\sigma \qquad (1)$$

where the integral is extended over the entire area S of the star occulted by the planet, I stands for the brightness distribution over the star, of surface element $d\sigma$, and γ is the angle between the surface normal and the line of sight.

In order to proceed with the evaluation of (1), a good knowledge of I is needed as given by the well-known effect of limb darkening. Though a linear limb-darkening law was used initially, and is the current approach for eclipsing binaries, in the case of planetary transits, a quadratic law is required to avoid systematic errors:

$$I(\mu) = I(1)[1 - u_a(1-\mu) - u_b(1-\mu)^2] \qquad (2)$$

where $\mu = \cos \gamma$, $I(1)$ is the intensity of radiation emerging normally to the surface, and u_a and u_b are the quadratic coefficients of limb-darkening. These coefficients have been extensively tabulated Claret (2000) for a wide range of temperatures and surface gravities using the best available model atmospheres.

Kopal (1979) showed that associated α_n functions could be defined, in such a way that the geometrical elements are decoupled from the radiation parameters, and the above integral (1) could be evaluated as a cross-correlation of two apertures; one representing the star undergoing eclipse, and the other the eclipsing disc (in our case, the transiting planet). This method links the fractional loss of light with the diffraction patterns of the two apertures as described in physical optics, allowing the use of well known mathematical tools. Thus, he found that the α_n functions can be expressed by means of a Hankel transform, and then written in terms of Jacobi polynomials of the type $G_n(p, q; x)$, with the final result,

$$\alpha_n(b, c) = \frac{(1-c^2)^{\nu+1}}{\nu \Gamma(\nu+1)} \sum_{j=0}^{\infty} (-1)^j (2j+\nu+2) \frac{\Gamma(\nu+j+1)}{\Gamma(j+2)}$$
$$\times \{b G_j(\nu+2, \nu+1; 1-b)\}^2 G_j(\nu+2, 1; c^2) \qquad (3)$$

where the relative radius of star, r_s and the planet, r_p, as well as the apparent separation between the centers of their projected discs δ, are expressed relative to the radius of the orbit, and the notation $b \equiv r_p/(r_s + r_p)$, $c \equiv \delta/(r_s + r_p)$ and $\nu \equiv (n+2)/2$ has been introduced. In the case of circular orbits we obviously have $\delta^2 = 1 - \cos^2 \theta \sin^2 i$. For $\delta > r_s + r_p$ there is no transit and $\alpha = 0$. For $\delta < r_s - r_p$ the transit will be annular, and for values $r_s + r_p > \delta > r_s - r_p$ it will be partial. Equation (3) was given by Kopal (1979), page 40, except for minor misprints, but recursion properties of the Jacobi polynomials have been used to ensure that they are all of the same degree, thus facilitating the computations. This is the most general expression for the associated α_n functions in algebraic form, valid for any degree of limb darkening, eccentricity and type of eclipse; be it total, annular, a transit or an occultation. Its right hand side is a rapidly convergent series of terms and the precision that can be achieved, under the adopted assumptions, is better than any need of existing or planned photometric surveys. This is a function of the degree adopted for the series of terms in the Jacobi polynomials but no problem has been found to evaluate α_n for large values of j, of the order of 2000. A good precision is obtained, below 5×10^{-5}, with values of j up to ~ 20, or below 5×10^{-7} for $j \sim 200$. The application of (3) for the analysis of a light curve is not only simple but also very fast. Further details on the test and validity of the involved equations are given by Giménez (2006).

3. The light curve solution

The relations between the elements of the transit and their observed characteristics are transcendental, and not capable of any type of analytic inversion. In solving the observed light curve for the system elements, five free parameters are involved: the relative radius of the star, r_s, the ratio of radii, $k = r_p/r_s$, the orbital inclination, i, and the limb darkening coefficients, u_a and u_b, for the case of a quadratic law. In practical cases, we have found that some of them can be substituted by other representative parameters. The first of them is the radius of the star for which we can use the, easily measured, phase of the transit ingress, θ_1. This parameter defines the sum of the relative radii of the two components as a function of the stellar latitude as given by the orbital inclination since, for adopted values of k and i, θ_1 is directly related to r_s by,

$$r_s = \left[\frac{1 - \cos^2 \theta_1 \sin^2 i}{1+k} \right]^{1/2}$$

On the other hand, it has been found that it is more convenient to use differential limb-darkening coefficients, $u_+ = u_a + u_b$ and $u_- = u_a - u_b$, rather than the individual values to avoid, or at least decrease, their correlation. While the information contained in the light curve can provide a good determination of u_+, u_- remains quite indeterminate.

We are not going to repeat here the possible methods to search for the best solution of the light curve, see Giménez (2006), but initial values of the parameters to be fitted are found necessary to speed up the adopted method or, at least, a range of validity consistent with the light curve to limit the search within reasonable boundaries. It is evident that the observed light curve itself provides a good initial value of θ_1 as well as a range of acceptable boundaries (or observational uncertainties). On the other hand, limb-darkening coefficients, both linear and quadratic, can be assumed from available model atmospheres for the expected effective temperature of the star, as derived from photometric or spectroscopic data. Otherwise, a good approximation for solar-type stars is the linear value $u = 0.6$ or, for a second-order limb-darkening law, the equivalent quadratic approximation $u_+ = 0.6$ and $u_- = 0$.

We have thus only two free parameters for which we need initial values: k and i and both can be obtained from the analytical expression derived in (3). For the case of the ratio of radii, we can get a good approximation from the depth of the loss of light during the transit. For no limb-darkening, we obviously have,

$$k = (1 - l(0))^{1/2}$$

and this can be used as an upper limit to k. Taking into account a linear law of limb-darkening with the value $u = 1$, then,

$$k = \left[1 - (1 - (1 - l(0)))^{2/3}\right]^{1/2}$$

which is a good approximation to the lower limit of k. Furthermore, the best initial value for a differential correction process is given by the solution of the equation valid for $i = 90$ degrees and a quadratic limb-darkening with coefficients, $u_+ = 0.6$ and $u_- = 0$. In this case,

$$k = \left[8.3(1 - l(0)) - 5.9(1 - (1 - k^2)^{3/2})\right]^{1/2}$$

which can be easily solved by iteration (another alternative is of course to use the average of the upper and lower values of k).

For the orbital inclination, an obvious upper value is $i = 90$ degrees while the lower value for the exploration range is given by $i = \arccos(r_s + r_p)$, indicating the limit in orbital inclination for eclipses to occur at all. In order to perform this estimation, the minimum sum of the radii is given by $(r_s + r_p)^2 = 1 - \cos^2 \theta_1$. Initial values for differential corrections may be obtained again by averaging the two estimations. The use of $i = 90$ degrees as an initial value does also work in general but may present numerical problems linked to the partial derivatives around it.

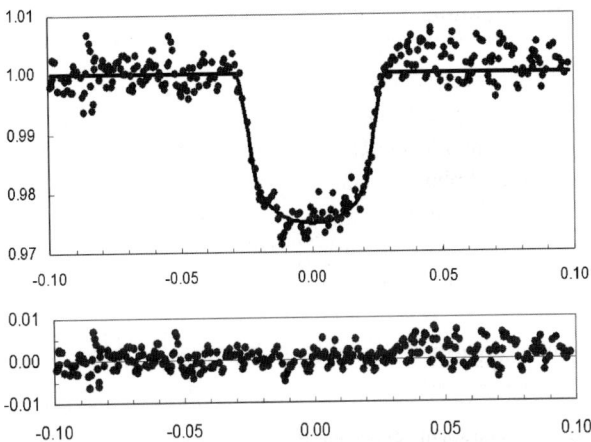

Fig. 1 Comparison of predicted and observed light curves of OGLE-TR-113 in the I band. Normalized luminosities versus orbital phase are shown in the upper plot and $O - C$ values in the lower one

In order to show the possibilities of the proposed equations in practical cases, two well known extra-solar planetary transits have been studied: the planet found in OGLE-TR-113, with a relatively low quality light curve obtained by means of ground-based telescopes, and the prototypical example of HD 209458, with a high-quality light curve obtained with *HST*.

In the first case, the same light curve analyzed by Konacki et al. (2004) has been adopted. The limb-darkening coefficients had to be derived from theoretical model atmospheres and only three parameters could be used to fit the light curve: θ_1, k and i. The range of exploration to find the best values of these parameters, following the above given guidelines, were $0.0275 - 0.0295$ for θ_1, $0.129 - 0.158$, for k, and $80 - 90$ for i. The adopted limb-darkening coefficients, for $T_{eff} = 4800$ K and I-band, are $u_+ = 0.64$ and $u_- = 0.17$. A final solution was found with parameters $r_s = 0.158 \pm 0.005$, $k = 0.145 \pm 0.002$ and $i = 88.2 \pm 1.8$. The result of the fit can be seen in Fig. 1 where the computed light curve is plotted together with the observational data. In the lower part, the $O - C$ curve is also shown. The obtained elements are in excellent agreement with previous studies Konacki et al. (2004) and lead to absolute radii of 0.78 ± 0.05 for the star in units of the solar radius and 1.10 ± 0.08 for the planet in units of the radius of Jupiter.

Similarly, the high-precision light curve of HD 209458, as given by Brown et al. (2001), was analyzed. In this case a parametric search was applied to θ_1 and u_- and a differential corrections method to the remaining k, i and u_+. A range for exploration $0.0180 - 0.0182$ was adopted as derived from the light curve for θ_1 and $-0.05 - 0.05$ for u_-, as indicated by model atmospheres. Initial values for the other three parameters could be estimated, as described above, to be $i = 86.7$ degrees, $k = 0.121$, and $u_+ = 0.6$. The best fit of the light curve was obtained for a quadratic limb darkening

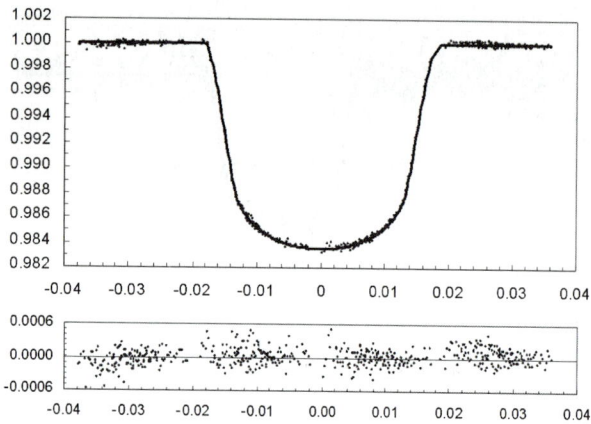

Fig. 2 Comparison of predicted and observed light curves of HD 209458. Normalized luminosities versus orbital phase are shown in the upper plot and $O - C$ values in the lower one

with $u_+ = 0.654 \pm 0.008$ and $u_- = -0.033 \pm 0.015$, i.e. $u_a = 0.310$ and $u_b = 0.343$. The geometrical elements were found to be $r_s = 0.1136 \pm 0.0006$, $k = 0.1208 \pm 0.0002$ and $i = 86.71 \pm 0.04$. The quality of the fit can be judged from the comparison of the computed and observed light curves as shown in Fig. 2. Finally, adopting the mass of the star to be 1.1 ± 0.1 solar masses, the relative radii can be expressed in physical units as 1.144 ± 0.040 solar radii for the star and 1.344 ± 0.050 in units of the radius of Jupiter for the planet. Again, an excellent agreement has been found with recent studies Wittenmyer et al. (2005) but the uncertainties in the absolute parameters could be limited to that of the adopted mass of the star.

4. Conclusions

Light curves produced by extra-solar planetary transits can be efficiently analyzed using tools from the study of eclipsing binaries. Contrary to this latter case, planetary transits present only one eclipse but the luminosity ratio is known to be negligible. In addition, transits always show an annular phase so that the ratio of radii can be well determined. This property, together with a good definition of the phase for beginning of ingress into eclipse, does also lead to an excellent determination of the sum of the relative radii. We should be aware nevertheless that the small value of k in planetary transits requires the use of second order laws of limb darkening rather than a linear approximation.

The well-detached nature of the planet-star system also allows for the adoption of simplified equations for spherical components and thus methods based on the direct evaluation of the α functions can be used efficiently. Giménez (2006) already discussed some of the possible deviations from the assumption of spherical components, particularly with respect to the parent star, which were found them to be negligible, but some further comments are necessary concerning the shape of the planet. Indeed, though tidal elongation may be negligible on the star, the effect on the planet could be large enough to be detected. Considering only first order terms and ignoring rotation, the shape of the planet due to tidal distortion should follow, $r = r_0(1 + \Delta_2 q r^3)$ in the direction of the perturbing potential, where r_0 is the radius of the sphere with the same volume as the actual configuration, q is the mass ratio in the sense star/planet, and Δ_2 is the distortion parameter (a function of the internal density concentration with value between 1 and 2.5). Even for $q \sim 1000$, with a relative radius of ~ 0.02, the expected difference is less than 2% of the equivalent radius. Since, during the transit, the elongation is perpendicular to the plane of the sky, the effect is negligible for the precision of the best light curves.

The planet Jupiter is known to have an oblateness factor of 7%, much larger than predicted for extra-solar planets. This is of course due to its fast rotation, not tidally locked with the orbital period. Most extra-solar planets are much closer to their parent stars and tidal effects are indeed inducing synchronized rotation. In this case, the rotational distortion of the planet should be less than 0.3%, smaller than the already negligible tidal effects. The relative radius actually derived from the observations will be an average between the polar and side radii, in the plane perpendicular to the tidal elongation axis. For a tidally distorted configuration with synchronized rotation, this radius can be easily converted into the sphere-equivalent one by means of $r_0 = r/(1 + (\Delta_2/4)(1 + q)r^3)$. This correction should not be necessary in normal cases due to its expected small value, less than 0.5%, for the above given values of r and q.

References

Brown, T.M., Charbonneau, D., Gilliland, R.L., Noyes, R.W., Burrows, A.: Astroph. J. **552**, 699 (2001)
Charbonneau, D., Brown, T.M., Latham, D.W., Mayor, M.: Astroph. J. **529**, L45 (2000)
Claret, A.: Astron. Astroph. **363**, 1081 (2000)
Deeg, H.J., Garrido, R., Claret, A.: New Astron. **6**, 51 (2001)
Giménez, À.: Astron. Astroph. **450**, 350 (2006)
Konacki, M., Torres, G., Sasselov, D.D., Pietrzyński, G., Udalski, A., Jha, S., Ruiz, M.T., Gieren, W., Minniti, D.: Astroph. J. **609**, L37 (2004)
Kopal, Z.: Language of the Stars (Reidel Publ., Dordrecht, Holland)
Mandel, K., Agol, E.: Astroph. J. **580**, L171 (2002)
Udalski, A., Żebruń, K., Szymanski, M., Kubiak, M., Soszyński, I., Szewczyk, O., Wyrzykowski, Ł., Pietrzyński, G.: Acta Astron. **52**, 115 (2002)
Wittenmyer, R.A., et al.: Astroph. J. **632**, 1157 (2005)

Photometric Studies of Twelve Deep, Low-mass Ratio Overcontact Binary Systems

Qian Shengbang · Yang Yuangui · Zhu Liying ·
He Jiajia · Yuan Jingzhao

Received: 1 November 2005 / Accepted: 1 February 2006
© Springer Science + Business Media B.V. 2006

Abstract The formations of the blue straggler stars and the FK Com-type stars are unsolved problems in stellar astrophysics. One of the possibilities for their formations is from the coalescence of W UMa-type overcontact binary systems. Therefore, deep ($f > 50\%$), low-mass ratio ($q < 0.25$) overcontact binary stars are a very important source to understand the phenomena of Blue Straggler/FK Com-type stars. Recently, 12 W UMa-type binary stars, FG Hya, GR Vir, IK Per, TV Mus, CU Tau, V857 Her, V410 Aur, XY Boo, SX CrV, QX And, GSC 619-232, and AH Cnc, were investigated photometrically. Apart from TV Mus, XY boo, and GSC 619-232, new observations of the other 9 binaries were obtained. Complete light curves of the 10 systems, FG Hya, GR Vir, IK Per, TV Mus, CU Tau, V857 Her, GSC 619-232, V410 Aur, XY Boo, and AH Cnc, were analyzed with the 2003 version of the W-D code. It is shown that all of those systems are deep ($f > 50\%$), low-mass ratio ($q < 0.25$) overcontact binary stars. We found that the system GSC 619-232 has the highest degree of overcontact ($f = 93.4\%$). The derived photometric mass ratio of V857 Her, $q = 0.0653$, indicates that it is the lowest-mass ratio system among W UMa-type binaries.

Of the 12 sample stars, long-term period changes of 11 systems were found. About 58% (seven) of the sample binaries show cyclic period oscillation. No cyclic period changes were discovered for the other 5 systems, which may be caused by the short observational time interval or by insufficient observations. Therefore, we think that all W UMa-type binary stars may contain cyclic period variations. By considering the long-term period changes (both increase and decrease) of those binary stars, we proposed two evolutionary scenarios evolving from deep, low-mass ratio overcontact binaries into Blue Straggler/FK Com-type stars.

Keywords Stars: binaries : close · Stars: binaries : eclipsing –Stars: individuals (FG Hya, GR Vir' IK Per, TV Mus, CU Tau, V857 Her, V410 Aur, XY Boo, SX CrV, QX And. GSC 619-232, and AH Cnc) · Stars: evolution

1. Introduction

Deep, low-mass ratio (DLMR) overcontact binary stars are a group of W UMa-type binaries with their mass ratios less than 0.25 ($q < 0.25$) and their degrees of overcontact larger than 50% ($f > 50\%$). They are a very important source to study in stellar astrophysics for the following reasons: (i) they are at the end evolutionary stage of overcontact binary stars and may be the progenitors of single rapid-rotating stars, therefore, they provide valuable information on the dynamical evolution of overcontact binaries and on the formation of the blue straggler (BS) stars and the FK Com-type stars, and (ii) they offer a good change to study the physics of the merge of overcontact binary and the formation of single rapid-rotating stars. Recently, some DLMR overcontact binaries were investigated photometrically. Here, we report some of the results.

Q. Shengbang (✉) · Y. Yuangui · Z. Liying · H. Jiajia ·
Y. Jingzhao
National Astronomical Observatories/Yunnan Observatory,
Chinese Academy of Sciences, P.O. Box 110, 650011 Kunming,
P.R. China
e-mail: qsb@ynao.ac.cn

2. Photometric solutions

Twelve W UMa-type binary stars, FG Hya, GR Vir, IK Per, TV Mus, CU Tau, V857 Her, V410 Aur, XY Boo, SX CrV,

Table 1 Photometric parameters of some DLMR overcontact binaries

Star name	Type	Sp.	Period	q	f%
FG Hya	A/W	G2V	0.32783	0.112	85.6
V410 Aur	AT	G0/2V	0.33653	0.142	66.5
GSC 619-232	WT	–	0.34396	0.104	93.4
GR Vir	AT	F7-8V	0.34698	0.112	78.6
AH Cnc	A/W	F5V	0.36044	0.168	58.5
XY BOO	AT	F5V	0.37055	0.143	59.8
V857 Her	AT	A6	0.38223	0.065	83.8
CU Tau	AP	G0	0.41254	0.177	50.1
TV Mus	AT	F8V	0.44586	0.166	74.3
IK Per	AT	A2	0.67603	0.191	52.0

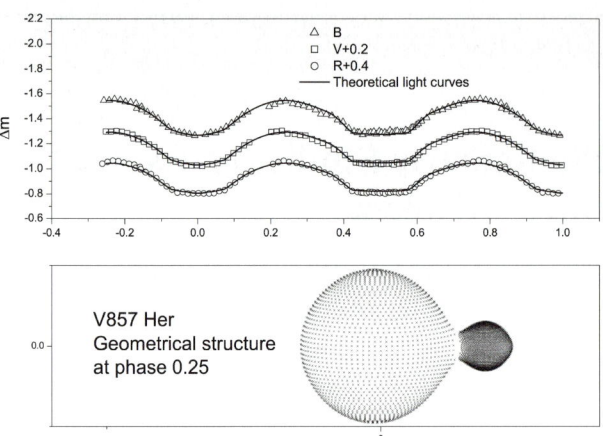

Fig. 1 Ligh curves and geometrical structure of the lowest-mass ratio overcontact binary V857 Her

QX And, GSC 619-232, and AH Cnc, have recently been included in our series photometric study. Apart from TV Mus, XY Boo, and GSC 619-232, new observations of 9 sample stars, FG Hya, GR Vir, IK Per, CU Tau, V857 Her, V410 Aur, SX CrV, QX And and AH Cnc, were obtained with the 1.0-m telescope in Yunnan observatory. Our purposes are: (1) obtain complete light curves and new times of light minimum of the sample stars, (2) analyze the variations of light curve and orbital period, (3) determine the photometric parameters and search for new deep, low-mass ratio overcontact binaries, and (4) finally, based on the long-term period changes and evolutionary states of those sample binary stars, study how do those deep, low-mass ratio overcontact binaries merge into single rapid-rotation stars. The variations of the light curves of several systems, e.g., FG Hya, CU Tau and AH Cnc, were discovered.

Photometric solutions of 10 systems, GR Vir (Qian and Yang, 2004), FG Hya (Qian and Yang 2005), IK Per (Zhu et al., 2005), TV Mus, CU Tau (Qian et al., 2005a), V857 Her (Qian et al., 2005b), GSC 619-232 (Yang et al., 2005a), V410 Aur, XY Boo (Yang et al., 2005b), and AH Cnc (Qian et al., 2006) were obtained with the 2003 version of the W-D code (Wilson and Devinney 1971; Wilson and Van Hamme, 2003). Of the 10 sample stars, only one system, CU Tau, is partial-eclipsing binary. The others show total eclipses. Their photometric parameters can be determined reliably. In order to check the agreement between photometric and spectroscopic mass ratios for total-eclipsing binaries, we analyze FG Hya in detail. It is shown that the derived photometric mass ratio ($q = 0.1115$) is almost the same as the spectroscopic one ($q_{sp} = 0.112$) determined by Lu and Rucinski (1999). Some photometric parameters of the 10 binaries are listed in Table 1 with the order of increasing period. Our solutions indicated that GSC 619-232 has the highest degree of overcontact ($f = 93.4\%$). The derived mass ratio of V857 Her suggests that it is the lowest-mass ratio system among W UMa-type binary stars ($q = 0.065$). Light curves and geometrical structure of V857 Her are shown in Fig. 1. Detail photometric analyses of the two systems, SX CrV and QX And, will be done.

3. Long-term and cyclic period changes

Orbital periods of W UMa-type binary stars are usually variable (e.g., Qian, 2001a,b, 2003) and their variations are usually very complex. The problem why some systems show a long-term period increase, while the periods of the others are decreasing is not unsolved. The possible connections between the long-term period change and the mass of the massive component M_1 and the mass ratio q were discussed by Qian (2001a,b; 2003), which may indicate that overcontact binary stars are oscillating around a critical mass ratio via the combination of TRO and the change of angular momentum loss. One of the purposes of our study is to determine new eclipse times and to analyze the period changes of the sample stars. Among the 12 systems, GSC 619-232 is a new discovered W UMa-type binary star (Gonzalez-Rojas, 2003). The observations are insufficient to reveal any period changes. The periods of the other 11 binaries are variable. The rates and the timescales of the long-term changes are displayed in Table 2.

After the long-term period variations were removed, the residuals of seven systems (XY Boo, AH Cnc, SX CrV, FG Hya, GR Vir, IK Per, and TV Mus) showed cyclic variations. The amplitudes and the periods of those cyclic chnages are listed in the fifth and the sixth columns of Table 2. The O-C curve of a typical sample star, FG Hya, is displayed in Fig. 2. A combination of a long-term period decrease and a cyclic oscillation were proposed to describe the general O-C change by Qian and Yang (2005). After the continuous period decrease and the cyclic change were removed, the residuals are plotted in the lowest panel of Fig. 2, where 4 eclipse times of Yang and Liu (2000) were not displayed. A small-amplitude

Table 2 Summary of long-term and cyclic period changes of 11 DLMR overcontact binaries

Star name	$dP/dt(10^{-7}$ d/y)	$\tau(10^5$ y)	Cyclic change	Amplitude (d)	Period (y)
V410 Aur	+8.22	4.09	–	–	–
XY Boo	+6.25	5.93	yes	0.0045	30.4
AH Cnc	+3.99	9.03	yes	0.0237	36.5
V857 Her	+2.90	13.1	–	–	–
CU Tau	−18.1	2.28	–	–	–
SX CrV	−9.76	3.45	yes	0.0218	32.1
FG Hya	−1.96	16.7	yes	0.0289	36.4
GR Vir	−2.05	16.9	yes	0.0140	19.3
IK Per	−2.50	27.0	yes	0.0203	50.5
QX And	−14.0	2.94	–	–	–
TV Mus	−2.16	20.6	yes	0.0240	29.1

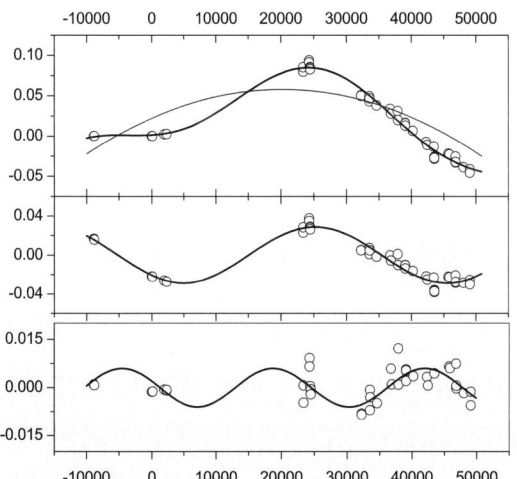

Fig. 2 Long-term and cyclic period changes of FG Hya. After the continuous period decrease and a cyclic change proposed by Qian and Yang (2005) were removed. The residuals in the lowest panel reveal a small-amplitude periodic variation

periodic variation is clearly seen in this panel. With least-squares method, the following equation,

$$R_e = -0.00006(22)$$
$$+ 0.0060(9)\sin[0.°0155E + 160.°1(\pm 3.°3)], \quad (1)$$

was obtained, which indicates a small-amplitude cyclic change with an amplitude of $0.^d0060$. A calculation with the relation $\omega = 360° P_e/T$ reveals a period of the periodic variation of about 20.9 years.

No cyclic period variations were found for the five other systems, V410 Aur, V857 Her, CU Tau, QX And and GSC 619-232. This may be caused by the reasons that they are observed in a short time interval (e.g., V410 Aur, V857 Her, CU Tau and GSC 619-232) or by insufficient observations (e.g., QX And). By considering the two reasons, it is expected that all W UMa-type binary stars may contain cyclic period variations. It should be pointed out here, the rates of long-term period decreases of two sample stars, CU Tau and QX And, are very large when comparing with those of the others. This may indicate that the present period decreases may be a part of a long-period cyclic variation or a combination of a cyclic change and a long-term variation as have been observed in other W UMa-type binaries (e.g., FG Hya).

4. Conclusions

From the series photometric studies of some DLMR overcontact binaries, the following conclusions can be drawn:

(1) At least part of the BS stars and the FK Com-type stars were formed from the merge of deep, low-mass overcontact binaries;
(2) Secular period changes may be very important to understand the merge of overcontact binary stars. Systems with a decreasing period will evolve into single rapid-rotation stars when the photospheric surface of the binary close the outer critical lobe, while systems with an increasing period may merge when they meets the more familiar criterion that the orbital angular momentum is less than 3 times of the total spin angular momentum (Hut, 1980);
(3) The time scale of the merge of the deep, low-mass ratio overcontact binary stars may be close to the time scale of the period change, i.e., $\tau \sim (2 \times 10^5 - 3 \times 10^6$ years);
(4) It is suggested that all W UMa-type binary stars may contain cyclic period variations, which provide valuable information on the multiplicity, magnetic activity cycle, formation and evolution of W UMa-type binary stars.

Acknowledgements This work is partly supported by Science and Technology Department of Yunnan Province (No. 2003RC19), Chinese Natural Science Foundation (No.10573032 and No.10433030) and by Chinese Academy of Sciences (No.KJCXZ-SW-T06).

References

Gonzlez-Rojas, D.J. et al.: IBVS No. 5437 (2003)
Hut, P.: A&A **92**, 167 (1980)
Lu, W., Rucinski, S.M.: AJ **118**, 515 (1999)
Qian, S.B.: MNRAS **328**, 635 (2001a)
Qian, S.B.: MNRAS **328**, 919 (2001b)
Qian, S.B.: A&A **400**, 649 (2003)
Qian, S.-B., Yang, Y.-G.: AJ **128**, 2430 (2004)
Qian, S.-B., Yang, Y.-G.: MNRAS **356**, 765 (2005)
Qian, S.-B., Yang, Y.-G., Soonthornthum, B. et. al.: AJ **130**, 224 (2005a)
Qian, S.-B., Zhu, L.-Y., Soonthornthum, B. et al.: AJ **130**, 1206 (2005b)
Qian, S.-B., Liu, L., Soonthornthum, B. et al.: AJ **131**, 3028 (2006)
Wilson, R.E., Devinney, E.J.: ApJ **166**, 605 (1971)
Wilson, R.E., Van Hamme, W.: Computing Binary Stars Observables, the 4th edition of the W-D program (2003)
Yang, Y.-G., Qian, S.-B., Gonzlez-Rojas, D.J. et al.: ApSS **300**, 337 (2005a)
Yang, Y.-G., Qian, S.-B., Yuan, J.-Z.: AJ **130**, 2252 (2005b)
Yang, Y.-L., Liu, Q.-Y.: A&AS **144**, 1 (2000)
Zhu, L.-Y., Qian, S.-B., Soonthornthum, B., Yang, Y.-G.: AJ **129**, 2806 (2005)

Updated UBV Light-Curve and Period Analysis of Eclipsing Binary HS Herculis

Zeynep Bozkurt · Ömer Lütfi Değirmenci

Received: 8 September 2005 / Accepted: 15 March 2006
© Springer Science + Business Media B.V. 2006

Abstract UBV light-curves of the eclipsing binary HS Herculis, obtained in 2002–2003 observational seasons, were analysed with Wilson-Devinney computer code. New absolute dimensions of the system were calculated using the results of the light-curve analysis. Period variation of the system was also investigated. Several new times of minima have been secured for this problematic system. An apsidal motion with a period of 80.7 years was confirmed and a third body in a pretty eccentric orbit ($e_3 = 0.90 \pm 0.08$) with a period of 85.4 years was found. The corresponding internal structure constants of the binary system, $\log k_2$, and the mass of the third body were derived.

Keywords HS Her: Period variation: apsidal motion · Light-time effect · Light-curve: WD solution

1. Introduction

The eclipsing binary HS Herculis (HD 174714 = BD + 24 3552; $P = 1^d.637$, $V = 8^m.61$, Sp = B5V + A4V) was discovered in 1934 by Martynov (1940). He classified the system as an Algol-type eclipsing binary. The variability and binary nature of HS Her were discovered independently by Jacchia (1940) using Harvard plates which were taken between 1898 and 1939. The first spectroscopic investigation was made by Cesco and Sahade (1945). They obtained a single-lined radial velocity curve and spectroscopic solution of the system. According to Cesco and Sahade (1945), the spectrum corresponds to that of a main-sequence star between B5 and B8 in the MK spectral classification system.

Z. Bozkurt (✉) · Ö. L. Değirmenci
Ege University, Faculty of Science, Department of Astronomy and Space Sciences, 35100, Bornova, İzmir, Turkey

Hall and Hubbard (1971) and Martynov (1974) made UBV photometric studies of HS Her. Hall and Hubbard (1971) pointed out a curious abnormality in the light-curve. In the ultraviolet, especially, it takes much longer for the system to recover maximum brightness after secondary mid-eclipse. As a consistent explanation of the photometric peculiarities, they offered a model including a cloud situated within the Roche lobe of the A4 star but not very close to its surface. They also showed that the orbital period of HS Her is variable and only rapid apsidal motion can explain this variation. Martynov (1971) emphasized that the apsidal motion period, obtained by Hall and Hubbard (1971), is too short. He found an apsidal motion with a period of 110 years using his own determination $\omega = 116°$ and taking into account the spectroscopic value $\omega = 37°$ (Cesco and Sahade, 1945) for the longitude of periastron. After this study, Scarfe and Barlow (1974) expressed that available photometric data do not put in evidence any effect of apsidal motion and another interpretation of the variation of period is required.

Todoran (1992) resumed the study of apsidal motion in the system and he found an apsidal motion with a period of 60 years. Bastian (1993) reported that O-C variation is not originated from the apsidal motion, but it can be explained by the light-time effect caused by a third body in the system. According to his study the orbital period of the third body is 60 years. Todoran and Agerer (1994) insisted on the apsidal motion effect. They noted that after 1955, the variation of O-C diagram can be explained by the effect of apsidal motion and there is not any indication about a third body. Wolf et al. (2002) made a detailed period analysis of the system and found that both apsidal motion and light-time effect cause the period variation. They derived an apsidal motion with the period of 78 years and a light-time effect with the period of 85 years. Bozkurt and Değirmenci (2005a,b) confirmed

Table 5 Comparison of the three different light curves solutions

Parameters	Hall and Hubbard (1971) (Russel-Merrill method)	Giuricin and Mardirossian (1981) (Wood method)	This study (WD method)
i	88°.7	87°.3 ± 1.3 (mean)	88°.1 ± 0°.1
r_h	0.259	0.251 ± 0.005 (mean)	0.2631 ± 0.0008 (mean)
r_c	0.142	0.141 ± 0.005 (mean)	0.1493 ± 0.0004 (mean)
e	–	0 (adopted)	0.016 ± 0.001
$f(m)$, Spectral type	0.1, B5	0.1, B5	0.1, B5
	Harris (1963)	Allen (1973)	Drilling and Landolt (2000)
$M_1(M_\odot)$	4.7	6.5	6.0 ± 0.5
$M_2(M_\odot)$	1.6	1.9	1.8 ± 0.2
$R_1(R_\odot)$	2.8	3.0	3.1 ± 0.2
$R_2(R_\odot)$	1.6	1.7	1.7 ± 0.1

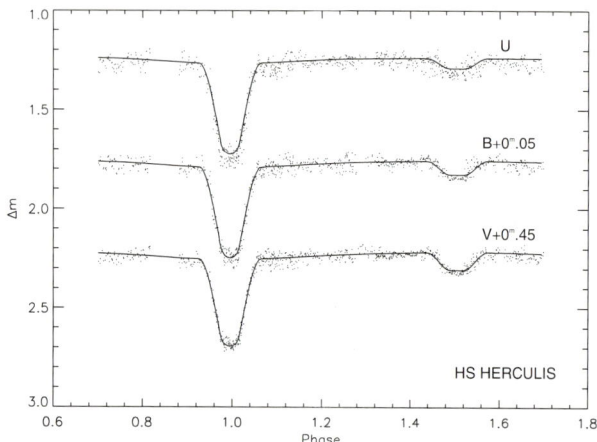

Fig. 2 U, B and V light curves of HS Her. The computed light curves (solid lines) formed with the parameters in Table 3

5. Conclusions

In this study we derived updated UBV light-curves and period analysis of HS Her. We confirmed the apsidal motion at a rate $\dot{\omega}_{obs} = 0.02001$ deg cycle^{-1} and a third body in an eccentric orbit with a period of 85.4 years. The derived parameters for the light-time effect allowed us to determine the mass function of the third body, $f(M_3) = 0.0485 M_\odot$, and its minimum mass, $M_{3,\min} = 1.64 M_\odot$. Observational and theoretical average values of internal structure constant of the binary system were computed as $\log k_{2,obs} = -2.32 \pm 0.01$ and $\log k_{2,theo} = -2.24$, respectively. These results show that HS Her is slightly centrally condensed than predicted by the theoretical models. The theoretical value of k_2 is for non-rotating models. The correction in the theoretical k_2 caused by rotation is not expected too large for the systems not excessively distorted (Claret, 1999). Claret (1999) showed that the correction due to rotation would be $\delta \log k_2 = -0.87 \lambda_s + 0.004$ where λ_s is $2v^2/(3gR)$, v being the rotational velocity, g the gravity and R the stellar radius. λ_s is an important parameter to evaluate the effects of rotation and it is computed at the surface. For HS Her, we found $\delta \log k_2 = -0.007$. We also calculated the rate of apsidal motion caused by the third body, $\dot{\omega}_{thrd} = -8 \times 10^{-6}$ deg/cycle, using the expression derived by Brown(1936). Since this value lies within the error of the observational rate of apsidal motion it can be neglected. Relativistic contribution to apsidal motion was also determined as $\dot{\omega}_{rel} = 0.00156$ deg cycle^{-1}. The values of eccentricity obtained independently from the light curves and O-C analyses are very close to each other as seen in Tables 2, 3. The time span covered by the available times of minima is longer than the period of apsidal motion, therefore the apsidal motion parameters presented in Table 2 are trustworthy. In the present case we think that our resulting e value determined from the O-C analysis is more reliable than that obtained from the light curves solution.

We analysed the U, B and V light-curves of the system simultaneously with the Willson-Devinney method and updated the absolute parameters. We compared this results with previous solutions. Our absolute parameters are based on the radial velocity curve of the primary component only (Cesco and Sahade, 1945). For more reliable results the double-lined spectra of the system are required.

Acknowledgements This study has been supported by the Ege University Research Foundation under contract 2002/FEN/003. Ege University Science and Technology Research Center also supported this investigation. We thank our referee for constructive comments.

References

Bastian, U.: AN **314**, 39 (1993)
Bastian, U.: IBVS **4822** (2000)
Bozkurt, Z., Değirmenci, Ö.L.: ASPC **335**, 277 (2005a)
Bozkurt, Z., Değirmenci, Ö.L.: XIV. National Astronomy Meeting, in press (2005b)
Brown, E.W.: Mon Not R Astron Soc **97**, 116 (1936)
Cesco, C., Sahade, J.: ApJ, **101**, 114 (1945)
Claret, A.: A&A **350**, 56 (1999)
Claret, A., Giménez, À.: A&AS **96**, 255 (1992)
Çolak, T., Müyyesseroğlu, Z.: IBVS **5619** (2005)

Díaz-Cordovés, J., Claret, A., Giménez, À.: A&A **110** 329 (1995)
Drilling, J.S., Landolt, A.U.: Allen's astrophysical quantities. In A.N. Cox (Ed.), New York: Springer-Verlag, p. 381 (2000)
Gillard, D.W., Holdaway, K.: The Astronomical Almanac for the Year 2005. Washington, DC, General Post Office (2005)
Giménez, A., Garcia-Pelayo, J.M.: Rev. R. Acad. Cienc. Exactes. Fis. Nat. Madr. **77**, 287 (1983)
Giménez, À., Bastero, M.: Ap&SS **226**, 99 (1995)
Giuricin, G., Mardirossian, F.: Ap&SS **76**, 111 (1981)
Hall, D.S., Hubbard, G.S.: PASP **83**, 459 (1971)
Harris, D.H.: in K. Aa. Strand (Ed.) Basic Astronomical Data, Chicago: University of Chicago Press, p. 263 (1963)
Irwin, J.B.: ApJ **116**, 211 (1952)
Irwin, J.B.: AJ **64**, 149 (1959)
Jacchia, L.: Harvard Bull. **912**, 20 (1940)
Kwee, K.K., van Woerden, H.: BAN **12**, 327 (1956)
Martynov, Ya.: Bull. Ast. Obs. V.P. Enge'gardta **18**, 38 (1940)
Martynov, Ya.: Astron. Circ. 651, 1 (1971)
Martynov, Ya.: Soobshch. Astron. Inst. Sternberga **185**, 3 (1974)
Russell, H.N., Merrill, J.E.: Princeton Contr. **26**, 54–92 (1952)
Scarfe, C.D., Barlow, D.Y.: PASP **86**, 181 (1974)
Todoran, I.: AN **313**, 183 (1992)
Todoran, I., Agerer, F.: AN **315**, 349 (1994)
Wehlau, W.: JRASC **54**, 164 (1960)
Wilson, R.E., Devinney, E.J.: ApJ **166**, 605 (1971)
Wolf, M., Harmanec, P., Diethelm, R., Hornoch, K., Eenens, P.: A&A **383**, 533 (2002)
Wood, D.B.: A Computer Program for Modeling Non-Spherical Eclipsing Binary Systems. Goddard Space Flight Center, Greenbelt, Maryland, U.S.A. (1972)

Investigation of Times of Minima of Selected Early-Type Eclipsing Binaries

Pavel Mayer · Marek Wolf · P. G. Niarchos · K. D. Gazeas · V. N. Manimanis · Drahomír Chochol

Received: 1 November 2005 / Accepted: 1 February 2006
© Springer Science + Business Media B.V. 2006

Abstract New precise times of minimum light for several early-type eclipsing binaries were obtained at three observatories. The changes of period of the following measured binaries are discussed: V1182 Aql, LY Aur, SZ Cam, FZ CMa, QZ Car, LZ Cen, V606 Cen, AH Cep and TU Mus.

Keywords Eclipsing binary · Period changes · Light-time effect

1. Introduction and observation

Change of periods of massive binaries might be observed due to the fast evolution of the components. Also number of multiple systems is high among these systems, and studies of periodicity in times of minima can reveal them. Several cases with the light-time effect (LITE) are also discussed below.

New times of minimum light were obtained during 2002–2004 at three observatories: Athens Observatory (40-cm reflector, CCD camera SBIG ST8), South Africa Astronomical Observatory (SAAO, 50-cm reflector, modular photometer) and Stará Lesná Observatory (60-cm reflector, photoelectric photometer). The precise moments are given in Table 1.

P. Mayer (✉) · M. Wolf
Astronomical Institute, Charles University Prague, Czech Republic

P. G. Niarchos · K. D. Gazeas · V. N. Manimanis
Dept. of Astrophysics, Astronomy and Mechanics, University of Athens, Greece

D. Chochol
Astronomical Institute, Slovak Academy of Sciences, Tatranská Lomnica, Slovakia

2. Remarks to individual systems

V1182 Aql. The minima are rather shallow and the measured times of minima show a large scatter; the new measurement fits the old ones quite well.

LY Aur. For this bright binary, there was a long gap since the last published minimum (Mayer, 1980). The $O - C$ in Fig. 1 were computed according to the ephemeris HJD $2439061.473 + 4.0024918 \cdot E$. Together with the time of minimum contained in the Hipparcos Catalogue (1997) and the time published by Krajci (2005) the new times allow to precise the ephemeris to

$$\text{Pri.Min.} = \text{HJD } 2439061.4646 + 4.0024932 \cdot E,$$

the period is constant. We measured also the brightness level in maximum; no change against older measurement has been detected. The UBV light curve during the minimum at JD 53382 was measured using the comparison star HD 35619.

SZ Cam. The new time of minimum fits well to the forecast of the $O - C$ curve published by Mayer et al. (1994); also minima published by Gorda (2000) fit the forecast. The solid curve represents the effect of the third-body in a highly eccentric orbit ($e \simeq 0.77$) with a period of 50.7 years and an amplitude of 0.086 days (Lorenz et al., 1998). The $O - C$ graphs for this and the other stars are presented in Fig. 2.

FZ CMa. New times of minima fit the LITE elements of Zasche (2005) quite well. The solid curve represents the effect of the third-body in a highly eccentric orbit ($e \simeq 0.8$) with one of the shortest known orbital period of 1.52 years and an amplitude of 14.5 min.

QZ Car. The measurements made at SAAO (JD 2453117, 2453120, 2453123) cover both minima, however only very sparse; nevertheless, they are compatible with phases of minima shifted to 0.03/0.53, as was observed for data by ASAS

Table 1 New times of minimum light

System	JD Hel.-2400000	Error [day]	Type	Method filter	Observatory
V1182 Aql	53129.6116	0.0002	pri	pe, UBV	SAAO
	53272.3302	0.0002	pri	CCD, V	Athens
LY Aur	53382.3826	0.0005	pri	pe, UBV	Stará Lesná
SZ Cam	53298.3965	0.0005	pri	CCD, R	Athens
	53325.37196	0.0003	pri	CCD, R	Athens
FZ CMa	53030.3201	0.0003	pri	CCD, V	Athens
	53037.3233	0.0003	sec	CCD, R	Athens
	53123.2502	0.0003	pri	pe, UBV	SAAO
	53336.4991	0.0002	sec	CCD, R	Athens
LZ Cen	53123.455	0.003	pri	pe, UBV	SAAO
V606 Cen	53124.5083	0.0003	sec	pe, UBV	SAAO
AH Cep	53001.2848	0.0034	pri	CCD, V	Athens
TU Mus	53119.27786	0.00025	sec	pe, UBV	SAAO

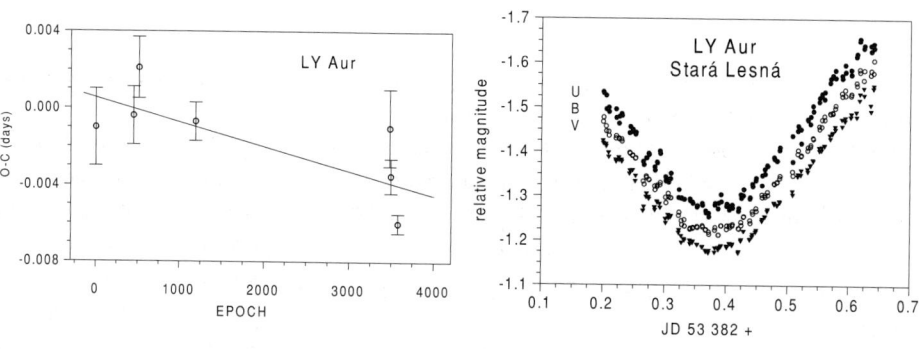

Fig. 1 *Left:* The $O - C$ diagram for the times of minimum of LY Aur. *Right:* The light curve of LY Aur obtained in Stará Lesná Observatory at JD 53382

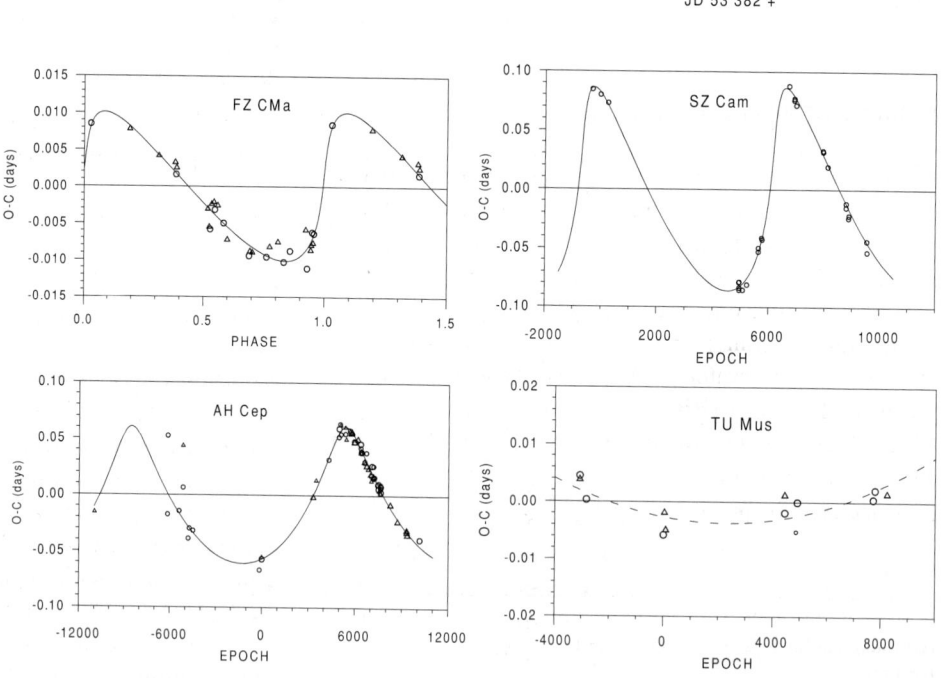

Fig. 2 The $O - C$ diagrams for FZ CMa (phased), SZ Cam, AH Cep and TU Mus. The individual primary and secondary minima are denoted by circles and triangles, resp

(Pojmanski, 2002), see Mayer (2005), and therefore confirm the light-time effect suggested by Mayer et al. (2001).

LZ Cen. No change of ephemeris is needed, new time fits very well the ephemeris by Vaz et al. (1995).

V606 Cen. No change of ephemeris is needed, new time fits the light elements by Lorenz et al. (1999).

AH Cep. New analysis of LITE confirms the third-body elements given in Kim et al. (2005). The solid curve represents

the effect of the third-body in an eccentric orbit ($e \simeq 0.56$) with a period of 67.6 years and an amplitude of 0.0608 days.
TU Mus. The photoelectric minima can be fitted by a linear function, however a better fit is obtained using a parabola (the dashed curve):

Pri.Min. = HJD $41699.8301 + 1\overset{d}{.}38728494 \cdot E + 1.96 \times 10^{-10} E^2$.

3. Conclusions

We obtained new times of minima for several eclipsing binaries to improve their orbital period. All these systems belong to the interesting group of early-type eclipsing binaries studied recently by many investigators. These important stellar systems deserve a continuous photometric, spectroscopic as well as astrometric monitoring.

Acknowledgements This investigation was supported by the Czech-Greek project of collaboration RC-3-18 of Ministry of Education, Youth and Sport and by the Grant Agency of the Czech Republic, grant No. 205/04/2063 and by the Science and Technology Assistance Agency under the contract No. APVT-20-014402. Based on observations secured at the South Africa Astronomical Observatory, Sutherland, South Africa. This research has made use of the SIMBAD database, operated at CDS, Strasbourg, France, and of NASA's Astrophysics Data System Bibliographic Services.

References

ESA: The Hipparcos and Tycho catalogues, ESA SP-1200, ESA (1997)
Gorda, S.Y.: IBVS 4939 (2000)
Kim, Ch.H., Nha, I.S., Kreiner, J.M.: AJ **129**, 990 (2005)
Krajci, T.: IBVS 5592 (2005)
Lorenz, R., Mayer, P., Drechsel, H.: A&A **345**, 531 (1999)
Mayer, P.: IBVS 1724 (1980)
Mayer, P., Chochol, D., Lorenz, R. et al.: A&A **288**, L13 (1994)
Mayer, P., Lorenz, R., Drechsel, H., Abseim, A.: A&A **366**, 558 (2001)
Mayer, P.: In: Drechsel, H.. Zejda, M. (eds.), Zdenek Kopal's Binary Star Legacy, Springer, p. 113 (2005)
Pojmański, G.: Acta Astron. **52**, 397 (2002)
Vaz, L.P.R., Andersen, J., Rabello-Soares, M.C.: A&A **301**, 693 (1995)
Zasche, P.: In: Drechsel, H., Zejda, M. (eds.), Zdenek Kopal's Binary Star Legacy, Springer, p. 127 (2005)

Photometric Studies of Bright Southern Binary Systems: ϵ Cra and ψ Ori

R. R. Shobbrook · S. Zola

Received: 1 November 2005 / Accepted: 1 February 2006
© Springer Science + Business Media B.V. 2006

Abstract We present results from detailed analyses of new data combined with previously published light curves of ϵ Cra and ψ Ori. Based on the shape of the secondary minimum of ϵ CrA we found that the discrepancy between the photometric and spectroscopic mass ratio, although marginal, is statistically significant. We propose a third light as a possible solution and derive the absolute parameters of components. The physical parameters of components of the early-type binary system ψ Ori were also obtained from the light curve modelling. Our solution indicate that ψ Ori is a detached, grazing-eclipse system.

Keywords Binaries: eclipsing, Binaries: close, Binaries: contact, Stars: Fundamental parameters

1. Introduction

Between 1991 and 2001 many observations were obtained of bright eclipsing or ellipsoidal binary stars which may also be observed with the Sydney University Stellar Interferometer, SUSI. The purpose of this program was to obtain photometry to enable some parameters to be fixed during the complex analysis of the SUSI data. The observations were taken using the 24-inch telescope of the Australian National University at Siding Spring Observatory (hereafter referred to as SSO).

R. R. Shobbrook
Research School of Astronomy and Astrophysics, Australian National University, Weston Creek P.O., ACT 2611, Australia

S. Zola (✉)
Astronomical Observatory of the Jagiellonian University, ul. Orla 171, 30-244 Cracow, Poland; Mt. Suhora Observatory, Pedagogical University, ul. Podchorazych 2, 30-084 Cracow, Poland

The new data will be referred to as *SSO V* magnitudes. The details of the observing programme and the equipment is given by Shobbrook (2004).

ϵ CrA (HD155813) is the brightest (4.73^m) W UMa-type system, with period $P = 0.59^d$ discovered by Cousins and Cox (1950). UBV observations were reported by Tapia (1969) and those in RI passbands by Hernandez (1972). The UBVRI set was analyzed with the Wilson-Devinney (W-D) code by Twigg (1979) who photometrically determined the mass ratio to be very small: $q = 0.113$. The spectroscopic orbit of ϵ Cra was published by Goecking and Duerbeck (1993). The following orbital elements were obtained: $K_1 = 34.5$ km/s, $K_2 = 266.9$ km/s and $\gamma = 57.9$ km/s.

ψ Ori (HD 35715, $V = 4.6^m$) was found to be a spectroscopic binary with period of 2.5 days by Frost and Adams (1903). The most recent orbital solution made by Telting et al. (2001) resulted in the following orbital elements: $K_1 = 144.6 \pm 0.5$ km/s, $K_2 = 237 \pm 4$ km/s, $\gamma = +19 \pm 5$ km/s, $e = 0.053 \pm 0.001$, and $\omega = 172 \pm 5°$. The light curve of ψ Ori, as shown by Percy (1969), is typical of an ellipsoidal variable. The depths of the minima in the SSO data are approximately 0.035^m and 0.020^m. The U and B light curves have been analyzed by Hutchings and Hill (1971). The authors found an orbital inclination of $58 \pm 8°$ and derived the absolute parameters: $M_1 = 14.2\ M_\odot$, $M_2 = 8.6\ M_\odot$, $R_1 = 6.1\ R_\odot$, $R_2 = 4.7\ R_\odot$.

2. Results from the light curve modelling

In order to obtain the physical parameters of each system we used the W-D code (Wilson, 1979) supplemented with the Monte Carlo search procedure. The theoretical values of both albedo and gravity darkening coefficients were used.

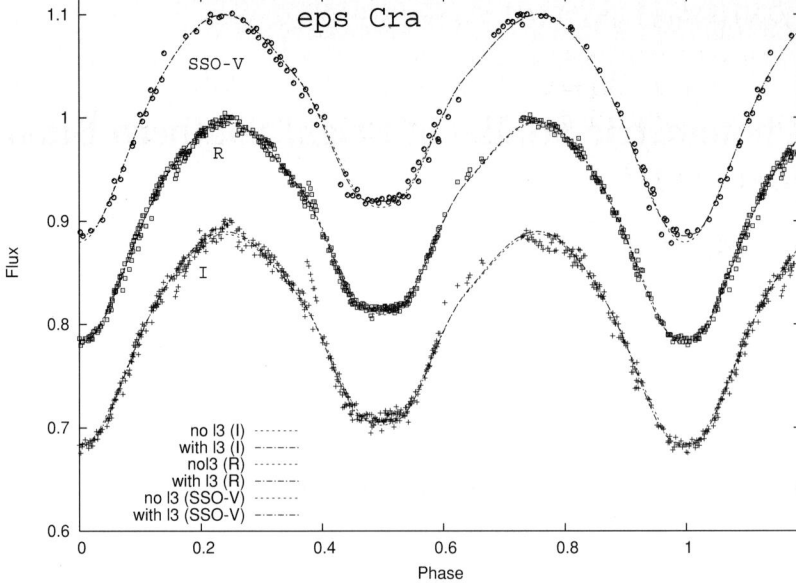

Fig. 1 Comparison between theoretical and observed light curves of ϵ Cra. Observations are shown by squares and theoretical curves by lines

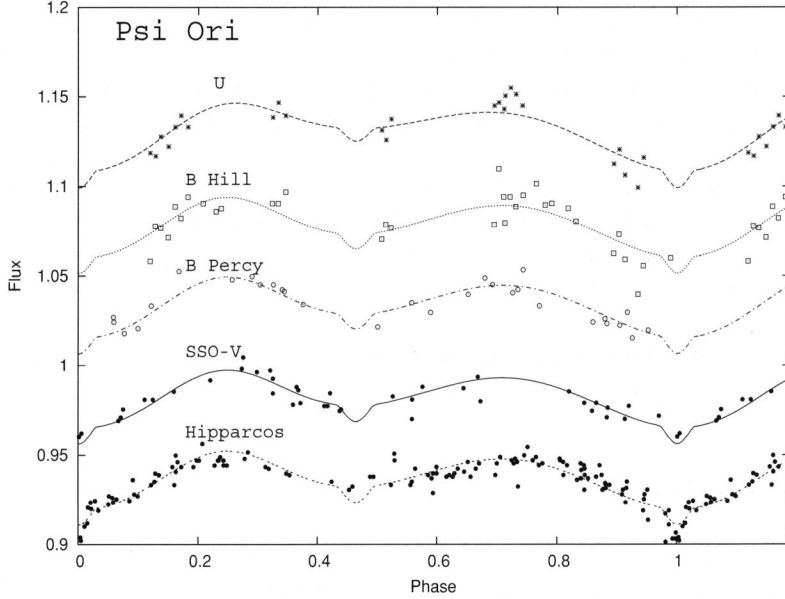

Fig. 2 Comparison between theoretical and observed light curves of ψ Ori

The limb darkening coefficients were adopted as functions of the temperature and wavelength from the Díaz-Cordovés et al. (1995) and Claret et al. (1995) tables. For both systems we fix two crucial parameters: the temperature appropriate for the spectral type and the mass ratio, as determined from the radial velocity curves.

The solution for ϵ CrA was made using the Hernandez (1972) RI light curves and the new SSO-V filter observations. We are obliged to point out that there appears to be a significant number of typing errors in the groups of Julian Dates of the Hernandez's data table. Most errors were able to be corrected by comparison with the Julian Dates in the Tapia (1969) UBV data tables, since their data were obtained on the same nights at the same observatory. However, there are some observations, which appear at phase close to 0.38 in Fig. 1 in the R and I curves, which are clearly in error but could not be corrected; these have not been included in the analysis discussed below.

For ϵ CrA, at first we fixed the mass ratio at $q = 0.129$, obtained by Goecking and Duerbeck (1993) and the temperature of the primary star at 6700. In our light curve modelling of ψ Ori we used the Toronto data (U and B filters), the Kitt Peak B data, the data collected by the Hipparcos mission and the new SSO V observations. We fixed the temperature of the primary component at 26000 K (B1 spectral type), the mass ratio value at $q = 0.61$ and the eccentricity $e = 0.053$, as obtained by Telting et al. (2001) . We assumed that there is no third light in this system.

Photometric Solution of the O-type Eclipsing Binary V1007 Sco

S. Nesslinger · H. Drechsel · R. Lorenz · P. Harmanec ·
P. Mayer · M. Wolf

Received: 1 November 2005 / Accepted: 1 December 2005
© Springer Science + Business Media B.V. 2006

Abstract Improved orbital and physical parameters of the eccentric O-type system V1007 Sco ($P = 5.8$ days) are presented. UBV light curves obtained over a period of 12 years were analyzed with a Wilson-Devinney-based solution code. Setting off from a set of parameters derived from spectroscopic investigations a consistent photometric solution was achieved. In accordance with the spectroscopic findings the eccentricity was determined as $e = 0.128$ and the apsidal motion with a spectroscopic period of about 150 days was confirmed.

Keywords V1007 Sco · O-type binaries · Eccentric

1. Overview

V1007 Sco (HD 152248) is one of the earliest eclipsing binaries known; both components are of spectral type O7. According to visual as well as IUE spectra a luminosity classification of class I is suggested; however the solution of light curves gives radii corresponding to giants. The binary is one of the brightest members (in maximum $V = 6.1$) of the open cluster NGC 6231. It has been known as an SB2 system since the work of Struve (1944); its eclipsing nature was discovered by Mayer et al. (1992), who also found the correct period of 5.814 days (later precised to 5.8160 days,

see Stickland et al., 1996). The UBV photometry by Mayer et al. showed that the orbit is eccentric.

Radial velocities of this star were published in several papers, the latest one being by Sana et al. (2001). Both values $K_{1,2}$ are close to 220 km/s. Penny et al. (1999) solved the V light curve by Mayer et al. and by HIPPARCOS (ESA, 1997). They found an inclination of $i = 72°$; the resulting component masses were under 30 M$_\odot$, i. e. smaller than would correspond to evolutionary models with T_{eff} about 37000 K.

The photometry of V1007 Sco is still rather scarce. Here we will discuss data consisting of the following sets of photoelectric observations and of far-UV magnitudes:

1. UBV observations obtained by Mayer et al. (1992) plus additional UBV observations secured by RL in 1993 and 1994.
2. Hipparcos H_p magnitudes, transformed into Johnson V using the transformation formula given by Harmanec (1998).
3. V observations (derived from Strömgren photometry published by Manfroid et al. (1994)).
4. UBV observations obtained by MW during 6 nights in 2004 with the 50 cm telescope at SAAO.
5. Fluxes from IUE spectra, which were also used for RV determination by Stickland et al. (1996) and Penny et al. (1999).

S. Nesslinger (✉) · H. Drechsel · R. Lorenz
Remeis Observatory Bamberg, Astronomical Institute of the University of Erlangen-Nürnberg, Germany
e-mail: nesslinger@sternwarte.uni-erlangen.de

P. Harmanec · P. Mayer · M. Wolf
Astronomical Institute of the Charles University,
V Holešovičkách 2, CZ-180 00 Praha 8, Czech Republic

2. Spectroscopic solution

In the binary spectra the He I and He II lines of both components are of nearly equal strength. This confines the light curve solution: both temperatures should not differ by more than about 2000 K. Using the temperature scale by

Fig. 1 U, B and V light curves obtained between 29 Feb 1992 and 10 Mar 1992 with best fits

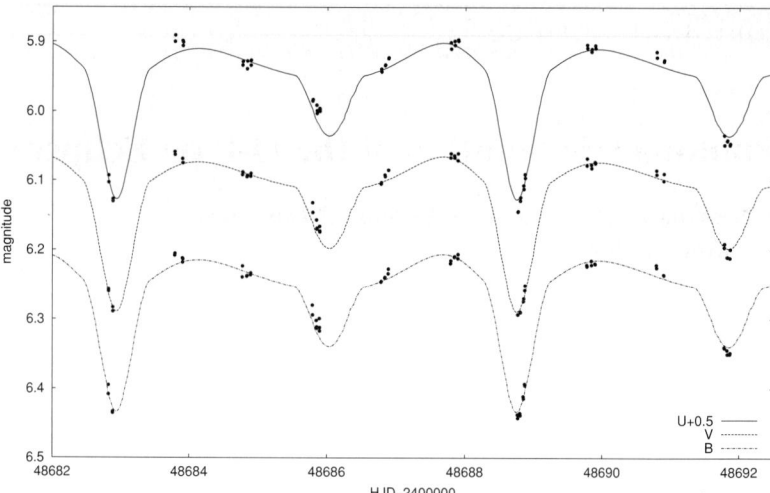

Table 1 Spectroscopically determined parameters

$P_{\text{sid.}}$	5.816051 ± 0.000034 d	K_1	213.4 ± 2.4 km/s
$T_{\text{per.}}$	2448502.35 ± 0.06	K_2	213.2 ± 2.4 km/s
e	0.128 ± 0.007	q	1.001 ± 0.016
$\omega(T_{\text{per.}})$	68.7 ± 3.7 deg	γ	-29.0 ± 4.3 km/s
$d\omega/dt$	2.336 ± 0.150 deg/year		

Table 2 Parameters obtained by light curve solution

	$i = 66.29 \pm 0.01$ deg		
	$T_{\text{eff},1} = 35300$ K (fixed)		
	$T_{\text{eff},2} = 38943 \pm 116$ K		
$r_{\text{pole},1}/a$	0.2856 ± 0.0007	$r_{\text{pole},2}/a$	0.3043 ± 0.0006
$r_{\text{point},1}/a$	0.3238 ± 0.0012	$r_{\text{point},2}/a$	0.3641 ± 0.0017
$r_{\text{side},1}/a$	0.2925 ± 0.0007	$r_{\text{side},2}/a$	0.3133 ± 0.0007
$r_{\text{back},1}/a$	0.3083 ± 0.0009	$r_{\text{back},2}/a$	0.3347 ± 0.0009

Schmidt-Kaler (1982), we expect the average temperature of the components to be about $T_{\text{eff}} = 37000$ K.

According to Mayer et al. (2001) and Sana et al. (2001) the radial velocities suggest an apsidal rotation with a period of about 150 years. This complicates the light curve solution. Therefore we first used the code FOTEL (Hadrava, 1990) to solve the radial velocity curves separately. In the achieved solution (see Table 1), also velocities obtained from unpublished spectra taken at La Silla in the years 1992–1995 were used. We believe that the period, eccentricity, longitude of periastron and its change can be more accurately found using only the radial velocities, since they cover a longer time span than the photometry.

3. Photometric solution

For the solution the code PHOEBE (Prša, 2005) was used, a user-friendly and much enhanced variant of the Wilson-Devinney approach. To reduce the effects of possible errors of the spectroscopic parameters, we used a time-limited data set; it covers a relatively short interval from JD 2448682 to 2449460. Multiple runs with different start parameters converged in the same region of parameter space. The best solution corresponds to the parameters given in Table 2 and the light curves in Fig. 1.

The difference in inclination between our solution and that of Penny et al. (1999) can be explained by several reasons:

1. Penny et al. considered a third light in the system; however the existence of this component was not confirmed (private note by Hartkopf).
2. Penny et al. did not consider the apsidal advance.
3. Penny et al. did not use a code suitable for solution of an eccentric orbit.

4. Absolute parameters

Combining the spectroscopic and photometric solutions, we derive the absolute parameters shown in Table 3 (mean stellar radii and semi-major axis given for time of periastron).

It should be noted that the values given here are still to be considered preliminary. Final results will be published in a subsequent paper.

Table 3 Absolute parameters of V1007 Sco

M_1	$30.60 \pm 0.71\ M_\odot$	R_1	$16.21 \pm 0.10\ R_\odot$
M_2	$30.63 \pm 0.71\ M_\odot$	R_2	$17.61 \pm 0.11\ R_\odot$
A	$53.64 \pm 0.29\ R_\odot$		

References

ESA: The Hipparcos and Tycho Catalogues, ESA SP-1200, Noordwijk, ESA (1997)

Hadrava, P.: Contr. Astron. Obs. Skalnaté Pleso **20**, 23 (1990)

Harmanec, P.: A&A **335**, 173 (1998)

Manfroid, J., Sterken, C., Cunow, B., de Groot, M., Jorissen, A., Kneer, R., Krenzin, R., Kruijswijk, M., Naumann, M., Niehues, M., Schneich, W., Sevenster, M., Vos, N., Vogt, N.: ESO Scientific Report **14**, 142 (1994)

Mayer, P., Lorenz, R., Drechsel, H.: IBVS **3765** (1992)

Mayer, P., Harmanec, P., Lorenz, R., Drechsel, H., Eenens, P., Corral, L.J., Morrell, N.: In: Vanbeveren, D. (ed.) *The Influence of Binaries on Stellar Population Studies*, Kluwer, Dordrecht, 567 (2001)

Penny, L.R., Gies, D.R., Bagnuolo, Jr., W.G.: ApJ, **518**, 450 (1999)

Prša, A.: http://www.fiz.uni-lj.si/~prsa/phoebe/ (2005)

Sana, H., Rauw, G., Gosset, E.: A&A **370**, 121 (2001)

Schmidt-Kaler, T.: In: Schaifers, K., Voight, H.H. (eds.), Landolt-Börnstein: Numerical Data and Functional Relationships in Science and Technology, Astronomy & Astrophysics, Group VI, Vol. 2b, Springer-Verlag, Berlin (1982)

Stickland, D.J., Lloyd, C., Penny, L.R., Gies, D.R., Bagnuolo, Jr., W.G.: The Observatory **116**, 226 (1996)

Struve, O.: ApJ **100**, 189 (1944)

First results of the Central-East-South European Binary Star Study Group (CESEB)

T. Hegedüs · J. Nuspl · N. Markova · H. Markov ·
H. Rovithis-Livaniou · I. Vince · J. Vinkó

Received: 1 November 2005 / Accepted: 1 February 2006
© Springer Science + Business Media B.V. 2006

Abstract Several eclipsing binary systems have been selected for combined spectral and photometric observations using the Bulgarian NAO 2 m telescope and several smaller telescopes located at various places in the CESE region. Preliminary results, based on a pilot study started in 2001, about radial velocity and light curve variations of the active W UMa system LS Del are presented here.

Keywords Eclipsing binaries · Radial velocity curve analysis · Individual: LS Del

1. The CESEB group

The concept of establishing a multilateral co-operation in the field of eclipsing binary stars in Central and East Europe is originated from a bilateral collaboration between Baja and Rhozen observatories in 2001. Its basis was the recognition of the necessity of more optimal utilization of the small- to mid-sized telescopes of the Central-East-South European region. Their importance is still obvious, since only a few space telescopes are available for carrying direct variable star astronomy measurements, and the larger ground-based telescopes are used for other (more "fashionable") researches like e.g. the extragalactic astronomy and cosmology. During 2003, this bilateral co-operation was expanded owing to the similar interest for several other teams in the same region. The abbreviation CESEB refers to the geographical region 'Central-East-South Europe' and the research area 'Binary systems'.

The supplementation of the existing photometric capabilities of smaller telescopes of the region with the spectroscopical facility of the 2 m telescope of Rhozen Observatory is based on the highly needed spectral data of many eclipsing binaries. Amongst these important astrophysical objects one can find very few spectral observations in the last decades. The experiences of the very first project reported here turned our attention to several practical problems of multi-site co-operations, like e.g. the sometimes large differences of the used data reduction methods, and of the generally too small CCD fields (since many telescopes of CESEB region were designed for PMT-based photometry).

2. Overview on the present program stars

CESEB group is mainly focusing its interest to the detached eclipsing SB2 binaries, having eccentric orbit, especially those suspected to exhibit non-radial pulsations (like e.g. V397 Cep). Moreover, several late-type detached systems are also investigated (e.g. WW Cep), as well as a few in-

T. Hegedüs (✉)
Baja Astronomical Observatory of the Bács-Kiskun County,
H-6500 Baja, Szegedi út, KT.766., Hungary

J. Nuspl
Konkoly Observatory, H-1525 Budapest, PF.67, Hungary

N. Markova · H. Markov
Institute of Astronomy, National Astronomical Observatory, P.O. Box 136, 4700 Smolyan, Bulgaria

I. Vince
Astronomical Observatory, Volgina 7, 11160 Belgrade, Serbia and Montenegro

H. Rovithis-Livaniou
Astronomical Institute, National Observatory of Athens, Zografos, Athens, Greece

J. Vinkó
Department of Optics & Quantum Electronics, University of Szeged, P.O. Box 406, Szeged H-6701, Hungary

teresting W UMa-type contact systems (e.g. UZ Leo). For testing the whole procedure of our group, a few well-known test systems have also been selected for checking the accuracy and compatibility of our instrumentation, reduction techniques and analysis methods. The crucial point of the project is to obtain new radial velocity and light curves for the target stars.

3. The very first target star: The W UMa type eclipsing binary LS Del

LS Del (= HD 199497 = BD +19°4574) was discovered in 1966 by Bond who also carried out the first photoelectric study of the system (Bond, 1976). The first light curve analysis of the system was presented by sezer et al. (1986). The inclination was found about $47 - 48°$, thus the eclipse should be very marginal, if any. They also derived a photometric mass ratio close to $q = m_c/m_h = 1.8$. They classified the system as contact one. Later, Demircan et al. (1991) extended the photometric investigation using new light curves, and they attempted to study the period variation of the system. They reported very fast (cycle-by-cycle) and complicate changes on the light curve. The shape and depth of both minima and the level of both maxima are continuously changing (the average value of the O'Connell effect is about 0.05 mag in V and 0.03 in B band). The spectroscopic orbit of LS Del was presented and analysed for the first time at Lu and Rucinski, amongst some other W UMa stars (1999). The resulted mass ratio: $q = m_2/m_1 = 0.375$.

3.1. The CESEB spectroscopy of LS Del

There were subsequent spectroscopic measurement of the system on 24 nights during 2001 at the Rhozen Observatory, using the 2 m RCC telescope. A few sample spectra (arranged by the orbital phases of LS Del) and the magnified part of 431–452 nm range of one of the measured spectra are shown

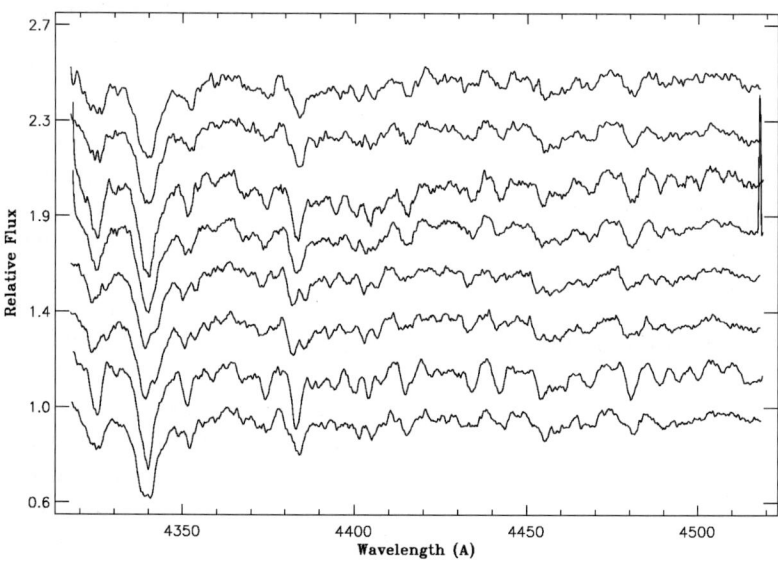

Fig. 1 A few Rhozen spectra of LS Del

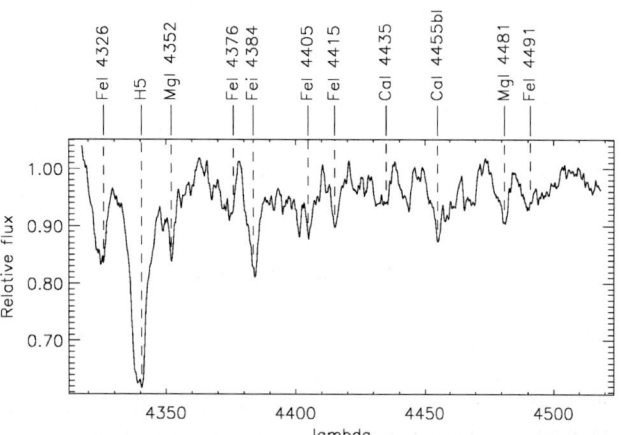

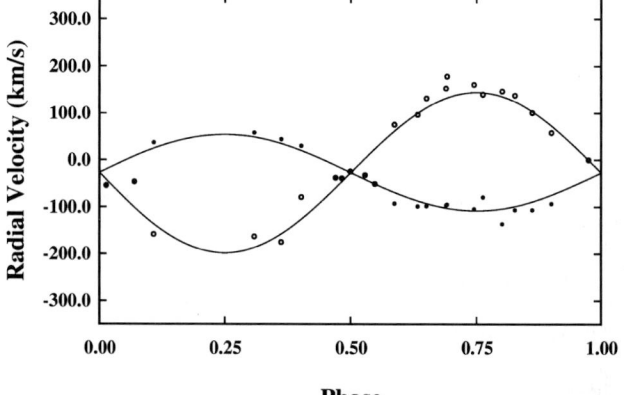

Fig. 2 (a) Part of LS Del spectrum; (b) The best fit to LS Del RV curve

Table 1 Comparison of two RV curves analysis of LS Del

Parameter	Lu and Rucinski (1999)	This work
$v_1 \sin i$ [km/s]	69.21 ± 1.41	81.02 ± 4.98
$v_2 \sin i$ [km/s]	184.78 ± 1.75	171.25 ± 6.52
q	0.3746	0.4731
$a \sin i$ [$R_\odot$]	1.83	1.81
$m_1 \sin^3 i$ [$M_\odot$]	0.45	0.41
$m_2 \sin^3 i$ [$M_\odot$]	0.17	0.20
γ [km/s]	−25.85 ± 1.89	−27.02 ± 8.56
corr. coeff.	0.995	0.945

on Figs. 1 and 2a. All spectra have been reduced by IRAF packages ONEDSPEC and TWODSPEC facilities, and corrected by VRCORRECT and DOPCORR for heliocentric time.

Both analyses have been performed using Binary Maker "velfit" code (Bradstreet, 1992). Our fit to the Lu and Rucinski original data is exactly reproducing their results. The respective new fit and the the observed original Rhozen RV data are shown on Fig. 2b.

3.2. The CESEB photometry of LS Del

There were numerous BVR measurement of the system during 2001 at the Baja Observatory, using the 50 cm RC telescope. Since these measurements show relatively large scatter and deeper minima in comparison with some previous studies, the system should have been in higher activity status. Thus, we did not make any attempt to use this data set for LC analysis. Up to now there were no spot analyses of

Table 2 Comparison of two solutions to Sezer et al. (1986) B filter measurements (with WD code, no spot)

Parameter	sezer et al. (1986)	This work
i [°]	47.62	47.78
T_c [K]	5487	5445
q	1.72	1.97
$L_h/(L_h + L_c)$	0.449	0.431

this system, but one spotted region can be a plausible working hypothesis for the forthcoming light curve analysis attempts. Reanalysing Sezer's B filter observations, we could get slightly different result, with a mass ratio closer to our spectroscopic value of 2.11 (see Table 1).

For the similar trial to Sezer's V light curve, we could not obtain fit better than the originally published one. The next trial was to fit Demircan et al. (1991) measurements, but that set exhibits strong spot activity and quick changes, so we did not yet obtain a satisfactory fit. CESEB shall observe this star seasonally. Parallel to the new measurements, the simultaneous spotted LC + RV analysis of all existing data will also be continued.

References

Bond, H.E.: IBVS **1214**, 1 (1976)
Bradstreet, D.H.: Binary Maker 2.0 (1992)
Demircan, O., Selam, S., Derman, E.: ApSS **186**, 57 (1991)
Lu, W., Rucinski, S.M.: AJ **118**, 515 (1999)
Sezer, C., Gülmen, Ö., Güdür, N.: ApSS **115**, 309 (1986)
Wang, R., Lu, W., Fan, Q.: IBVS **2982**, 1 (1986)

Spot Modelling of ζ Andromedae

Zsolt Kővári · Katalin Oláh · János Bartus · Klaus G. Strassmeier · Thomas Granzer

Received: 1 November 2005 / Accepted: 1 February 2006
© Springer Science + Business Media B.V. 2006

Abstract The photometric light modulation of ζ Andromedae originates from the distorted geometry of the primary, and additionally, from spots of which parameters (temperature, size, location) are variable in time. We present spot modelling results for six two-colour light curves which show that spots preferably appear on the stellar surface towards the companion star and opposite to it, where the distortion also causes dimming. Therefore, simple fitting of the measured data for the ellipticity effect does not yield correct result. Instead, ellipticity calculated from exact stellar parameters should be removed from the data to get reliable spotted light curves.

Keywords Stars: Activity of, Stars: Imaging, Stars: individual: ζ Andromedae, Stars: late-type, Starspots

1. Spots and ellipticity

The primary component of the RS CVn-type single line spectroscopic binary ζ And (HD 4502) is well-known of its ellipsoidal shape, which is the main reason of its photometric light variability. Apart from that, spot activity is also present on the star (see Strassmeier et al., 1989 and references therein).

In this paper we analyse those observations which were suitable for spot modelling simultaneously in two colours for determining spot temperature together with spot positions. Our B and V data were gathered from the literature

Z. Kővári (✉) · K. Oláh
Konkoly Observatory, Budapest, Hungary

J. Bartus · K. G. Strassmeier · T. Granzer
Astrophysical Institute Potsdam, Germany

(Strassmeier et al., 1993, Zhang et al., 2000) for 1984 (2), 1985 and 1988, while new observations were made in 1996–97 with the Wolfgang 0.75-m APT of the University of Vienna (Strassmeier et al., 1997) in Strömgren b and y at Fairborn Observatory, Arizona. All observations are phased using the equation $HJD = 2449992.781 + 17.769426 \times E$ (Fekel et al., 1999).

The ellipticity effect of ζ And has been studied before by fitting the observational data with appropriate functions, summarized in Kaye et al. (1995). However, if the star has also spots facing the companion star and opposite to it, as is observed on several other giants, e.g., on UZ Lib (Oláh et al., 2002a,b) or on IM Peg (Ribárik et al., 2003), then a simple $\sin\theta$ plus $\sin 2\theta$ fit to the data cannot separate the ellipticity effect from effects of spots situated near the two elongated parts of the star. To avoid a possible misinterpretation, we simply predict the ellipsoidal light curve using the accurately determined stellar and orbital parameters from our Doppler imaging study (see our other paper in this volume by Kővári et al., hereafter paper II). The resulting double-wave curve with unequal minima has amplitudes smaller than quoted before by Kaye et al. (1995) (see Table 1). The left panel of Fig. 1 shows the y, V data suitable for modelling (the picture is similar for the b, B data). For the light curve correction we matched the mean level of the ellipsoidal light curve at the unspotted light level of $V = 4^m.00$, which is the maximum brightness observed in 2001.

2. Spot modelling

Hipparcos lists $B - V = 1^m.100$ and that is the same as observed in 1984–85 by Strassmeier et al. (1989). Thus, using $V_{\rm unsp} = 4^m.00$ we obtain $B_{\rm unsp} = 5^m.1$. In 1996–97 the observations were made in Strömgren b, y colours. Although

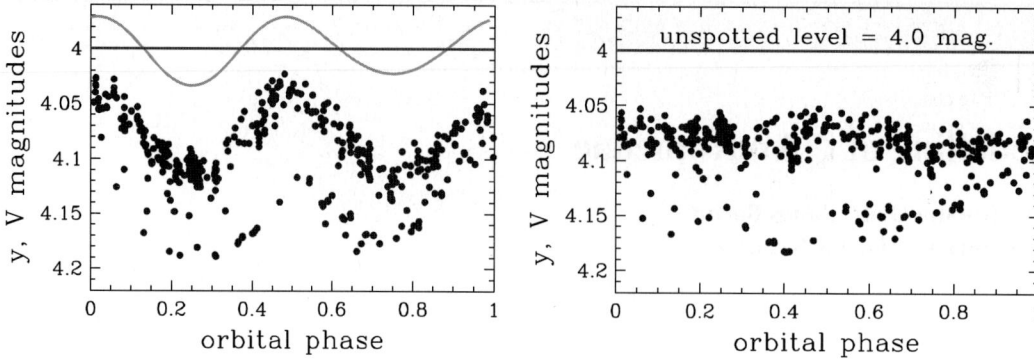

Fig. 1 *Left*: Calculated ellipsoidal light curve from the stellar and orbital parameters using NIGHTFALL (www.lsw.uni-heidelberg.de/~rwichman/Nightfall.html), and observations in y, V bandpasses. Right: the ellipticity-corrected light curves

y is almost exactly the same as V, we had to transform both the B magnitude of the comparison star and the unspotted B magnitude of ζ And to b. Using the method of Harmance and Božic (2001) we get $b = 4\overset{m}{.}679$ as unspotted b light. The error of the B-to-b transformation is expected to be less than $0\overset{m}{.}01$. Limb darkening coefficients are from the tables of van Hamme (1993).

For spot modelling we use SPOTMODEL by Ribárik et al. (2003). The code treats two-colour light curves simultaneously, and thereby fits the spot temperature together with

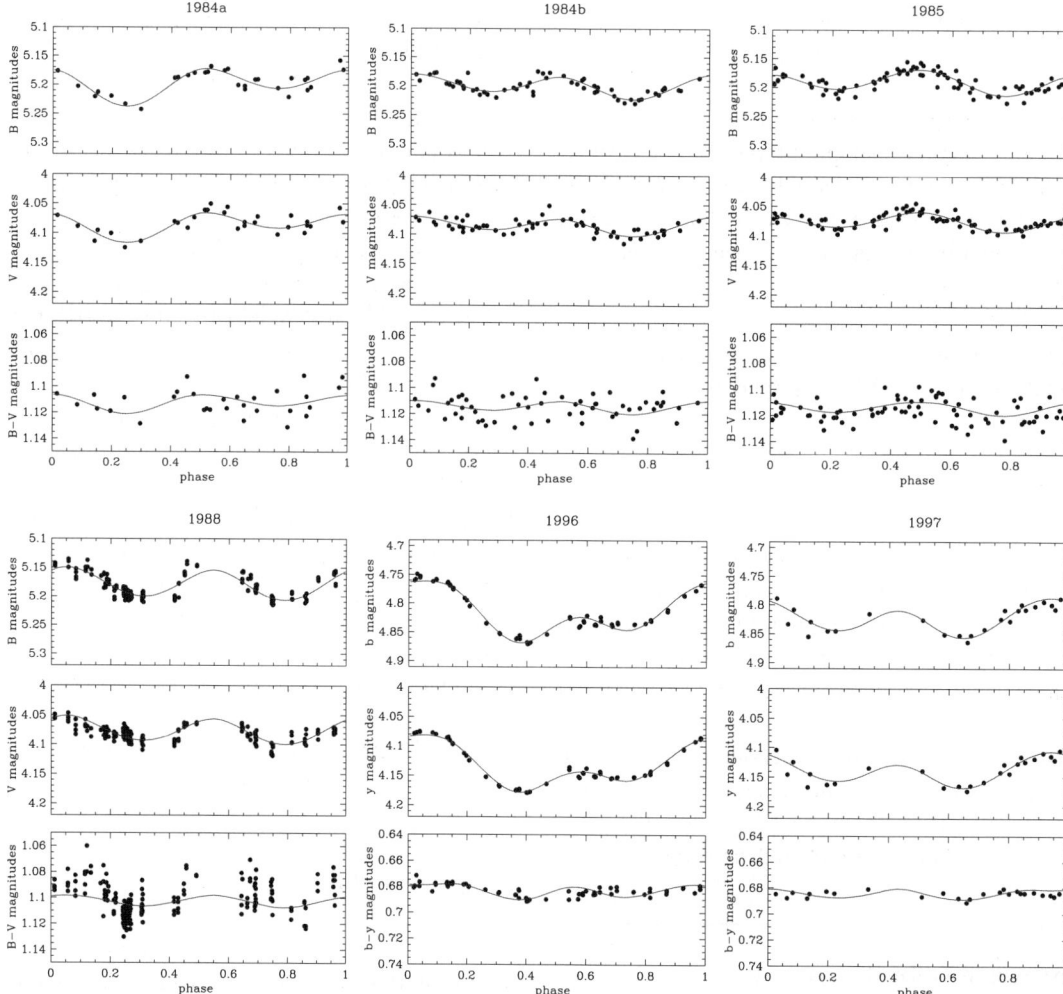

Fig. 2 Simultaneously fitted B, V and b, y light curves of ζ And. Blocks corresponding to six observing seasons (1984a, 1984b, 1985, 1988, 1996, 1997) show the data and the simultaneous fits in both colours, and the colour indices with the difference fits

Fig. 3 Resulting spot temperature (top) and spot coverage (bottom) showing significant temperature changes with simultaneous spot coverage variations

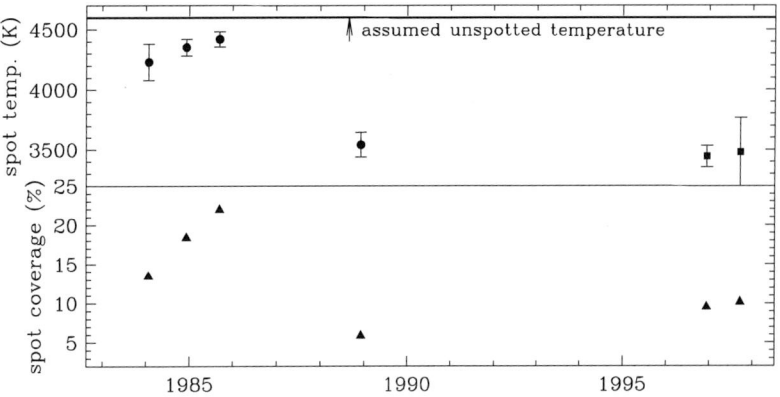

Table 1 Amplitudes of the ellipsoidal light curve

Minimum	A(B)	A(b)	A(V)	A(y)
Main	0.^{m}0681	0.^{m}0674	0.^{m}0625	0.^{m}0638
Secondary	0.^{m}0557	0.^{m}0550	0.^{m}0510	0.^{m}0518

the spot position and size. Because the latitude information in our small-amplitude light curves is very low, we use two spots, both with fixed latitudes of 25° (i.e. at center of the stellar disk at $i = 65°$) and one spot on the pole which accounts for the variable mean light level. Results are plotted in Fig. 2. The light curves are always double-humped showing two minima near phases 0.25 and 0.75, i.e., spots are facing the companion and opposite to it.

The changes of spot temperature and spot coverage are plotted in Fig. 3. Considering the errors, significant spot temperature and corresponding spot coverage variability is found. During 1984–85 the spot coverage was between 14–23%, and the temperature was about 4200–4400 K, i.e., ζ And had large, not too cool spotted areas, whereas in 1989 and 1996–97 the spots were much smaller (5–10%) with temperatures of $\approx$3500 K. The spot temperature results for the years 1996–97 agree well with the value of $\approx$1000 K below the photosperic temperature obtained from Doppler imaging (see paper II).

We think that spot temperature changes could mean different plage-to-spot ratio, since active regions are usually combinations of cool and hot areas. From the solar analogy this would mean, that an old active region slowly dissolves with growing plage-to-spot ratio and coverage, and later new activity starts with cool spots and only a few (if any) plages.

Acknowledgements ZsK and KO are grateful to the Hungarian Science Research Program (OTKA) for support under grants T-038013, T-043504 and T-048961.

References

Fekel, F.C., Strassmeier, K.G., Weber, M., Washüttl, A.: A&A Suppl. **137**, 369 (1999)
Harmanec, P., Božić, H.: A&A **369**, 1140 (2001)
Kaye, A.B., Hall, D.S., Henry, G.W., et al.: AJ **109**, 2177 (1995)
Oláh, K., Strassmeier, K.G., Weber, M.: A&A **389**, 202 (2002)
Oláh, K., Strassmeier, K.G., Granzer, T.: Astron. Nachrichten **323**, 453 (2002)
Ribárik, G., Oláh, K., Strassmeier, K.G.: AN **324**, 202 (2003)
Strassmeier, K.G., Boyd, L.J., Epand, D.H., Granzer, T.: PASP **109**, 697 (1997)
Strassmeier, K.G., Hall, D.S., Boyd, L.J., Genet, R.: ApJ Suppl. **69**, 141 (1989)
Strassmeier, K.G., Hall, D.S., Fekel, F.C., Scheck, M.: A&A Suppl. **100**, 173 (1993)
van Hamme, W.: AJ **106**, 2096 (1993)
Zhang, Zh., Yulan, Li, Huisong, Tan, Hongguang, S.: IBVS 4935 (2000)

Chromospheric Activity and Orbital Solution of Six New Late-type Spectroscopic Binary Systems

M.C. Gálvez · D. Montes · M.J. Fernández-Figueroa · J. López-Santiago

Received: 1 November 2005 / Accepted: 1 December 2005
© Springer Science + Business Media B.V. 2006

Abstract We present here the results of our high resolution echelle spectroscopic observations of six recently identified spectroscopic binary systems with late-type stellar components (HD 82159 (BD +11 2052 A); HIP 63322 (BD +39 2587); HD 160934 (RE J1738 +611); HD 89959 (BD +41 2078); HD 143705 (BD +29 2752); HD 138157 (OX Ser)). The orbital solution has been obtained using precise radial velocities determined by cross-correlation with radial velocity standard stars as well as previous values reported by other authors. These multiwavelength optical observations allow us to study the chromosphere of these active binary systems using the information provided by several optical spectroscopic features (from Ca II H & K to Ca II IRT lines) that are formed at different heights in the chromosphere. The chromospheric contribution in these lines has been determined using the spectral subtraction technique. In addition, we have determined rotational velocities (vsin i), lithium (Li I λ 6707.8 Å) abundance, and kinematic properties (membership in representative young disk stellar kinematic groups).

Keywords Binary · Cromospheric activity · Orbital solution

1. Observations and data reduction

1.1. Observations

High resolution spectroscopic observations of six near spectroscopic binaries were obtained from 29 March to 7 April 2004 using the 2.2 m Telescope at the German Spanish Astronomical Observatory (CAHA) in Almeria (Spain), with the Fibre Optics Cassegrain Echelle Spectrometer (FOCES). HD 160934 was observed in similar way in 5 observing runs from 2000 to 2004 in several telescopes.

1.2. Radial velocities

Heliocentric radial velocities of the components have been determined by using the cross-correlation technique, where the stars were cross-correlated order by order, using the routine fxcor in IRAF, against spectra of radial velocity standards of similar spectral types. The radial velocity of the components is derived from the position of the cross-correlation peak, one peak from each component in the case of the SB2 systems. In BD +39 2587 case, we observe the cross-correlation function of the primary, but we have obtained the secondary radial velocities using wavelengh differences between emission lines (see Fig. 1).

1.3. Chromospheric activity indicators

The contribution in the different optical chromospheric activity indicators has been determined using the spectral subtraction technique Montes et al. (1998). The synthesized spectrum was constructed using the program STARMOD developed at Penn State. We have obtained the subtracted spectra for all the optical indicators (Ca II IRT, Hα, Hβ, Hγ, Hδ and H & K Ca II lines).

M.C. Gálvez (✉) · D. Montes · M.J. Fernández-Figueroa · J. López-Santiago
Universidad Complutense de Madrid, Spain
e-mail: mcz@astrax.fis.ucm.es

Table 1 Unspotted light curve solutions of V387 Cyg; all errors are standard deviations

Parameter	Mode 2	Mode 4	Mode 5
ϕ_0	−0.0008(1)	−0.0008(1)	−0.0008(1)
i (degrees)	84.162(990)	81.241(610)	83.517(860)
T_1 (K)	8250*	8250*	8250*
T_2 (K)	5972(4)	5790(4)	5977(4)
g_1, g_2	1.00*, 0.32*	1.00*, 0.32*	1.00*, 0.32*
A_1, A_2	1.0*, 0.5*	1.0*, 0.5*	1.0*, 0.5*
Ω_1	3.0049(54)	3.4525	2.9708(48)
Ω_2	3.0254(118)	3.7690(151)	2.8935
$q = M_2/M_1$	0.5461(38)	0.8208(45)	0.5092(21)
$L_1/(L_1+L_2)$ (B)	0.9984(14)	0.9953(11)	0.9911(14)
$L_1/(L_1+L_2)$ (V)	0.8536(12)	0.8491(9)	0.8453(12)
$L_1/(L_1+L_2)$ (R)	0.7778(11)	0.7739(8)	0.7704(11)
$L_1/(L_1+L_2)$ (I)	0.7094(10)	0.7059(8)	0.7027(10)
r_1 (volume)	0.4242(28)	0.3959(29)	0.4230(27)
r_2 (volume)	0.3137(64)	0.3137(66)	0.3208(64)
χ^2	0.06103	0.06422	0.06045

*Assumed

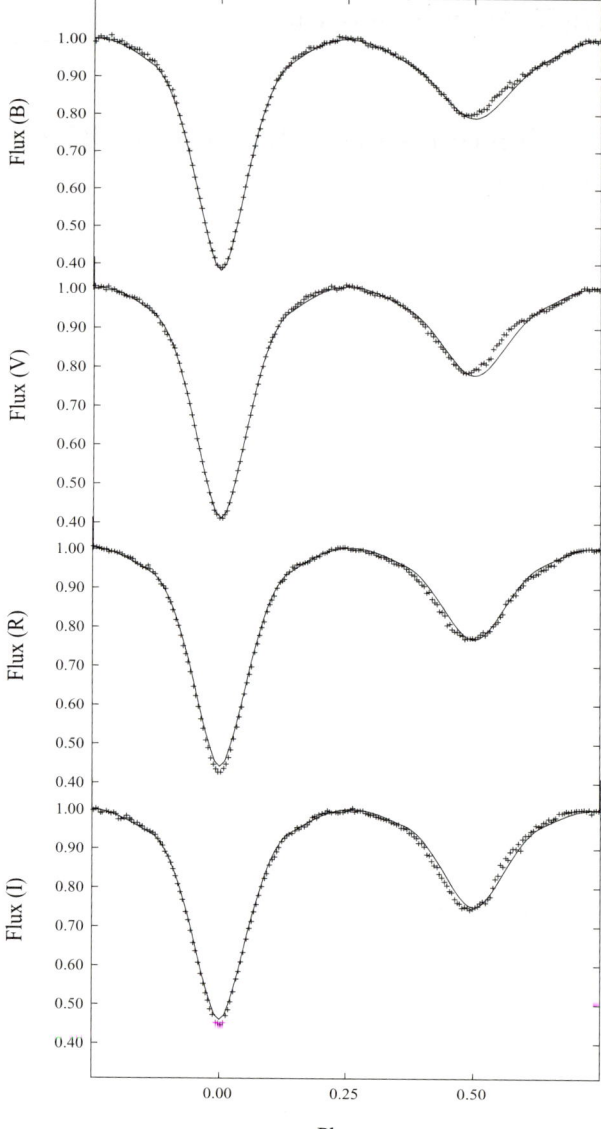

Fig. 1 The observational points and the theoretical light curve fitting (solid line) for our unspotted model (Mode 5)

points for each color were used in the analysis. The subscripts 1 and 2 refer to the star being eclipsed at primary and secondary minimum, respectively. The PHOEBE 0.26 version (Prša and Zwitter, 2005) of the Wilson-Devinney program was used in the light curve analysis. Since no double-line spectroscopy was available, a "q-search" conducted separately in Modes 2, 4 and 5 gave a starting value of $q = 0.6$ in all three Modes. A fixed value of the temperature T_1 was obtained from the spectral type given by Shaw (1994); also, assumed values for gravity, limb darkening coefficients and bolometric albedos according to the spectral types of the components were used. The results converged with the Modes 2, 4 and 5 of the program, and are shown in Table 1; the best fit (minimum χ^2 value averaged for all filters) was achieved with Mode 5, indicating that the secondary fills its Roche lobe. This corresponds to the FO Virginis subclass of near-contact binaries as defined by Shaw (1994). The parameters found in the unspotted solution were subsequently used in order to model the light curves with spots. We placed only one cool spot on the surface of the secondary, since T_1 is too high to allow spots in theory. However, the program did not converge for any parameters of the spot, although combinations of both a cool and a hot spot, or a single hot spot, were tried. The theoretical light curves of the unspotted solution, along with the observed ones, are shown in Fig. 1.

4. Discussion and conclusions

From the results obtained using the W-D program it seems more probable that the system of V387 Cygni is a semi-detached one. There are two hints that this near-contact binary undergoes a mass transfer phase. The first is the clear evidence for period change and the second is the asymmetry immediately after phase 0.5, which defies a spot explanation, thus suggesting either a temporary feature (flare) or the presence of circumstellar matter. We plan to observe again this asymmetrical part of the light curve; irrespective of the reappearance of the asymmetry, we believe that a proper quantitative modelling of the photometric effect of circumstellar material should be developed and tried for this system, and for any similar systems to be discovered in the future.

Acknowledgements This research was included in the *PYTHAGORAS* project for the support of research groups in the universities, co-funded by the EPEAEK program and the European Social Fund (ESF).

References

Baldinelli, L., Maitan, A.: IBVS No. 5220 (2002)
Berthold, T.: IBVS No. 1203 (1976)
Brancewicz, H. K., Dworak, T. Z.: AcA **30**, 501 (1980)
Manimanis, V.N., Niarchos, P.G.: IBVS No. 5624 (2005)
Morgenroth, V.O.: AN **255**, 425 (1935)
Nelson, R.H.: IBVS No. 5040 (2001)
Nelson, R.H.: IBVS No. 5371 (2001)
Prša, A., Zwitter, T.: ApJ **628**, 426 (2005)
Shaw, J.S.: Mem. S.A. Ital. **65**, 95 (1994)

Monitoring Secular Orbital Period Variations of Some Eclipsing Binaries at the Ankara University Observatory

K. Yüce · S. O. Selam · B. Albayrak · T. Ak

Received: 1 November 2005 / Accepted: 1 February 2006
© Springer Science + Business Media B.V. 2006

Abstract This study concerns the long-term monitoring of the secular variation character in the orbital period of some short-period eclipsing binaries observed at the Ankara University Observatory. Among the systems of our observing list are CK Boo, V502 Oph and V836 Cyg that show long-term secular variations in their orbital periods. We use classical O-C diagram analysis technique as a tool to reveal the character of the period variations of these binary systems.

Keywords Binaries · close - Binaries · eclipsing - CK Boo · V502 Oph and V836 Cyg

1. Introduction

CK Boo (HD 128141, BD+09°2916) was discovered to be a variable star by Bond (1975). Aslan et al. (1981) have noticed for the first time the presence of the O'Connell effect and the interchanging depths of primary and secondary minima in the light curves of CK Boo, and improved the ephemeris of the system. The first period study of the system was performed by Pajdosz and Zola (1988). Recently, Shengbang and Quingyau (2000) have studied the period variation of CK Boo.

V502 Oph (HD 150484, BD+00°3562) was discovered to be a variable star by Hoffmeister (1935). The spectral type of the system was assigned as G2V and G0V by Struve and

K. Yüce (✉) · S. O. Selam · B. Albayrak
Ankara University Faculty of Science, Dept. of Astronomy and Space Sciences, TR-06100, Tandoğan, Ankara, Turkey

T. Ak
İstanbul University, Faculty of Science, Dept. of Astronomy and Space Sciences, TR-34452, Üniversite, İstanbul, Turkey

Gratton (1948) and Pych et al. (2004), respectively. Kwee (1958) has first noticed the orbital period variation of the system. Derman and Demircan (1992) have suspected the existence of an additional component in the system based on their (O-C) analysis of the eclipse timings.

V836 Cyg (HD 203470, BD+35°4496) was discovered and classified as an Algol type binary by Strohmeier et al. (1956). Later, Deinzer and Geyer (1959) and Cester (1963) classified the type of the system as Beta Lyrae and overcontact, respectively.

2. Observations and (O-C) analyses

We have collected all available minima times for these three binary systems from the literature. For the (O-C) analyses, we had a data set of 1 ptg, 79 pe, 4 CCD timings for CK Boo, 1 vis, 1 ptg, 76 pe, 16 CCD timings for V502 Oph and 19 vis, 88 pe, 4 CCD timings for V836 Cyg. The data contains our new photoelectric times of minima (three in 2005 for CK Boo, two in 2005 for V502 Oph, one in 2004 for V836 Cyg) determined by using the Kwee and van Woerden (1956) method from the photoelectric observations obtained at the Ankara University Observatory. (O-C) diagrams for each of the systems were constructed using these data and displayed in Fig. 1a, d, g. All of them have obvious quadratic character, which is the indication of the monotonic change in the orbital periods of the systems. Parabolic least-squares fits to the (O-C) values yield the following quadratic ephemerides for our systems as:

CK Boo: $HJDMinI = 2442898.0760 + 0^d.35515081 \times E + 1.21 \times 10^{-10} \times E^2$

V502 Oph: $HJDMinI = 2439639.9572 + 0^d.45339408 \times E - 1.79 \times 10^{-10} \times E^2$

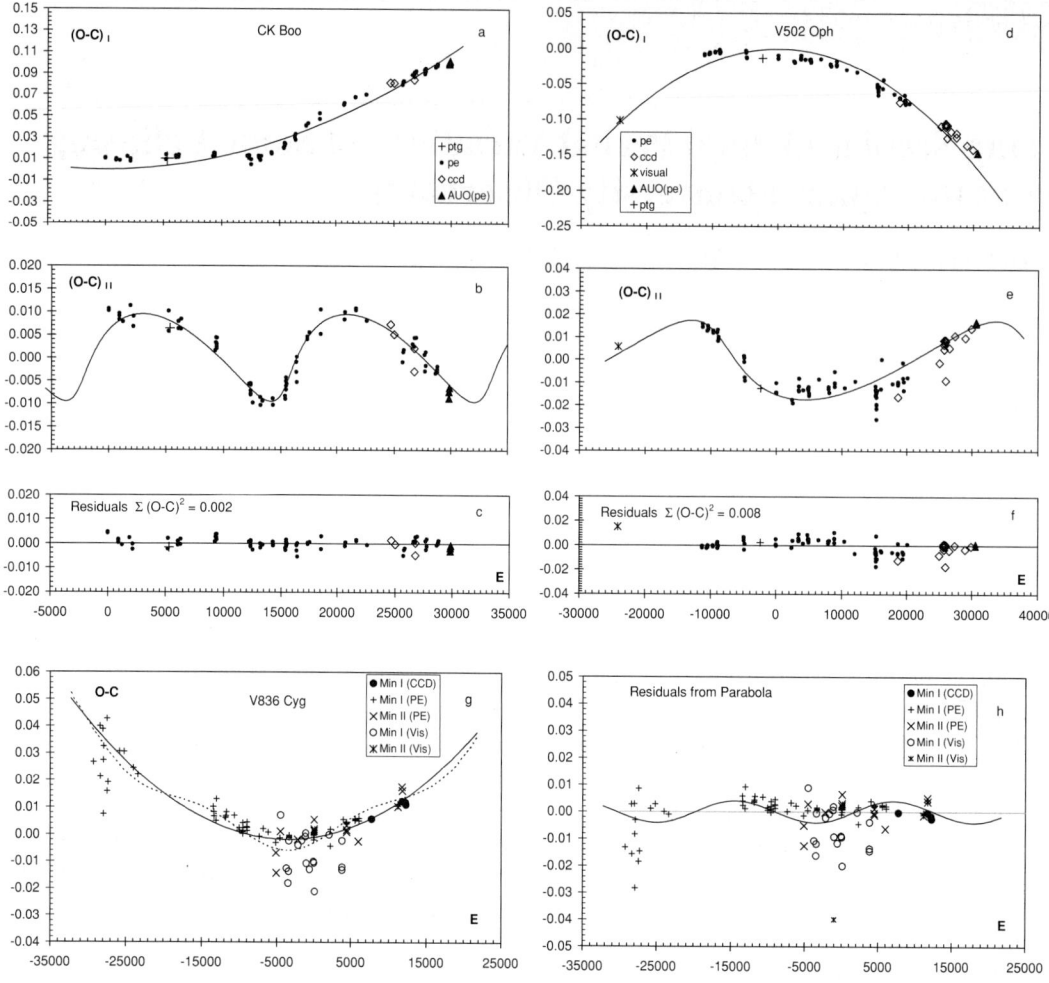

Fig. 1 (a, d) The O-C diagram with the parabolic representation for CK Boo and V502 Oph, respectively. (b, e) The $(O-C)_{II}$ residuals from the parabolic best fits with a third body approximation. (c, f) Final residuals from the overall fits for two binaries. (g) The O-C diagram for V836 Cyg and its represantation with a best parabolic fit (solid line) and a periodic term. (h) The residuals from the parabolic fit

V836 Cyg: $HJDMinI = 2444853.4903 + 0^d.65341220 \times E + 6.28 \times 10^{-11} \times E^2$

It can be seen from the (O-C) diagrams and the sign of the quadratic elements derived from the analysis, that the orbital period of the CK Boo and V836 Cyg is increasing with a dP/dt rate of 2.49×10^{-7} and 7.02×10^{-8} days yr^{-1}, respectively, while the orbital period of V502 Oph is decreasing with a rate of -2.88×10^{-7} days yr^{-1}.

The long term orbital period decrease or increase can be attributable to mass exchange/loss mechanisms in those systems. Under the assumption of conservative mass transfer between the components of the system, the calculated rate of the mass transfer, dM/dt in $M_\odot$ yr^{-1}, and their directions are 3.92×10^{-8} (less massive to the more massive) for CK Boo, 1.64×10^{-7} (more massive to the less) for V502 Oph and 3.94×10^{-8} (less massive to the more) for V836 Cyg.

Residuals from the quadratic ephemerides were also plotted in each figure (Fig. 1b, e, h). Although it is less evident in the case of V836 Cyg, these residuals show some evidence of cyclic changes. These cyclic changes can be produced by gravitationally bound additional components in the systems or magnetic activity cycle effect of their active components. Under the assumption of the existence of third bodies in CK Boo and V502 Oph, the residuals of the (O-C) diagrams were fitted with the formulation given by Irwin (1952) and the third body specific parameters were derived (see Table 1). The magnetic activity cycle effect is more favourable to the (O-C) residuals of V836 Cyg, because of its quasi-periodic nature.

3. Discussion

An analysis of the (O-C) diagrams of three eclipsing binaries, CK Boo, V502 Oph and V836 Cyg, was performed. The (O-C) variations of the systems suggest that the orbital periods of CK Boo and V836 Cyg show secular increases which may be due to mass transfer from the less massive to the more massive component for each system while the orbital period

Table 1 Parameters derived from the (O-C) analyses

Parameter	3rd body CK Boo	3rd body V502 Oph	Parameter	Magnetic activity cycle V836 Cyg
$a'_{12} \sin i'$ [AU]	1.81 ± 0.11	3.33 ± 0.12	ΔP (sn cycle^{-1})	0.101
e'	0.55 ± 0.03	0.49 ± 0.03	ΔJ (cgs)	-1.49×10^{47}
ω' [°]	321.0 ± 2.3	150.0 ± 1.7	ΔE (cgs)	5.60×10^{40}
T' [HJD]	2441977 ± 26	2437970 ± 32	ΔL (cgs)	1.45×10^{32}
P_{12} [years]	17.41 ± 0.14	57.88 ± 0.17	B (kG)	6.0
A [days]	0.0095 ± 0.0004	0.0174 ± 0.0003	Δm (mag.)	0.001
$f(m_3)$ [$M_\odot$]	0.0197 ± 0.0035	0.011 ± 0.001	P_{cyc} (year)	38.46
m_3 [$M_\odot$]	0.48 ± 0.03	0.37 ± 0.02		

of V502 Oph shows a decrease. For CK Boo and V502 Oph, the periodic variations of orbital period suggest the presence of a third-body in the systems. The masses of the third stars are found to be 0.48 and 0.37 $M_\odot$, respectively. In CK Boo the mass of the third body turns out to be larger than the secondary component and could be a compact object or a close binary. The residual (O-C) variation of V836 Cyg can be well arisen from the magnetic activity cycle effect of its late type active component. Under these circumstances, we calculated the activity related parameters using the Applegate's (1992) formulation. They are listed in Table 1.

Acknowledgements This work has been supported by the research fund of Ankara University (BAP) under the research projects no : 20040705089. We would like to thank M. Helvaci for his help during the analysis of V836 Cyg.

References

Applegate, J.H.: ApJ **385**, 621 (1992)
Aslan, Z., Gören, M., Derman, E.: IBVS No. 2043 (1981)
Bond, H.E.: PASP **87**, 877 (1975)
Cester, B.: Oss. Astron. Tiresto, No. 317 (1963)
Deinzer, W., Geyer, E.: ZfA **47**, 211 (1959)
Derman, E., Demircan, O.: AJ **103**, 1658 (1992)
Hoffmeister, C.: AN **255**, 403 (1935)
Irwin, J.B.: ApJ **116**, 211 (1952)
Kwee, K.K., van Woerden, H.: BAN **12**, 327 (1956)
Kwee, K.K.: Bull. Astron. Inst. Neth. **14**, 131 (1958)
Pajdosz, G., Zola, S.: IBVS No. 3251 (1988)
Pych, W., Rucinski, S.M., DeBond, H., Thomson, J.R., Capobianco, C.C., Melvin Blake, R.: AJ **127**, 1712 (2004)
Shengbang, Q., Quingyau, L.: Ap&SS **331**, 271 (2000)
Strohmeier, W., Kippenhahn, R., Geyer, E.: Kl. Veroff. Remeis-Sternwarte Bamberg Nr. 15 (1956)
Struve, O., Gratton, L.: ApJ **108**, 497 (1948)

The W UMa-type Stars Program: First Results, Current Status and Perspectives

J. M. Kreiner · S. Zola · W. Ogloza · B. Pokrzywka ·
M. Drozdz · G. Stachowski · B. Zakrzewski ·
P. G. Niarchos · K. Gazeas · S. M. Rucinski · M. Siwak ·
D. Koziel · D. Kjurkchieva · D. Marchev

Received: 1 November 2005 / Accepted: 1 February 2006
© Springer Science + Business Media B.V. 2006

Abstract We present the results of a statistical investigation of the period-color and period-bolometric magnitude relations using a carefully selected sample of 120 contact systems with known physical parameters.

Keywords Eclipsing binaries · W UMa stars · Physical parameters

1. Aims of the program

The primary aim of this program is the discussion of certain relations between the most important parameters of contact binary systems. Our approach is novel in that the absolute parameters (masses, radii and luminosities) of the systems have been derived for a homogeneous sample obtained by observation according to the following criteria: 1) The star has a mass ratio q_{sp} determined spectroscopically. 2) The complete light curve for a given system is obtained at one observatory using the same telescope and the same instrument (a CCD camera

J. M. Kreiner (✉) · S. Zola · W. Ogloza · B. Pokrzywka · M. Drozdz · G. Stachowski · B. Zakrzewski
Mt. Suhora Observatory, Cracow Pedagogical University, Poland

P. G. Niarchos · K. Gazeas
Department of Astrophysics, Astronomy and Mechanics, Faculty of Physics, University of Athens, Greece

S. M. Rucinski
David Dunlap Observatory, University of Toronto, Canada

M. Siwak · D. Koziel
Astronomical Observatory of the Jagiellonian University, Cracow, Poland

D. Kjurkchieva · D. Marchev
Department of Physics, Shoumen University, Shoumen, Bulgaria

or a multichannel photoelectric photometer) with an accuracy of at least ±0.01 mag. 3) The light curves are obtained in the shortest possible time to avoid any possible changes in the light curve due to, for example, spot activity. In total, 41 systems were observed with the results from light curve modelling available in the series of papers: Kreiner et al. (2003), Baran et al. (2004), Zola et al. (2004), Gazeas, Baran et al. (2005), Zola et al. (2005), Gazeas, Niarchos et al. (2005). In all cases, the light curve modelling was performed using the Wilson-Devinney code (Wilson, 1979, 1993) supplemented by a Monte Carlo search method. The effective temperature was derived from the spectral type, while the mass ratio q was deduced from spectroscopic measurements. The separation a and fill-out factor f were obtained as a result, allowing calculation of the effective surface $S(q, f)a^2$ of the binary (see Mochnacki, 1984). The above quantities, together with the temperature difference between the components, was used to compute the bolometric absolute magnitude M_{bol} of the system. The mean color index $(B-V)$, bolometric correction BC and absolute visible magnitude M_V were derived assuming that the radiative properties of the common envelope can be described using formulae adequate for main sequence stars (e.g. Popper, 1980).

The sample of 41 contact systems obtained in this way was augmented using data from the literature. Only parameters derived from models with spectroscopically-determined mass ratios were taken into account.

2. Relations between some parameters

W UMa stars are interesting objects as distance tracers. Extensive work concerning the $M_V(\log P, B-V)$ calibration has been done in the series of papers by Rucinski and co-workers e.g. Rucinski (2000) or Rucinski and Duerbeck

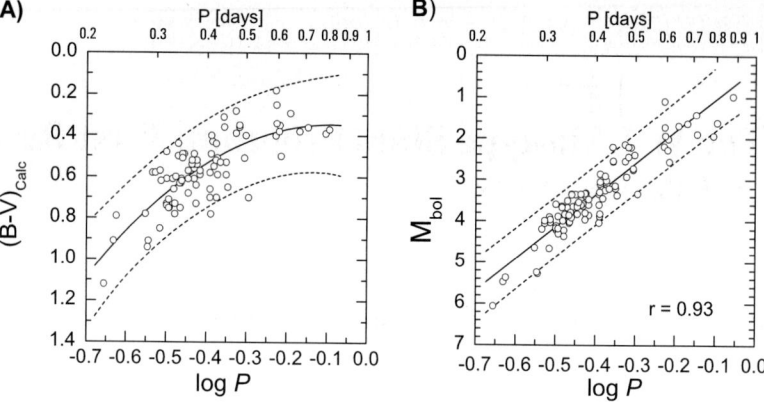

Fig. 1 Left (A): period – color diagram. Right (B): period – bolometric magnitude diagram. Continuous line denotes the best fit while dashed lines denote prediction bands at $\alpha = 0.95$

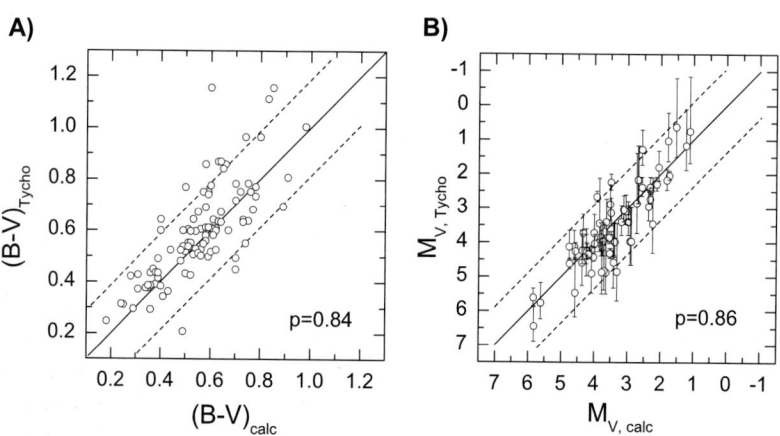

Fig. 2 Left (A): comparison between $(B-V)_{\text{calc}}$ and $(B-V)_{\text{Tycho}}$. Right (B): comparison between $M_{V,\text{calc}}$ and $M_{V,\text{Tycho}}$. Continuous line denotes $y = x$ dependence while dashed lines denote prediction bands at $\alpha = 0.95$. p denotes probability that relation between parameters is described by the $y = x$ function

(1997). The investigation of the $M_V(\log P, B-V)$ relation is complicated because both parameters are not independent. The set of physical parameters of contact binaries obtained both within our program and the literature was used to check the relation between the color index $(B-V)$ and the absolute bolometric magnitude versus the orbital period. Our sample is more numerous and homogeneous than those considered in earlier works. Two types of correlation can be derived from our data set. The first is the period-color diagram presented in Fig. 1A. Statistical analysis of the data leads to the conclusion that the best formula describing the $(B-V) - \log P$ relation is the second order polynomial:

$$(B-V)_{\text{calc}} = 0.363 + 0.362 \log P + 1.99 (\log P)^2.$$

The above relation propagates into M_{bol} versus $\log P$. Color index is related to temperature, which in the fourth power is the first factor determining the luminosity of a binary. However, the surface of the common envelope also affects luminosity. The size of the system described by component separation a is related to the period according to Kepler's law, and this dependence is a second source of the $M_V - \log P$ correlation. The dependence of M_{bol} on $\log P$ is presented in Fig. 1B. In this case a linear function is sufficient:

$$M_{\text{bol}} = 0.30 - 7.70 \log P$$

The self-agreement between absolute magnitudes M_V and color indices $(B-V)$ derived directly from observations and those derived from modelling are presented in Fig. 2A and 2B. The values of $(B-V)_{\text{Tycho}}$ and $M_{V,\text{Tycho}}$ were taken from Hipparcos and Tycho catalogues and corrected for interstellar extinction (assuming 1.6 mag/kpc).

3. Perspectives

Our set of physical parameters of contact binaries is far from complete, even for stars brighter than 9th magnitude. Therefore, we will continue spectroscopic and photometric observations to obtain more radial velocity and photometric light curves. Subsequently, we will perform light curve modelling using the method described above. We hope to obtain physical parameters for most contact binaries brighter than 9th magnitude for which trigonometric parallaxes were obtained by the Hipparcos mission. This data set will be very useful for investigation of the spatial distribution of contact systems near the Sun and for testing the relations between physical parameters.

This project was made possible by very good international cooperation, and we hope that observers from the southern hemisphere will join us and perform observations of low-declination stars which are not accessible from the northern hemisphere.

Acknowledgements This project was supported by the Polish National Committee grant No 2 P03D 006 22. The support from the Special Account for Research Grants (Nos. 70/4/5806 and 70/3/7187) of the University of Athens is kindly acknowledged.

References

Baran, A., Zola, S., Rucinski, S.M., et al.: Acta Astron. **54**, 195 (2004)
Gazeas, K., Baran, A., Niarchos, P., et al.: Acta Astron. **55**, 121 (2005)
Gazeas, K., Niarchos, P., Zola, S., et al.: Acta Astron. **56**, 127 (2006)
Kreiner, J.M., Rucinski, S.M., Zola, S., et al.: Acta Astron. **412**, 465 (2003)
Mochnacki, S.W.: J. Suppl. Ser. **55**, 551 (1984)
Popper, D.M.: Ann. Rev. Astron. Astrophys. **18**, 115 (1980)
Rucinski, S.M.: Astron. J. **120**, 319 (2000)
Rucinski, S.M., Duerbeck, H.W.: PASP **109**, 1340 (1997)
Wilson, R.E.: Astrophys. J. **234**, 1054 (1979)
Wilson, R.E.: Documentation of Eclipsing Binary Computer Model (1993)
Zola, S., Rucinski, S.M., Baran, A., et al.: Acta Astron. **54**, 299 (2004)
Zola, S., Kreiner, J.M., Zakrzewski, B., et al.: Acta Astron. **55**, 389 (2005)

A Mechanism for Producing Short-Period Binaries

Peter P. Eggleton · Ludmila Kisseleva-Eggleton

Received: 1 November 2005 / Accepted: 1 February 2006
© Springer Science + Business Media B.V. 2006

Abstract When three point masses form a hierarchical triple system, the short-period orbit can be severely modified by the long-period orbit if the two orbits are inclined to each other by more than about 39°deg ($\sin^{-1}\sqrt{2/5}$). Such an inclination can induce 'Kozai cycles' (Kozai, 1962), in which the eccentricity of the inner orbit cycles by a large amount while its period and therefore semimajor axis remains roughly constant. During those periastra when the eccentricity is largest, tidal friction may become important, and this can result in a secular shrinkage of the orbit, until it becomes circularised at a period of a few days.

However, apsidal motion due to either GR or to the quadrupolar distortion of the components (if they are no longer treated as point masses) can reduce the range of eccentricity. We explore the limits on outer and inner orbital period that these perturbations imply.

If the components are F/G/K/M dwarfs, then rotationally-driven dynamo activity can become important at the short periods that can occur in the right circumstances. It can cause the period to shorten further. The result may be a contact binary, and/or a merger in which the two stars of the inner pair coalesce to form a single rapidly rotating star. We suggest that this may be the origin of AB Dor, a very rapidly rotating K dwarf that is probably about 50 Myr old.

Keywords Binary stars · Triple stars

P. P. Eggleton (✉)
Lawrence Livermore National Laboratory, Livermore, CA 94550, USA

L. Kisseleva-Eggleton
California State University Maritime Academy, Vallejo, CA 94590-8181, USA

1. Introduction

Binary systems constitute probably ~50% of stellar systems, and triple systems may be ~10% more. Among the 60 systems nearer than ~5.5 pc (61 including the Sun), about 20 are binary (including three with massive planets) and 8 are triple, but the census of multiples continues to increase slowly (e.g. McCaughrean et al., 2004). Nearby stars are usually of low mass, and massive stars appear to be even more likely to be multiple.

It is not clear how either single or multiple stars form, although it *is* fairly clear that most or all form in dense 'Star Forming Regions' (SFRs). It may well be that multiple systems form most readily, and later break up, rather than that single or binary stars form first and then agglomerate into higher multiples. The multiple system T Tau was recently observed to break up (Loinard et al., 2003; Furlan et al., 2003); one component was apparently ejected from a previous orbit of ~20 yr. The system was presumably at least quadruple before ~1998, and is now at least triple with the fourth member ejected. However, this interpretation has been challenged by others (Tamazian, 2004; Johnston et al., 2004).

Because most stellar formation is seen to take place in SFRs, and yet most stars in the solar neighbourhood are *not* in dense clusters, we infer that clusters typically disintegrate. This is partly due to gravitational scattering, but is probably more strongly due to the ejection of a considerable fraction of the original mass of gas by energetic processes: supernovae, and jets from young stellar objects. If a half or more of the primordial gas is ejected rapidly by such processes, the remaining fraction can be gravitationally unbound. As the systems move apart, their chance of interacting with each other diminishes, and after ~10 Myr there should be little such interaction. In about the same time interval, subclusters on various scales will also be disintegrating, if they are non-hierarchical. Thus the outcome should be a large number of independent hierarchical multiples, in which one might suppose there is practically no scope for further dynamical evolution.

However, even if an SFR *has* disintegrated into a large number of independent hierarchical multiples, there still exists a prospect of dynamical interaction on a long timescale within some of these, the ones which are triple or higher-multiple. Kozai cycles by themselves can lead to ejection events, including possibly the ejection of T Tau Sc from the T Tau multiple. The combination of Kozai cycles and tidal friction can lead to a secular hardening of the innermost members of systems, and only stops when, at an inner period of a few days, the quadrupolar distortion of each star by the other leads to apsidal motion which is comparable in magnitude to the apsidal motion induced by the distant third body. Even then there is the possibility of further secular hardening, because in such close binaries, at least if they are of type F/G/K/M, magnetic braking may continue the period decrease. On a timescale that may be many Myr, or even Gyr, the system may be reduced to a contact binary and/or a merger into a single rapidly-rotating star. We suggest that AB Dor, a very rapidly rotating K dwarf which is a member of a loose association ∼50 Myr old (Zuckerman et al., 2004), and is also a member of a hierarchical quadruple sub-system of this association, may have originated fairly recently from a merger in what was previously a hierarchical quintuple system.

2. Kozai cycles

If the outer orbit of a triple is inclined at more than $39°$ to the inner, the parameters of the inner orbit, in particular the eccentricity and angular momentum but *not* the period or semimajor axis, are forced to cycle between two values (Kozai, 1962). The larger value of eccentricity can be quite close to unity. The maximum is given implicitly by

$$\cos^2 \eta = \frac{2e_{\min}^2(1 - e_{\max}^2)}{5e_{\max}^2(1 - e_{\min}^2)}, \qquad (1)$$

where η is the angle between the two orbits and $e_{\min}$ is the minimum eccentricity. The maximum is unity if the orbits are exactly perpendicular. An analysis of the Kozai mechanism, in the quadrupole approximation, was given by Kiseleva et al. (1998).

If the mutual inclination of the two orbits is random, as can be expected if triples are subjected to considerable dynamical interaction with nearby stars during the formation process, then the median inclination should be $60°$. This is quite enough to drive the peak eccentricity to 0.764, even if the orbit is circular to start with. But with the same random distribution of inclinations, 17% will have an inclination of over $80°$ and then the peak eccentricity is in excess of 0.975. Thus it is not improbable that the periastron separation may decrease by a factor of 40 in the course of a Kozai cycle.

Remarkably, the range of eccentricity fluctuation is independent of (a) the masses, (b) the periods, and (c) the *outer* eccentricity. These quantities only affect the *period* P_K of the cycle, according to

$$P_K \sim \frac{M_1 + M_2 + M_3}{M_3} \frac{3 P_{\text{out}}^2}{2\pi P_{\text{in}}} \left(1 - e_{\text{out}}^2\right)^{3/2}. \qquad (2)$$

This is much the same period as for precession and apsidal motion. Even a brown dwarf or major planet might cause a Kozai cycle of large amplitude, with a period of $\lesssim 10$ Myr if the outer period is $\lesssim 100$ yr.

If at the peak of a Kozai cycle the periastron separation is less than ∼6 times the sum of the radii, it is likely that tidal friction will be important. This is *not* time-reversible, and tends to circularise the orbit at roughly constant periastron separation, so that the period can be substantially reduced. It may take many Kozai cycles to do this, but if the cycle time is ∼1 Myr there is still scope for the process to work in say $10 - 10^4$ Myr.

Three physical processes may, however serve to reduce the maximum eccentricity that is predicted by Table 1. They are (a) the effect of GR on the inner orbit, (b) the effect of quadrupolar distortion, due to rotation, in each of the inner pair, and (c) the effect of quadrupolar distortion of each star by the other in the inner pair. All three of these processes produce apsidal motion, and if this is comparable to the apsidal motion produced by the third body then they interfere with the Kozai cycle. The eccentricity still cycles (because all these processes are time-reversible) but over a range which may be much more limited.

For example take three equal (and solar) masses, and $P_{\text{in}} = 10$ d, $P_{\text{out}} = 10$ yr. Start with both orbits circular, and with a mutual inclination of $80°$. GR, stellar spin ($P_{\text{rot}} = 1$ d), and mutual distortion reduce the peak eccentricity from 0.975 to 0.78; mutual distortion was dominant. Thus the periastron separation is reduced by a factor of 4.5 rather than 40, but this is still enough to give tidal friction a good chance to operate on a reasonably short timescale.

Figure 1 shows the effect of tidal friction over a longer timescale. The behaviour appears to be quite complex, and we cannot say that we understand every aspect of it yet, but a longer-term effect is that the inner orbit is finally circularised at a period of 2.5 d, after about 50 Myr, and the inclination reduced from $80°$ to $74°$. Although the pattern is fairly complex, we find much the same pattern independent of starting conditions, provided those conditions are such as to bring the stars close enough together at their periastra for tidal friction to be significant.

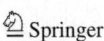

Fig. 1 Kozai cycles and tidal friction for a system with initial parameters $((1 + 1\,M_\odot; 10\,\mathrm{d}) + 1\,M_\odot; 10\,\mathrm{yr})$, $\eta = 80°$. (a) Eccentricity: the Kozai fluctuations diminish, but e tends first to a high value of ~ 0.7 before decreasing in ~ 100 Myr to near zero. (b) The mutual inclination η: it fluctuates before settling at $\sim 74°$

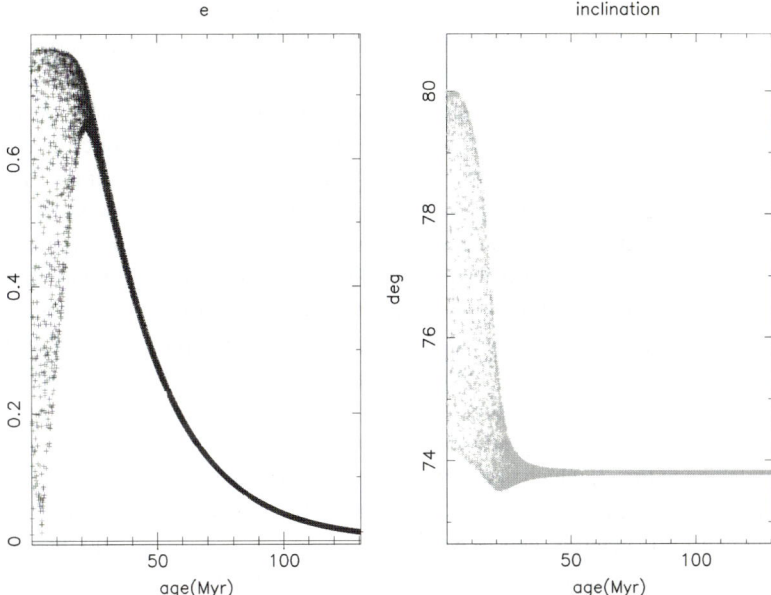

Two additional physical processes are included in Fig. 1, although they only become important at later stages than shown: a crude model of stellar evolution, and a crude model of the combination of mass loss by dynamo-driven stellar wind and angular momentum loss by magnetic braking. Stellar evolution, of course, makes the components evolve to larger radii, so that interaction by Roche-lobe overflow (RLOF) is almost inevitable. The increasing size increases the quadrupole moment, and so would quench the Kozai cycles, but almost certainly not before they, combined with tidal friction (which also increases with size), have circularized the orbit at a period substantially shorter than the initial period.

Figure 2 shows, on the plane of $\log P_{\mathrm{out}}$ versus $\log P_{\mathrm{in}}$, the area where Kozai cycles can be expected to operate. There is an excluded area to the lower right, where P_{in} and P_{out} are sufficiently close together that the system is non-hierarchical, and can be expected to break up very rapidly. The sloping upper boundary is dictated by GR towards the right and by quadrupole distortion towards the left. We computed this figure three times, each time for three equal masses which were 0.1, 1 and 10 $M_\odot$, and found much the same upper boundary. Very crudely, the region for Kozai cycles is $P_{\mathrm{out}} \gtrsim 4 P_{\mathrm{in}}$, $P_{\mathrm{out}}(\mathrm{y}) \lesssim 0.5 P_{\mathrm{in}}^{1.5}(\mathrm{d})$. We find the existence (or otherwise) of Kozai cycles to depend rather little on the mutual inclination of the orbits, provided it is above 39°, although their amplitude does.

The region for Kozai cycles is quite substantial, but the region where tidal friction is significant at periastra, the darker area in Fig. 2, is smaller, and rather harder to quantify. On the one hand, the dissipation coefficient in the equilibrium-tide model is by no means precise, and on the other it is not clear whether the equilibrium-tide model is good enough

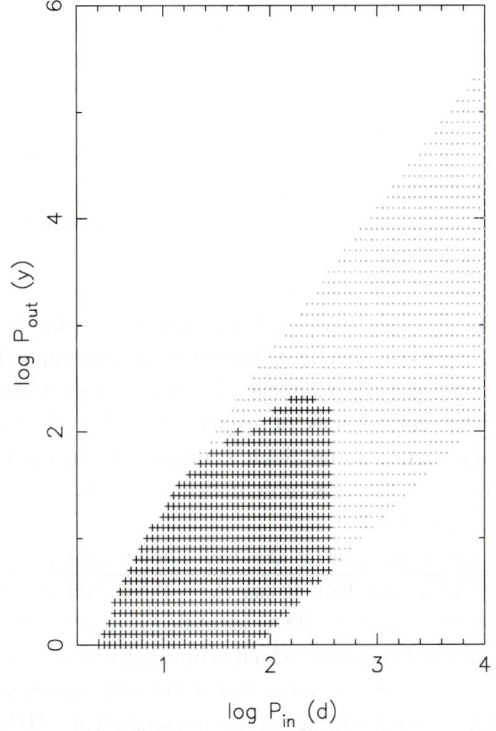

Fig. 2 The entire shaded area is where Kozai cycles, starting from $e = 0.01$ and $\eta = 80°$, are able to increase e to above 0.5 cyclically. The darkly shaded area is where the timescale of tidal friction at the peak of eccentricity is enough to reduce the eccentricity on a timescale of $\lesssim 1$ Gyr. The three masses were all $1\,M_\odot$

for high-eccentricity encounters. Further, the effectiveness of tidal friction depends strongly on the maximum eccentricity, and hence on the inclination. In Fig. 2, the initial mutual inclination was taken to be 80°. Thus we cannot say with certainty what fraction of systems will be subject to the combination

Evolutionary Status of Late-type Contact Binaries: Case of a $1.2 + 1\, M_\odot$ Binary

Kazimierz Stępień

Received: 1 November 2005 / Accepted: 1 February 2006
© Springer Science + Business Media B.V. 2006

Abstract It is argued that typical W UMa-type stars are old, advanced evolutionary objects, similar to Algols, in the sense that they are past mass exchange resulting in a mass ratio inversion. Their secondaries are oversized due to depletion of hydrogen, and in many cases they possess small helium cores. An alternative evolutionary scenario leading to such a configuration is presented. Differences between the evolution of binaries with the initial mass ratio far and close to unity are discussed.

Keywords Contact binaries · Stellar activity · Evolution of binary stars

1. Introduction

Cool contact binaries, or W UMa-type stars, are systems in which both components share a common envelope between the inner and outer critical surface in the Roche model. Most systems have components with spectral types F-K and nearly equal surface temperatures. Early analyses of these stars assumed that binaries are young, with both components on, or close to the Zero Age Main Sequence (ZAMS). This posed a fundamental problem with explaining their basic physical parameters. There exists a well known mass-radius relation for young, MS stars whereas the Roche geometry requires a different mass-radius relation which cannot be satisfied by two MS stars of different mass. Furthemore, the effective temperature of MS stars increases with mass, so the two stars of different mass should have different temperatures. The observations of contact binaries with unequal components contradict both relations obeyed by single stars. The solution to this paradox was suggested by Lucy (1968a,b) and later on elaborated by other authors. It consists of the following basic points:

1. Both components are MS stars
2. When a primary component reaches its critical surface, mass transfer begins
3. After mass transfer of $\sim 0.1 M_\odot$ the secondary (lying initially well inside its Roche lobe) expands and a common envelope is formed
4. Further mass transfer is blocked by the swollen secondary which fills its Roche lobe
5. The convective energy transfer from the primary to the secondary results in equal entropies of convective envelopes of both components – this keeps the secondary oversized
6. The components are out of thermal equilibrium, which forces thermal relaxation oscillations (TRO) making the binary to alternate between the contact and semi-detached configuration
7. Luminosity increase of the primary due to its (faster) nuclear evolution drives a secular mass transfer from the secondary to the primary
8. When the mass ratio reaches ≈ 0.1 the secondary is tidaly disrupted and a fast rotating single star is formed.

Points 2–4 apply to initially detached binaries. This paradigm stands behind most of the analyses aimed at explainig the evolutionary status of W UMa-type stars (e.g. Kähler, (2004) and references therein). Arguments will be given in the next Section that the paradigm is not correct. In Section 3 an alternative evolutionary scenario is shortly described and the results of more detailed calculations are presented for a binary with initial masses $1.2 + 1.0\, M_\odot$.

K. Stępień
Warsaw University Observatory, Al. Ujazdowskie 4, 00-478 Warszawa, Poland

Pribulla, T., Kreiner, J.M., Tremko, J.: Contr. Astr. Obs. Skalnate Pleso **33**, 38 (2003)
Rucinski, S.M.: AJ **116**, 2998 (1998)
Rucinski, S.M.: AJ **120**, 319 (2000)
Rucinski, S.M., Lu, W.: MNRAS **315**, 587 (2000)
Shaw J.S.: Mem. S. A. It. **65**, 95 (1994)
Stępień K.: MNRAS **274**, 1019 (1995)
Stępień K.: In Dupree, A.K., Benz, A.O. (eds.), Stars as suns: activity, evolution and planets. IAU Symp. No. 219, Astr. Soc. of Pacific, pp. 967 (2004)
Szymański, M., Kubiak, M., Udalski, A.: Acta Astron. **51**, 259 (2001)
Wood, B.E., Müller, H.-R., Zank, G.P., Linsky, J.L.: ApJ **574**, 412 (2002)

Wind Ionization Structure of the Short-Period Eclipsing LMC Wolf-Rayet Binary BAT99-129: Preliminary Results*

C. Foellmi · A. F. J. Moffat · S. V. Marchenko

Received: 1 November 2005 / Accepted: 1 December 2005
© Springer Science + Business Media B.V. 2006

Abstract BAT99-129 is a rare, short-period eclipsing Wolf-Rayet binary in the Large Magellanic Cloud. We present here medium-resolution NTT/EMMI spectra that allow us to disentangle the spectra of the two components and find the orbital parameters of the binary. We also present VLT/FORS1 spectra of this binary taken during the secondary eclipse, i.e. when the companion star passes in front of the Wolf-Rayet star. With these data we are able to extract, for the first time in absolute units for a WR + O binary, the sizes of the line emitting regions.

Keywords Binaries: close · Binaries: Wolf-Rayet · Binaries: eclipsing · Binaries: individual: BAT99-129 · Binaries: evolution · Stars: massive stars

1. Presentation

The Wolf-Rayet (WR) star BAT99-129 (a.k.a. Brey 97, see Breysacher et al., 1999, hereafter BAT129) has been discovered to be a short-period eclipsing binary by Foellmi et al. (2003b). It consists of a nitrogen-rich WN4 Wolf-Rayet component and a companion of unknown spectral-type. Along with BAT99-19 (Brey 16) in the LMC and HD 5980 in the SMC, BAT129 belongs to the very small group of extragalactic eclipsing Wolf-Rayet binaries.

Absorption lines are visible in the spectrum of BAT129, especially in the blue part of the optical range ($\lesssim 4000$Å), while WR emission lines strongly dominate everywhere else. The origin of these absorption lines was unclear: they are either produced by an O-type companion, and/or they originate in a WR wind, as has been found in all the *single early-type WN (WNE) stars* in the SMC and, to a lesser extent, in the LMC (Foellmi et al., 2003a,b, 2004).

Previous data did not allow us to obtain the orbital motion of the companion star. Moreover, being eclipsing (see Fig. 1), the system deserved much more study. Thus, we acquired medium-resolution spectra with the EMMI spectrograph, mounted on the New Technology Telescope (NTT, La Silla Observatory, Chile). These data allowed us to resolve the orbital motion of the companion star, and find a relatively large hydrogen content on the WR star. In addition, we obtained long-slit spectra with the instrument FORS1 at the Very Large Telescope (VLT, Paranal Observatory, Chile). These spectra allow us to study the wind ionization structure of BAT129.

2. The orbit of BAT129

The EMMI spectra have wavelength coverage of 3920–4380 Å and a resolving power of 3500 (resolution of 2.6 pixels). The typical 30-min exposure resulted in S/N ratio between 50 and 90. The spectra were shifted to the heliocentric rest frame, and normalized to unity using continuum regions common to all spectra (see Foellmi et al. 2005) for details).

We then performed an iterative method, first described in Demers et al. (2002), to reconstruct the individual mean

---
*Based on observations obtained at the La Silla and Paranal Observatories, European Southern Observatory (Chile).

C. Foellmi (✉)
European Southern Observatory, 3107 Alonso de Cordova, 19 Vitacura, Santiago, Chile
e-mail: cfoellmi@eso.org

A. F. J. Moffat
Département de physique, Université de Montréal, C.P. 6128, Succ. Centre-Ville, Montréal, QC, H3C 3J7, Canada

S. V. Marchenko
Department of Physics and Astronomy, Thompson Complex Central Wing, Western Kentucky University, Bowling Green, KY 42101-3576, USA

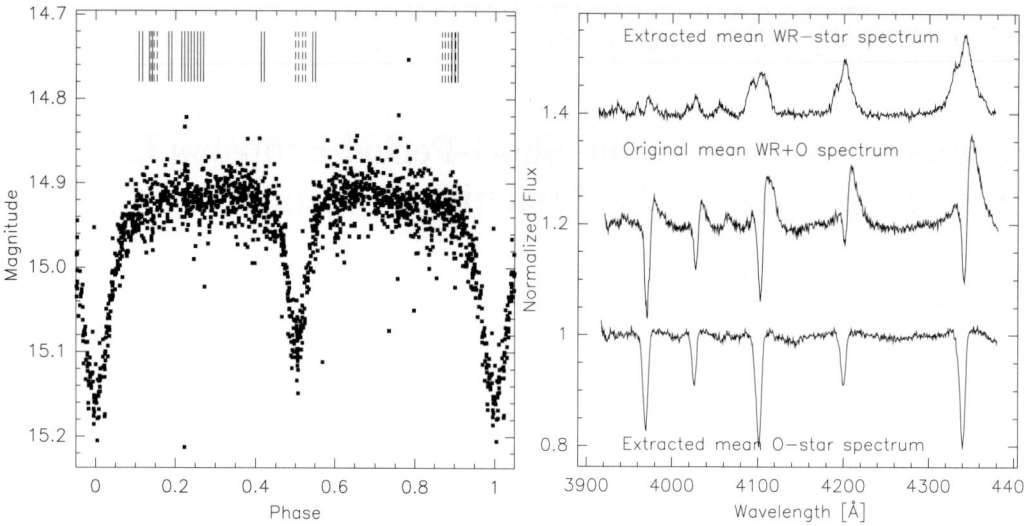

Fig. 1 Left panel: MACHO light curve of BAT129, as described in Foellmi et al. (2003b). The phases at which EMMI spectra were obtained are indicated by vertical lines. Right panel: result of the separation procedure. The original mean spectrum and the extracted WR spectrum have been shifted by 0.2 and 0.4 continuum units for clarity

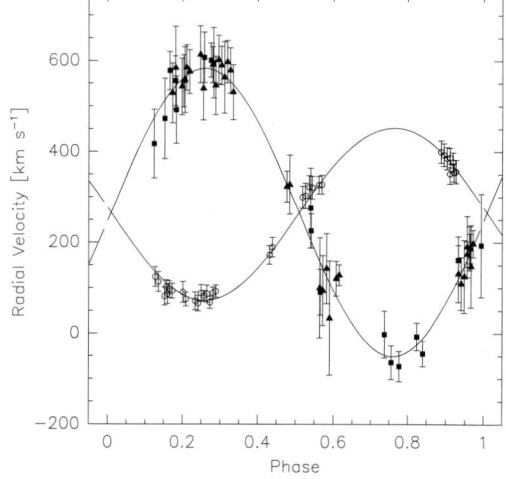

Fig. 2 The final orbit of the system. The orbital parameters (with their usual denominations) are the following: Period $= 2.7689 \pm 0.0002$ days (from MACHO photometry), $e = 0$ (assumed), $K_{\rm WR} = 316 \pm 5\,{\rm km\,s^{-1}}$, $K_{\rm O} = 193 \pm 6\,{\rm km\,s^{-1}}$, $E_0 = 2451946.4274 \pm 0.0054$ (HJD), $\gamma = 265 \pm 5\,{\rm km\,s^{-1}}$

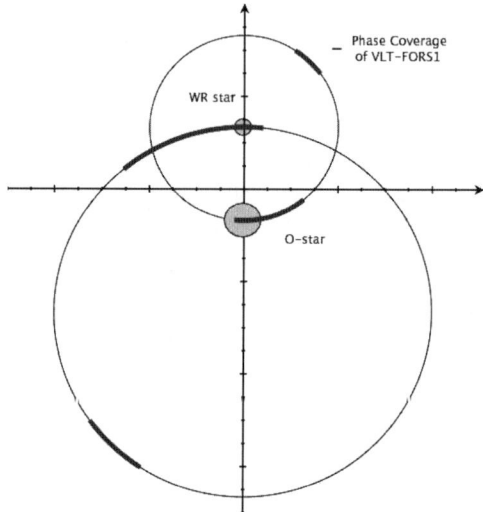

Fig. 3 Sketch of the observations obtained with VLT/FORS1. A mass of the O star roughly twice that of the WR star has been assumed. The two large circles indicate the respective orbital trajectories of the stars. Bold parts on these circles indicates when, according to the ephemeris of our observations, FORS1 spectra were obtained. The longest bold parts correspond to the data of the first night, while the shortest ones correspond to the second half-night

spectra. Basically, this hypothesis-free method consists of shifting the spectra in the radial-velocity (RV) space to a common rest frame of, say, the first component and then combining them to build a high S/N spectrum of this component. This latter spectrum is shifted back to the original positions and subtracted from the original spectra. This provide a first guess of the spectrum of the second component. This procedure is started again for the second component with its respective RVs, and the new mean spectrum is again subtracted from the original spectra. This completes one full iteration. Between 3 and 5 iterations are necessary to obtain stable results.

Interestingly, a blue-shifted absorption profile is discovered in the emission lines of the WR spectrum. These absorption lines are certainly related to the WR star itself, similarly to what was found in other SMC and the LMC WR stars (see e.g. Foellmi et al., 2003a). We were then able to compute a full orbital solution (summarized in Fig. 2), confirm the WN4ha spectral type of the WR star and assign an O5V spectral type to the companion (Foellmi et al., 2005). The minimum masses of the two stars are $M_{\rm WR} \sin^3 i = 14 \pm 2\,M_\odot$ and $M_O \sin^3 i = 23 \pm 2\,M_\odot$, respectively, and the minimum separation $a \sin i = 27.9 \pm 0.5\,R_\odot$.

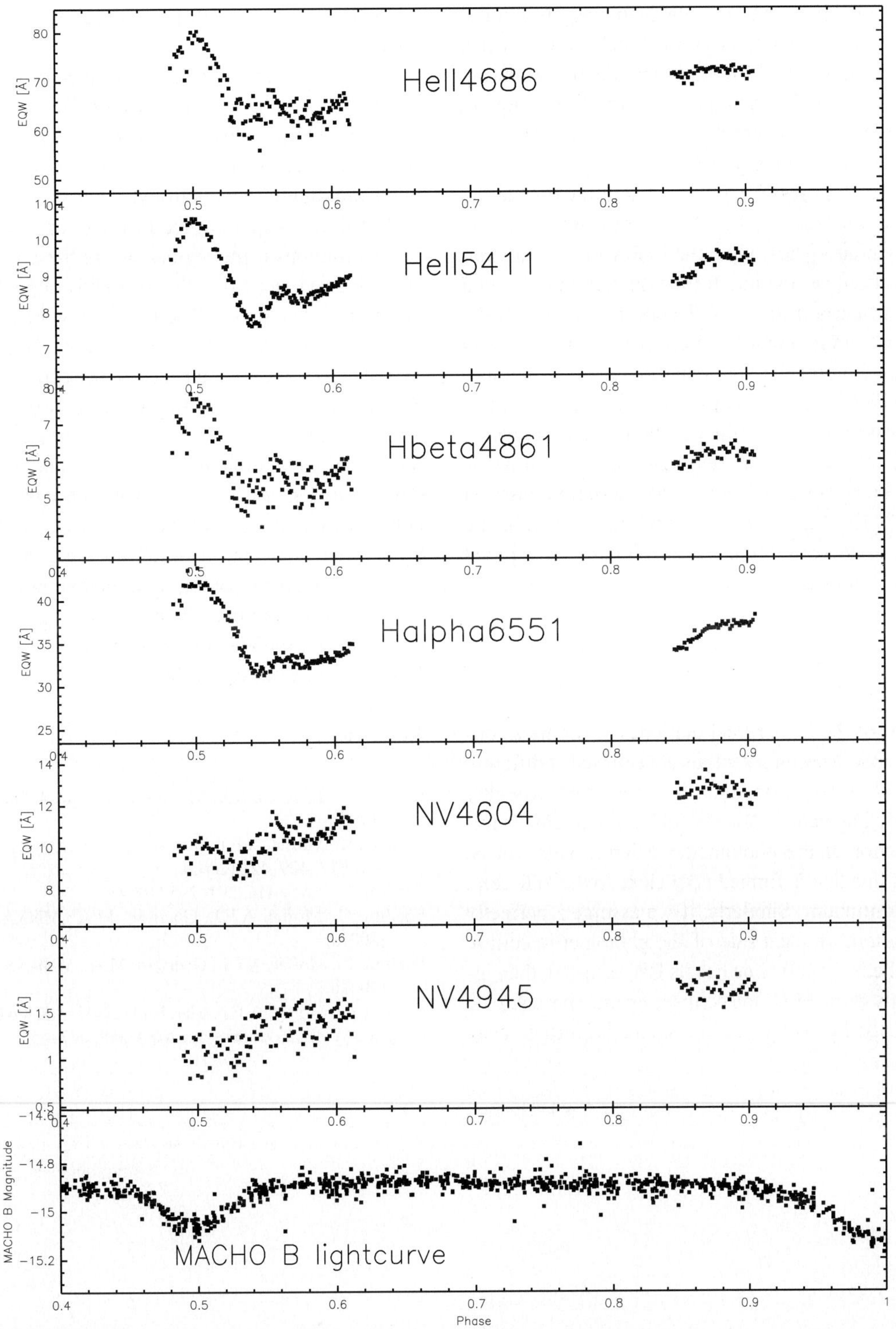

Fig. 4 Variations of the *absolute values* of the equivalent widths of various lines in the FORS1 spectra of BAT129 versus the orbital phase. The bottom panel shows the MACHO lightcurve. It can be easily seen that lines of different ions and/or different ionization potentials are not eclipsed at the same time

3. The ionization structure of the wind

Armed with the full set of orbital parameters of the system (except the inclination angle), we are able to use our VLT/FORS1 spectra to sketch the ionization structure of the wind of the WR component in BAT129. Our FORS1 spectra were taken in December 2003, and have a wavelength range from 3700 to 8900Å, and a resolving power of about 500.

One spectrum of S/N ~ 200 was obtained every 6 minutes during one night through the secondary eclipse (i.e. when the O star is in front of the WR star), and identically during the next half-night. A sketch of the geometry is presented in Fig. 3. It shows the orbital plane seen from above, and where the spectra were obtained during the orbit.

The idea is as follows. We assume that in the optical the O component can be treated as a geometrically-occulting disk (i.e. a windless star); plus, the WR wind has a clear wind stratification, i.e. the line-formation zones are located at different distances from the WR core, depending on the ion and its ionization potential. Therefore, following the O star passing in front of the WR star should reveal, minute-by-minute, which region is eclipsed and during how much time. This can be measured by the variations of the equivalent widths (EWs) of the different emission lines present in the spectra (after being corrected for the varying continuum level using the MACHO lightcurve). The preliminary results are shown in Fig. 4, where the normalized EWs have been plotted against the orbital phase.

4. Discussion

The variations of the EWs reveal that lines of different ions and/or different ionization potentials are eclipsed at different times, *and with different patterns*. It is possible to roughly draw the following picture. The NV $\lambda 4945$ line seems to follow the behavior of the photometric eclipse. This can be understood if this line is formed very close to the WR core, as does the continuum. Similarly, if we assume a perfectly symmetric pattern on each side of the photometric eclipse core (PEC), the NV $\lambda 4604$ line has its EW eclipsed, then increased slightly at the PEC, and eclipsed again. This could be interpreted as the line is formed in a shell a bit further away from the WR core, and its projected appearance is essentially an annulus.

The interpretation of the other lines is more difficult. At first sight, a strong increase seems to occur precisely during the photometric eclipse. However, it can be seen that the reference level, as shown in the EW values of the second half night, are slightly changing, and their mean level is *lower* than that of the maximum at the PEC. It is possible to interpret these variations if the regions where these lines are emitted as having an inner radius that is slightly larger than that of the O star projected radius. Therefore, the eclipsing effect of the O star is at its *minimum* when the O star is aligned with the WR star. When the O star moves away from this alignment, the line emitting regions start to be eclipsed.

The situation is probably not as clear cut as this, since the orbital inclination of the system is not known. Although close to 90 degrees, it is likely that even at the PEC, the O star eclipses a (small) part of the regions where the above lines are formed. Interestingly, small wiggles in the EW variations (around phase 0.57, see Fig. 4) seem to show that line emitting regions are not simple shells. These measurements have a direct impact on WR atmosphere models, such as those of Hamann and Koesterke (2000).

References

Breysacher, J., Azzopardi, M., Testor, G.: A&A Supl. **137**, 117–145 (1999)
Demers, H., Moffat, A.F.J., Marchenko, S.V., Gayley, K.G., Morel, T.: ApJ **577**, 409–421 (2002)
Foellmi, C.: A&A **416**, 291–295 (2004)
Foellmi, C., Moffat, A.F.J., Guerrero, M.A.: MNRAS **338**, 360–388 (2003a)
Foellmi, C., Moffat, A.F.J., Guerrero, M.A.: MNRAS **338**, 1025–1056 (2003b)
Foellmi, C., Moffat, A.F.J., Marchenko, S.V.: A&A **447**, 667 (2006)
Hamann, W.-R., Koesterke, L.: A&A **360**, 647–655 (2000)

Masses and Radii of Low-Mass Stars: Theory Versus Observations

I. Ribas

Received: 1 November 2005 / Accepted: 1 February 2006
© Springer Science + Business Media B.V. 2006

Abstract Eclipsing binaries with M-type components are still rare objects. Strong observational biases have made that today only a few eclipsing binaries with component masses below 0.6 $M_\odot$ and well-determined fundamental properties are known. However, even in these small numbers the detailed comparison of the observed masses and radii with theoretical predictions has revealed large disagreements. Current models seem to predict radii of stars in the 0.4–0.8 $M_\odot$ range to be some 5–15% smaller than observed. Given the high accuracy of the empirical measurements (a few percent in both mass and radius), these differences are highly significant. I review all the observational evidence on the properties of M-type stars and discuss a possible scenario based on stellar activity to explain the observed discrepancies.

Keywords Binaries: eclipsing · Binaries: spectroscopic · Stars: fundamental parameters · Stars: late-type

1. Introduction

Most of the stars in the Galaxy have masses well below that of the Sun. In spite of the numerous population, detailed investigation of the properties of low-mass stars has been often difficulted by their intrinsic faintness. However, the observation and study of low-mass stars is now a field in rapid development mostly because of the increasing number of deep photometric surveys and the advent of powerful instrumentation able to obtain spectroscopy of these faint stars. But also renewed interest arises from one of the "hot topics" of this past decade: exoplanets. Very low mass stars, brown dwarfs, and giant planets share many physical characteristics and their study and modeling is often intimately related.

Efforts in the theoretical description of low mass stars have been intense in recent years. Current stellar structure models of low mass stars have reached a high level sophistication and maturity (e.g., Chabrier and Baraffe, 2000). However, theoretical progress has not been matched by observational developments because of the difficulty in obtaining accurate determinations of the physical properties of low-mass stars. The comparison of model predictions with observations is a central point. Only by limiting the number of free parameters can a stringent test of stellar models be carried out. Therefore, there is a strong need for stars with well-determined properties such as masses, radii, effective temperatures, metallicities and ages. Models will pass the test *only* if they are able to reproduce *all* of the observed stellar properties.

The best source of such high-quality stellar properties is the analysis of double-lined eclipsing binaries (EBs) in which the components are detached. Optimum results are achieved when the system components are similar (i.e., deep eclipses and two radial velocity sets). Unfortunately, the number of known EBs with M-type components is small because of the faintness of the stars and the often strong intrinsic variations due to magnetic activity. To complement the dataset, in recent years it has become possible to determine the radii of nearby M-type stars directly from IR interferometry. The current precision does not match that of eclipsing systems but the prospects are bright. Furthermore, planetary transit research has also contributed to our database of low-mass stars. Follow-up of OGLE transit candidates has

I. Ribas
Institut de Ciències de l'Espai – CSIC, Campus UAB, Facultat de Ciències, Torre C5 – parell – 2a planta, 08193 Bellaterra, Spain;
Institut d'Estudis Espacials de Catalunya (IEEC), Edif. Nexus, C/Gran Capità, 2–4, 08034 Barcelona, Spain
e-mail: iribas@ieec.uab.es

uncovered a number of EBs with F-G primaries and M-type secondaries.

In this paper I review the current data on masses and radii of low-mass stars, including both EBs and stars with direct radius measurements, and compare them with the predictions of stellar models. As already pointed out by several authors (e.g., Torres and Ribas, 2002), a highly significant discrepancy exists between observation and theory. Here I analyze possible reasons for such discrepancy.

2. Eclipsing M-type systems

Eclipsing binaries with similar components yield the stellar physical properties potentially to an accuracy of 1–2%. Such data have often served as valuable benchmarks for the validation of structure and evolution models. For two decades only two bona-fide EBs with M-type components were known: The member of the Castor multiple system YY Gem (Torres and Ribas, 2002), with components of spectral type M1 V, and CM Dra (Lacy, 1977; Metcalfe et al., 1996), composed of two M4.5 Ve stars. These were the only two M-type EBs that had been well studied until Delfosse et al. (1999) reported the discovery of eclipses in the M3.5 star CU Cnc and Ribas (2003) carried out accurate determinations of the components' physical properties. Very recently, three new M-type EBs have been studied in detail. These are BW3 V38 (Maceroni and Montalbán, 2004), TrES-Her0-07621 (Creevey et al., 2005), and GU Boo (López-Morales and Ribas, 2005). Unfortunately, the quality of the available observations for BW3 V38 and TrES-Her0-07621 does not permit high-accuracy determinations of both masses and radii, which have uncertainties of up to 10–15%. GU Boo has well-determined physical properties that make it twin system of YY Gem. Table 1 gives the masses and radii of the components of these M-type binaries. Also listed in the table are the components of two eclipsing systems with masses below 0.8 $M_\odot$: The Hyades EB V818 Tau (Torres and Ribas, 2002) and RXJ0239.1-1028 (López-Morales et al., in prep.). These K-type stars can be useful to better understand the results of the comparison with stellar models.

Stringent tests of stellar models can only be carried out if the chemical compositions and ages of the EBs can be constrained. YY Gem and CU Cnc are interesting cases because their kinematic properties indicate that they belong to the so-called Castor moving group, with an age of approximately ~300 Myr and solar metallicity. Using also kinematic criteria, it appears that GU Boo is an intermediate-age star in the galactic disk and probably its metallicity is not far from the solar value. CM Dra is most likely a Population II EB with sub-solar metallicity and an old age (of the order of 10 Gyr). The ages and metallicities of BW3 V38 and TrES-Her0-07621 have not been estimated. An important re-

Table 1 Masses and radii of the components of double-lined EB systems with masses below 0.8 $M_\odot$

Name	Mass ($M_\odot$)	Radius ($R_\odot$)	Ref.
V818 Tau B	0.7605 ± 0.0062	0.768 ± 0.010	1
RXJ0239.1-1028 A	0.736 ± 0.009	0.735 ± 0.018	–
RXJ0239.1-1028 B	0.695 ± 0.006	0.710 ± 0.016	–
GU Boo A	0.610 ± 0.007	0.623 ± 0.016	2
GU Boo B	0.599 ± 0.006	0.620 ± 0.020	2
YY Gem AB	0.5992 ± 0.0047	0.6191 ± 0.0057	1
TrES-Her0-07621 A	0.493 ± 0.003	0.453 ± 0.060	3
TrES-Her0-07621 B	0.489 ± 0.003	0.452 ± 0.050	3
BW3 V38 A	0.44 ± 0.07	0.51 ± 0.04	4
BW3 V38 B	0.41 ± 0.09	0.44 ± 0.06	4
CU Cnc A	0.4333 ± 0.0017	0.4317 ± 0.0052	5
CU Cnc B	0.3890 ± 0.0014	0.3908 ± 0.0094	5
CM Dra A	0.2307 ± 0.0010	0.2516 ± 0.0020	6,7
CM Dra B	0.2136 ± 0.0010	0.2347 ± 0.0019	6,7

Ref.: 1. Torres and Ribas (2002); 2. López-Morales and Ribas (2005); 3. Creevey et al. (2005); 4. Maceroni and Montalbán (2004); 5. Ribas (2003); 6. Lacy (1977); 7. Metcalfe et al. (1996)

mark on the significance of stellar ages is that M-type stars have long evolutionary timescales once they have reached the main sequence. Thus, the only relevant point to model comparisons is whether any of the studied EBs could be pre-main sequence (i.e., an age <100 Myr). Available evidence indicates that this is not the case.

3. Other M-type stars with masses and radii

Besides double-lined EBs, other sources of masses and radii of low-mass stars have emerged in recent years. Spectacular developments in interferometry (such as the PTI or VLTI instruments) have made it possible to resolve nearby M-type stars and determine their angular diameters with uncertainties of just a few hundredths of a milliarcsecond. From those measurements and trigonometric distances, determinations of stellar radii can be carried out (Lane et al., 2001; Ségransan et al., 2003). The drawback of this technique is that the masses cannot be determined directly (unless the resolved M-type star belongs to a visual binary) but have to be inferred from calibrations. Fortunately, the empirical mass-luminosity relationship in the infrared K band is well defined and has little intrinsic scatter (Delfosse et al., 2000).

Follow-up of OGLE planetary transit candidates has uncovered a number of eclipsing systems consisting of main sequence F-G stars with M dwarf companions (Bouchy et al., 2005; Pont et al., 2005). Because of selection effects, their light curves have shallow and flat-bottom eclipses corresponding to the transit of the M-type star (the occultation is not observable). Also, only the lines of the F-G components are visible in the spectra due to the large contrast. These

Table 2 Other low-mass stars with well-determined masses and radii

Name	Mass ($M_\odot$)	Radius ($R_\odot$)	Ref.
OGLE-TR-114	0.82 ± 0.08	0.72 ± 0.09	1
GJ 105A	0.790 ± 0.039	0.708 ± 0.050	2,3
GJ 380	0.670 ± 0.033	0.605 ± 0.020	2,3
GJ 205	0.631 ± 0.031	0.702 ± 0.063	3
OGLE-TR-34	0.509 ± 0.038	0.435 ± 0.033	4
GJ 887	0.503 ± 0.025	0.491 ± 0.014	3
OGLE-TR-120	0.47 ± 0.04	0.42 ± 0.02	1
GJ 15A	0.414 ± 0.021	0.383 ± 0.020	2,3
GJ 411	0.403 ± 0.020	0.393 ± 0.008	2,3
OGLE-TR-18	0.387 ± 0.049	0.390 ± 0.040	4
OGLE-TR-6	0.359 ± 0.025	0.393 ± 0.018	4
GJ 191	0.281 ± 0.014	0.291 ± 0.025	3
OGLE-TR-7	0.281 ± 0.029	0.282 ± 0.013	4
OGLE-TR-5	0.271 ± 0.035	0.263 ± 0.012	4
OGLE-TR-78	0.243 ± 0.015	0.240 ± 0.013	1
OGLE-TR-125	0.209 ± 0.033	0.211 ± 0.027	1
GJ 699	0.158 ± 0.008	0.196 ± 0.008	2,3
GJ 551	0.123 ± 0.006	0.145 ± 0.011	3
OGLE-TR-106	0.116 ± 0.021	0.181 ± 0.013	1
OGLE-TR-122	0.092 ± 0.009	0.120 ± 0.018	1

Ref.: 1. Pont et al. (2005); 2. Lane et al. (2001); 3. Ségransan et al. (2003); 4. Bouchy et al. (2005).

restrictions imply that the masses and radii of the M-type stars have to be determined through some assumptions (some of which are model dependent). The resulting accuracies are in the range 5–20%. The masses and radii of M-type stars resulting from both interferometry and OGLE transit follow-up are listed in Table 2.

4. Models versus observations

An obvious test of the performance of low-mass stellar models is to compare the observational mass-luminosity diagram with theoretical predictions. Most of the checks of state-of-the-art models using the absolute magnitude in the V band have indicated good overall agreement but significant scatter in the measurements. Further works (e.g., Delfosse et al., 2000) have shown that such scatter is most likely caused by starspots since the same mass-luminosity relationship is much better defined in the infrared K band. From those tests, one may naively conclude that models are successful at predicting the properties of low-mass stars. However, this is a very restrictive comparison that uses only two of the several independent properties that define a star.

The accurate masses and radii of the stars described above offer an excellent opportunity to carry out critical tests to evaluate the performance of low-mass stellar models. Such tests have been carried out by a number of authors in the past (Popper, 1997; Clausen et al., 1999; Torres and Ribas, 2002; Ribas, 2003), who have systematically pointed out a (rather serious) discrepancy between the stellar radii predicted by theory and the observations. Model calculations appear to underestimate stellar radii by ∼10%, which is a highly significant difference given the observational uncertainties. Furthermore, the comparisons in some cases were made with virtually no free parameters since the ages and metal contents of the stars could be constrained independently.

With the extended stellar sample in this paper, the question of the comparison between theory and observation can be revisited. Empirical mass-radius diagrams are shown in Fig. 1 showing both the entire sample (top) and a subsample including those stars with masses and radii determined to better than 3% (bottom), which all happen to be EB members. The line represents a 300 Myr isochrone (i.e., main sequence) calculated with the models of Baraffe et al. (1998).

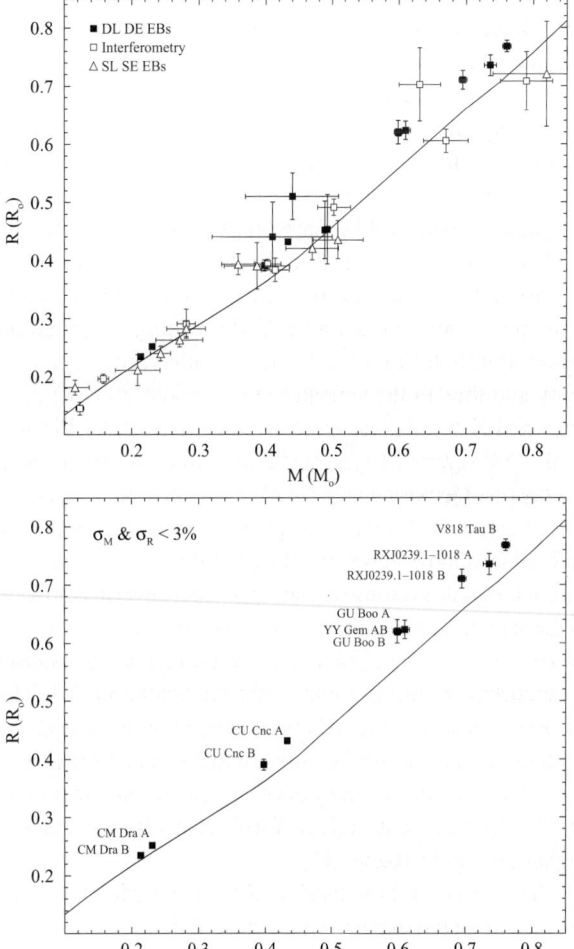

Fig. 1 Top: Mass-radius plot for stars in the lower main sequence with empirical determinations. The solid line represents a theoretical 300 Myr isochrone calculated with the Baraffe et al. (1998) models. Bottom: Same as above for those stars with determinations of masses and radii better than 3% (double-lined EBs)

Inspection of the top panel shows two mass intervals with different characteristics. Stars with masses below $\sim$0.30–0.35 $M_\odot$ seem to show small scatter and good agreement with stellar models, while the more massive have larger scatter and radii that tend to fall systematically above the theoretical line. These distinct mass regimes are not well established yet but it is tantalizing that the apparent division occurs near the limit between fully convective stars and stars with radiative cores.

The high-accuracy sample in the bottom panel of Fig. 1 leaves no doubt that a significant discrepancy exists between models and observations with regards to stellar radii. Other detailed comparisons have also shown that the stellar effective temperatures appear to be overestimated by $\sim$5%. This, together with the good agreement in the mass-luminosity plot, argues in favor of a scenario in which the stars have larger radius and cooler temperature than predicted by models but just in the right proportions to yield identical luminosities. A $\sim$10% radius undererestimation is compensated by a $\sim$5% temperature overestimation to yield identical luminosities. What would explain such coincidence? The answer to this question is not clear yet, but there are some hints pointing in certain directions.

Perhaps the first question to address is whether the EB sample used to compare with models is representative of the low-mass star population. These systems are all detached and should have evolved like single stars. However, as members of close binaries (with periods less than 2.8 days) the components have undergone tidal interactions forcing them to spin up in orbital synchronism. The resulting high rotational velocities (10–60 km s^{-1}) give rise to enhanced magnetic activity and thus to the appearance of surface spots, emission lines and X-ray fluxes. As shown in the work by Pizzolato et al. (2003), any M-type star with a rotation period below 10 days will experience these phenomena at their peak (saturated activity). It might be speculated that the larger radii and lower temperatures could be a reflex of such enhanced activity. Perhaps the significant spot areal coverage observed in these eclipsing systems has the effect of lowering the overall photospheric temperature, which the star compensates by increasing its radius to conserve the total radiative flux. Thus, there may be a correlation between the radius and the activity level of an M-type star. A similar conclusion of stellar activity causing the discrepancy between models and theory was reached in the recent study of Torres et al. (2006) for stars of higher masses ($\sim$0.9 $M_\odot$).

The sample we have used in our comparison may be representative of the population of active M-type stars only. This does not diminish the relevance of the discrepancy between models and observations. Low-mass stars with ages younger than a few Gyr are very active because they are generally fast rotators. Therefore, not only a star in a close binary system but any active M-type star (e.g., in a stellar cluster) may have its radius severely underestimated if computed from stellar models. A definitive test of the magnetic activity hypothesis will have to wait for further observational data. In particular, EBs with periods $\gtrsim$10 days (i.e., not synchronized) and components of visual binaries resolved interferometrically should provide the necessary proof. Ongoing large scale surveys and future space missions, such as COROT or Kepler, are expected to increase the number of EBs significantly. If the activity correlation is firmly established, it will be time for theory to catch up by introducing magnetic activity in stellar evolution codes as a major ingredient influencing the observable properties of low-mass stars.

Acknowledgements I am grateful to E. Guinan, M. López-Morales and G. Torres for a number of fruitful discussions. Support from the Spanish MCyT through a Ramón y Cajal fellowship and grant AyA2003-07736 is acknowledged.

References

Baraffe, I., Chabrier, G., Allard, F., Hauschildt, P.H.: A&A **337**, 403 (1998)
Bouchy, F., Pont, F., Melo, C., Santos, N.C., Mayor, M., Queloz, D., Udry, S.: A&A **431**, 1105 (2005)
Chabrier, G., Baraffe, I.: ARA&A **38**, 337 (2000)
Clausen, J.V., Baraffe, I., Claret, A., VandenBerg, D.A.: In: Giménez, À., Guinan, E.F., Montesinos, B. (eds.) In theory and tests of convection in stellar structure, ASP Conf. Ser. 173, (San Francisco: ASP), 265 (1999)
Creevey, O.L. et al.: ApJ **625**, L127 (2005)
Delfosse, X., Forveille, T., Mayor, M., Burnet, M., Perrier, C.: A&A **341**, L63 (1999)
Delfosse, X., Forveille, T., Ségransan, D., Beuzit, J.-L., Udry, S., Perrier, C., Mayor, M.: A&A **364**, 217 (2000)
Lacy, C.H.: ApJ **218**, 444 (1977)
Lane, B.F., Boden, A.F., Kulkarni, S.R.: ApJ **551**, L81 (2001)
López-Morales, M., Ribas, I.: ApJ **631**, 1120 (2005)
Maceroni, C., Montalbán, J.: A&A **426**, 577 (2004)
Metcalfe, T.S., Mathieu, R.D., Latham, D.W., Torres, G.: ApJ **456**, 356 (1996)
Pizzolato, N., Maggio, A., Micela, G., Sciortino, S., Ventura, P.: A&A **397**, 147 (2003)
Pont, F., Bouchy, F., Melo, C., Santos, N.C., Mayor, M., Queloz, D., Udry, S.: A&A **438**, 1123 (2005)
Popper, D.M.: AJ **114**, 1195 (1997)
Ribas, I.: A&A **398**, 239 (2003)
Ségransan, D., Kervella, P., Forveille, T., Queloz, D.: A&A **397**, L5 (2003)
Torres, G., Ribas, I.: ApJ **567**, 1140 (2002)
Torres, G., Lacy, C.H.S., Marschall, L.A., Sheets, H.A., Mader, J.A.: ApJ **640**, 1018 (2006)

Light-Time Effect in the Eclipsing Binaries GO Cyg, GW Cep, AR Aur and V505 Sgr

D. Chochol · T. Pribulla · M. Vaňko · P. Mayer ·
M. Wolf · P. G. Niarchos · K. D. Gazeas ·
V. N. Manimanis · L. Brát · M. Zejda

Received: 30 September 2005 / Accepted: 19 December 2005
© Springer Science + Business Media B.V. 2006

Abstract Orbital period changes of the eclipsing binaries GO Cyg and GW Cep are explained by the light-time effect for the first time. New minima of the eclipsing binary AR Aur improve the predicted light-time orbit. The light-time orbit with the quadratic ephemeris of the binary matches the new observations of V505 Sgr better than the linear one. As the light-time effect fits in corresponding $O - C$ diagrams of all four systems have been reaching extreme values, the observations of minima times in forthcoming years are highly desirable.

Keywords Eclipsing binary · Period changes · Light-time effect

1. Multiplicity and light-time effect

The close binaries are often members of multiple systems (triple, quadruple or systems with higher multiplicity). The third and further bodies of the binaries can be found: (a) astrometrically by visual, speckle or direct interferometric observations; (b) spectroscopically by detection in broadening functions, composite spectra signatures, radial-velocity changes of the mass center of the binary; (c) photometrically in eclipsing systems by a light-time effect (LITE), the third light found by analysis of their light curves, changes of the orbital inclination due to the tidal interaction of the third body. Tokovinin (1997) catalogue of physically multiple stars contains 728 systems of multiplicity 3 to 7. Their incidence provides important clues on the evolution of close binaries.

The LITE in eclipsing binaries is a periodic change of the orbital period caused by the movement of the binary in space due to the presence of the third or multiple bodies. It can be studied by the analysis of the periodic shifts of minima times. The basic formulae for the LITE computation were published by Irwin (1959). The time of the minimum light, caused by the eclipse, can be predicted as follows:

$$\text{Min I} = JD_0 + P \times E + Q \times E^2 + \\ + \frac{a_{12} \cdot \sin i}{c} \left[\frac{1-e^2}{1+e\cos\nu} \sin(\nu+\omega) + e\sin\omega \right], \quad (1)$$

where c is a speed of light, $JD_0 + P \times E + Q \times E^2$ is the quadratic (to include internal causes) ephemeris of the eclipsing pair, a_{12}, i, e, ω are orbital parameters and ν is the true anomaly of the binary on the orbit around the center of mass of the triple system. From the observed Heliocentric JDs of minima times and corresponding epochs it is possible to optimize the linear or quadratic ephemeris of the eclipsing pair (optimized parameters are JD_0, P, Q) and the parameters of the LITE orbit: P_3 – orbital period of the third body, $a_{12} \sin i, \omega, e, T_0$ – time of the periastron passage.

early photographic minima times published by Hoffmeister (1934).

V505 Sgr was shown (Mayer, 1997) to be one of eclipsing binaries where the LITE can be combined with the visual orbit to obtain more parameters of the system (namely the parallax). After updating the interferometric data by new measurements (see http://ad.usno.navy.mil) we found that to explain them, the visual orbit has to be about 60 years. Therefore, no satisfactory explanation for the behaviour of the times of minima can be offered at present. Probably the LITE caused by the third body is combined with another cause (fourth body?).

Acknowledgements This investigation was supported by the Science and Technology Assistance Agency, contract No. APVT-20-014402, the Czech-Greek project of collaboration RC-3-18 and research plan J13/98: 113200004 of Ministry of Education, Youth and Sport and by the Grant Agency of the Czech Republic, grant No. 205/04/2063. This paper uses observations made at the South African Astronomical Observatory (SAAO).

References

Aksu, O., Özavci, I., Yüce, K. et al.: Inf. Bull. Var. Stars, No. 5588 (2005)
Albayrak, B., Ak, T., Elmasli, A.: Astron. Nachr. **324**, 523 (2003)
Chambliss, C.R., Walker, R.L., Karle, J.H. et al.: Astron. J. **106**, 2058 (1993)
Chochol, D., Juza, K., Mayer, P. et al.: Bull. Astron. Inst. Czechosl. **39**, 69 (1988)
Cook, J.M., Divoky, M., Hofstrand, A. et al.: Inf. Bull. Var. Stars, No. 5636 (2005)
Edalati, M.T., Atighi, M.: Astrophys. Space Sci. **253**, 107 (1997)
Hoffmeister, C.: Astron. Nachr. **251**, 321
Ibanoglu, C., Cakirh, Ö., Degirmenci, Ö. et al.: Astron. Astrophys. **354**, 188 (2000)
Irwin, J.B.: Astron. J. **64**, 149 (1959)
Khokhlova, V.L., Zverko, Y., Žižňovský, J. et al.: Astron. Lett. **21**, 818 (1995)
Kreiner, J.M., Kim, C., Nha, I.: An atlas of $(O-C)$ diagrams of eclipsing binary stars, Wydawnictvo Naukove AP, Krakow (2001)
Mayer, P.: Astron. Astrophys. **324**, 988 (1997)
Maceroni, C., van 't Veer, F.: Astron. Astrophys. Suppl. Ser. **311**, 523 (1996)
McAlister, H.A., Hartkopf, W.I., Hutter, D.J. et al.: Astron. J. **93**, 183 (1987)
Nordström, B., Johansen, K.T.: Astron. Astrophys. **282**, 787 (1994)
Pribulla, T., Vaňko, M., Chochol, D. et al.: Contrib. Astron. Obs. Skalnaté Pleso **31**, 26 (2001)
Pojmanski, G.: Acta Astron. **52**, 397 (2002)
Tokovinin, A.A.: Astron. Astrophys. Suppl. **121**, 71 (1997)
Tomkin, J.: Astrophys. J. **387**, 631 (1992)
Zverko, J., Chochol, D., Juza, K. et al.: Inf. Bull. Var. Stars, No. 1997 (1981)

X-ray Observations of Binary and Single Wolf-Rayet Stars with XMM-Newton and Chandra

Stephen Skinner · Manuel Güdel · Werner Schmutz · Svetozar Zhekov

Received: 1 November 2005 / Accepted: 1 February 2006
© Springer Science + Business Media B.V. 2006

Abstract We present an overview of recent X-ray observations of Wolf-Rayet (WR) stars with *XMM-Newton* and *Chandra*. These observations are aimed at determining the differences in X-ray properties between massive WR + OB binary systems and putatively single WR stars. A new *XMM* spectrum of the nearby WN8 + OB binary WR 147 shows hard absorbed X-ray emission (including the Fe Kα line complex), characteristic of colliding wind shock sources. In contrast, sensitive observations of four of the closest known single WC (carbon-rich) WR stars have yielded only non-detections. These results tentatively suggest that single WC stars are X-ray quiet. The presence of a companion may thus be an essential factor in elevating the X-ray emission of WC + OB stars to detectable levels.

Keywords Stars:Wolf-Rayet · Stars:X-rays

1. Wolf-Rayet stars

Wolf-Rayet (WR) stars are luminous objects that are losing mass at very high rates $\dot{M} \sim 10^{-5} - 10^{-4}$ $M_\odot$ yr^{-1}. WR stars are the evolutionary descendants of massive O-type stars and are in advanced nuclear burning stages, approaching the end of their lives as supernovae. They are broadly classified as either WN or WC subtypes according to whether N or C lines dominate the optical spectra. WC stars show chemical evidence of He burning and are thought to be more evolved than WN stars. The powerful winds of WR stars enrich the ISM with heavy elements that will ultimately be recycled into future generations of stars. X-ray studies of WR stars provide crucial information on physical conditions in their winds and outer atmospheres. Detailed comparisons of X-ray properties with shock models can be used to infer information on mass-loss properties. High-resolution X-ray grating spectra also offer the potential of measuring chemical abundances in their metal-rich winds.

2. X-rays from Wolf-Rayet binaries

Strong X-ray emission has been detected from several massive WR + OB binary systems (e.g. Skinner et al., 2001). It is believed that the X-rays originate mainly in a colliding wind (CW) shock region between the two stars. However, other mechanisms may be involved. Intrinsic emission from the stars themselves may be present, and processes such as radiative recombination far out in the winds can give rise to X-ray spectral lines (Schild et al., 2004).

WR 147 is an excellent target for testing the predictions of CW models because it lies nearby at a distance of 0.63 kpc and its stellar properties are relatively well-known. *HST* observations show that the system consists of a WN8 star lying ≈0.62″ south of a O5-7 star (Lépine et al., 2001), but other studies assign a later B0 type for the companion. Several high-resolution radio and IR observations indicate that a CW shock is present near the surface of the OB star. Strong X-ray emission was detected by *ASCA* (Skinner et al., 1999), and

S. Skinner (✉)
CASA, 389 UCB, Univ. of Colorado, Boulder, CO 80309, USA

M. Güdel
Paul Scherrer Inst., Würenlingen und Villigen, CH-5232 Villigen PSI, Switzerland

W. Schmutz
Physik-Met. Observatorium Davos (PMOD), Dorfstrasse 33, CH-7260 Davos Dorf, Switzerland

S. Zhekov
Space Research Inst., Moskovska Str. 6, Sofia 1000, Bulgaria, and JILA, 440 UCB, Univ. of Colorado, Boulder, CO 80309, USA

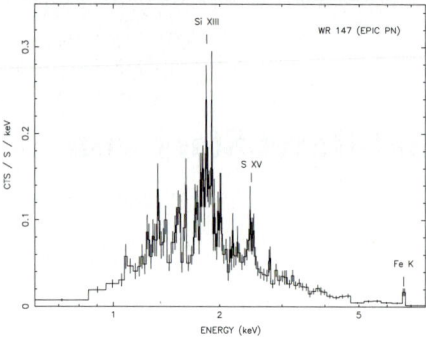

Fig. 1 X-ray spectrum of the composite WR 147 N + S system binned to a minimum of 15 counts per bin. Data are from the *XMM* EPIC-PN camera using the MED optical blocking filter. Prominent emission lines are identified

Chandra shows the emission is extended on arcsecond scales (Pittard et al., 2002).

Our sensitive 22 ksec *XMM* observation of WR 147 reveals a strongly absorbed X-ray spectrum and high-temperature emission lines (Fig. 1). The Fe Kα complex at 6.67 keV is detected for the first time in this system. Acceptable spectral fits with two-temperature thin plasma models give an absorption column density log $N_H = 22.4 \, cm^{-2}$ with plasma temperatures $kT_{cool} \approx 0.8$ keV (9 MK) and $kT_{hot} \approx 3 - 4$ keV (35 – 46 MK). The derived absorption is consistent with estimates based on the known large visual extinction $A_V = 11.5$ mag. Detailed modeling of the X-ray spectrum is underway to determine if the observed X-ray properties are consistent with CW shock predictions based on currently accepted mass-loss parameters of WR 147.

3. X-rays from single Wolf-Rayet stars

In contrast to WR + OB binaries, pointed X-ray observations of single WR stars are almost non-existent and little is known about their X-ray properties. Characterizing their emission is important for interpreting the X-ray spectra of WR + OB systems, whose components are usually too closely spaced to be resolved in X-rays. Single OB stars have been detected in X-rays and almost always show soft emission at characteristic temperatures kT < 1 keV (Berghöfer et al., 1997). Theoretical models generally attribute the soft emission to shocks that are formed in their radiatively-driven winds, but other processes involving magnetic phenomena may also be at work in young OB stars.

Given that single OB stars are X-ray emitters, one expects that single WR stars should be as well. To investigate this question, we are obtaining sensitive X-ray observations of selected WR stars. To date, we have observed four of the closest known single WC stars (Table 1), but single WN stars have not yet been studied. Surprisingly, none of the four WC stars were detected! Our most stringent upper limit is based on a 20 ksec *Chandra* observation of WR 135, which detected only one X-ray count in a nominal source extraction region centered on the star's optical position (consistent with background predictions). This implies a remarkably low intrinsic X-ray luminosity $L_X \leq 10^{29.1}$ ergs s^{-1}, but we quote a more conservative upper limit in Table 1 based on a nominal 5-count *Chandra* detection threshold. The L_X/L_{bol} value of WR 135 is at least two orders of magnitude below what is typically seen for O stars (Fig. 2). These initial results suggest that single WC stars may be X-ray quiet, but additional observations are needed to enlarge the sample. The reason for the apparent suppression of their X-ray emission is not yet known, but X-ray absorption in their dense metal-rich winds may be a factor.

Table 1 X-ray data for WR stars

Star	Sp. type	Dist. (kpc)	Log L_X (erg/s)	Log[L_X/L_{bol}]
WR 147	WN8 + OB	0.63	32.60	−6.7
WR 5	WC6	1.91	≤30.97	≤−7.7
WR 57	WC8	2.37	≤30.95	≤−8.1
WR 90	WC7	1.64	≤30.29	≤−8.7
WR 135	WC8	1.74	≤29.82	≤−9.1

Note: Upper limits on the unabsorbed X-ray luminosity in the 0.5 – 7 keV band are from the *PIMMS* mission simulator, assuming a 1 keV thermal plasma with mean extinctions and distances from van der Hucht (2001). The *XMM* upper limits for WR 5, 57, and 90 are for 68% encircled energy (15″ extraction radii) based on mean background MOS count rates (MED filter). Time intervals during background flares were excluded; the WR5 observation was severely affected by high background. The *Chandra* upper limit for WR 135 assumes a 5 count detection threshold in the ACIS-S detector. L_{bol} values are from Koesterke & Hamann (1995); the value for WR 90 is an average for WC7 stars

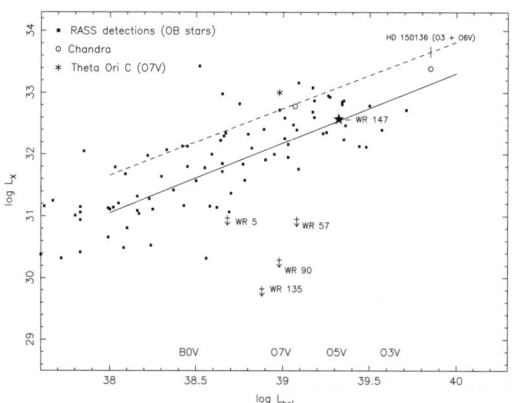

Fig. 2 X-ray versus bolometric luminosity for O stars detected in RASS (solid squares), with Wolf-Rayet stars discussed in the text overlaid for comparison. The solid line is a regression fit for O stars based on RASS detections and non-detections (Berghöfer et al., 1997). The dashed line is a similar regression fit based on *Einstein* data for O stars (not shown). The two circles are based on *Chandra* data for the O stars HD 150135 (O6.5V) and HD 150136 (O3 + O6V) from Skinner et al., (2005). The downward arrows denote X-ray upper limits for undetected WC stars

Acknowledgements This work was supported by NASA grants NNG05GA10G, NNG05GB48G, and GO 5003-X.

References

Berghöfer, T.W., Schmitt, J.H.M.M., Danner, R., Cassinelli, J.P.: A&A **322**, 167–174 (1997)
Koesterke, L., Hamann, W.-R.: A&A **299**, 503–519 (1995)
Lépine, S., Wallace, D., Shara, M.M.: AJ **122**, 3407–3418 (2001)
Pittard, J.M., Stevens, I.R., Williams, P.M.: A&A **388**, 335–345 (2002)
Schild, H.R., Güdel, M., Mewe, R.: A&A **422**, 177–191 (2004)
Skinner, S.L., Itoh, M., Nagase, F., Zhekov, S.: ApJ **524**, 394–405 (1999)
Skinner, S.L., Güdel, M., Schmutz, W., Stevens, I.R.: ApJ **558**, L113–L116 (2001)
Skinner, S.L., Zhekov, S.A., Palla, F., Barbosa, C.: MNRAS **361**, 191–205 (2005)
van der Hucht, K.A.: New Ast. Rev. **45**, 135–232 (2001)

Libration Points in Schwarzschild's Circular Restricted Three-Body Problem

Roman Rodica · Mioc Vasile

Received: 1 November 2005 / Accepted: 1 February 2006
© Springer Science + Business Media B.V. 2006

Abstract The restricted three-body problem in Schwarzschild's gravitational field is analyzed. The existence of the equilibrium points in the orbital plane is discussed and the corresponding positions are established. There are three collinear libration points, and, if they exist, two triangular libration points (situated in the orbital plane of the primaries). If triangular points exist, they may not form equilateral triangles; the triangles are isosceles for equal masses of the primaries, and scalene else.

Keywords Restricted three-body problem · Schwarzschild's field

1. Introduction

As it is well-known, in 1916 Schwarzschild gave the solution of Einstein's field equations in the relativistic central force problem (see Misner, Thorne and Wheeler, 1973). The Schwarzschild metric leads to a Binet-type equation which describes the motion as governed by a force originating in a potential of the form $A/r + B/r^3$, where the constants A and B have well established expressions. Mioc and Stoica (1997) use the name of Schwarzschild problem for the two-body problem associated to such a potential in which A and B may take any real value. We shall use the Schwarzschild potential in a restricted three-body problem, using for the constants A and B the expressions: $A = GMm$ and $B = \frac{GMmC^2}{c^2}$ (see Mioc and Stavinschi, 1998), so the Schwarzschild potential is:

$$U = G\frac{Mm}{r}\left(1 + \frac{C^2}{c^2 r^2}\right), \qquad (1)$$

where: $G =$ Newtonian gravitational constant; $M, m =$ masses of two interacting particles in this field; $C =$ constant angular momentum; $r =$ distance between M and m; $c =$ speed of light.

2. Equations of motion and Jacobian first integral

Consider the circular restricted three-body problem, with primaries of masses M and m, in Schwarzschild's gravitational field. Adopting a rotating frame (with Keplerian angular velocity ω) in the motion plane of the primaries, originated in the common mass center of M and m, the motion equations of the infinitesimal mass read (see Roman, 2003):

$$\ddot{X} - 2\omega\dot{Y} = \omega^2 X - \frac{GM}{r_M^3}\left(1 + \frac{3C_M^2}{c^2 r_M^2}\right)(X + R_M)$$
$$- \frac{Gm}{r_m^3}\left(1 + \frac{3C_m^2}{c^2 r_m^2}\right)(X - R_m)$$

$$\ddot{Y} + 2\omega\dot{X} = \omega^2 Y - \frac{GM}{r_M^3}\left(1 + \frac{3C_M^2}{c^2 r_M^2}\right)Y$$
$$- \frac{Gm}{r_m^3}\left(1 + \frac{3C_m^2}{c^2 r_m^2}\right)Y \qquad (2)$$

$$\ddot{Z} = -\frac{GM}{r_M^3}\left(1 + \frac{3C_M^2}{c^2 r_M^2}\right)Z - \frac{Gm}{r_m^3}\left(1 + \frac{3C_m^2}{c^2 r_m^2}\right)Z$$

R. Rodica (✉) · M. Vasile
Astronomical Institute of the Romanian Academy, Romania
E-mail: rroman@math.ubbcluj.ro

where (X, Y, Z) are the spatial coordinates in the above specified frame (r_M, r_m) are the distances of the infinitesimal mass from (M, m), respectively, (R_M, R_m) are the distances of (M, m) from their common mass center, and (C_M, C_m) are the constants of area in the two-body problem (M, test particle), respectively (m, test particle). Here $C_M^2 = a_M GM(1 - e_M^2)$ and $C_m^2 = a_m Gm(1 - e_m^2)$, where a_M, a_m, e_M, e_m are the usual notations for these two-body problems.

Remark 1. In Equation (2), the terms in $\dot{X}$ and $\dot{Y}$ in the left-hand side represent the Coriolis acceleration. As regards the right-hand side, the gravitational terms (containing G) stand for the attracting Schwarzschild force, whereas the terms that contain ω^2 represent the centrifugal force.

Let us denote

$$U(X, Y, Z) = \frac{\omega^2}{2}(X^2 + Y^2) + \frac{GM}{r_M}\left(1 + \frac{C_M^2}{c^2 r_M^2}\right) + \frac{Gm}{r_m}\left(1 + \frac{C_m^2}{c^2 r_m^2}\right) \quad (3)$$

It is clear that (2) + (3) lead to

$$\ddot{X} - 2\omega \dot{Y} = U_X, \quad \ddot{Y} + 2\omega \dot{X} = U_Y, \quad \ddot{Z} = U_Z, \quad (4)$$

where $U_W := \partial U / \partial W$, $W \in \{X, Y, Z\}$.

Theorem 1. *Schwarzschild's circular restricted three-body problem admits the Jacobian first integral.*

3. Libration points

To determine the libration points of the problem and to locate them, we need some preliminary results.

Proposition 1. *In Schwarzschild's circular restricted three-body problem, the zero-relative-velocity surfaces are equipotential surfaces.*

Corollary 1. *In Schwarzschild's circular restricted three-body problem, the equipotential surfaces are given by*

$$\frac{\omega^2}{2}(X^2 + Y^2) + \frac{GM}{r_M}\left(1 + \frac{C_M^2}{c^2 r_M^2}\right) + \frac{Gm}{r_m}\left(1 + \frac{C_m^2}{c^2 r_m^2}\right) = \text{const.} \quad (5)$$

Let us now pass to the $(Mxyz)$ coordinate system, originated in M, via the transformations: $x = R_M + X$, $y = Y$, $z = Z$, under which we have: $r_M = \sqrt{x^2 + y^2 + z^2}$, $r_m = \sqrt{(R_M + R_m - x)^2 + y^2 + z^2}$. This coordinate system will be useful below for a better positioning of the libration points. Equation (5) becomes:

$$\frac{\omega^2}{2}(x^2 + y^2) - \omega^2 x R_M + \frac{GM}{r_M}\left(1 + \frac{C_M^2}{c^2 r_M^2}\right) + \frac{Gm}{r_m}\left(1 + \frac{C_m^2}{c^2 r_m^2}\right) = \text{const.} \quad (6)$$

Next, we choose the following system of units: $R_M + R_m = 1$ as unit of length, $M + m = 1$ as unit of mass, ω^{-1} as unit of time. Therefore we have $G = 1$ and $P = 2\pi$, where P stands for the orbital period. This system of units leads to:

$$M = \frac{1}{(1+q)}, \quad q = \frac{m}{M}, \quad m = \frac{q}{(1+q)},$$

$$R_M = \frac{q}{(1+q)}, \quad R_m = \frac{1}{(1+q)}.$$

These expressions make (6) become:

$$x^2 + y^2 - \frac{2q}{1+q}x + \frac{2}{1+q}r_M^{-1} + \frac{2q}{1+q}r_m^{-1} + \frac{2C_M^2}{c^2(1+q)}r_M^{-3} + \frac{2qC_m^2}{c^2(1+q)}r_m^{-3} = \text{const.}$$

Now we are in the position to tackle the main topic of our note: the libration points and their location. Recall that they are double points situated on the zero-relative velocity surfaces, their position being determined by the condition: $U_x = U_y = U_z = 0$.

Proposition 2. *In Schwarzschild's circular restricted three-body problem, the libration points are located in the orbital plane of the primaries.*

Theorem 2. *In Schwarzschild's circular restricted three-body problem, there are three collinear libration points (L_1, L_2, L_3).*

Theorem 3. *The coordinates of the collinear libration points in Schwarzschild's circular restricted three-body problem are given by*

$$\left(x_{L_1} - x_{L_1}^{-2}\right) - q\left[(1 - x_{L_1}) - (1 - x_{L_1})^{-2}\right] - \frac{3C_M^2}{c^2}x_{L_1}^{-4} + \frac{3qC_m^2}{c^2}(1 - x_{L_1})^{-4} = 0, \quad (7)$$

$$\left(x_{L_2} - x_{L_2}^{-2}\right) + q\left[(x_{L_2} - 1) - (x_{L_2} - 1)^{-2}\right]$$

$$-\frac{3C_M^2}{c^2}x_{L_2}^{-4} - \frac{3qC_m^2}{c^2}(x_{L_2}-1)^{-4} = 0, \quad (8)$$

$$(x_{L_3} + x_{L_3}^{-2}) - q[(1-x_{L_3}) - (1-x_{L_3})^{-2}]$$
$$+ \frac{3C_M^2}{c^2}x_{L_3}^{-4} + \frac{3qC_m^2}{c^2}(1-x_{L_3})^{-4} = 0. \quad (9)$$

As regards the triangular libration points, for which $y \neq 0$, we state

Theorem 4. *If Schwarzschild's circular restricted three-body problem admits triangular libration points, the properties of these ones are:*
(i) their coordinates depend on q;
(ii) there is only one couple of triangular libration points;
(iii) the respective triangles are never equilateral and are isosceles for $\frac{m}{M} = \frac{a_M(1-e_M^2)}{a_m(1-e_m^2)}$ and scalene else.

Remark 2. If $q=1$, we have $M=m$ and consequently $a_M = a_m$, and $e_M = e_m$, so the above relation is fulfilled.

Corollary 2. *Schwarzschild's circular restricted three-body problem always admits isosceles libration points.*

Theorem 5. *Schwarzschild's circular restricted three-body problem admits scalene triangular libration points if and only if*

$$\frac{C_m^2}{C_M^2} > \frac{r_M^5 + r_m^5 - r_M^4 - r_m^4 - r_M^3 r_m - r_m^3 r_M - r_M^2 r_m^2 - r_M^2 - r_m^2 + r_M + r_m}{r_M^5 + r_m^5 + r_M^4 + r_m^4 + r_M^3 r_m + r_m^3 r_M + r_M^2 r_m^2 - r_M^2 - r_m^2 + r_M + r_m}.$$

With this, we have offered a first insight into Schwarzschild's circular restricted three-body problem.

References

Mioc, V., Stoica, C.: Rom. Astron. J. **7**, 19 (1997)
Mioc, V., Stavinschi, M.: Rom. Astron. J. **8**, 125 (1998)
Misner, Ch., Thorne, K., Wheeler, J.A.: Gravitation W.H. Freeman and Company, San Francisco (1973)
Roman, R.: Modelul Roche la stele duble, Ed. Casa Cărții de Știință, Cluj-Napoca, Romania (2003)

Symbiotic Stars in the Context of Binary Evolution and Metallicity

Laurits Leedjärv

Received: 1 November 2005 / Accepted: 1 February 2006
© Springer Science + Business Media B.V. 2006

Abstract A brief description of symbiotic binary stars is given, followed by a discussion on their evolutionary status. It is evident that different subtypes of symbiotic stars belong to the different Galactic populations. Discussion of some properties of the hot and cool components of symbiotic stars is presented.

Keywords Stars: binaries: symbiotic · Galactic populations

1. Classification schemes

Symbiotic stars (SyS) are among the widest interacting binaries. The pairs consisting of a red giant and a hot companion (usually white dwarf) are embedded into an extended nebula which owes its presence to the strong stellar wind from the red giant or asymptotic giant branch (AGB) star and which is partially ionized by the radiation from the hot component. About 200 known symbiotic and suspected symbiotic stars (Belczyński et al., 2000) actually form a rather heterogeneous class.

The most common classification scheme divides SyS into two main types, according to the spectral energy distribution in the infrared: S (stellar) type and D (dust) type. S-type stars form about 80% of the total number of SyS, they contain normal red giants, usually have orbital periods $P_{orb} \sim 200 - 1000$ days, and possess dense ($n_e \sim 10^8 - 10^{10}$ cm^{-3}) compact (~ 1 AU) nebulae. D-type SyS contain Mira type pulsating AGB stars and an extended (≥ 10 AU) rarified ($n_e \sim 10^6$ cm^{-3}) nebulae. Orbital periods of D-type SyS mostly are supposed to be well over 20 years (Mikołajewska, 2003).

Munari (1994) has proposed a more detailed scheme:

- Classical SyS (cool component M or late K giant, ~ 70 % of total number, corresponds to S type);
- Mira SyS (late M Mira, ~ 20 %, corresponds to D type);
- Carbon SyS (carbon star, ~ 5 %);
- Yellow SyS (F-G giant or supergiant, ~ 5 %).

Yellow SyS, in turn, can be divided into S- and D-type (the latter containing dust) yellow SyS.

Based on the outburst characteristics, a discrimination into classical SyS, symbiotic novae and symbiotic recurrent novae can be made.

2. Evolutionary status

A pair of low-mass stars with an orbital period in a suitable range, in principle, can appear as symbiotic in two different evolutionary stages:

(i) When an initially more massive star becomes a red giant, and transfers matter to the main sequence companion. High mass accretion rates ($\sim 10^{-5} M_\odot$yr^{-1}) required for sustaining symbiotic star brightness are attainable through Roche lobe overflow.

(ii) When an initially more massive star has become a white dwarf (WD), and secondary star is a RGB or AGB star. Hot component is powered by accretion from stellar wind of the giant.

In practice, there are only a couple of SyS (CI Cyg, AX Per) suspected to be of type (i). Even those stars do not entirely contradict the RGB+WD interpretation. Thus, practically all SyS are detached pairs of a red (yellow) giant and a

L. Leedjärv
Tartu Observatory, Tõravere, EE 61602 Estonia
e-mail: leed@aai.ee

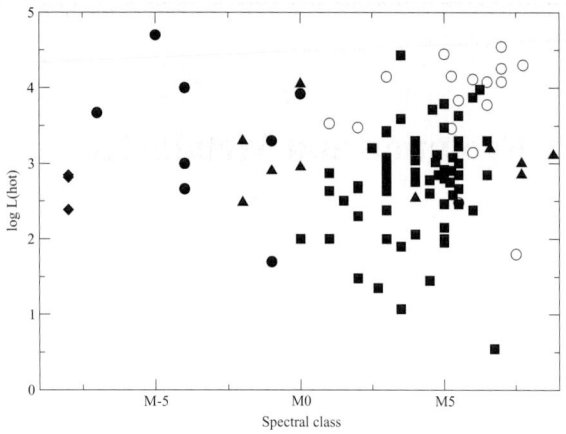

Fig. 1 L_{hot} depending on spectral class of the cool component. Filled rectangles denote red S-type SyS, open circles D-type, filled circles yellow S-type, and diamonds – yellow D-type. Triangles tentatively mark carbon SyS in Magellanic Clouds. M–5 corresponds to the spectral subclass K3

WD. Such binaries should have been able to avoid common envelope stage in their evolution.

A question remains: where are the symbiotic stars of type (i)? According to Kenyon (1994) such stars are easy to "make", there should be $\sim 500 - 5000$ RG + MS symbiotics in the Galaxy. Given the total number of SyS from the same paper, 30 000, this makes about $1.5 - 15\%$ of total number. Two possible RG + MS SyS out of about 200 known hardly approach this lower limit.

3. Cool and hot components of different symbiotic stars

A priori, one should not expect any correlation between the characteristics of the cool and hot components of SyS. However, assuming that the masses of the WDs in SyS are not spread over very broad range, we can expect some correlation between the luminosity of the hot component L_{hot} and the spectral class of the cool component, as mass loss rate and thus also mass accretion rate by the WD increases with the spectral class. Mikołajewska et al. (1997) have found a weak correlation of the above mentioned quantities for a large sample of southern SyS. L_{hot}, in general, is difficult to find, as distances to SyS are not well known. Moreover, this quantity usually is variable, especially in symbiotic novae. However, not pretending to very high accuracy, data for L_{hot} in about 110 SyS can be found in the literature. Being aware of big uncertainties and possible caveats, some conclusions from Fig. 1 can be made:

- In S-type SyS with cool components of spectral class M, L_{hot} weakly correlates with the spectral subclass.
- L_{hot} in D-type SyS tends to be higher than that in S-type SyS.
- In yellow S-type SyS, L_{hot}, in general, is higher than could be expected from the correlation for red S-type SyS; the same is true for yellow D'-type SyS.
- At least some SyS with the lowest L_{hot} (CH Cyg, EG And, R Aqr) are among the nearest known SyS ($\sim 200 - 300$ pc).

4. Discussion

Heterogeneity of SyS implies that actually we have to deal with different Galactic populations and possibly with different evolutionary states. According to Kenyon (1994), Munari (1994), and Jorissen (2003), a following picture can be drawn:

(i) Classical red SyS belong to the bulge/thick-disk (old) population rather than to the thin-disk (young) population.
(ii) Yellow SyS belong to two different populations: yellow S-type SyS belong to the Galactic halo, while yellow D'-type SyS are rather young Galactic disk objects.

It is problematic, how giants of spectral class G or early K can provide strong enough stellar wind for sustaining the high L_{hot}. Although at low metallicity (Z) mass loss rates usually are lower, lower velocity of the wind may enhance accretion rate which depends on the relative velocity of the accreting object. Here may be a clue to understand why yellow S-type SyS contain relatively luminous hot components. Yellow D-type SyS are more difficult to understand. They might actually be similar to planetary nebulae where the nebula originates from the AGB star just evolved to WD.

SyS are sometimes considered as precursors to the supernovae of type Ia (SN Ia). Munari (1994) estimates that only 4% of all classical SyS (assuming that there are $\sim$300 000 such stars in the Galaxy) would be able to create the Galactic SN Ia population. Other authors often are more pessimistic. The main problem is which fraction of the mass accreted by the WD will really accumulate on the WD, and which will be ejected from the binary system. Outflows from hot components (fast winds) certainly exist in symbiotic novae and recurrent novae. Extended nebulae have been detected in many D-type SyS. More or less collimated bipolar jets have been found in about 10 SyS (Brocksopp et al., 2004; Leedjärv 2004) – all those facts show that ejection of matter is an important phenomenon in the life of SyS. Relations to other close binaries with WD components, like CVs and SSXSs, but also links to peculiar red giants like Ba-stars, CH stars, S-giants etc. should be useful to understand the accretion phenomena in binary stars of different Galactic populations and metallicities.

Acknowledgements The present study and the attendance of the Syros conference was financially supported by the Estonian Science Foundation grant No 5003 and by the project "Structure, chemical composition and evolution of stars" financed by the Estonian Ministry of Education and Research.

References

Belczyński, K., Mikołajewska, J., Munari, U., Ivison, R.J. Friedjung, M.: A&ASS **146**, 407 (2000)

Brocksopp, C., Sokoloski, J., Kaiser, C., Richards, A.M., Muxlow, T.W.B., Seymour, N.: MNRAS **347**, 430 (2004)

Jorissen, A.: In: J. Mikołajewska et al. (eds.), Symbiotic Stars Probing Stellar Evolution, ASP Conf. Ser. **303**, 25 (2003)

Kenyon, S.J.: Mem. Soc. Astron. It. **65**, 135 (1994)

Leedjärv, L.: Baltic Astronomy **13**, 109 (2004)

Mikołajewska, J.: In: J. Mikołajewska et al. (eds.), Symbiotic Stars Probing Stellar Evolution, ASP Conf. Ser. **303**, 9 (2003)

Mikołajewska, J., Acker, A., Stenholm, B.: A&A **327**, 191 (1997)

Munari, U.: Mem. Soc. Astron. It. **65**, 157 (1994)

Properties of Components in Contact Systems

S. Zola · K. Gazeas · J. M. Kreiner · B. Zakrzewski

Received: 1 November 2005 / Accepted: 1 February 2006
© Springer Science + Business Media B.V. 2006

Abstract A new sample of contact systems, consisting of more than 100 stars, was created for binaries for which the physical parameters have been determined using both photometric light curves and radial velocity measurements of both components. Properties of components are discussed including their evolutionary status.

Keywords Binaries: eclipsing · Binaries: close · Binaries: contact

1. Introduction

Evolutionary status of contact binaries was investigated by several authors (Mochnacki, 1981) and references therein, (Kałużny, 1985; Maceroni and van't Veer, 1996; Csizmadia and Klagyivik, 2004), mostly by comparing the basic parameters of components with theoretical ones for single stars. They were based on a sample of a few dozen of the best known systems. In theoretical work, the greatest progress was made by Lucy (1968a,b). The Lucy theory explained the properties of contact systems by energy transport within a common convective envelope surrounding the two components. The solutions in hydrostatic equilibrium were successful only for

S. Zola (✉)
Astronomical Observatory of the Jagiellonian University, ul. Orla 171, 30-244 Cracow, Poland

K. Gazeas
Department of Astrophysics, Astronomy and Mechanics, Faculty of Physics, University of Athens, Panepistimiopolis, GR-15784 Zografos, Athens, Greece

J. M. Kreiner · B. Zakrzewski
Mt. Suhora Observatory, Pedagogical University, ul. Podchorazych 2, 30-084 Cracow, Poland

hotter systems; the majority of W UMa-type systems were explained by introducing a thermal instability on a thermal timescale, leading to periodical breaking of the contact configuration with the systems appearing as semidetached at this phase. It was argued that the primaries, more massive components, were still in core hydrogen burning phase (Eggen, 1967; Hilditch and King, 1988).

As it has been shown (Rucinski et al., 2005) and references therein), the determination of absolute parameters of contact systems only from the LC modeling, without knowing the spectroscopic mass ratio, can often result in spurious solutions, thus leading to unreliable physical parameters. We created a new sample, consisting of more than 100 contact systems, for which the parameters have been determined using both photometric light curves and radial velocity measurements for both components see (Kreiner et al., 2005) for a description).

2. Temperature of components versus ΔT relations

The results from the light curve modeling show that the temperature difference between the components (ΔT) in contact systems is small, however for several systems it can be a few hundred K, while for some it can reach more than 1000 K (the subgroup B introduced by Lucy and Wilson (1979). Throughout this paper we define the more massive component as the primary while less massive as the secondary. In Fig. 1 we show relations between the temperature of the primary (T_1) versus ΔT and that for the secondary (T_2) versus ΔT. The whole sample was divided into systems with a small (q < 0.35) and large mass ratio. As it can be seen in Fig.1 (left panel), there is a correlation between T_1 and ΔT: the cooler primaries tend to have negative ΔT while hotter positive ΔT. Such a correlation is stronger for systems

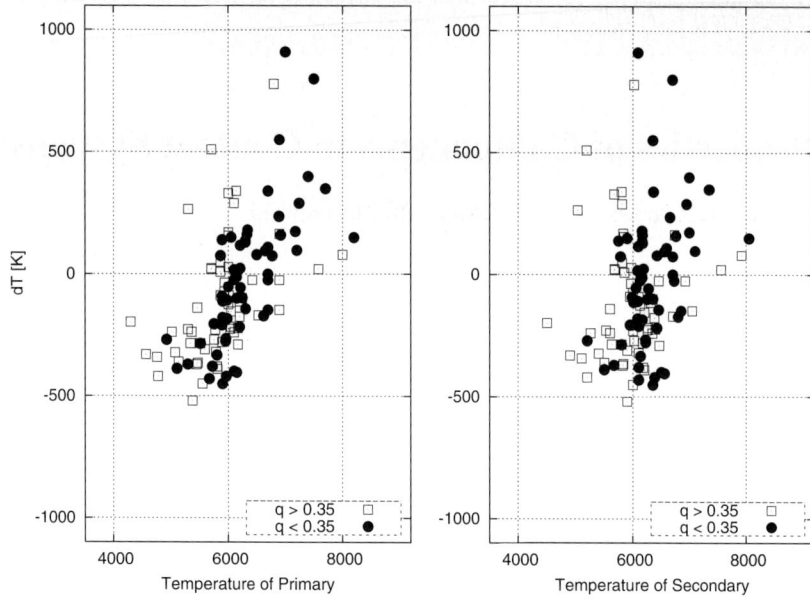

Fig. 1 Temperature of components versus ΔT

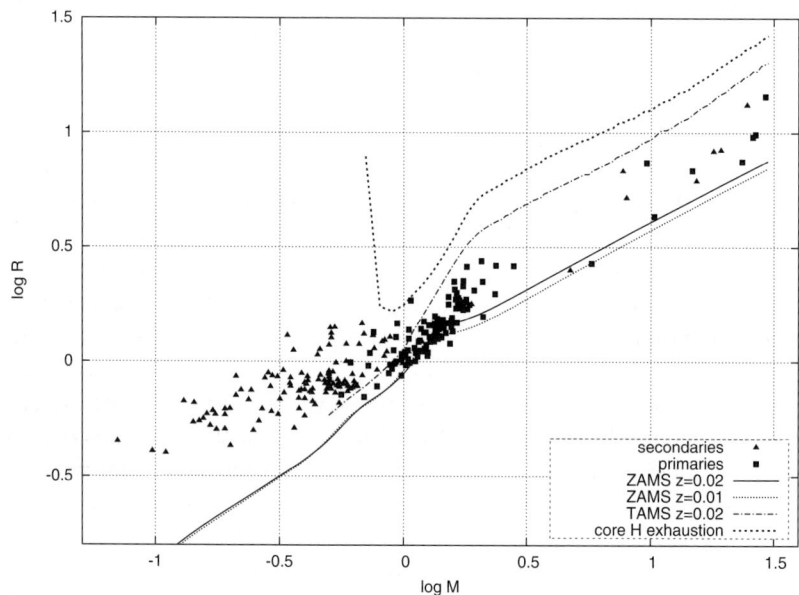

Fig. 2 Mass-Radius relation for our sample of contact systems

with a low mass ratio. There is no correlation between T_2 and ΔT regardless of the mass ratio (see Fig. 1, right panel). Another interesting feature is that the lowest temperature systems seem to avoid having components with very close temperatures.

3. Mass-Luminosity and Temperature-Luminosity relations

To investigate the evolutionary status of components we created a diagram of the Mass-Radius (M-R) relation for both components in a contact system and compared it with theoretical computations of the ZAMS for single stars (Fig. 2). In this figure, the parameters for more massive (squares) and less massive (triangles) stars are shown together with results from theoretical ZAMS computations for two metalicities: $z = 0.02$ and $z = 0.01$. Also shown are parameters for the TAMS ($z = 0.02$) and the evolutionary stage when hydrogen exhaustion in the core occurs. Paxtons's **EZ** code (astro-ph/0405130) was used for computations of the theoretical lines. We applied corrections for components to acount for (1) surface distortion (making use of the Wilson-Devinney code (Wilson, 1979) and (2) energy transfer from the more massive star to the less massive one (Mochnacki, 1981). Except for the temperature correction (formula [7] in Mochnacki (1981)) we additionaly introduced a factor ($\Delta T/T_2 + 1$) changing the energy transfer rate when there is a temperature difference between the components (Mochnacki's formula assumed equal temperatures).

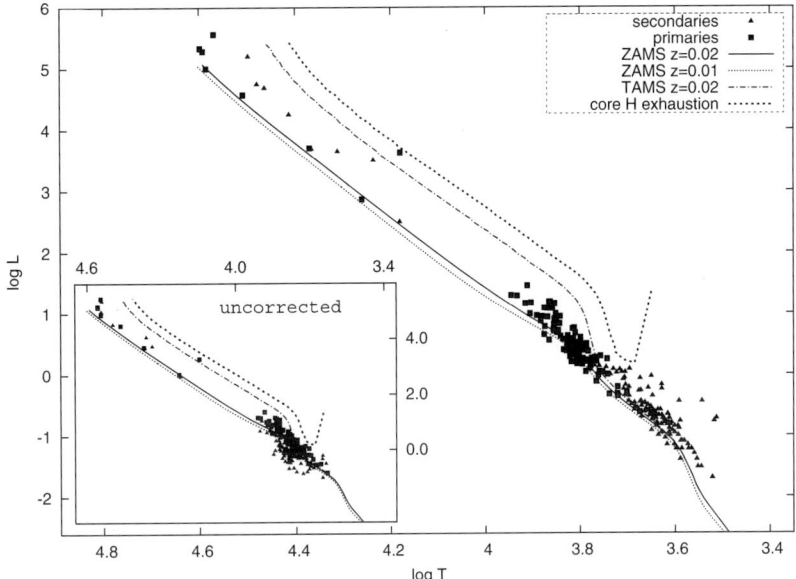

Fig. 3 Temperature-Luminosity relation for contact systems

As it can be seen from the (M-R) relation (Fig. 2), all components of more massive systems ($\log M_\odot > 0.5$) are to be found in the stage between the ZAMS and TAMS. The less massive systems show different properties. While the primaries are not far from the ZAMS, the secondaries are significantly oversized, which is interpreted as the result of energy transfer from the more massive star. The Mass-Luminosity (M-L) relation **without the corrections** confirms the same properties: the primary components of lower temperature systems are near the MS, some are somewhat underluminous, while secondaries are significantly more luminous than single MS stars with equal masses. In Fig. 3 we show the Temperature-Luminosity (T-L) relation for components of contact systems when the above corrections have been accounted for. It turned out that both components are close to the Main Sequence.

Acknowledgements This work was supported by the KBN grant No. 2 P03D 006 22. We would like to thank S. Rucinski and S. Mochnacki for their comments.

References

Csizmadia, Sz., Klagyivik, P.: A&A **426**, 1001 (2004)
Eggen, O.J.: MNRAS **70**, 111 (1967)
Hilditch, R.W., King, D.J.: MNRAS **231**, 397 (1988)
Kaluźny, J.: Acta Astron. **35**, 311 (1985)
Kreiner, J.M., Rucinski, S.M., Niarchos, P. et al.: in press (2005)
Lucy, L.: ApJ **151**, 1123 (1968a)
Lucy, L.: ApJ **153**, 877 (1968b)
Lucy, L., Wilson, R.E.: ApJ **231**, 502 (1979)
Maceroni, C., van't Veer, F.: A&A **311**, 523 (1996)
Mochnacki, S.W.: ApJ **245**, 650 (1981)
Rucinski, S.M., Pych, W., Ogloza, W. et al.: AJ **130**, 767 (2005)
Wilson, R.E.: ApJ **234**, 1054 (1979)

On $(\dot{m} - \dot{P})$ and $(\dot{J} - \dot{P})$-Type Relations for Close Binaries

A. Kalimeris · H. Livaniou-Rovithis

Received: 1 November 2005 / Accepted: 1 February 2006
© Springer Science + Business Media B.V. 2006

Abstract In this study we develop relations connecting the rate of orbital period variations $\dot{P}$ of binary stars with the significant parameters of their evolution, under non-conservative conditions in a fundamental level of description. In this framework, a new generalized $\dot{J} - \dot{P}$-type relation is derived, that can be used in estimation of parameters connected with basic evolutionary processes, when $\dot{P}$ is known from $O - C$ time-series analysis.

Keywords Binary stars : Close binary evolution · Physical processes : Mass transfer

1. Introduction

Orbital period variations (hereafter OPVs) are connected with (and modulated by) physical processes playing a central role in the evolution of close binaries. This connection can be described by equations relating the (observed) rate $\dot{P}$ of OPVs with the basic parameters of binary evolution, such as the rate of mass transfer $\dot{m}$, the rate of mass and/or angular momentum loss (to be referred as $\dot{m} - \dot{P}$ or $\dot{m} - \dot{P}$-type relations hereafter). Equations of this kind have been broadly used in estimations of the mass transfer rate $\dot{m}$ when the rate of period changes $\dot{P}$ are known by observations. The most well known relation of this kind is that proposed by Kruszewski (1966):

$$\frac{\dot{P}}{P} = 3\frac{M_1 - M_2}{M_1 M_2}\dot{m} = 3\frac{1 - q^2}{qM}\dot{m} \quad (1)$$

A thorough review of such relations has been given by Hilditch (2001). The $\dot{m} - \dot{P}$-type relations proposed so far, have been developed under special conditions or particular phases of binary evolution. For completeness and accuracy, we develop a new $\dot{J} - \dot{P}$ relation valid for a broader variety of evolutionary processes and general enough to reproduce most of the preceding relations of this type, as special applications of it.

In the following we assume that the orbital period changes between an initial orbital cycle E_0 and a final E_f, as well as the masses M_i and the radii R_i, $(i = 1, 2)$ of the components are known.

2. A general $(\dot{J} - \dot{P})$-type relation

We consider that, when mass transfer (*of any origin*) is established in a binary, then the rate $\dot{m}_d$ of mass outflow from the donor star and the rate $\dot{m}_g$ of accumulation by the gainer are connected by $\dot{m}_g = -p_m \dot{m}_d$ (where p_m is the portion of the mass that is finally captured by the gainer, $0 < p_m \leq 1$). We further assume that the rest mass escapes from the system (presumably through L_2) at a rate $\dot{m}_{L2} = (1 - p_m)\dot{m}_d$. As a consequence $\dot{m}_d + \dot{m}_g = \dot{m}_1 + \dot{m}_2 = \dot{m}_{L2}$ is valid, irrespective of which component is the donor or the gainer star. Further, if $\dot{m}_w = \dot{m}_{w,1} + \dot{m}_{w,2}$ and $\dot{J}_w = \dot{J}_{w,1} + \dot{J}_{w,2}$ are the total mass and AML due to a stellar wind, then $\dot{M}_i = \dot{m}_i + \dot{m}_{w,i}$ is valid for the total rate of change of the mass of the i-component. And, the variation of the total mass M is given by $\dot{M} = \dot{M}_1 + \dot{M}_2 = \dot{m}_{w,1} + \dot{m}_{w,2} + \dot{m}_{L2} = \dot{m}_w + \dot{m}_{L2}$.

A. Kalimeris (✉)
Technological & Educational Institute of Ionian Islands,
Zakynthos, Greece
e-mail: taskal@teiion.gr

H. Livaniou-Rovithis
Dept. of Astrophysics, Astronomy & Mechanics, Physics Fac.,
Athens Univ., Greece

Radiation Pressure and Surface Gravity of Close Binaries: First Results from Observational Data Analysis

S. Tsantilas · H. Rovithis-Livaniou · G. Djurasevic

Received: 1 November 2005 / Accepted: 1 February 2006
© Springer Science + Business Media B.V. 2006

Abstract We present the results from a sample of early type binaries, for which the surface gravity of their primary components was computed using two different ways. From a comparison of the derived gravity values, is found that the existing difference between them turns to vanish or even to inverse as the luminosity and temperature ratios increase. This implies the presence of radiation pressure, which strongly affects the calculations. A conclusion that has to be further investigated with a larger sample.

Keywords Stars: close binaries · Stars: early type

1. Introduction

Radiation pressure in binaries can be treated using various ways. In most of the cases, the presence of radiation force is assumed to exert from the primary component only (e.g. Schuerman, 1972; Kondo and McCluskey, 1976; Vanbeveren, 1977, 1978). On the other hand, Drechsel et al. (1995), and Djurasevic (1986) considered the case of coming from both components, although they used different models. The latter, made a hydrodynamic and magnetic approach. The former as well as Bauer (2005) analyzed the light curves of IU Aur & AB Cru and V606 Cen, respectively, and using a computer code based on a modified model to include radiation pressure, they achieved a better fit than that of the classical Roche model. One of the main features of the modified models is that for lobes of the same size the value of potential C is smaller than that of the classical Roche.

2. The method

In our project, we used Schuerman's (1972) approach, a simple but well-established model that refers to a binary with an early type primary and a rather cool secondary component. A comparison between this and the classical model reveals that for a given primary component, the value of the potential (and therefore of the surface gravity, g) should be larger in the classical case, increased by $\frac{\delta \cdot \mu}{r_1}$, where $\mu = \frac{m_1}{m_1+m_2}$, $r_1 = \sqrt{(x+\mu-1)^2+y^2}$ and $\delta = F_{\rm rad}/F_{\rm grav}$ is the ratio of the radiation pressure to the gravitational force.

And since the surface potential cannot be measured directly, another physical parameter must be chosen to investigate the existence of any possible increase (i.e. departure from the classical Roche model). We believe that surface gravity is the most promising candidate for this purpose, because it can be measured in three different ways: spectroscopically, from the systems' absolute parameters and from the surface potential. The first two manners are independent of the model used, while the later is model-dependent through the value of the potential, because $g_c = -\nabla C$. Different models yield to different potential C, and therefore to different gravity g_c. But the surface gravity computed from the absolute parameters of the system, g_{ap}, provides an independent on the model value, which can be used to test the model's validity. So, we considered the difference

$$\Delta \log(g) = \log(g_c) - \log(g_{ap}) \qquad (1)$$

where: g_c is the primary's surface gravity computed from the potential according to the classical Roche model. g_{ap} is

S. Tsantilas (✉) · H. Rovithis-Livaniou
Dept. of Astrophysics, Astronomy & Mechanics, Faculty of Physics, Athens University
e-mails: stsant@phys.uoa.gr, elivan@cc.uoa.gr

G. Djurasevic
Astronomical Observatory of Belgrade, Volgina 7, 11160
Belgrade, Serbia & Montenegro
e-mail: gdjurasevic@aob.bg.ac.yu

Table 1 Sample of 16 close binaries

Name	Group	T_1	T_2	References
AW Cam	A	9900	6381	Russo, G. and Milano, L. 1983, A&AS **52**, 311
TW Cas	A	12825	5005	Mardirossian et al., 1980, A&AS **39**, 235
U Cep	A	13600	5454	Markworth, N. 1979, MNRAS **187**, 166
RZ Cas	A	8720	5150	Narusawa et al., 1994, AJ **107**, 1141
IM Aur	A	10350	4953	Gulmen et al., 1985, A&AS **60**, 389
SX Aur	A	19200	14800	Champliss, C. & Leung, K. C. 1979, ApJ **228**, 828
68 Her	A	21700	13800	Jabbar et al., 1987, Ap&SS **135**, 377
CQ Cep	A	43600	37000	Demircan et al., 1997, AN **5**, 267
VV Ori	A	25000	15579	Sarma, M. & Vivekananda Rao P. 1995, JApA **16**, 407
HU Tau	A	13600	5740	Parthasarathy et al., 1995, A&A **297**, 359
RW Tau	A	11750	4271	Van Hamme, W. & Wilson, R. E. 1990, AJ **100**, 1981
BF Aur	B	15600	15400	Demircan, O. 1997, RMxAA **33**, 131
AH Cep	B	29900	28570	Bell et al., 1986, MNRAS **223**, 513
AO Cas	B	34700	35033	Schneider, D. & Leung, K. C. 1978, ApJ **223**, 202
OO Aql	C	5754	5623	Hrivnak, B. J. 1989, ApJ **340**, 458
TY Pup	C	7800	7609	Maceroni et al., 1982, A&AS **49**, 123

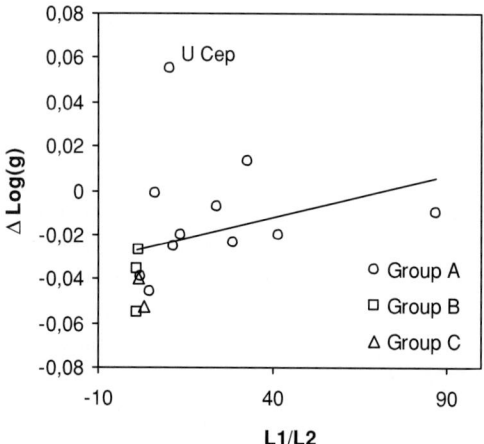

Fig. 1 $\Delta log(g)$ as a function of the luminosity ratio L_1/L_2

the primary's surface gravity computed from the system's absolute parameters selected from the literature.

And as it was mentioned above, in the presence of the radiation pressure, $\log(g_c)$ – hence $\Delta \log(g)$ – should take larger value as the luminosity L_1 increases.

3. Data analysis

The foregoing mentioned were applied to a sample of 16 close binaries that can be divided in three groups (Table 1):

Group A (main group): 11 systems with one component significantly hotter than the other.

Group B: 3 systems with almost similar hot components.

Group C: 2 systems with almost similar cool components. The binary systems of groups B and C were included to test the results of the main group. The obtained results are presented in Fig. 1.

4. Conclusion-Discussion

Two significant results come from our analysis. First, the correlation between $\Delta \log(g)$ and the ratio L_1/L_2 shows that $\log(g_c)$ gets larger values than these of $\log(g_{ap})$ as a result of the radiation pressure effect. This strongly implies that the radiation pressure effect plays an important role in the formation of binaries with one early-type component and therefore should be included in the models. Second, the behavior of groups B and C is similar, which is a rather unexpected result. Because it seems that binaries with two hot components are not affected by radiation pressure, like binaries with two cool stars (in which indeed the effect is negligible); as if radiation pressure forces neutralize each other. Although these are very preliminary results, the general trend seems to be the same, and it does not change significantly, even if different absolute values are chosen, (as they are given by the various investigators). Hence, although Schuerman's (1972) model describes well the case of one hot star, it cannot be generalized in the case of both hot components. So, new models should be implemented, especially in the case of two hot components.

Moreover, more accurate analysis can be achieved if we include non-synchronous rotation. That's because this phenomenon can affect our results by enlarging the measured surface gravity, if the used model is the classical one. For example, we think that the especially large $\Delta \log(g)$ value for *U* Cep (Fig. 1) is partly due to its well-established non-synchronous rotation and partly due to the radiation pressure effect. Another interesting approach is to analyze a larger sample, in groups divided not only according to their temperature, but to their geometrical configuration as well (i.e. contact, detached, etc.). This may reveal a possible diminishing of the radiation pressure consequences in over-contact binaries, due to shadowing effect. This will be the task of further investigation.

References

Bauer, M.: Ap&SS **296**, 255 (2005)
Djurasevic, G., Ap&SS **124**, 5 (1986)
Drechsel, H., Haas, S., Lorenz, R., Gayler, S. A&A, **294**, 723 (1995)
Kondo, Y., McCluskey, G., Proc. IAU Symp. 73, Structure and evolution of close binary systems. Reidel, Dordrecht, 277 (1976)
Schuerman, D., Ap&SS **19**, 351 (1972)
Vanbeveren, D., A&A **54**, 877 (1977)
Vanbeveren, D., Ap&SS **57**, 41 (1978)

The Spatial Distribution of W UMa-Type Stars

Waldemar Ogloza · Bartlomiej Zakrzewski

Received: 1 November 2005 / Accepted: 1 February 2006
© Springer Science + Business Media B.V. 2006

Abstract The aim of this project is to study the spatial distribution of contact binaries and to compare the sample from the project of study the W UMa-type stars, with existing databases such as Hipparcos, GCVS or ASAS (ESA, 1997; GCVS, 2005; Pojmanski, 2002).

Keywords Contact binary stars

1. Sky distribution

As shown in Fig. 1, the distribution of contact systems on the sky (α, δ) is much more homogenous than that of detached. The Milky Way is visible, although weakly. The figure was based on data from only one survey (ASAS-3) to avoid selection effects arising from varying observational methods. Figures. 2 and 3 show a histogram of stars classified as contact systems in the GCVS catalogue. The distributions are of course biased by the higher density of observatories in the northern hemisphere and by seasonal effects. The solid line on the declination histogram represents constant surface density of stars on the celestial sphere.

2. Brightness distribution

Using a variety of methods of discovering contact systems leads to departures of the observed brightness histogram from the expected power law. Left panel of Fig. 3 shows a histogram of brightness based on GCVS and Hipparcos data. The dashed curve represents the theoretical number of contact systems under the assumptions of constant spatial density and perfect effectiveness of discovery. Of course, the probability of detecting an eclipsing binary depends on the inclination of its orbit; however, such tight systems as contact binaries can be detected even for very small values of the inclination. Right panel of Fig. 3 shows a histogram of orbital inclinations based on models found in the literature. In practice, the detection of an eclipsing binary is also determined by the amplitude of the brightness variations. Central panel shows a histogram of the depths of minima of stars classified as contact systems in the GCVS catalogue.

3. Period distribution

Figure 4 shows the distributions of the number of systems as a function of orbital period for samples from the GCVS, Hipparcos and the "W UMa-type stars program" sample (Kreiner et al., 2006). The double maximum (at periods 0.4 d and 0.65 d) could perhaps be the result of an unknown selection effect, however it does not appear in the distribution of periods of RR Lyrae-type, which have similar periods, brightness and amplitude (see insert picture).

The positions of these systems in three-dimensional galactic coordinates are shown in Fig. 5. The values of velocity vectors in space for those systems with known distances and proper motions are shown in Fig. 7. Most of these systems are field stars, however some do belong to moving clusters: 44i Boo, FN Cam, TW Cet, YY CrB, SW Lac, AM Leo, V351 Peg, NN Vir, V776 Cas, UX Eri (Bilir et al., 2005).

W. Ogloza (✉) · B. Zakrzewski
Mt. Suhora Observatory, Cracow Pedagogical University, Poland
e-mail: ogloza@ap.krakow.pl

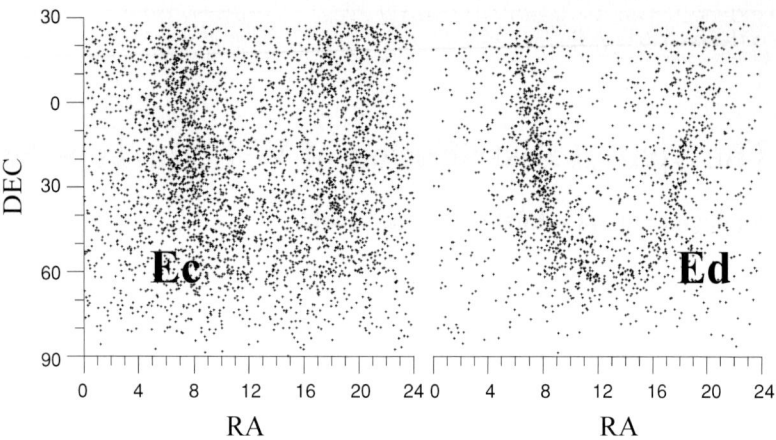

Fig. 1 Sky positions of W UMa-type and detached systems based of ASAS database

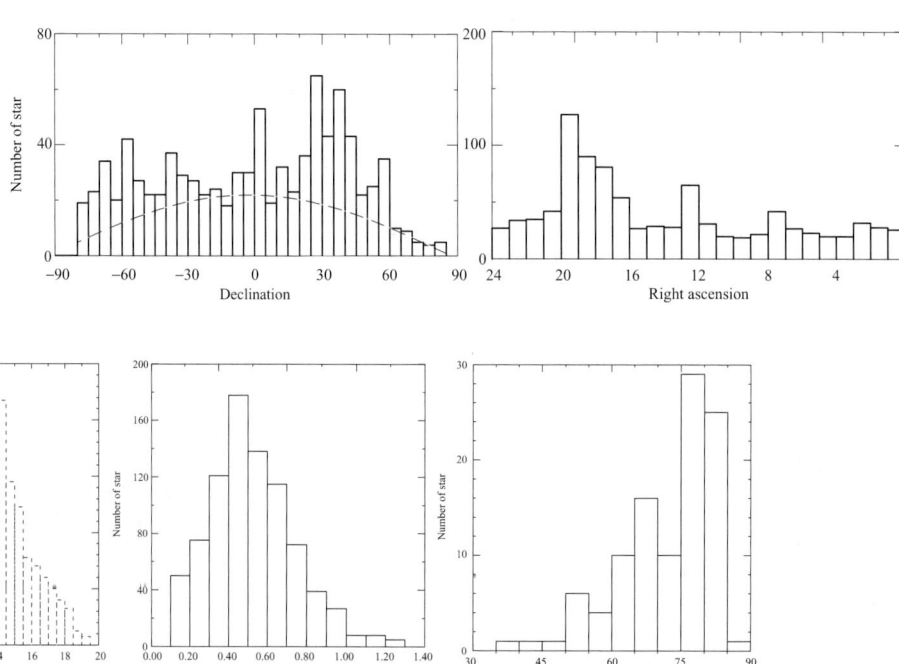

Fig. 2 Sky distribution of W UMa-type stars from GCVS catalogue

Fig. 3 Brightness, amplitude and inclination distribution

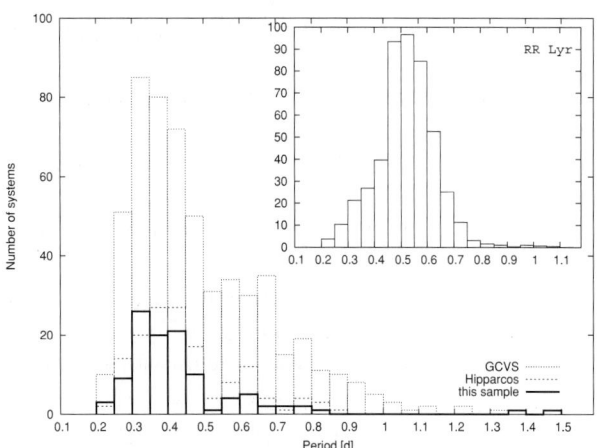

Fig. 4 Period distribution

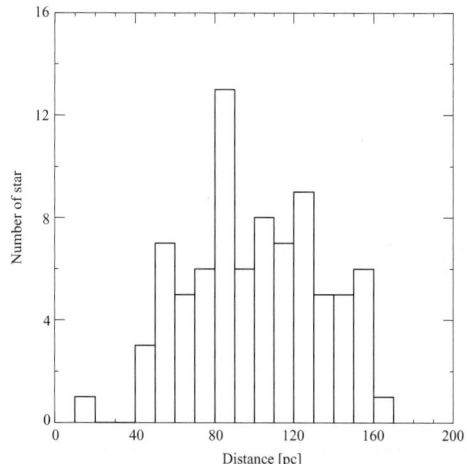

Fig. 5 Distance distribution

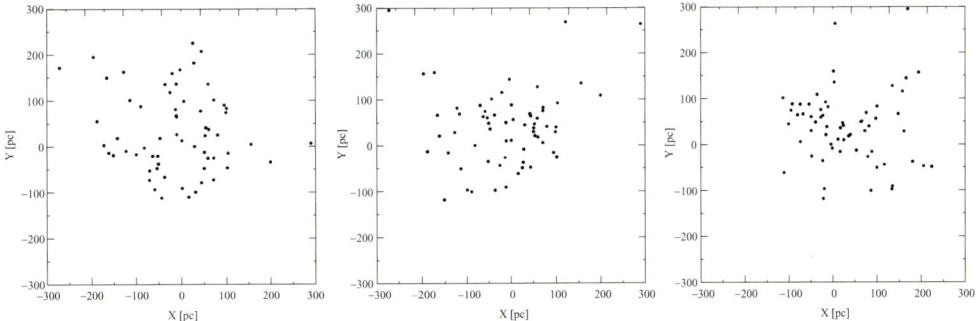

Fig. 6 Spatial distribution

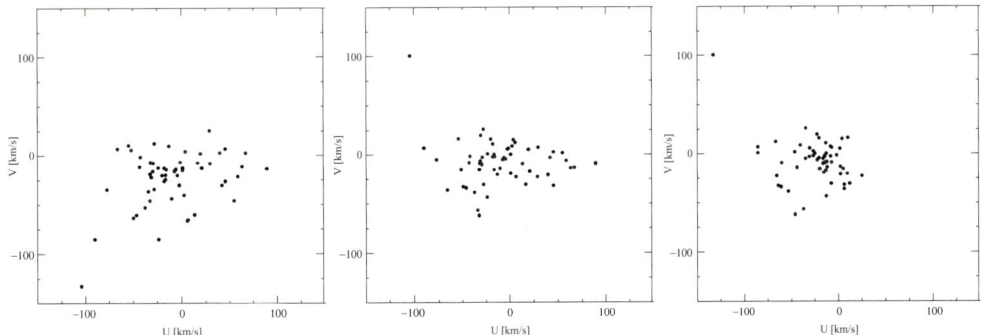

Fig. 7 Velocity distribution

Acknowledgements This project was supported by Polish National Committee grant No. 2 PO3D 006 22.

References

Bilir, S., Karatas, Y., Demircan, O., Eker, Z.: MNRAS **357**, 497 (2005)
ESA, 1997, ESA Publications Division, Noordwijk. ESA SP-1200, (1997)
Kholopov, P.N. et al.: The Combined General Catalogue of Variable Stars, electronic version
Kreiner, J.M., Niarchos, P.G., Rucinski, S.M. et al.: Poster on the conference "Close Binaries in the 21st Century: New Opportunities and Challenges", Syros Island, Greece (2005)
Pojmanski, G.: Acta Astron. **52**, 397 (2002)
Rucinski, S.M.: Publ. Astr. Soc. Pacific **114**, 1124 (2002)
Selam, S.O.: Astron. and Astroph. **416**, 1097 (2004)

Table 1 Parameters derived from $(O-C)$ analyses for both stars

Parameter	TZ Boo value	Std. error	CF Tau value	Std. error
T_0 [HJD]	2443655.492	0.0002	2430651.247	0.0010
P_{orb} [days]	0.297160446	7.4×10^{-9}	2.7558773	0.00000051
$\frac{dP}{dE}$ [days/cycle]	-1.76×10^{-11}	0.02×10^{-11}	–	–
$a'_{12} \sin i'$ [AU]	6.0495	0.0294	3.728	0.329
e'	0.37	0.012	0.51	0.04
ω' [°]	277	1.1	306	4.0
T' [HJD]	2459051.0	22.2	2450879.0	139.0
P_{12} [years]	28.07	0.02	44.42	0.75
A [days]	0.035	0.0002	0.0206	0.0015
$f(m_3)$ [$M_\odot$]	0.28101	0.00367	0.02627	0.00648

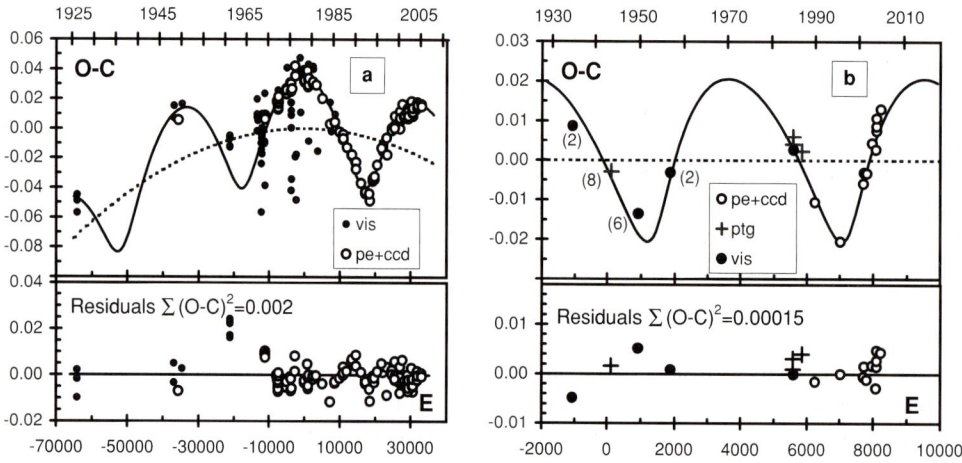

Fig. 1 (a) The $O-C$ diagram of TZ Boo. (b) The $O-C$ diagram of CF Tau. The solid curves representing the least-square fits to the data. Lower panels in each diagram display the final residuals from the overall fits

As seen in Fig. 1a, the $O-C$ variation of TZ Boo has an obvious cyclic character superimposed on a downward curving parabolic structure. In the case of CF Tau, the $O-C$ variation has an obvious cyclic character too, as seen from the Fig. 1b. We have analysed these cyclic characters under the assumption of the existence of additional components in both systems following the formulation given by Irwin (1952). The resulting parameters of the analyses for both systems are listed in Table 1.

3. Interpretations and results

The secular character in the $O-C$ diagram of TZ Boo which was represented with a quadratic term in the light elements during the analysis corresponds to a rate of orbital period decrease by $dP/dt = -4.33 \times 10^{-8}$ days yr^{-1} and can be attributed to mass exchange/loss mechanism in the system. If the period decrease is originated from the conservative mass transfer phenomenon, then the direction of the mass transfer should be from the more massive to the less massive component with a rate of about $dM/dt = 5.07 \times 10^{-9}$ $M_\odot$ yr^{-1}. Under the assumption of presence of a gravitationally bound third body in the system as a cause of observed cyclic character in the $O-C$ diagram we calculated the third body specific parameters. The eclipsing pair completes a revolution on that wide orbit in 28.07 ± 0.02 yrs. The projected distance of the mass center of the eclipsing pair to the center of mass of the triple system should be 6.050 ± 0.029 AU. These values lead to a relatively large mass function of $f(m_3) = 0.28101 \pm 0.00367$ $M_\odot$ for the hypothetical third body. If the third body orbit is co-planar with the systemic orbit (i.e., $i' = i = 78°$), its mass would then be 0.95 ± 0.14 $M_\odot$ which is larger than the total mass of the eclipsing pair ($M_{tot} = 0.74$ $M_\odot$). Then, Kepler's third law gives the semi-major axis of the orbit to be 11.00 ± 0.37 AU. By adopting the distance to TZ Boo as $d = 147.93$ parsecs (ESA, 1997), we get the maximum angular separation of the third body from the eclipsing pair to be $0''.0744 \pm 0.0224$. Thus, the third star, if it exists, should be a very under-luminous compact object in comparison to the main sequence stars or it may also be a low-mass close binary or a multiple system. McLean and Hilditch (1983) have not mentioned a trace of a third component in their spectroscopic work. But the complications in obtaining radial velocity data for TZ Boo which was mentioned by several authors, may be an indication of the existence of the additional component in the system.

The $O - C$ deviations for CF Tau also resemble the light-time effect, orbiting around a third body. By the analysis of the $O - C$ curve, the eclipsing pair completes a revolution on this orbit in 44.42 ± 0.75 yrs. The projected distance of the mass center of the eclipsing pair to the center of mass of the triple system should be 3.728 ± 0.329 AU. These values lead to a mass function of $f(m_3) = 0.02627 \pm 0.00648$ $M_\odot$ for the hypothetical third body. We estimated the mass of the third body under the co-planarity assumption (i.e. $i' = i = 90°$) to be 0.552 ± 0.056 $M_\odot$. In this computation, the masses of the components of CF Tau were adopted as $M_1 = 1.11$ $M_\odot$, $M_2 = 0.87$ $M_\odot$ (Brancewicz and Dworak, 1980). Under the co-planarity assumption of the third body orbit with the systemic orbit, Kepler's third law gives the semi-major axis of the third body orbit to be 13.386 ± 0.052 AU. By adopting the distance of CF Tau $d = 200$ parsecs (Kashyap and Drake, 1999), we get the maximum angular separation of the third body from the eclipsing pair to be $0''.0770 \pm 0.0102$. Using the mass-luminosity relation for main-sequence stars given by Demircan and Kahraman (1991), we can estimate the bolometric absolute magnitude of the third body for the given distance to be $M_{\text{bol}} \cong 15^m.7$ which is too fainter than the binary system. Thus, the third star, if it exists, will be difficult to detect with various observational techniques.

Acknowledgements This work has been supported by the Research Fund of Ankara University (BAP) under the research projects nos. 20040705089 and 20040705090.

References

Brancewicz, H.K., Dworak, T.Z.: AcA **30**, 501 (1980)
Demircan, O., Kahraman, G.: ApSS **181**, 313 (1991)
ESA: The Hipparcos and Tycho catalogs. ESA-SP-1200 (1997)
Guthnick, P., Prager, R.: AN **228**, 331 (1926)
Irwin, J.B.: ApJ **116**, 211 (1952)
Kashyap, V., Drake, J.J.: ApJ **524**, 988 (1999)
McLean, B.J., Hilditch, R.W.: MNRAS **203**, 1 (1983)
Morgenroth, V.O.: AN **252**, 391 (1934)
Popper, D.M.: ApJS **106**, 133 (1996)
Qian, S., Liu, Q.: A&A **355**, 171 (2000)
Szafraniec, R.: AcA **10**, 99 (1960)
Wood, D.B., Forbes, J.E.: AJ **68**, 257 (1963)

Distributions of Geometrical and Physical Parameters of Contact Binaries

M. Vaňko · J. Tremko · T. Pribulla · D. Chochol · Š. Parimucha · J. M. Kreiner

Received: 1 November 2005 / Accepted: 1 February 2006
© Springer Science + Business Media B.V. 2006

Abstract The distributions of geometrical and physical parameters from the CCBS (Catalogue of Contact Binary Stars) and the ASAS-3 (The All Sky Automated Survey) are discussed. The distributions of orbital periods of light curves for 374 contact binaries from the CCBS as well as 3590 contact binaries, selected by Fourier decomposition of 4216 eclipsing binaries from the ASAS-3 database, are similar. The maxima of the period distributions are between 0.31–0.40 days (0.25–0.32 days for W-type and 0.35–0.40 days for A-type) and 0.40–0.45 days for the CCBS and ASAS-3 dataset, respectively.

Keywords Contact binaries · Geometrical and physical parameters · Distribution

1. Introduction

The present paper analyzes the distributions of geometrical and physical parameters of contact binaries (CBs), based on improved version of the catalogue containing 374 contact binaries (hereafter CCBS), published by Pribulla et al. (2003). However, the CCBS catalogue is partly inhomogeneous, because for different systems original authors use different model assumptions. The parameters such as: filling factor, total mass, inclination, photometric mass ratio (derived from partial eclipses), temperature of the secondary component, etc. depend on model assumptions which are in many cases different. Therefore, we have chosen for our study 6 parameters not dependent on the method of analysis: orbital period (P), amplitude of the light curve (A), spectroscopic mass ratio (q_{sp}), photometric mass ratio (derived from systems with total eclipses) (q_{ph}) and temperature of the primary component (T_1). Each distribution (except the distribution of temperatures) is divided into 3 subdistributions for the types of CBs: (i) CBs with the primary minimum being the transit are of A-type, (ii) CBs with the primary minimum being the occultation are of W-type (Binnendijk, 1957) and (iii) CBs with hot components of the OB spectral type are assigned as E type. Systems denoted as B (components are in physical but not in thermal contact, reminding of β Lyrae LC) were excluded from this treatment. The data were not corrected for selection effects and therefore we are dealing with the apparent distribution. The data used to study the distributions of orbital periods and amplitudes of CBs were taken from the ASAS-3 (The All Sky Automated Survey)[1] database (Pojmanski, 2002). The final goal of ASAS survey is photometric monitoring of $\approx 10^7$ stars brighter than 14 magnitude all over the sky. Selection of the EW (W UMa LCs, denoted in ASAS-3 as Ec) systems from the ASAS-3 database was performed by Fourier decomposition of the LCs of all eclipsing variables (see Fig. 1).

2. Selection of systems from the ASAS-3 database

The light curves of possible eclipsing CBs were extracted from the database of the ASAS-3 (the part of the catalogue

M. Vaňko · J. Tremko · T. Pribulla · D. Chochol
Astronomical Institute, Slovak Academy of Sciences, 059 60 Tatranská Lomnica, Slovakia

Š. Parimucha
Institute of Physics, Faculty of Natural Sciences University of P.J. Šafárik, 040 01 Košice, Slovakia

J. M. Kreiner
Mt. Suhora Observatory, Pedagogical University, ul. Podchorazych 2, 30-084 Cracow, Poland

[1] http://www.astrouw.edu.pl/~gp/asas/asas.html.

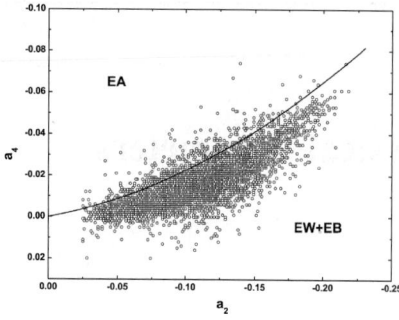

Fig. 1 Relations between the Fourier coefficients a_2 and a_4 for 4216 eclipsing binaries from the ASAS-3 database. The solid curve presents the boundary $a_4 = a_2(0.125 - a_2)$ of Fourier classificator corresponding to the theoretical position of the systems at inner contact. The detached binaries EA are located above the curve and EB + EW types below the curve, respectively (left). Relation between the Fourier coefficients a_2 and a_1 have been used to separate 3590 genuine EW binaries from EB binaries. The boundary $a_1 = -0.02$, is represented by a solid line (right)

available on Feb. 24, 2005). The type of variability is often ambiguous, so the systems were often preliminary classified by more than one type. Hence, we have chosen only the variables, where the catalogue classification string of possible types contains Ec (eclipsing contact) binary type. This sample contains 4216 possible Ec systems. The separation of the CBs from other eclipsing binaries can be done by Fourier decomposition of their light curves using the automatic classifier constructed in the same way as described originally by Rucinski (1997). Note that the Fourier coefficients a_1, a_2 and a_4 are important for the selection process. The a_1 coefficient is sensitive to the difference in depth of eclipses, the a_2 coefficient is proportional to the overall amplitude of the light variations and the a_4 coefficient expresses the degree of contact. For EW binaries with good energy exchange between the components, the a_1 term is expected to be very small, reflecting almost identical depth of the primary and secondary minima, in accordance with almost constant

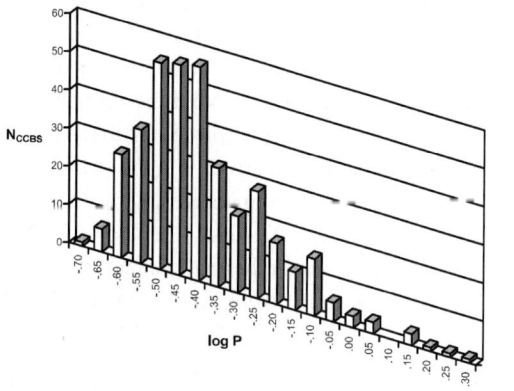 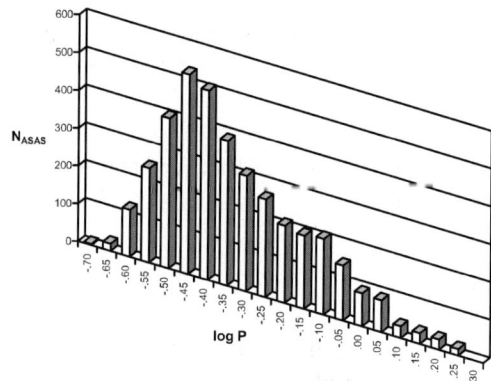

Fig. 2 Orbital periods distribution of CBs from the CCBS (left) and ASAS-3 database (right). The orbital periods are given in log P with the bins of 0.05

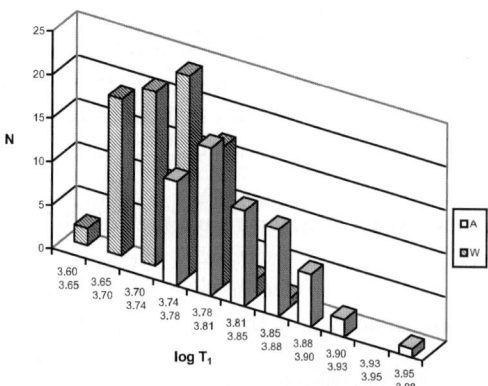 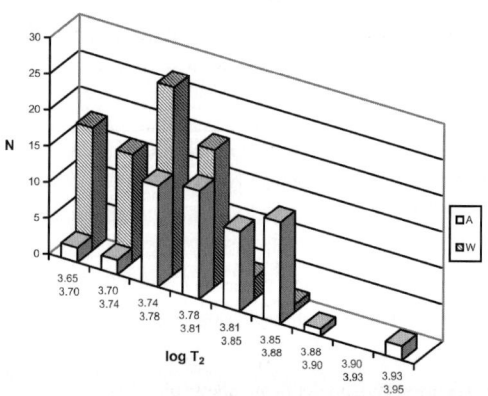

Fig. 3 The log T_1 (left) and log T_2 (right) distributions for CBs from CCBS

effective temperature over the whole contact configuration. Finally, we have selected 3590 EW binaries.

3. Results and discussion

The distributions of orbital periods and orbital light curve amplitudes of CBs from the CCBS were compared with the distributions from the ASAS-3 database. To distinguish the CBs from other eclipsing binaries from ASAS-3 dataset, the "Fourier filter" was used. The distribution extracted from CCBS (Fig. 2) shows that the normal sequence of classical W UMa systems is extended to the range 1.3–1.5 days and then suddenly disappears. The similar behaviour is in the case of period histogram from the ASAS-3. The maximum of period occurrence is between 0.40–0.45 days and 0.31–0.40 days for ASAS-3 and CCBS, respectively. The difference in the period distribution for A and W types in CCBS is conspicuous in the short period domain and could be explained by a different internal constitution at least of one component of a CB. The surface temperatures of A-type CBs are higher than in W-type. Therefore, it is often found in the literature that the A-type and W-type systems have the radiative and convective envelopes, respectively. The thermal boundary between these two types of energy transfer is considered to be $T \approx 7200$ K (Lucy, 1967). This result is in agreement with our distribution of T. As seen in Fig. 3, there are no W-type systems which exceed the temperature of 7080 K. The $\log T_1$ distribution shows the maxima of occurrence of W-type and A-type in the range of 3.74–3.78 and 3.78–3.81, respectively. The value of T_2 is usually calculated during the LCs analysis (resulting from various model assumptions varying from author to author for a given star), so it is an inhomogeneous parameter. Thus, we cannot make firm conclusions from the distribution. On the other hand, reliability of the T_1 parameter, which is usually fixed from spectral type, is in many cases not sufficient.

Acknowledgements This work was supported by Science and Technology Assistance Agency, contract No. APVT-20-014402, by the VEGA grant No. 2/4014/4 and by the Polish KBN grant No 2 P03D 006 22. The authors are grateful to Prof. Rucinski for valuable comments.

References

Binnendijk, L.: JRASC **51**, 81 (1957)
Lucy, L.B.: Z. Astroph. **65**, 89 (1967)
Pojmanski, G.: Acta Astron. **52**, 397 (2002)
Pribulla, T., Kreiner, J.M., Tremko, J.: Contrib. Astron. Obs. Skalnaté Pleso **33**, 38 (2003)
Rucinski, S.M.: Astron. J. **113**, 407 (1997)

to non-axisymmetric plasma streams in TTS binaries Petrov et al. (2001).

Finally, in the present contribution we have discussed several cases of extremely active classical T Tauri stars where close binarity appears to be the engine of the enhanced activity. Even though there are several such CTTS which are known or suspected to be close binaries, it can not be concluded that all active CTTS are close binaries.

The fox is a red animal – but maybe not all red animals are foxes.

References

Alencar, S. H. P., Basri, G., Hartmann, L., Calvet, N.: A&A, in press
Artymowicz, P.: private communication (2005)
Artymowicz, P., Lubow, S. H.: in Beckwith, S., Staude, J., Quetz, A., Natta, A. (eds.), Disks and outflows around young stars, Lecture Notes in Physics **465**, 115 (1996)
Basri, G., Johns-Krull, C. M., Mathieu, R. D.: AJ **114**, 781 (1997)
Covino, E., Melo, C., Alcalá, J. M., Torres, G. et al.: A&A **375**, 130 (2001)
Gahm, G. F., Liseau, R.: in Havnes, O., et al. (eds.) Proc. Activity in cool star envelopes, (Kluwer Academic Publ.), p. 99 (1988)
Gahm, G. F., Petrov, P. P., Duemmler, R., Gameiro, J. F., Lago, M. T. V. T.: A&A **352**, L95
Gahm G.. F., Petrov, P. P., Stempels, H. C.: in Favata, F., et al. (eds.), Proceedings of the 13th Cambridge Workshop on Cool Stars, Stellar Systems and the Sun, p. 563 (2005)
Herbig G. H., Bell K. R.: Lick Obs. Bull., **1111** (1988)
Herczeg, G. J., Walter, F. M., Linsky, J. L., Gahm, G. F., et al.: AJ, in press
Lopéz-Martin, L., Cabrit, S. and Dougados, C.: A&A **405**, L1 (2003)
Martin, E. L., Magazzù, A., Delfosse, X., Mathieu, R. D.: A&A **429**, 939 (2005)
Melo C. H. F.: A&A **410**, 269 (2003)
Petrov, P. P., Gahm, G. F., Gameiro, J. F., Duemmler, R., et al.: A&A **369**, 993 (2001)
Stempels H. C., Gahm, G. F.: A&A **421**, 1159 (2004)
Stempels H. C., Piskunov, N.: A&A **391**, 595 (2002)

Modelling Eclipsing Binaries with Dense Spot Coverage

S. V. Jeffers

Received: 1 November 2005 / Accepted: 1 February 2006
© Springer Science + Business Media B.V. 2006

Abstract To synthesise images of stellar photospheres with high spot filling factors, we model an extrapolated solar size distribution of spots on an immaculate SV Cam. These models of starspot coverage show that the primary star is peppered with a large number of subresolution spots. Using these model starspot distributions we generate a photometric lightcurve, which is then used as input to an maximum-entropy eclipse mapping code, that is based on chi-squared minimisation. I solve for the system parameters to show the effect of dense spot coverage on the derived system parameters, and show that surface brightness distributions reconstructed from these lightcurves have distinctive spots on the primary star at its quadrature points. It is concluded that two-spot modelling or chi-squared minimisation techniques are more susceptible to spurious structures being generated by systematic errors, arising from incorrect assumptions about photospheric surface brightness, than simple Fourier analysis of the light-curves.

Keywords Binaries: eclipsing · Stars: individual (SV Cam) · Late-type · Activity · Spots

1. Introduction

It is well established that rapidly rotating RS CVn binary systems show long lived polar caps. A successful theoretical model of the formation of polar caps assumes that magnetic flux from decaying active regions is swept towards the pole by meridional flows (Schrijver and Title, 2001). In order to produce polar spots the bipolar active regions have to emerge at a rate approximately 30 times faster than on the Sun, implying that the photospheres of active stars should be peppered with a large number of small starspots. Motivated by the work of Jeffers et al. (2006) and Eaton et al. (1996), we apply the extrapolated solar size distributions of Solanki (1999) to a hypothetically immaculate SV Cam. I show how different degrees of spot coverage influences the shape of the lightcurve, and how accurately these lightcurves are reconstructed into surface brightness distributions using a maximum entropy χ^2 minimisation technique (Collier Cameron, 1997).

2. Models

I use the binary eclipse-mapping code DoTS (Collier Cameron, 1997) to synthesize spot maps that follow a log-normal size distribution (Solanki, 1999) on the surface of an immaculate SV Cam. The model parameters we use are for the primary star: $R_1=1.24 R_\odot$, $T_{\rm eff}=6038$ K, and for the secondary star: $R_2 = 0.79 R_\odot$, $T_{\rm eff} = 4804$ K (Jeffers et al., 2005). The input parameters to DoTS for modelling spots are, where x is the output from the random number generator ($0 \leq x \leq 1$);

(i) **longitude:** randomly distributed between 0° & 360°
(ii) **latitude:** $-\frac{\pi}{2} < \theta < \frac{\pi}{2}$, following $\theta = \arcsin(2x+1)$ with $0 \leq x \leq 1$, to eliminate an artificial concentration of spots at the pole
(iii) **spot radius:** computed using a log-normal distribution as described by Jeffers (2006)
(iv) **spot brightness & spot sharpness:** modelled to obtain an umbral to penumbral ratio of 1:3.

---

*Marie Curie Intra-European Fellow

S.V. Jeffers*
Laboratoire d'Astrophysique de Toulouse-Tarbes (UMR 5525), Observatoire Midi-Pyrenees, 14 ave Edouard Belin, 31000 Toulouse, France; School of Physics and Astronomy, University of St Andrews, North Haugh, St Andrews, KY16 9SS, Scotland

discrete Fourier transformation to get the spectrum of ΔV in order to find the periodicities it contains. The main spectral component corresponds to a periodicity of 15.3 ± 0.7yr.

6. Concluding remarks

CG Cyg is a detached, short period RS CVn type system. In short-period RS CVn stars the magnetic activity of one or both components is the main cause of the period variations. In CG Cyg both components show magnetic activity (Kjurkchieva et al., 2003). But the contribution of the secondary to total light of the system in V filter is only 27 percent. So, its magnetic activity can not produce the light variation with amplitude of ~ 0.09. Therefore we considered the primary component to be the main cause of period and brightness variations. An Applegate mechanism was assumed. We estimated the parameters of this mechanism, on the assumption that the absolute parameters of CG Cyg are: $M_1 = 0.85 M_\odot$, $M_2 = 0.54 M_\odot$, $R_1 = 0.77 R_\odot$, and $R_2 = 0.73 R_\odot$. Moreover, we adopted the values of $\Delta P = 1.4 \times 10^{-7}$ as the amplitude of the orbital period modulation, and $P_{\text{mod}2} = 15.9$yr as the modulation periodicity. The magnetic induction was found to be ~ 2.1 KG. The O-C diagram also contained a variation with $P_{\text{mod}1} = 52$yr and amplitude of $\Delta P \sim 1.55 \times 10^{-6}$day. But no correlation was found between $P_{\text{mod}1}$ and mean brightness variations outside eclipse. So magnetic activity cycles of 52yr appear unlikely and it can be attributed to the presence of a third component.

References

Afsar, M., Heckert, P.A., Ibanoglu, C.: A&A **420**, 595–604 (2004)
Applegate, J.H.: AJ **385**, 621–629 (1992)
Beckert, D., Cox, D., Gordon, S., Ledlow, M., Zeilik, M.: IBVS No. 3398 (1989)
Bedford, D.K., Dapergolas, A., Kontizas, E., Kontizas, M.: IBVS No. 3322 (1989)
Dapergolas, A., Kontizas, E., Kontizas, M.: IBVS No. 3609 (1991)
Hall, D.S., Kreiner, J.M.: Acta Astron. **30**, 387 (1980)
Hall, D.S.: ApJ **380**, L85 (1991)
Herkert, P.A.: IBVS No. 4127 (1994)
Herkert, P.A.: IBVS No. 4627 (1998)
Jassur, D.M.Z.: Ap&SS **67**, 191 (1980)
Kalimeris, A., Livaniou-Rovithis, H., Rovithis, P.: A&A **282**, 775 (1994)
Kjurkchieva, D.P., Marchev, D.V., Ogloza, W.: A&A **400**, 623–631 (2003)
Milone, E., Ziebarth, K.: PASP **86**, 684 (1974)
Milone, E., Castle, K., Robb, R., Hall, D.S., Zissell, R.: AJ **84**, 417 (1979)
Naftilan, S., Grillmair, C., Trager, G.: AJ **92**, 210 (1987)
Sowell, J.R., Wilson, J.W., Hall, D.S., Peyman, P.E.: PAPS **99**, 407 (1986)
Wiliams, A.: MNRAS **82**, 300 (1992)
Yü, C.: ApJ **58**, 75 (1923)
Zeilik, M., Elston, R., Henson, G., Schmolke, P., Smith, P.: IBVS No. 2090 (1982)
Zeilik, M., Cox, D., DeBlasi, C., Rhodes, M., Budding, E.: ApJ **345**, 991 (1989)
Zeilik, M., Heckert, P., Ledlow, M. et al.: IBVS No. 3663 (1991)

Short Time-Scale Variability in the Light Curve of TW Draconis

Miloslav Zejda · Zdeněk Mikulášek · Marek Wolf · Ondřej Pejcha

Received: 1 November 2005 / Accepted: 1 February 2006
© Springer Science + Business Media B.V. 2006

Abstract We have re-analyzed old photometric *UBV* data obtained by Papoušek et al. at Skalnaté Pleso, Slovakia (1967–1980) and Brno, Czech Republic (1976–1979) observatories and analyzed new own CCD measurements (Brno, 2003–2005). In both set of photometric data the oscillations in the vicinity of the primary minimum were found. The detected periods are compatible with the previously published ones.

Keywords Eclipsing binary · Short-time variations · Star: Individual: TW Dra

M. Zejda (✉)
N. Copernicus Observatory and Planetarium, Kraví hora 2, 616 00 Brno, Czech Republic; Institute of Theoretical Physics and Astrophysics, Masaryk University, Kotlářská 2, 611 37 Brno, Czech Republic; Charles University Prague, Faculty of Mathematics and Physics, V Holešovičkách 2, 180 00 Prague 8, Czech Republic

Z. Mikulášek
N. Copernicus Observatory and Planetarium, Kraví hora 2, 616 00 Brno, Czech Republic; Institute of Theoretical Physics and Astrophysics, Masaryk University, Kotlářská 2, 611 37 Brno, Czech Republic

M. Wolf
Charles University Prague, Faculty of Mathematics and Physics, V Holešovičkách 2, 180 00 Prague 8, Czech Republic

O. Pejcha
N. Copernicus Observatory and Planetarium, Kraví hora 2, 616 00 Brno, Czech Republic; Charles University Prague, Faculty of Mathematics and Physics, V Holešovičkách 2, 180 00 Prague 8, Czech Republic

1. Introduction

TW Draconis ($\alpha = 15^h33^m51.1^s$, $\delta = 63°54'26''$ (2000.0); = HD 139 319 = BD + 64 1077 = HIP 76196 = GSC 04184-00061; Sp. A8V + K0III) belongs among well-known and often observed Algol-type eclipsing binaries. This star is an A-component of visual binary ADS 9706. The light changes of TW Dra are caused predominantly by eclipses of the hot main sequence star A8V by the cooler and fainter giant component K0III. The deep primary minimum 8.0–10.5 mag (spanning about 11.5 hours) and shallow secondary minimum (0.1 mag) are repeating with the period of 2.8068 days. The O-C diagram illustrates large changes of the light variation's period in the last 150 years as well as shorter oscillation of it (see Fig. 1). The system has been studied photometrically and spectroscopically using both ground based observations and observations from various satellites. However, the satisfactory solution of this unique system has not been published up to now. Different authors have explained the observed properties by the presence of gas stream between components, accretion disk, light time effect, third body, planets orbiting the basic pair, changes of internal structure of secondary component, changes of magnetic fields in the system, pulsations etc. It seems this very interesting system could serve as an important astrophysical laboratory indeed.

2. Observations

The first observations of this star are dated back to 1845! We have obtained 12 photographic observations (plates weakening), 393 visual observations (since 1910), 66 photometric observations (photoelectric since 1947 and CCD since 1996). The first solution of the system based on photometric observations was published by Baglow (1952). Later did it

Table 1 Comparison of detected oscillations

Observations	Period [day]	Semi-Ampl. [mmag]	Pass-band	Notes
Papoušek et al. (1984, this paper)	0.0501(13)	9(2)	B	31 nights (4 processed); possible 2nd $P = 0.034$ d
Kusakin et al. (2001)	0.0556	2.1	B	3 nights
Kim et al. (2003)	0.0527	5.3	B	2 nights; possible 2nd $P = 0.037$ d
Zejda et al. (2005, this paper)	0.0519	10	BVRI	35 nights (1 processed)

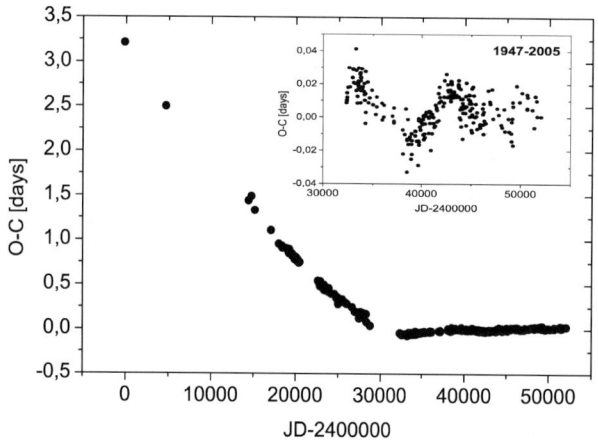

Fig. 1 O-C diagram of TW Dra. The inside figure shows a detail of time interval after 1947 (JD 24320000)

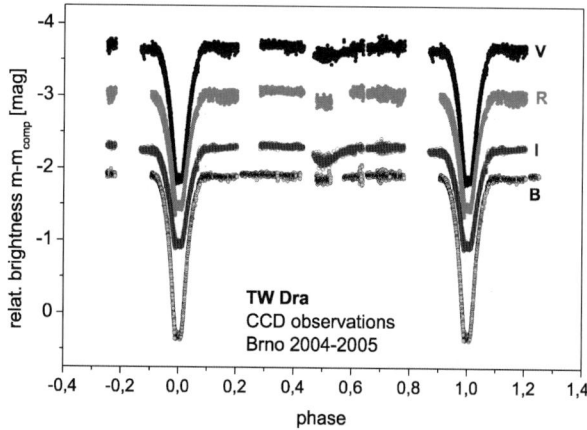

Fig. 2 CCD light curve of TW Dra in *BVRI* passbands (comparison star GSC 4184 0227)

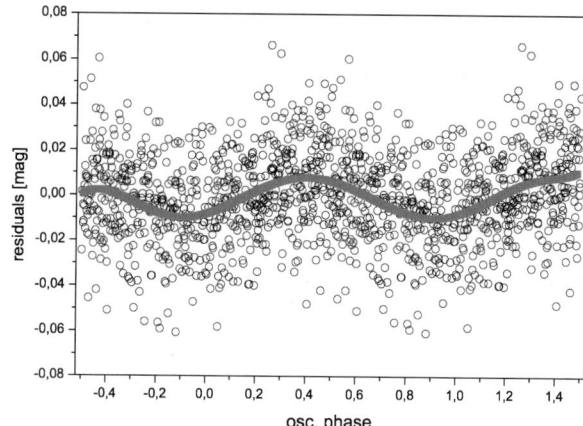

Fig. 3 Phased curve of residuals from CCD observations (*V* passband)

Kopal and Shapley (1956), Brancewicz and Dworak (1980), Al-Naimiy (1983) and others. The first spectroscopic observations and estimate of mass ratio were published by Pearce (1937). Later, TW Dra was studied spectroscopically by Smith (1949), Batten (1967) or Popper (1989).

TW Dra has been observed since 2003 using the CCD cameras SBIG ST7 and ST8 attached to several telescopes (RL400, RL200, RF80) at N. Copernicus Observatory and Planetarium Brno. For the photometric monitoring of the star was used the standard set of $BV(RI)_C$ filters. We obtained 5245 CCD frames in B, 7572 frames in V, 8693 in R and 10792 in I passbands, totally 32 302 CCD frames up to May 2005.

3. Oscillations

The light curves in *BVRI* filters based on new CCD observations are shown in Fig. 2. It seems the scatter of some parts of light curve is larger than one could expect. This is partly due to the small size of CCD frames and the large difference between variable and comparison stars and partly it is very likely a real effect. Kusakin et al. (2001) and later Kim et al. (2003) reported short periodic oscillations in the light curve interpreted as δ Scuti pulsations of primary component. They observed TW Dra in B passband for three and two nights, respectively. The semi-amplitude of variations found by them is only 5 mmag and the period 0.053 day.

Although the amplitude of described oscillation is close to observational scatter, we tried to analyze our data with the aim to confirm or refuse the pertinent oscillation in set of our data in different filters. We have also decided to re-analyze older photometric observations made by Papoušek et al. at Skalnaté Pleso and Brno observatories in 1967–1980 and 1976–1979, respectively.

The short-time oscillations in the light curve become more distinct if one subtracts the mean light curve. For the purpose of finding of the mean light curve we have used sophisticated mathematical method briefly described in Mikulášek et al. (2005). The method is based on the expression of fitting function by linear combination of set of orthonormal trigonometric polynomials. Phased curve of residuals with found period are shown in Fig. 3.

4. Conclusion

The new thorough analyses of old photometric data as well as new CCD measurements have proven the occurrence of oscillations in the light curve of TW Dra on the time-scale of about hours. We have found two periods in Papoušek's data set of about 0.0501 day and 0.0330 day (see Table 1) and one period in the new CCD data 0.0519 day. In both datasets are semi-amplitudes of oscillations larger then reported ones.

However, the nature of these oscillations remains mysterious up to now. The published hypotheses, namely the possible δ Scuti pulsations of one of components or some processes in gas stream between components have not been satisfactorily confirmed yet. Moreover they are not able to explain found oscillation in time of the primary minimum (totally eclipse).

The further, more precise photometric and spectroscopic observations of TW Dra are needed in the future to enlarge the time span for better analysis of all phenomena in this interesting binary.

Acknowledgements This work was supported by the grants GA ČR 205/04/2063 and 205/06/0217 and by the Czech-Greek project of collaboration RC-3-18 of Ministry of Education, Youth and Sport of Czech Republic.

References

Al-Naimiy, H.M.K., Al-Sikab, A.O.: Astrophys. & Space Sci. **103**, 115 (1984)
Baglow, R.L.: Publ. David Dunlop Observatory **2**(1) (1952)
Brancewicz, K., Dworak T.Z.: Acta Astr. **30**(4), 501 (1980)
Batten, A.H.: Publ. Dominion Astroph. Obs. **13**(8) (1967)
Kim, S.-L., Lee, J.W., Kwon, S.-G., Youn, J.-H., Mkrtichian, D.E., Kim, C.: A&A **405**, 231 (2003)
Kopal, Z., Shapley, M.B.: Jodrell Bank Ann. **1**, 141 (1956)
Kusakin, A.V., Mkrtichian, D.E., Gamarova, A.Yu.: IBVS 5106 (2001)
Mikulášek, Z., Wolf, M., Zejda, M., Pecharová, P.: DOI 10.1007/s10509-006-9158-0 (2005)
Papoušek, J., Tremko, J., Vetešník, M.: Folia Fac. Scientiarium Natur. Univ. Purkynianae Brunensis **25**, Physica 39, opus 4 (1984)
Pearce, J.A.: Publ. Am. Astron. Soc. **9**, 131 (1937)
Popper, D.M.: Astrophys. J. Suppl. **71**, 595 (1989)
Smith, B.: Astrophys. J. **110**, 63 (1949)
Walter, K.: Astron. Astrophys. Suppl. **32**, 57 (1978)

A Search for Pulsating, Mass-Accreting Components in Algol-Type Eclipsing Binaries

D. Mkrtichian · S. -L. Kim · A. V. Kusakin ·
E. Rovithis-Livaniou · P. Rovithis · P. Lampens ·
P. van Cauteren · R. R. Shobbrook · E. Rodriguez ·
A. Gamarova · E. C. Olson · Y. W. Kang

Received: 1 November 2005 / Accepted: 1 February 2006
© Springer Science + Business Media B.V. 2006

Abstract We present a status report on the search for pulsations in primary componants of Algols systems (oEA stars). Analysis of 21 systems with A0-F2 spectral type primaries revealed pulsations in two systems suggesting that of the order of ten persent of Algols primaries in this range are actually pulsators.

Keywords Binaries: close · Binaries: eclipsing · R CMa · TZ Eri · TZ Dra

1. Introduction

The pulsating components of semi-detached Algol-type systems (oEA stars) form a group of pulsators attractive for asteroseismic studies (Mkrtichian et al., 2004). The short-period ($P \geq 22$ min) δ Scuti oscillations of A-F spectral class prime components coexist with mass accretion on their surfaces. We present here a status report on the results of an international survey to search for new pulsators in eclipsing binaries (Mkrtichian et al., 2002; Kim et al., 2003).

2. A search for pulsations in eclipsing binaries

The Table 1 lists the stars that have been analyzed, the spectral classes of their main components, whether the data are already published or have been collected for this program, the detection threshold and comments. In total, we analyzed

Table 1 List of analyzed eclipsing binary stars

System	Sp(pri)	Publ./Obs.	Detection threshold mmag	Comment
KO Aql	A0 V	P	3.5	No var.
KP Aql	A3	P	2.8	No var.
RY Aqr	A8	O	5	Uncertain
WW And	A5	P	4.4	No var.
WW Aur	A3	P	4	No var.
R CMa	F0	O	2.7	No var.
BR Cyg	A5 V	P	2.2	No var.
SW Cyg	A2	O	1.2	No var.
RW CrB	A8-F2 V	P	2.1	No var.
V477 Cyg	A3 V	P	3.3	No var.
TZ Dra	A7 V	P/O		var., oEA
TZ Eri	A5 V	P/O		var., oEA
V1647 Sgr	A3 III	P	1.28	No var.
CM Lac	A2 V	P	2.6	No var.
RZ Tau	A7 V	P	5.4	Uncertain
SW Lyn	F1 V	P	3.6	No var.
TX Her	A9 V	P/O	5	Uncertain
BH Dra	A 0	O	9	Uncertain
δ Lib	A0 IV	O	2	No var.
UX Her	A3 V	O	6.0	Uncertain
X Tri	A5-A7 V	P/O	1.9	No var.

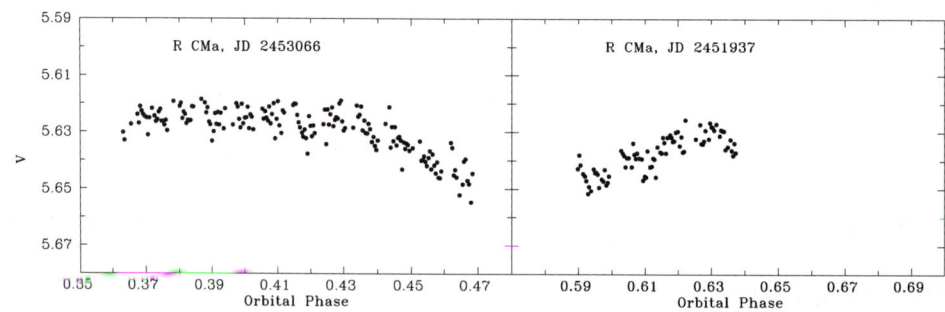

Fig. 1 The JD 2451937 and JD 2453066 V filter observations of R CMa. The data for JD 2453066 does not show the short-periodic pulsational variability

the published data for 17 eclipsing variables and obtained follow-up photometric observations for four of them. For six systems we obtained new observations. For 19 systems we obtained negative or uncertain results, two new systems TZ Eri and TZ Dra were recognized as pulsating and will be discussed together with R CMa in the next subsection.

2.1. R CMa, TZ Eri and TZ Dra

The orbital period of R CMa system is 1.13594 days. The cooler, Roche lobe filling star has the small mass of 0.17 $M_\odot$, while the mass of the accreting star is 1.07 $M_\odot$. Pulsational variability in RCMa with the period of 0.047 day (21.21 c/d) was announced by Mkrtichian and Gamarova (2000) based on the analysis of published photometric observations of Koch (1960) carried out by using the comparison star HD 56405 ($V = 5.454$, A1V). To check whether the pulsational variability belongs to R CMa we carried out additional observations using comparison stars HD 56316 and HD 56971, these being different from those used for Koch's observations. The data were obtained on 27 January 2001 (JD 2451937) and 1 March 2004 (JD 2453066) on the 0.48-m telescope of Tien-Shan Astronomical Observatory. The instrumental V-filter observations vs the orbital phase are shown in Fig. 1. The observations correspond to the orbital phases just before and after secondary minimum. We used for analysis the longer night on JD 2453066. As can be seen in Fig. 1, no signatures of 0.047 day period, nor any short period variability, is visible in the data. The periodogram analysis of these observations, after the removal of the trend due to the orbital variations, does not show any variability with an semi-amplitude above 2.7 mmag in the range of frequencies 10–75 c/d.

So, we cannot yet firmly disprove or confirm of existence of pulsational variability of R CMa. Faced with uncertainty over the earlier results (Mkrtichian and Gamarova, 2000) we cannot now exclude the possibility that the comparison star HD 56405 ($V = 5.454$, A1V) used by Koch (1960) is

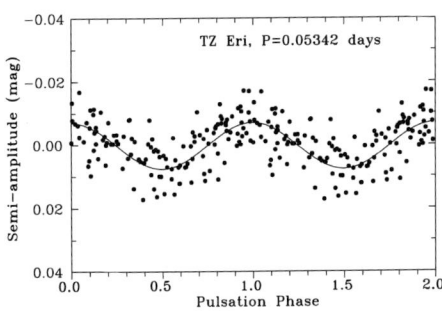
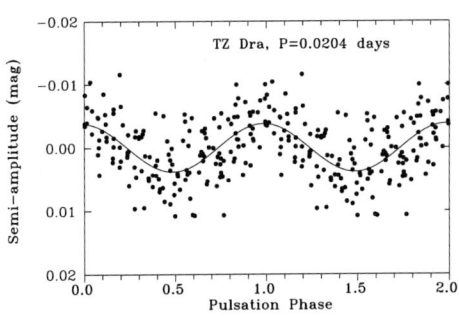

Fig. 2 The out-of-eclipse pulsational B filter light curves of TZ Eri (left panel) and TZ Dra (right panel) phased to periods of 0.05342 d and 0.0204 d respectively

actually a low-amplitude variable star and this caused the 0.047 day variability in (R CMa-HD 56405) magnitude difference determined by Mkrtichian and Gamarova (2000). The variability of HD 56405 should be checked by further observations.

The pulsational variability of TZ Eri and TZ Dra was found by authors during the re-analysis of published photometry obtained by Burblan et al. (1998) and unpublished observations of E. Rovithis-Livaniou and P. Rovithis, respectively. The additional observations of these binaries carried out by authors in 2004 confirmed with high confidence the pulsations in these stars. In more detail, the results will be shown in further publications. Here we show (see Fig. 2) the phase curves of pulsational light variability of TZ Eri and TZ Dra folded with periods of pulsations 0.05342 day and 0.0204 day, respectively.

3. Conclusion

From the analysis of the out-of-eclipse light curves of 21 stars belonging to the Algol class of eclipsing binaries we have found two new pulsators: TZ Dra and TZ Eri. We do not confirm the pulsations of the primary component of R CMa clamed by Mkrtichian and Gamarova (2000). Based on this small sample of A0-F2 class primary components, we find that about 10% of analyzed stars are actually oEA pulsators. The results of the survey of another sample of 15 Algol-type stars carried out by Kim et al. (2005) shows that about 20% are variable. Both results show that oEA stars are promising targets for asteroseismic studies. The progress in field depends on finding and detail studying the pulsations in many Algols.

References

Barblan, F., Bartholdi, P., North, P., Burki, G., Olson, E.C.: A&AS **132**, 367 (1998)
Kim, S.-L., Lee, J.W., Kwon, S.-G., Youn, J.-H., Mkrtichian, D.E., Kim, C.: A&A **405**, 231 (2003)
Kim, S.-L., Kim, S.H., Lee, D.-J., Lee, J.A., Kang, Y.B., Koo, J.-R., Mkrtichian, D., Lee J.W.: PASPC **333**, 217 (2005)
Koch, R.H.: AJ **65**, 326 (1960)
Mkrtichian, D.E., Gamarova, A.Yu.: IBVS **No. 4836**, 1 (2000)
Mkrtichian, D.E., Kusakin, A.V., Gamarova, A., Yu., Rodriguez, E., Kim, S.L., Kim, C., Janiashvili, E.B., Kuratov, K.S., Mukhamednazarov, S.: PASPC **259**, 102 (2002)
Mkrtichian, D.E., Kusakin, A.V., Rodriguez, E., Gamarova, A., Yu., Kim, C., Kim, S.-L., Lee, J.W., Youn, J.-H., Kang, Y.W., Olson, E.C., Grankin, K.: A&A **419**, 1015 (2004)

Astrophys Space Sci (2006) 304:171–173
DOI 10.1007/s10509-006-9102-3

ORIGINAL ARTICLE

HD 172189, a Cluster Member Binary System with a δ Scuti Component in the Field of View of COROT

Pedro J. Amado · Susana Martín-Ruíz ·
Juan Carlos Suárez · Armando Arellano Ferro ·
Andrés Moya · Ignasi Ribas · Ennio Poretti

Received: 1 November 2005 / Accepted: 1 February 2006
© Springer Science + Business Media B.V. 2006

Abstract Photometric and spectroscopic results for the star HD 172189, member of the open cluster IC 4756 in the summer field of the space mission COROT, are presented. From photometric observations in the Strömgren system carried out at various epochs, its binary nature as well as the presence of a δ Scuti-type pulsating component have been discovered. The frequency analysis of the whole dataset confirms a dominant frequency of 19.5974 c d^{-1} with a maximum amplitude near 0.02 mag plus other frequencies in the range 18–20 c d^{-1}. A preliminary orbital solution from the light curve and from four FEROS spectra reveals two similar components of around 1.5 $M_\odot$ orbiting with a period of 5.702 d.

P. J. Amado (✉)
European Southern Observatory, Alonso de Cordova 3107, Santiago 19, Chile

S. Martín-Ruíz · Juan Carlos Suárez
Instituto de Astrofísica de Andalucía, CSIC, Apdo 3004, 18080 Granada, Spain

A. Arellano Ferro
Observatoire de Paris-Meudon, LESIA, 92195 Meudon, France

A. Moya
Instituto de Astronomía, UNAM, Apdo. 70-264, 04510 Mexico D.F., Mexico

I. Ribas
Institut dEstudis Espacials de Catalunya/CSIC, Campus UAB, Facultat de Ciéncies, Torre C5-parell-2a planta,08193 Bellaterra, Spain

E. Poretti
INAF-Osservatorio Astronomico di Brera, Via Bianchi 46, 23807 Merate (LC), Italy

Keywords Stars: Binaries: Eclipsing · Stars: Oscillations · Stars: Variables: δ Sct · Galaxy: Open clusters · Association: IC 4756

1. Introduction

HD 172189 was discovered as a binary system in the summer of 1997 during systematic photometric observations carried out for detecting γ Doradus variables in the open cluster IC 4756 Martín (2003). The Strömgren photometer attached to the 90-cm telescope at the Observatorio de Sierra Nevada (OSN) was used to perform this study. The time span of 10 days was not enough for determining the orbital period. Nevertheless, an eclipse-like minimum was detected in the light curve.

The star has also been classified as a member of the cluster IC 4756 (IC 4756 93) by Sanders (1971), and Missana and Missana (1995) from proper motion studies.

2. Observations

To verify the binary status of the star, a second campaign was performed in June 2003. The new light curves taken by this author showed, however, a short and small variation of, at least, one of the components. These results were confirmed with new observations in August-September of the same year using the same instrument. For both campaigns, July 1997 and August-September 2003, the comparison stars were HD 172365 (IC 4756 145) and SAO 123720 (IC 4756 48). In June the 2003 observations, HD 181414 and HD 173369 were used as check stars.

A frequency analysis has been performed using all out-of-eclipse data. In addition to the frequency $f_1 = 0.17541$ c d^{-1},

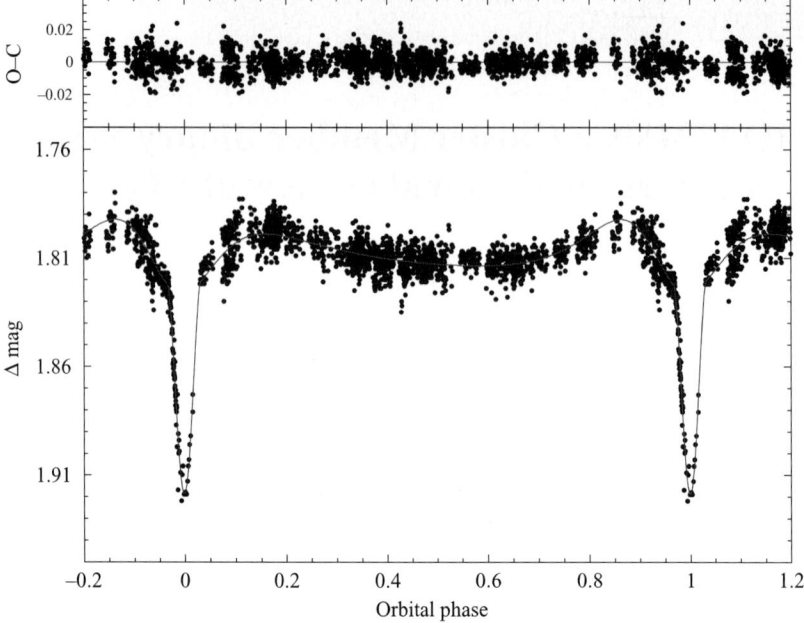

Fig. 1 Strömgren v light curve observed in the summer of 2004 in phase with a period of $P = 5.702$ d. The solution from a preliminary model for the system is overplotted and the $(O–C)$ residuals shown in the upper panel

corresponding to the orbital period, we detected a set of peaks in the region of the spectrum between 18–20 c d^{-1}, typical of δ Sct stars, with the frequencies with the highest amplitudes corresponding to 19.59740 and 18.87939 c d^{-1}.

Moreover, during the summer of 2004, a multisite campaign was carried out at Sierra Nevada and San Pedro Mártir observatories. The main goal of these coordinate efforts was to confirm binarity and thereby better determine the orbital period. In a recent analysis, a value of its orbital period of 5.70198 days and an ephemeris of T(min I) = HJD 2452914.644(3) + 5.70198 (4) d were obtained. In Fig. 1, these data are shown phased with the orbital period.

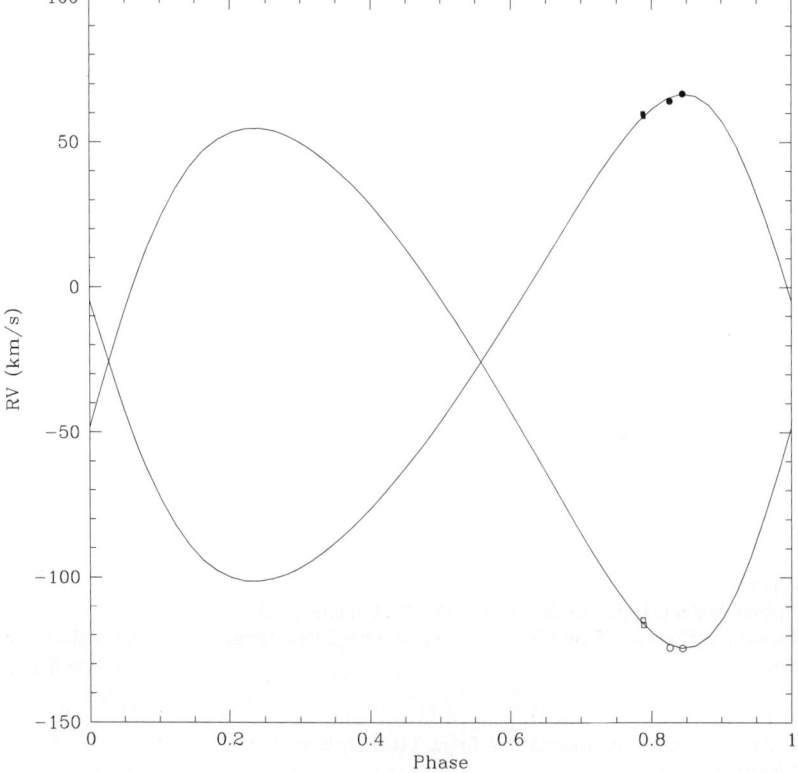

Fig. 2 Radial velocity curve in phase with the period derived form the photometric data ($P = 5.702$ d). Symbols represent the observed data taken with FEROS at the ESO 2.2m telescope

Table 1 Orbit of HD 172189

Period (d)	5.70198(4)
γ (km s^{-1})	−26.0
K_p (km s^{-1})	89.0
K_s (km s^{-1})	101.0
ω (°)	68.0
e	0.2
i (°)	78.0
Mp/Ms	≈ 0.9

The observed primary eclipse depth is of approximately 0.12 mag.

Observations taken with FEROS at the 2.2-m telescope in La Silla (ESO, Chile) at two different epochs (2004 and 2005) produced four spectra. We used the IRAF task FXCOR to determine the radial velocities of the binary components. These data helped, together with the parameters determined from the light curve, obtaining a preliminary orbit of the binary (Fig. 2), whose parameters are given in Table 1.

3. Conclusions

Photometric measurements collected over several years have allowed us to discover the binary nature and δ Sct pulsation of HD 172189. The first minimum in brightness was observed during the summer of 1997. To detect a second eclipse, new campaigns were carried out in 2003, which also unveiled small amplitude oscillations in the light curves. The orbital period of the binary system could not be determined until the two-site campaign in 2004 took place. The most important preliminary results of these observations are: 1. An orbital period of 5.702 d and an eclipse depth in the primary minimum of about 0.12 mag have been estimated. 2. Preliminary simulations of the binary solution suggest that this object is a classic eclipsing binary system with an eccentric orbit. The components likely have similar masses and might be at a rather evolved evolutionary stage. 3. Frequencies have been detected in the region of 18–20 c d^{-1} when analyzing the out-of-eclipse data. Within this range, the frequencies of 19.59740 and 18.87939 c d^{-1} are those with higher amplitudes (less than 5 mmag). With high probability, HD 172189 is a binary belonging to the IC 4756 open cluster and also showing δ Sct-type pulsations. We have presented a preliminary analysis of the currently available data, but more photometric and spectroscopic observations are required to verify the physical parameters, since this object might become a target for the COROT space mission.

References

Herzog, A.D., Sanders, W.L., Seggewiss, W.: A&AS **19**, 211 (1975)
Martín, S.: ASP Conf. Series **292**, 59 (2003)
Missana, M., Missana, N.: AJ **109**, 1903 (1995)
Sanders, W.L.: A&A **15**, 226 (1971)

Eclipsing Binaries with Possible Light-Time Effect

Petr Zasche · Miloslav Zejda · Luboš Brát

Received: 1 November 2005 / Accepted: 1 February 2006
© Springer Science + Business Media B.V. 2006

Abstract The period changes of six eclipsing binaries have been studied with focus on the light-time effect. With the least squares method we also calculated parameters of such an effect and properties of the unresolved body in these systems. With these results we discussed the probability of presence of such bodies in the systems with respect to possible confirmation by another method. In two systems we also suggested the hypothesis of fourth body or magnetic activity for explanation of the "second-order variability" after subtraction of the light-time effect of the third body.

Keywords Stars: Binaries: Eclipsing · Stars: Individual · Multiple stars · Period variations

1. Introduction

Eclipsing binary systems provide a good opportunity for studying the presence of an unresolved third body by observing their minima times affected by the light-time effect (hereafter LITE). This effect was explained by Irwin (1959) and its necessary criteria have been mentioned by Frieboes and Hertzeg (1973) and also by Mayer (1990). Presence of a third body in the system is possible only if the times of minima

P. Zasche (✉)
Astronomical Institute, Charles University, Prague, Czech Republic
e-mail: petr.zasche@email.cz

M. Zejda
Nicolas Copernicus Observatory and Planetarium, Brno, Czech Republic

L. Brát
Velká Úpa 193, Pec pod Sněžkou, Czech Republic

behave in agreement with a theoretical LITE curve, the resultant mass function has reasonable value and corresponding variations in radial velocities are measured. In the last decade a confirmation by astrometry seems to be also possible.

In each case we have calculated new light elements of the eclipsing pair and also the parameters of the predicted third body orbit. The Tables 1 and 2 present the results for each system, where the M_i are the masses of the components, p_i the computed period of the unresolved body, A_i the semi-amplitude of LITE, e_i the eccentricity, ω_i the length of periastron, $f(M_i)$ the mass function and $M_{i,\min}$ the minimum mass (for $i = 90°$) of predicted body, respectively. The subscripts 3 and 4 refer to the parameters of the third and fourth body, respectively.

2. Investigated systems

We have derived new linear light elements for the following systems:

OO Aql: Min I = HJD 24 39322.6902+0^d.50679191 · E,

V338 Her: Min I = HJD 24 41945.3324+1^d.30574608 · E,

T LMi: Min I = HJD 24 23856.4200+3^d.01988799 · E,

RV Lyr: Min I = HJD 24 45526.4003+3^d.59901148 · E,

TW Lac: Min I = HJD 24 24387.2547+3^d.03749292 · E,

V396 Mon: Min I= HJD 24 34769.4234+0^d.39634326 · E.

In the case of OO Aql and T LMi the hypothesis of the fourth body was also suggested. The results are given in Table 2. In the system OO Aql the components are very similar to the Sun (sp. G5V, T $\simeq$ 5700 K), so the second order variations could be also caused by the the magnetic cycles similar to those of Sun (so-called Applegate mechanism, (see eg. Applegate, 1992).

Table 1 The results for six investigated LITE systems

Name of star	Spectrum	$M_1 + M_2$ [$M_\odot$]	p_3 [years]	A_3 [days]	e_3	ω_3 [deg]	$f(M_3)$ [$M_\odot$]	$M_{3,\min}$ [$M_\odot$]	Ref.
OO Aql	G5V+G5?	1.05+0.88	71.72	0.0369	0.13	0.0	0.052	0.714	[1]
V338 Her	F2V+K0	1.42+0.40	26.35	0.0150	0.59	128.7	0.031	0.562	[2]
T LMi	A0+G5III	6.10+0.60	136.98	0.1023	0.29	24.6	0.329	3.177	[3]
RV Lyr	A5+K4III	3.70+1.30	67.99	0.0331	0.31	138.5	0.044	1.192	[4]
TW Lac	A2IV	2.80+1.87	105.98	0.0844	0.70	213.1	0.523	3.184	[5],[2]
V396 Mon	?F8+K8?	1.20+0.47	133.71	0.0519	0.41	114.2	0.042	0.603	[6]

Ref.: [1] – Al-Naimiy et al. (2000); [2] – Budding (2004); [3] – Cester (1979); [4] – Budding (1984); [5] – Halbedel (1984); [6] – Yang and Liu (2001)

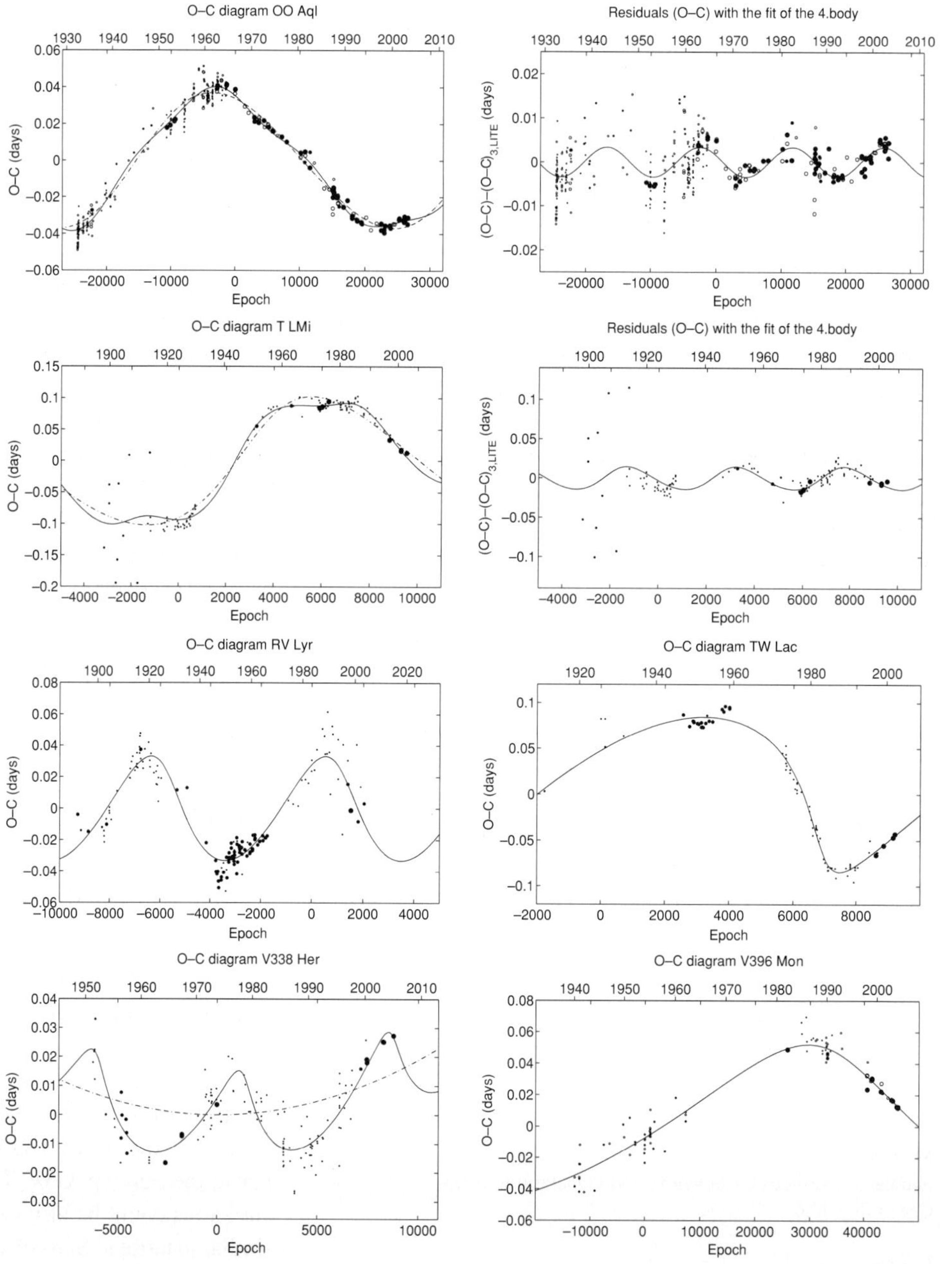

Fig. 1 The $O-C$ diagrams of selected eclipsing binaries. The curves represent proposed light-time effect caused by the third and fourth bodies with periods and semi-amplitudes given in the Tables 1 and 2. The individual primary and secondary minima are denoted by dots and circles, respectively. Larger symbols correspond to the photoelectric or CCD measurements which were given higher weights in our computations. In OO Aql and T LMi cases the LITE caused by the third body is represented by the dashdotted line together with the LITE caused by the fourth body by the solid line. The residuals after subtraction of LITE caused by the third body are shown in figures on the right

Table 2 Parameters for the proposed fourth components

Name of star	p_4 [years]	A_4 [days]	e_4	ω_4 [deg]	$f(M_4)$ [$M_\odot$]	$M_{4,min}$ [$M_\odot$]
OO Aql	19.71	0.0034	0.01	31.7	0.001	0.159
T LMi	37.42	0.0144	0.17	44.8	0.011	1.109

In the system V338 Her the quadratic term was also applied, so the mass transfer in the system could be present. From this hypothesis (with mass transfer parameter $q = 2 \cdot 10^{-10}$) we have derived the mass transfer rate to $\simeq 4 \cdot 10^{-8} M_\odot/\mathrm{yr}$.

In a few cases here the potential third body would be detectable in the detailed light curve analysis. Especially the cases where the third body has not negligible mass, comparing with the eclipsing pair. Regrettably we have no information about the distance of individual systems and also about the absolute magnitudes, so the determination of the angular separation of the possible third body is very uncertain.

3. Conclusion

We have derived new LITE parameters for six eclipsing binaries by means of an $O-C$ diagram analysis. In two cases, OO Aql and T LMi, another variation was found, so there is a possibility of a presence of the fourth body in the system, or magnetic activity in them. But we have not enough data to make a final decision. So the consequence is, that for the confirmation of the presence of LITE in these systems, we need detailed photometric, spectroscopic or astrometric data of these binaries.

Acknowledgements This research has made use of the SIMBAD database, operated at CDS, Strasbourg, France, and of NASA's Astrophysics Data System Bibliographic Services. This investigation was supported by the Czech-Greek project of collaboration RC-3-18 of Ministry of Education, Youth and Sport and by the Grant Agency of the Czech Republic, grant No. 205/04/2063

References

Applegate, J.H.: ApJ **385**, 621 (1992)
Budding, E.: BICDS **27**, 91 (1984)
Budding, E.: A&A **417**, 263 (2004)
Cester, B. et al.: A&A **36**, 273 (1979)
Frieboes-Conde, H., Hertzeg, T.: A&AS **12**, 1 (1973)
Halbedel, E.M.: IBVS **2549**, 1 (1984)
Irwin, J.B.: AJ **64**, 149 (1959)
Mayer, P.: BAICz **41**, 231 (1990)
Naimiy, H.M.K., Al-Masharfeh, T.H.S.: ApSS **273**, 83 (2000)
Yang, Y., Liu, Q.: AJ **122**, 425 (2001)

Eccentric Eclipsing Binary YY Sagittarii

Marek Wolf · P. G. Niarchos · K. D. Gazeas ·
V. N. Manimanis · Lenka Kotková · Anton Paschke ·
Miloslav Zejda

Received: 1 November 2005 / Accepted: 1 February 2006
© Springer Science + Business Media B.V. 2006

Abstract Seven new precise times of minimum light have been gathered for the triple eccentric eclipsing binary YY Sgr ($P = 2^d\!.63$, $e = 0.16$). Its O–C diagram is presented and improved elements of the apsidal motion and the light-time effect are given. We found a new short period of the third body of about 18.5 years in an eccentric orbit ($e_3 \simeq 0.4$).

Keywords Eclipsing binary · Apsidal motion · Light-time effect · YY Sgr

1. Introduction

Eccentric eclipsing binaries with apsidal motion are useful sources of our knowledge about the internal structure of stars. The rate of motion of the apsis is dependent on the internal structure of each component. The study of the apsidal motion thus provides an important observational test of the theoretical models of stellar structure and evolution. The combination of apsidal motion with a light-time effect (LITE) in such multiple systems serves as an excellent laboratory of celestial mechanics to study a wide variety of processes in stellar astrophysics. Here we re-analyse the observational data and rate of apsidal motion for YY Sgr, a southern-hemisphere object, well-known with eccentric orbit and that exhibit apsidal motion.

The detached eclipsing binary YY Sgr (HD 173140; $V_{max} = 10^m\!.03$; Sp. B5+B6) is an early-type binary with an eccentric orbit ($e = 0.16$) and a period of about 2.63 days. This binary is a classical example of apsidal motion and its observational time span covers practically one century. It was discovered to be a variable by Ms. Cannon (Pickering, 1908, 1909), the first photoelectric light-curve was obtained by Keller and Limber (1951). The history of work on this binary was summarized by Woodward and Koch (1992). Photometric orbit, apsidal motion parameters and absolute dimensions were derived by Lacy (1993, 1997), who found an apsidal motion period of $U = 297$ years and eccentricity $e = 0.1575$. The following linear light elements were also given

$$\text{Pri. Min.} = \text{HJD } 24\,48059.68793 + 2^d\!.62846355 \cdot E.$$

In our previous paper (Wolf 2000) we improved the apsidal period and predicted a third body orbiting with a probable period $P_3 = 44.3$ years in an eccentric orbit ($e = 0.44$).

New times of minimum light, covering about 300 epochs, were obtained during 2002–2005 at several observatories: Ondřejov (0.65-m reflector + CCD camera Apogee AP7), South Africa Astronomical Observatory (SAAO, 0.5-m reflector + modular photometer), Athens (0.4-m reflector + CCD camera SBIG ST8) and Hakos IAS, Namibia (Celestron 11 + CCD camera SBIG ST8). The precise moments are given in Table 1. A total of 77 times of minimum light were used in our analysis, with 31 secondary eclipses among them.

M. Wolf (✉)
Astronomical Institute, Charles University Prague, V. Holesovickah 2, Praha 8 Czech Republic
e-mail: wolf@cesnet.cz

P. G. Niarchos · K. D. Gazeas · V. N. Manimanis
Dept. of Astrophysics, Astronomy and Mechanics, University of Athens,
Greece

Lenka Kotková
Astronomical Institute, Academy of Sciences, Ondřejov,
Czech Republic

A. Paschke · M. Zejda
Brno Regional Network of Observers, Brno, Czech Republic

Table 1 New times of minimum light of YY Sgr

JD Hel. - 2400000	Error [day]	Epoch	Method filter	Observatory
52810.4734	0.0003	1807.5	CCD, R	Ondřejov
52839.3853	0.0001	1818.5	CCD, R	Athens
53128.5158	0.0001	1928.5	pe, UBV	SAAO
53178.4556	0.0003	1947.5	CCD, R	Ondřejov
53517.5283	0.0001	2076.5	CCD, R	Ondřejov
53550.5446	0.0002	2089.0	CCD, –	Hakos
53579.4577	0.0002	2100.0	CCD, R	Ondřejov

2. Apsidal motion and LITE analysis

The apsidal motion and LITE in YY Sgr were solved simultaneously by an O-C diagram analysis. The computed apsidal motion elements and their internal errors of the least squares fit (in brackets) are given in Table 2, where P_a denotes the anomalistic period, e represents the eccentricity and $\dot\omega$ is the rate of periastron advance (in degrees per cycle or in degrees per year). The position of periastron at zero epoch is represented by ω_0. The O–C residuals for all times of minimum with respect to the linear part of the apsidal motion equation are shown in Fig. 1. The sinusoidal solution is plotted as continuous and dashed lines for primary and secondary eclipses, resp. New elements of the LITE are also given in Table 2, where P_3 is orbital period of the third body (in days or in years), A the semiamplitude of LITE, e_3 eccentricity of the third body orbit and ω_3 the lenght of periastron. Assuming a coplanar orbit ($i_3 = 90°$) and a total mass of the eclipsing pair $M_1 + M_2 = 7.38\,M_\odot$ (Lacy, 1997) we can obtain a value of the mass function $f(m)$ and a lower limit for the mass of the third component $M_{3,\min}$. The third component could be a star of spectral type K3 with the bolometric magnitude of about +6.1 mag, practicaly invisible in the system with a B5 primary ($M_{\rm bol} \simeq -2$ mag, Harmanec 1988).

The observed average value of the internal structure constant (ISC) $k_{2,\rm obs}$ can be derived using the well-known equation (Kopal, 1978). Taking into account the value of the eccentricity and the masses of the components, we have to subtract from $\dot\omega$ a relativistic correction $\dot\omega_{\rm rel}$. For YY Sgr we have $\dot\omega_{\rm rel} = 0.0011$ deg/year and $\dot\omega_{\rm rel}/\dot\omega = 12.6\,\%$. The resulting mean ISC is log $k_{2,\rm obs} = -2.286$. The theoretical value of ISC according to available models for the internal stellar structure computed by Claret and Giménez (1992) gives for YY Sgr the value of log $k_{2,\rm theo} = -2.23$. The agreement between the theoretical and observed value of ISC is relatively good.

Table 2 Apsidal motion and LITE parameters of YY Sgr

Element	Unit	Value	Element	Unit	Value
T_0	HJD	24 48059.5782 (5)	P_3	days	6783 (20)
P_s	days	2.6284739 (3)	P_3	years	18.57 (5)
P_a	days	2.6285381 (5)	T_3	JD	24 53040 (100)
e	–	0.158 (1)	e_3	–	0.404 (8)
$\dot\omega$	deg/cycle	0.0088 (2)	A	days	0.00674 (15)
$\dot\omega$	deg/yr	1.22 (0.03)	ω_3	deg	166.8 (1.2)
ω_0	deg	214.6 (0.2)	$f(m)$	$M_\odot$	0.0059
U	years	294.7 (1.4)	$M_{3,\min}$	$M_\odot$	0.73

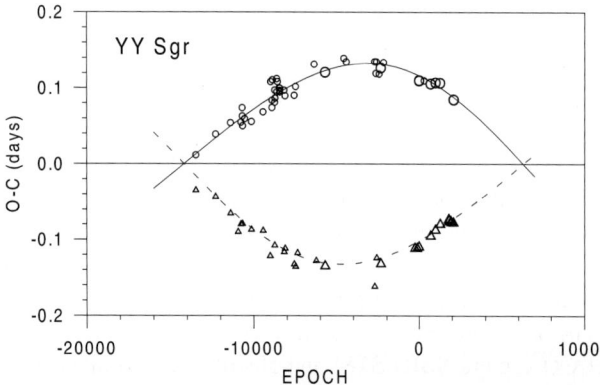 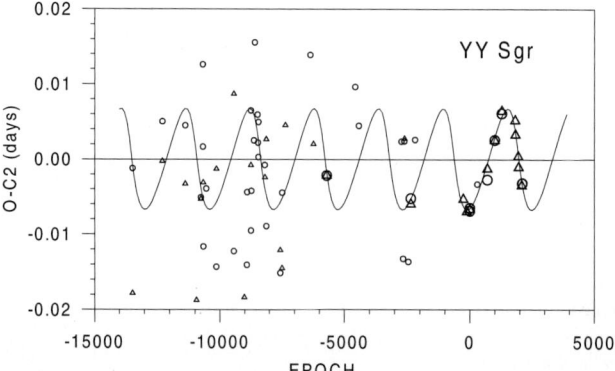

Fig. 1 Left: The O–C graph for the times of minimum of YY Sgr. The individual primary and secondary minima are denoted by circles and triangles, resp. Larger symbols correspond to the photoelectric or CCD measurements which were given higher weights. Right: The $O-C_2$ diagram in detail after subtraction the terms of apsidal motion. The solid curve represents LITE for the third body orbit with a period of 18.5 years and an amplitude of about 0.007 days

3. Conclusions

We improved apsidal motion parameters and derived new LITE elements with the short period for the eccentric eclipsing binary YY Sgr. This system belongs to the interesting group of triple eccentric eclipsing binaries showing the apsidal motion and the LITE together (V889 Aql, V539 Ara, AS Cam, TV Cet, RU Mon, U Oph, AO Vel, DR Vul) described in Wolf et al. (2001), Bozkurt and Degirmenci (2005) and Wolf and Zejda (2005). This category of eclipsing binaries deserves a continuous photometric, spectroscopic as well as astrometric monitoring.

Acknowledgements This investigation was supported by by the Czech-Greek project of collaboration RC-3-18 of Ministry of Education, Youth and Sport and by the Grant Agency of the Czech Republic, grant No. 205/04/2063. Based on observations secured at the South Africa Astronomical Observatory, Sutherland, South Africa.

References

Bozkurt, Z., Degirmenci, O.L.: ASP Conf. Series **335**, 277 (2005)
Claret, A., Giménez, À.: A&AS **96**, 255 (1992)
Harmanec, P.: Bull. Astr. Inst. Czech **39**, 329 (1988)
Keller, G., Limber D.N.: ApJ **113**, 637 (1951)
Kopal, Z.: Dynamics of Close Binary Systems, Reidel, Dordrecht, Holland (1978)
Lacy, C.H.S.: AJ **105**, 637 (1993)
Lacy, C.H.S.: AJ **113**, 1091 (1997)
Pickering, E.C.: HCO Circ. No. 137 (1908)
Pickering, E.C.: ANAS **179**, 7 (1909)
Woodward, E.J., Koch, R.H.: AJ **104**, 796 (1992)
Wolf, M.: A&A **356**, 134 (2000)
Wolf, M., Diethelm, R., Hornoch, K.: A&A **374**, 243 (2001)
Wolf, M., Zejda, M.: A&A **437**, 545 (2005)

Photospheric Abundance Peculiarities in RS CVn Binaries

Thierry Morel · Giuseppina Micela · Fabio Favata

Received: 1 November 2005 / Accepted: 1 February 2006
© Springer Science + Business Media B.V. 2006

Abstract We discuss the results of a LTE abundance study of 14 single-lined RS CVn binaries. Increasingly peculiar abundance ratios are observed for the cooler and more active stars in this sample (this is best illustrated in the case of oxygen). This may arise from the existence of large spot groups, departures from LTE much larger than anticipated and/or inadequacies in the Kurucz model atmospheres for these objects.

Keywords Stars: fundamental parameters · Stars: abundances · Stars: activity · Line: formation

1. Observations and analysis

Spectra of 14 single-lined RS CVn binaries (G8-K2 IV-III) were obtained at ESO using the echelle spectrograph FEROS (3600–9200 Å; $R = 48\,000$). Only slow rotators were selected ($v \sin i \lesssim 10$ km s^{-1}). A control sample made up of 7 single stars of similar spectral type, but with much lower X-ray luminosities was also observed.

The abundances of 13 chemical species were determined using the measured EWs of about 90 carefully-selected spectral lines, along with a set of 1-D line-blanketed LTE Kurucz atmospheric models, as input for the MOOG software. The oscillator strengths were calibrated with a very high-quality FEROS solar spectrum. The effective temperatures and surface gravities were determined from the excitation and ionization equilibrium of the Fe lines, while the

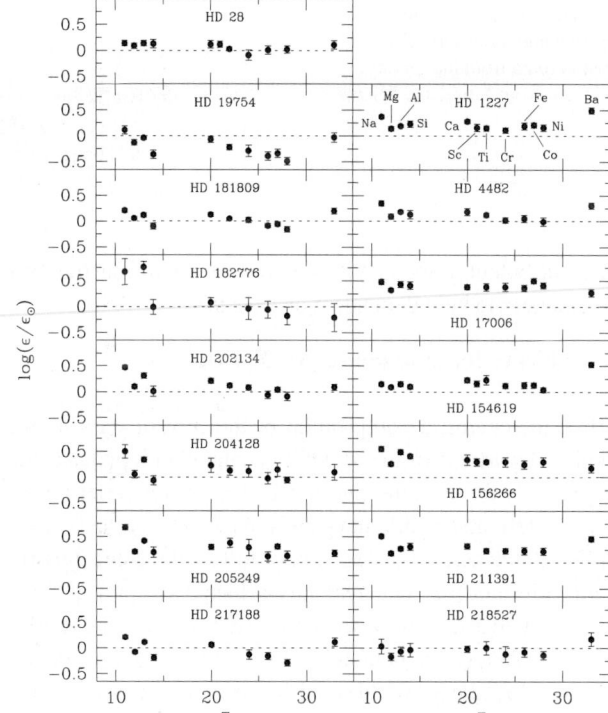

Fig. 1 Abundance patterns of the RS CVn binaries (*left-hand panels*) and the stars in the control sample (*right-hand panels*). The chemical elements are identified in the upper, right-hand panel. The position of barium has been shifted for the sake of clarity to $Z = 33$. The dashed line corresponds to the solar abundance pattern. Adapted from Morel et al. (2004)

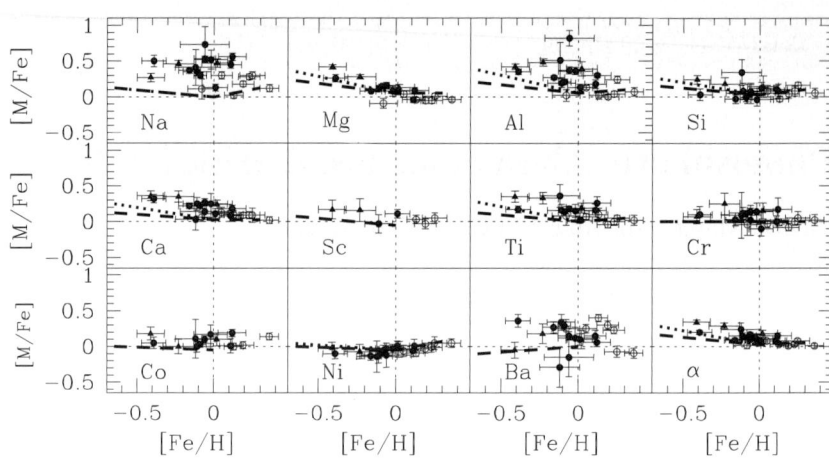

Fig. 2 Abundance ratios as a function of [Fe/H]. The active binaries and stars in the control sample are indicated by filled and open circles, respectively (from Morel et al., 2004). The thick dashed and dotted lines show the characteristic trends of kinematically-selected samples of thin and thick disk stars, respectively (Bensby et al., 2003)

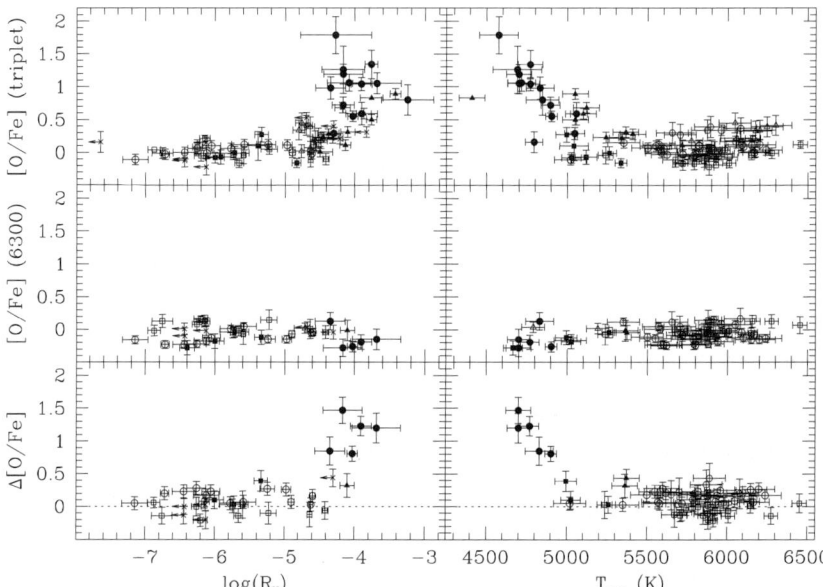

Fig. 3 Oxygen abundances as a function of the activity index, R_X, derived from X-ray data (*left-hand panels*) and the excitation temperature, T_{exc} (*right-hand panels*). The bottom panels show the difference between [O/Fe] yielded by the O I 7774-Å triplet and the [O I] λ6300 line. Filled circles: RS CVn binaries, filled squares: field subgiants (Morel et al., 2003, 2004), filled triangles: Pleiades stars, open triangles: Hyades stars, open circles, squares and hexagons: disk dwarfs (data from the literature). Adapted from Morel and Micela (2004)

microturbulent velocity was derived by requiring the Fe I abundances to be independent of the EWs.

2. Evidence for abundance peculiarities

While the chemical composition of the control stars shows modest departures from a scaled solar abundance pattern, this is generally far to be the case for the RS CVn binaries (compare the left- and right-hand panels of Fig. 1). As can be seen in Fig. 2, the active binaries do not follow the characteristic trends presented by kinematically-selected samples of inactive, FG field dwarfs between the abundance ratios and [Fe/H] (e.g. Al). At a given metallicity, a significant overabundance of several elements is observed in the active stars (see also Katz et al., 2003).

3. Activity and/or temperature effects?

Peculiar abundance ratios are preferentially observed for the cooler and more active objects in our sample. This phenomenon is particularly well illustrated by oxygen (see Fig. 3). In this particular case, the distinct behaviour of the O I triplet and [O I] λ6300 points to unexpectedly large NLTE effects affecting the permitted lines. Although a similar interpretation might apply to other elements, the same analysis carried out on synthetic, composite Kurucz spectra with a varying spot coverage shows that cool spots generally lead to an overestimation of the abundance ratios, in qualitative agreement with the observations. Temperature effects (e.g. inadequacies in the Kurucz models for K-type subgiants) may also play a significant role.

4. Conclusion

This first systematic study of the chemical composition of tidally-locked active binaries raises some concerns about the reliability of classical LTE abundance analyses in cool, chromospherically active stars. The combined action of presumably strong NLTE effects, cool spots and inadequate atmospheric models (e.g. which neglect the chromospheric component) may potentially have a significant impact on the final results. These limitations must be kept in mind when comparing the photospheric and coronal abundance patterns of active binaries, for instance (see, e.g. Sanz-Forcada et al., 2004). In a broader context, the chemical peculiarities reported here are strongly reminiscent of recent observations of young open cluster members (e.g. Schuler et al., 2004). It is thus likely that the abundances derived in both classes of objects are affected by similar processes. We expect our forthcoming abundance analysis of a large sample of inactive K-type stars to help assessing the relative importance of temperature and activity effects (Morel et al., in preparation).

References

Bensby, T., Feltzing, S., Lundström, I.: A&A **410**, 527 (2003)
Katz, D., Favata, F., Aigrain, S., Micela, G.: A&A **397**, 747 (2003)
Morel, T., Micela, G., Favata, F., Katz, D., Pillitteri, I.: A&A **412**, 495 (2003)
Morel, T., Micela, G., Favata, F., Katz, D.: A&A **426**, 1007 (2004)
Morel, T., Micela, G.: A&A **423**, 677 (2004)
Sanz-Forcada, J., Favata, F., Micela, G.: A&A **416**, 281 (2004)
Schuler, S.C., King, J.R., Hobbs, L.M., Pinsonneault, M.H.: ApJ **602**, L117 (2004)

Possible Light-Time Effect in the Eclipsing Binary System TY Boo

A. Elmaslı · S. O. Selam · B. Albayrak · D. Özuyar

Received: 1 November 2005 / Accepted: 1 February 2006
© Springer Science + Business Media B.V. 2006

Abstract The $O - C$ diagram for the eclipsing binary system TY Boo was constructed with the new minima times observed at the Ankara University Observatory along with the collected ones from the literature. The $O - C$ diagram shows a cyclic variation superimposed on a quadratic variation. The quadratic variation can be explained in terms of mass loss/exchange mechanism in the system while the cyclic variation is attributed to a possible light-time effect caused by a third body revolving around the close binary.

Keywords Binaries: Close · Binaries: Eclipsing · TY Boo

1. Introduction

TY Boo ($V_{max} = 10^m.81$) was discovered and classified as a W UMa type eclipsing binary by Guthnick and Prager (1926). Carr (1972) obtained the first UBV photoelectric light curves of the system and analyzed them using Russell and Merrill (1952) method. He also determined the spectral types of both components as G3 and G7, and comfirmed the W UMa type variability. Subsequent photometric observations of TY Boo were obtained by Samec and Bookmyer (1987), Samec et al. (1989), Rainger et al. (1990) and Milone et al. (1991). Samec et al. (1989) analyzed the Samec and Bookmyer's (1986) light curves using the Wilson-Devinney code and determined the photometric mass ratio as 0.447. Rainger et al. (1990) obtained and analyzed the first radial velocity curves of both components. They also analyzed the Samec and Bookmyer's (1986) B-band light curve using the LIGHT2 code by Hill (1989). They concluded that TY Boo belongs to the shallow contact W sub-type of the W UMa type contact binaries with a

A. Elmaslı (✉) · S. O. Selam · B. Albayrak · D. Özuyar
Ankara University Faculty of Science, Dept. of Astronomy and Space Sciences, TR-06100, Tandoğan, Ankara, Turkey

spectroscopic mass ratio of $q = 0.437$. Milone et al. (1991) performed a simultaneous analysis of their new light and radial velocity curves using Wilson-Devinney code and derived the following parameters for the system; $M_1 = 1.14$ $M_\odot$, $M_2 = 0.53$ $M_\odot$, $i = 77°.5$. Their analysis yielded two spotted regions on the larger component and they noted that the system may contain a chromospherically active star due to the CaII flare observed in the spectrum. They also classified the component stars as G3 and G8 which are almost in agreement with Carr's (1972) results. The period variation of TY Boo was first announced by Szafraniec (1953). Kreiner (1971), Samec and Bookmyer (1987), and Rainger et al. (1990) presented the $O - C$ diagram of the system. Samec and Bookmyer (1987) emphasized that the early visual minima times indicate two period changes in a 19-years interval. Later Qian (2001) indicated that the $O - C$ diagram of TY Boo has an oscillation around a quadratic trend. The latest period study was performed by Li et al. (2005) and they found two periodic variatons (caused by light-time effect and magnetic activity cycle) superimposed on a continuous increase in the systems period.

2. Observations and $O - C$ Diagrams

TY Boo was observed at the Ankara University Observatory during the nights of 3 April, 10 May, and 7 June 2005. From the observations we obtained one primary and two secondary minima (see IBVS 5649). During the observations the 30 cm Maksutov telescope equipped with a SSP-5A photometer head which consists of a Hamamatsu R1414 photomultiplier tube was used. PPM 78302 was used as the comparison star. The new minima were calculated using Kwee and van Woerden's (1956) method. All available times of minima of TY Boo were collected from the literature. Thus, we have a data set of 190 visual, 9

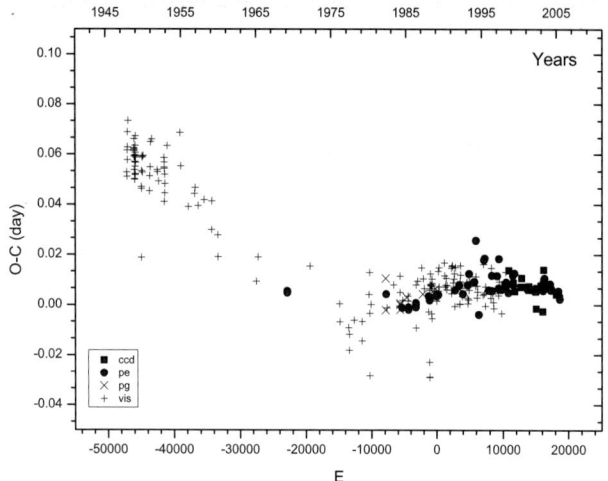

Fig. 1 The $O - C$ diagram of TY Boo obtained with all minima times

photographic, 88 photoelectric, and 34 ccd timings resulting to a total of 321. A quadratic least-square fit to the data (weights for ccd, photoelectric, photographic, and visual minimum times are 10, 10, 4, and 1, respectively) yields the ephemeris as Min I = HJD 2447611.6475(4) + $0^d.317149901(2) \times E + 2.470(7) \times 10^{-11} \times E^2$. Figure 1 shows the $O - C$ diagram constructed with the light elements given above. It indicates an increasing period which can be explained with a mass exchange/loss in the system. The rate of period change arisen from the quadratic term of the fit is: $dP/dE = 4.94 \times 10^{-11}$ day/cycle.

Under the assumption of conservative mass transfer phenomena, the direction of the transfer should be from the less massive component to the more massive one with a rate of $\Delta M = 1.79 \times 10^{-7} \, M_\odot \mathrm{yr}^{-1}$. Residuals from the parabolic fit indicate a continuous oscillatory behaviour covering one evident minimum and two maxima. The cyclic variation could be produced by the light-time effect due to a third body in the system. Since the visual and photographic data show large scatter from the general trend, only the photoelectric and ccd minima were used to calculate the theoretical $O - C$ curve. The resulting fit can be seen in Fig. 2 (top panel). During the analysis, Irwin's (1952) equations were used in order to derive the light-time orbit and the parameters of the third body. The results are presented in Table 1.

3. Conclusions

An O-C analysis for TY Boo based on all available eclipse timings together with newly determined ones is presented. The orbital period of the binary star shows cyclic variation superimposed on a quadratic variation. The periodic O-C variation can be explained in terms of the light-time effect due to the presence of an additional third body. Our results yield the third body completing its revolution in 41.07 ± 0.17 years with a rather low mass of $0.178 \pm 0.003 \, M_\odot$ (M_1, M_2

Table 1 Parameters of the light-time orbit and the third-body

Parameters	Value	Standard deviation
$a'_{12} \sin i' \, (AU)$	1.41	0.02
e'	0.22	0.02
$\omega' \, (°)$	90	2
T' HJD	2419500	36
P_{12} (year)	41.07	0.17
A (day)	0.00814	0.00011
$f(m_3) M_\odot$	0.00165	0.00005
$\sum()^2 \, (\mathrm{day}^2)$	0.0243	
$a'_3 (AU)$ (for $i' = 77.5$)	14.91	0.02
$\alpha^{(\prime\prime)}$ (for $d = 125$, Hipparcos)	0.119	0.008

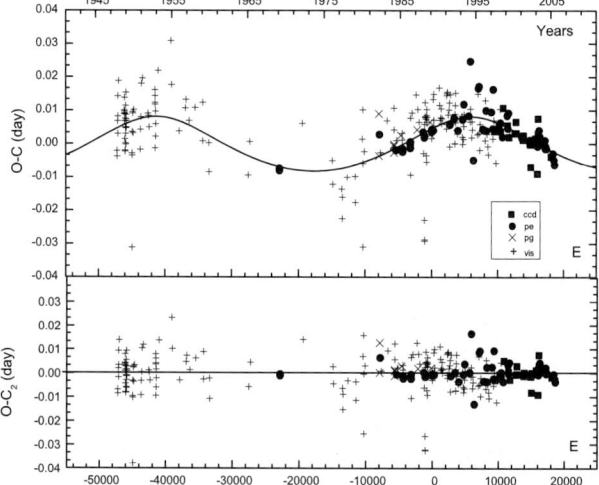

Fig. 2 Top panel: The $O - C$ residuals from the quadratic fit. The continuous curve represents the light-time effect fit. Bottom panel: The $O - C$ residuals from the light-time effect fit

and i adopted from Milone et al. (1991) and accepting third bodies' orbit co-planar with the the systemic orbit).

Acknowledgements This work has been supported by the research fund of Ankara University (BAP) under the research projects no: 20040705089 and 20040705090.

References

Carr, R.B.: AJ **77**, 155 (1972)
Guthnick, P. Prager, R.: AN **228**, 99 (1926)
Hill, G.: Publ. Dom. Astrophys. Obs. **15**, 297 (1989)
Irwin, J.B.: AJ **116**, 211 (1952)
Kreiner, J.M.: AcA **21**, 365 (1971)
Kwee, K.K., van Woerden, H.: BAN **12**, 32 (1956)
Li, L., Han, Z., Zhang F.: PASJ **57**, 187 (2005)
Milone, E.F., Groisman, G., Fry, D.J.I., Bradstreet, D.H.: ApJ **370**, 677 (1991)
Qian, S.: MNRAS **328**, 635 (2001)
Rainger, P.P., Hilditch, R.W., Bell, S.: MNRAS **246**, 42 (1990)
Russell, H.N., Merrill, J.E.: Contr. Princeton Obs. **26**, 50 (1952)
Samec, R.G., Bookmyer, B.B.: IBVS **2931** (1986)
Samec, R.G., Bookmyer, B.B.: PASP **99**, 842 (1987)
Samec, R.G., Van Hamme, W., Bookmyer, B.B.: AJ **98**, 2287 (1989)
Szafraniec, R.: AcA **2**, 101 (1953)

Preliminary Photometric Evidence of Starspot Umbrae and Penumbrae on Late-Type Active Stars

Sergio Messina · Edward F. Guinan

Received: 10 October 2005 / Accepted: 1 February 2006
© Springer Science + Business Media B.V. 2006

Abstract We constrain the properties of the spotted regions on the photosphere of the active late-type star DX Leonis by comparing the observed amplitudes of light and color variations with synthetic amplitudes obtained by means of Dorren's spot model and computed for a grid of values of spot temperatures, areas and latitudes.

Keywords Stars: Late-type · Stars: Spots · Stars: Photometry · Stars: DX Leonis

1. Introduction

Light and color variations observed in chromospherically active stars are due to the presence of surface brightness inhomogeneities which have a temperature different than the unperturbed photosphere and whose visibility is modulated by the stellar rotation. The amplitudes of the light and color variations depend on the properties and distribution of such regions over the stellar surface. We have modelled simultaneously the light and color variations (see Fig. 1) shown by the magnetically active K0 dwarf DX Leonis (HD 82443) by using the spot modelling approach of Dorren (1987) and found the spot temperature to vary from season to season.

S. Messina (✉)
INAF-Catania Astrophysical Observatory, Via S. Sofia 78, I-95123 Catania, Italy

E. F. Guinan
Dept. of Astronomy and Astrophysics, Villanova Univ., Villanova 19085, PA USA

2. The model

Our aim is to model the amplitude of the V magnitude and $B-V$ color variations to derive information on the spot area

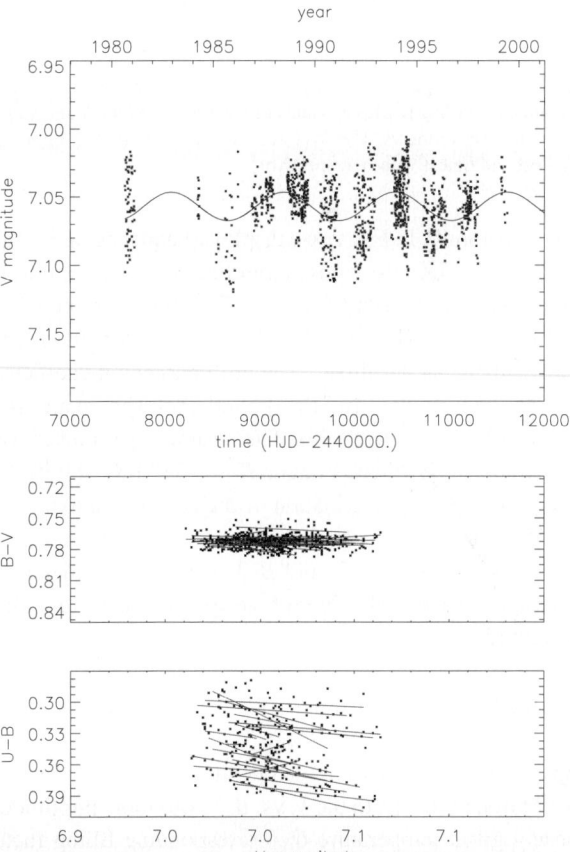

Fig. 1 Upper plot: V mag vs. time. The continuous curve is a sinusoidal fit with the starspot cycle period of $P_{cyc} = 3.21$ yr. Lower plots: The $B-V$ and $U-B$ colors vs. V magnitude along with linear fits to the individual light curves

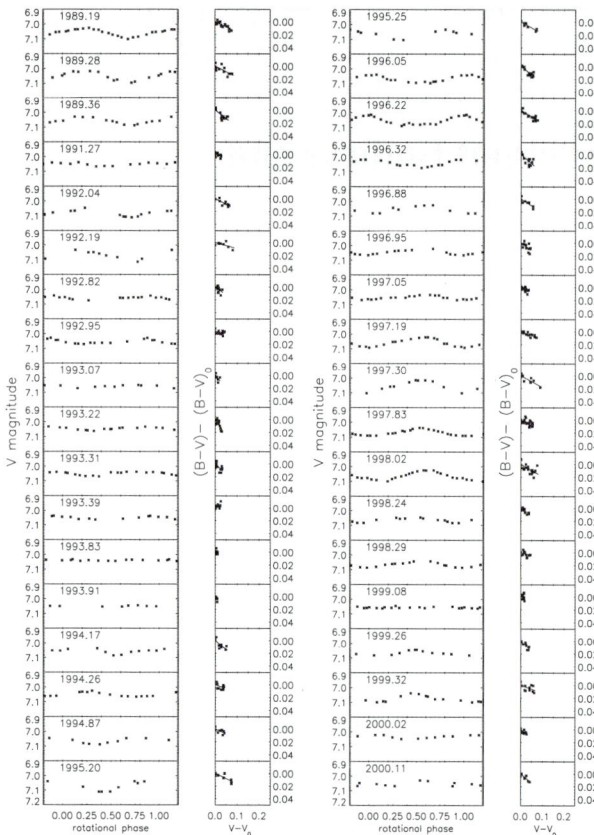

Fig. 2 *Left-hand panels*: V mag of DX Leo vs. rotational phase as computed according to Messina and Guinan (2002). *Right-hand panels*: B-V color variations vs. V mag variations, V_0 and $(B-V)_0$ being the brightest and bluest values, respectively

and temperature. In Fig. 2, we display a sequence of 36 light curves of DX Leonis, whose amplitude variations are to be compared with the synthetic ones. The NextGen model atmosphere grid of Hauschildt et al. (1999) is adopted to compute synthetic stellar fluxes. The limb-darkening coefficient for both the unspotted and the spotted photosphere are from Diaz-Cordoves et al. (1995). The assumed parameters are the stellar effective temperature $T_{\text{eff}} = 5370$ K, gravity log $g = 4.5$ cm s^{-2}, the inclination of the stellar rotational axis $i = 55°$, and the unspotted magnitude $V = 7.01$ and color $B-V = 0.75$. Synthetic V and B-V curve amplitudes were generated for a grid of values of the spot temperature, radius and latitude.

3. Results and conclusions

Figure 3 shows, in latitude-temperature plots, the residual distribution of the fits to the V vs. B-V variation amplitudes. For any given temperature the corresponding filling factor is also plotted. As expected, solutions are not unique and the higher the spot temperature and latitude, the higher their filling factor values. However, the solutions tend to cluster around a finite range of values for the spot temperature. Gray

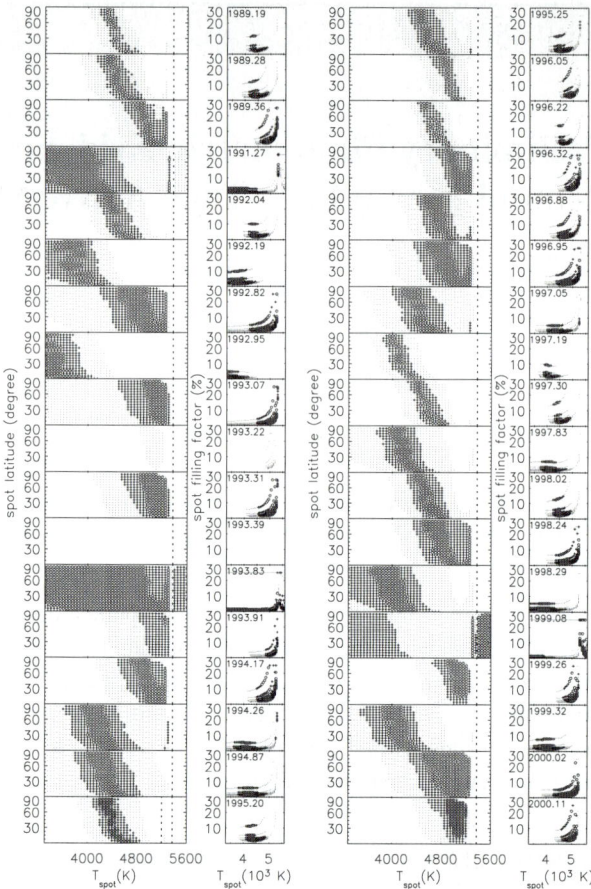

Fig. 3 The best fit solutions in latitude-temperature plots with corresponding filling factors (%). Heavy gray color denotes solutions with minimum χ^2 (0–0.5) while light gray denotes solutions with χ^2 up to 2.0

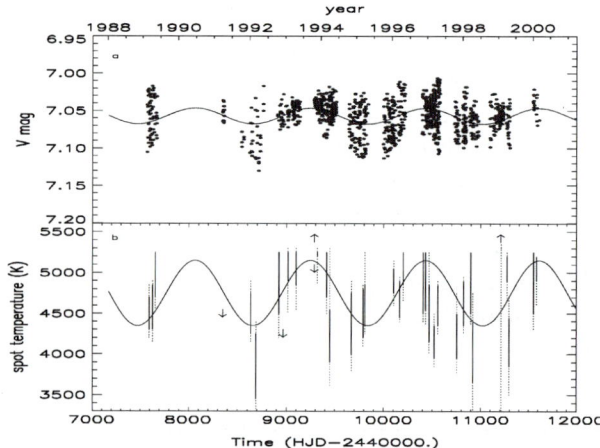

Fig. 4 Spot temperature vs. time. Vertical solid and dotted lines represent the solution in the $\chi^2 = 0.0$–0.5 and $\chi^2 = 0.5$–1.0 range, respectively. Downward and upward arrows represent upper and lower limits. The brightness variations are plotted in the upper panel to better show the correlation between temperature variations and starspot cycle phase

levels indicate solutions with different χ^2: heavy gray $\chi^2 =$ 0.0–0.5, light gray $\chi^2 = 1.0$–2.0.

The spot temperature is found to vary significantly from epoch to epoch (see Fig. 4). The V and $B - V$ data allow us to derive the average temperature of those active regions which are asymmetrically distributed along the stellar longitudes. The inferred seasonal variations of spot temperature are likely to be attributed to the two-component (two-temperature) nature of active regions. These are expected to be composed, as guided by the solar analogy, of a dark umbra surrounded by a less dark penumbra. The observed variation of the average temperature arises from the time variation of their areal ratio. In another reasonable scenario, we can imagine that active regions consist of dark spots and bright faculae, and the changing ratio of spot/plage is responsible for the observed average temperature variation. Figure 4 shows the tendency of the average spot temperature to be higher at the epochs when the star is less spotted. Thus, according to our interpretation, penumbrae or faculae are dominant when the star is at or near the spot activity minimum.

References

Díaz-Cordovés, J., Claret, A., Giménez, A.: A&A **110**, 329 (1995)
Dorren, J.D.: ApJ **320**, 756 (1987)
Hauschildt, P.H., Allard, F., Baron, E.: ApJ **512**, 377 (1999)
Messina, S., Guinan, E.F.: A&A **393**, 225 (2002)

Late-Type Active Stars: Rotation & Companions

T. H. Dall · K. G. Strassmeier · H. Bruntt

Received: 1 November 2005 / Accepted: 1 February 2006
© Springer Science + Business Media B.V. 2006

Abstract Rapid rotation has been established empirically as the controlling factor for the magnetic field strength and the magnetic activity level of single late-type (F-M) stars. The dynamo theories explain this fact as due to interaction between differential rotation and helical motion in the transition layer between the convective envelope and the radiative interior. The presence of a close companion even down to the size of a "hot Jupiter" could alter the physical processes responsible for the activity, by introducing longitude- and latitude-dependencies and inhomogenious chemical abundances – effects that cannot be ignored in the attempt to understand magnetic activity on late type stars. Here we present first results for the well-known, single, active star, HD 27536. Binarity is established by very precise radial velocity (RV) measurements using HARPS spectra. The spectral line bisectors are examined for correlations between RV and bisector shape to distinguish between the effects of stellar activity and unseen companions.

Keywords Binaries: general · Stars: individual: HD 27536 · Stars: activity

T. H. Dall (✉)
European Southern Observatory, Casilla 19001, Santiago 19, Chile

K. G. Strassmeier
Astrophysical Institute Potsdam, An der Sternwarte 16, 14482 Potsdam, Germany

H. Bruntt
Niels Bohr Institute, Juliane Maries Vej 30, 2100 Copenhagen Ø, Denmark

1. Introduction

There is currently still no single theory that adequately describes magnetic activity on cool main-sequence stars with outer convective envelopes. Although rotation is thought to be the necessary condition for magnetic activity, it is not clear to what extent binarity influences the generation and morphology of magnetic fields and the corresponding chromospheric and coronal emission. The differential gravitational pull from a companion may cause a longitude- and latitude-dependent relationship between rotation rate and activity level. It may also contribute to an inhomogeneous chemical abundance by effectively lowering/enhancing the chemical stratification process, e.g. due to diffusion. If such relationships exist, the models of the evolution of close binaries would then need to be reconsidered.

We are currently conducting a search for true single stars among the known active stars, in order to study the activity-rotation relation in a sample that is not "polluted" by any type of binaries. Our sample consist of about 30 known active G–M stars, with known photometric variations attributed to rotational modulation of star spots, including both binaries and – supposedly – single stars.

Several aspects need to be investigated to properly analyze the sample, including (1) time-resolved radial velocity monitoring and (2) line bisector analysis, (3) monitoring of activity indexes, (4) determination of fundamental atmospheric parameters, and (5) accurate photospheric abundance analysis (Dall et al., 2005).

2. HARPS observations

HD 27536 is a G8 giant with $v \sin i < 2$ km/s and an inclination $i \sim 90°$. Its rotationally modulated brightness variation

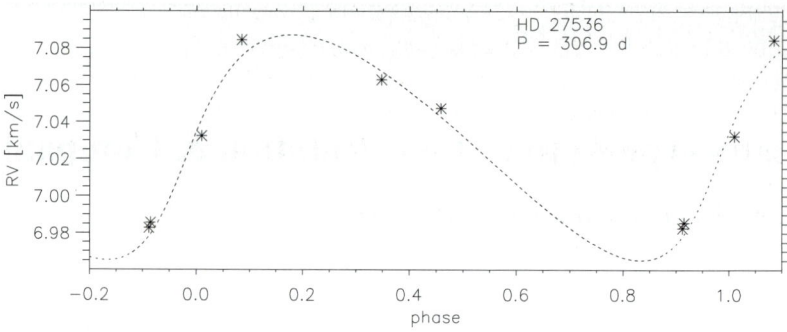

Fig. 1 The RV curve of HD 27536 with a binary orbit fit (dotted line)

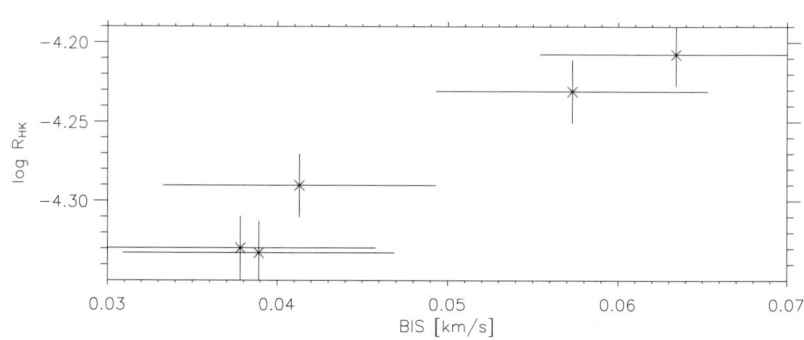

Fig. 2 Chromospheric activity index log R_{HK} versus the photospheric indirect activity indicator BIS. A tight correlation seem to exist between chromosphere and photosphere

has a period of $P = 306.9\,d$. The activity is believed to be due to a fossil Ap type field viewed equator-on (Stępień, 1993).

All spectra for this program were taken with the HARPS spectrograph at the ESO 3.6m telescope at La Silla Observatory. The reduction process of HARPS spectra is fully automatic and extremely accurate due to the high intrinsic stability of the spectrograph. The final data products is a wavelength calibrated 1D spectrum, and a cross-correlation function averaged over all 72 spectral orders. The formal error on the wavelength calibration is $\sim 0.02\,\mathrm{m\,s^{-1}}$, while the total RV uncertainty is ~ 1–$2\,\mathrm{m\,s^{-1}}$. The spectral resolution is $R \sim 100,000$.

3. Determining stellar parameters

The first step in the analysis is the measurement of the EWs, which is accomplished using DAOSPEC (Stetson and Pancino, in preparation), which uses an iterative Gaussian fitting and subtraction procedure to fit the lines and the effective continuum. The lines are identified using a list of lines from the VALD database (e.g. Kupka et al., 1999). With this, we calculate abundances and determine T_{eff}, $\log g$ and microturbulence by matching with ATLAS9 models (Kurucz, 1993). The results for HD 27536 yields $T_{\mathrm{eff}} = 5240 \pm 35\,\mathrm{K}$, $\log g = 3.55 \pm 0.06$, $\xi_t = 1.2 \pm 0.1\,\mathrm{km\,s^{-1}}$.

We calculate the absolute emission fluxes in the Ca II H and K lines following the method of Linsky et al. (1979), integrating the flux f_K between the K_{1V} and K_{1R} points, and nor-

malizing by the flux f_{50} between 3925 and 3975 Å. Similarly for f_H. Using $V - R$ colors (Strassmeier et al., 2000) this is transformed and corrected for photospheric flux to the index R_{HK}. We use our own derived T_{eff}, and values of $V - R$ derived from $B - V$ where R measurements are not available.

3.1. Radial velocities

In Fig. 1 we show the RV measurements for HD 27536, along with a binary orbit fit, using the known 306.9 d photometric period. To investigate whether this variation may be due to a low-mass companion, we measured the bisector inverse slope (BIS) of the mean cross-correlation function of the spectra, and compared with the measured RV to check for correlations. We also compared with the activity index $\log R_{HK}$.

We find that both BIS and $\log R_{HK}$ correlate with the RV, indicating that the variation is internal to the star and not caused by a orbiting companion. Furthermore, the correlation between the photospheric variation (BIS) and chromospheric activity ($\log R_{HK}$) seem to be very strict, as can be seen in Fig. 2. This is in accordance with expectations (e.g. Messina et al., 2001).

4. Discussion

Part of the RV variation of HD 27536 may be due to an unseen companion, hiding in the activity induced jitter. Queloz et al. (2001) found a negative correlation between BIS and

RV for HD 166435, and concluded that the RV variation was induced by activity. We find a positive correlation between BIS and RV, as did Santos et al. (2002) for the visual binary HD 41004. Later, Zucker et al. (2004) identified the RV signature of a giant planet around the A component. Hence, in the case of HD 27536 we cannot exclude the possibility of companion induced RV, activity and BIS variations, although the variation is consistent with purely activity induced effects.

Note also, that the photometric variation is sinusoidal, as already suggested by Strassmeier et al. (1999) and confirmed by continuous monitoring (Strassmeier et al., unpublished), which is hard to reconcile with the eccentric "orbit" suggested by the RV data.

The sinusoidal photometric lightcurve and the asymmetric RV curve may hint that the geometric configuration of HD 27536 is not a settled issue: Are we really seeing it equator-on? Or is it really a fast rotator viewed pole-on and the photometric and RV variations reflecting activity cycles and not rotation?

To disentangle any companion-induced pull from the activity jitter would require several years of monitoring, since the period is so long and since the orbital pull and the induced activity may have the same period. Hence, we will keep monitoring this very interesting object in the years to come.

References

Dall, T.H., Bruntt, H., Strassmeier, K.G.: A&A **444**, 573 (2005)
Kupka, F., Piskunov, N., Ryabchikova, T.A. et al.: A&AS **138**, 119 (1999)
Kurucz, R.: CD-ROM No. 13. Cambridge, Mass.: SAO (1993)
Linsky, J.L., McClintock, W., Robertson, R.M., Worden, S.P.: ApJS **41**, 47 (1979)
Messina, S., Rodonò, M., Guinan, E.F.: A&A **366**, 215 (2001)
Queloz, D., Henry, G.W., Sivan, J.P. et al.: A&A **379**, 279 (2001)
Santos, N.C., Mayor, M., Naef, D. et al.: A&A **392**, 215 (2002)
Stępień, K.: ApJ **416**, 368 (1993)
Strassmeier, K.G., Stępień, K., Henry, G.W., Hall, D.S.: A&A, 343, 175 (1999)
Strassmeier, K., Washuettl, A., Granzer, et al.: A&AS **142**, 275 (2000)
Zucker, S., Mazeh, T., Santos, N.C., Udry, S., Mayor, M.: A&A **426**, 695 (2004)

Eclipsing Binaries in Open Clusters

John Southworth · Jens Viggo Clausen

Received: 1 November 2005 / Accepted: 1 February 2006
© Springer Science + Business Media B.V. 2006

Abstract The study of detached eclipsing binaries in open clusters can provide stringent tests of theoretical stellar evolutionary models, which must simultaneously fit the masses, radii, and luminosities of the eclipsing stars and the radiative properties of every other star in the cluster. We review recent progress in such studies and discuss two unusually interesting objects currently under analysis. GV Carinae is an A0 m + A8 m binary in the Southern open cluster NGC 3532; its eclipse depths have changed by 0.1 mag between 1990 and 2001, suggesting that its orbit is being perturbed by a relatively close third body. DW Carinae is a high-mass unevolved B1 V + B1 V binary in the very young open cluster Collinder 228, and displays double-peaked emission in the centre of the Hα line which is characteristic of Be stars. We conclude by pointing out that the great promise of eclipsing binaries in open clusters can only be satisfied when both the binaries and their parent clusters are well-observed, a situation which is less common than we would like.

Keywords Stars: fundamental parameters · Stars: binaries: eclipsing · Stars: binaries: spectroscopic · Open clusters and associations: general

1. Eclipsing binaries in open clusters

Detached eclipsing binary stars (dEBs) are of fundamental importance to stellar physics because they are, apart from the few closest objects to the Earth, the only stars for which we can accurately measure basic quantities such as mass, radius and surface gravity (Andersen, 1991). The realism and reliability of the current generation of theoretical stellar evolutionary models ranks as one of the great achievements of modern astrophysics, but this success would have been much more difficult without the ability to check the effects of particular physics against the accurate physical properties of stars in dEBs.

Given good photometric and spectroscopic data, it is possible to derive masses and radii of stars in a dEB to accuracies better than 1% and surface gravities to within 0.01 dex (Southworth et al., 2005b). However, the predictions of theoretical models can usually match even properties as accurate as this, both because of their sophistication and because there are several important unconstrained parameters, e.g., metal and helium abundance and age. More constraints are needed to investigate the success or otherwise of a number of physical parameters which are only very simplistically treated in theoretical models, e.g., convective core overshooting, mass loss, mixing length and rotational effects. For example, small changes in the mixing length can change the derived ages of the oldest globular clusters, which constrain the age of the Universe, by 10% (Chaboyer, 1995).

An answer to this problem is to study dEBs which are members of stellar clusters (Southworth et al., 2004a; Thompson et al., 2001). Because the dEB and the other cluster members have the same age and chemical composition, theoretical models must be able to simultaneously match the masses, radii and luminosities of the two stars in the dEB and the radiative parameters of every other cluster member, for one age and chemical composition. This allows much more detailed tests to be made of the success or otherwise of different physical ingredients in models. Alternatively, if the cluster is poorly studied, a comparison of the properties of the dEB with model predictions allows the cluster metal abundance and age to be derived.

J. Southworth (✉) · J. V. Clausen
Niels Bohr Institute, Copenhagen University, Denmark
e-mails: jkt@astro.keele.ac.uk, jvc@astro.ku.dk

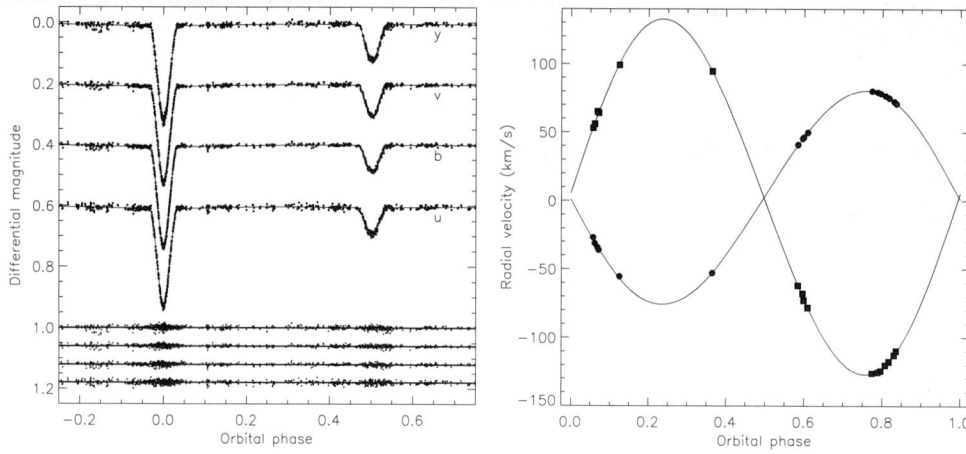

Fig. 1 The observed light curves (left) and radial velocity curves (right) of GV Car with the best-fitting models shown using solid lines

Detached eclipsing binaries are also excellent distance indicators (Clausen, 2004), through a variety of methods such as using bolometric corrections or surface brightness calibrations (see Southworth et al., 2005a, for a detailed analysis). dEBs in clusters therefore give an accurate distance to the cluster without the problems and theoretical dependence which affect the main sequence fitting method.

An unique advantage of studying dEBs in clusters is that it can be possible to place four or more stars with the same age and chemical composition onto one mass–radius or T_{eff}–$\log g$ plot, if several dEBs are members of one cluster (Southworth et al., 2004b, c). Some Galactic open clusters are known to contain four or more dEBs, e.g., NGC 7086 (Robb et al., these proceedings).

2. GV Carinae in NGC 3532

GV Car ($m_V = 8.9$, $P = 4.29$ d) is a member of the nearby open cluster NGC 3532 and contains two metallic-lined A stars. It displays apsidal motion with a period of $U \approx 300$ yr.

Complete Strömgren $uvby$ light curves, with 775 observations in each passband, were obtained at the Strömgren Automated Telescope (ESO La Silla) in 1987–1991, with additional data from 2002–2004 (Fig. 1). These light curves have been analysed using the EBOP code (Popper and Etzel, 1981; Etzel, 1975) and the Monte Carlo error analysis algorithm implemented in JKTEBOP (Southworth et al., 2004b,c).

Spectroscopic observations of GV Car were obtained in 2001–2004 using the FEROS échelle spectrograph at the 1.5 m and 2.2 m telescopes at ESO La Silla. Radial velocities have been derived by cross-correlating spectra of GV Car from the 4360–4520 Å échelle order against a spectrum of GV Car taken at the midpoint of a secondary eclipse. They have been fitted with an eccentric orbit using SBOP (written by P. B. Etzel), which is shown in Fig. 1.

The spectra have been modelled using the UCLSYN synthesis code (see Southworth et al., 2004a, for references), giving $T_{\mathrm{eff A}} = 10\,100 \pm 300$ K and $T_{\mathrm{eff B}} = 7750 \pm 350$ K, consistent with the $uvby\beta$ colours of the system and the flux ratios found in the light curve analysis.

The masses and radii of GV Car are $M_{\mathrm{A}} = 2.51 \pm 0.03\,\mathrm{M}_\odot$, $M_{\mathrm{B}} = 1.54 \pm 0.02\,\mathrm{M}_\odot$, $R_{\mathrm{A}} = 2.57 \pm 0.05\,\mathrm{R}_\odot$ and $R_{\mathrm{B}} = 1.43 \pm 0.06\,\mathrm{R}_\odot$. These are well fitted by the Cambridge theoretical models (Pols et al., 1998) for an age of 360 ± 20 Myr and a metal abundance of $Z = 0.01$ (Fig. 2). This age is in good agreement with, and more precise than, main-sequence-fitting estimates (González and Lapasset, 2002). Our value for the metal abundance is the first published estimate for NGC 3532.

But we do not yet understand GV Car fully; whilst its eclipses were 0.33 and 0.12 mag deep in 1987, they had shallowed to 0.22 and 0.08 mag in depth in 2004. This change can be explained by either an increase in third light (implying a companion which is itself variable) or a decrease of about 3° in orbital inclination (which suggests a perturbed orbit). The latter explanation seems more likely, but because there is no other evidence of a third star in the system it must have a low mass or be a compact object. Further observations will be required to fully understand this interesting system.

3. DW Carinae in Collinder 228

DW Car ($m_V = 9.7$, $P = 1.33$ d) is a high-mass dEB in the young open cluster Cr 228. Strömgren $uvby$ light curves, 518 points in each passband, were obtained as with GV Car and modelled using the 2003 version of the Wilson-Devinney code (Wilson and Devinney, 1971) (Fig. 3).

The radial velocity analysis of DW Car is difficult because the spectra have very few features. Apart from the hydrogen lines, which do not give reliable velocities (Andersen, 1975),

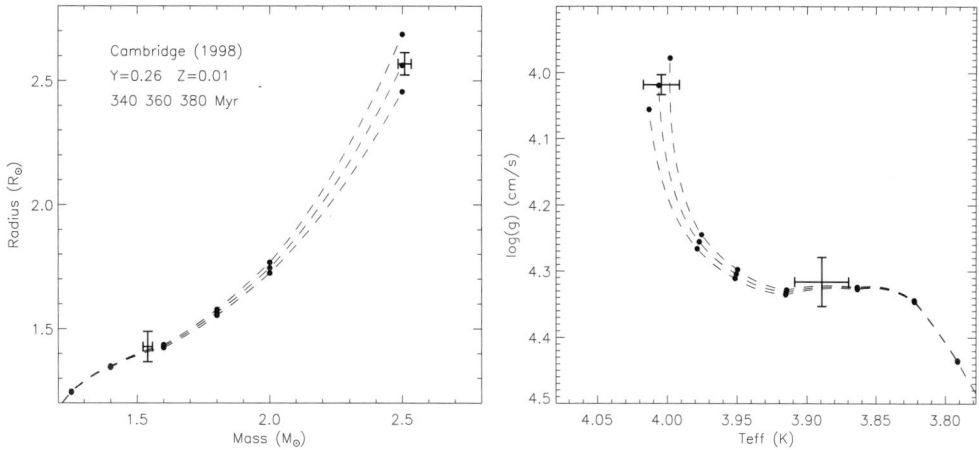

Fig. 2 Mass–radius and T_{eff}–log g comparison plots between the properties of the components of GV Car and the predictions of the Cambridge theoretical models

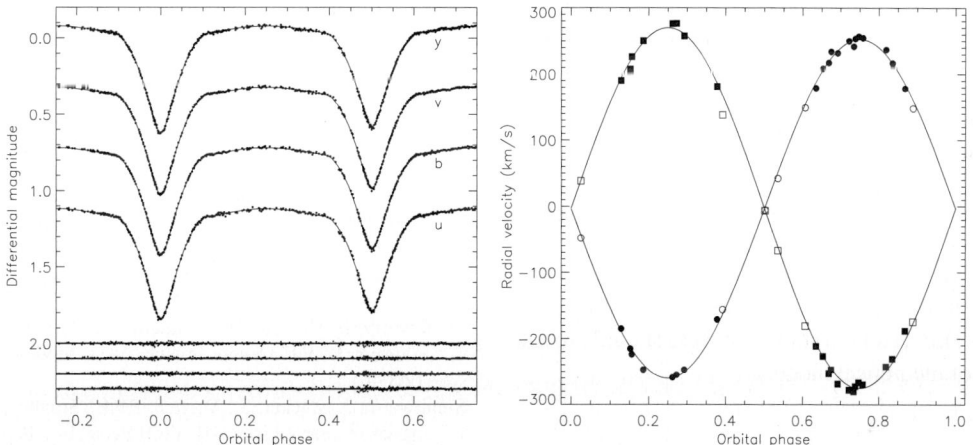

Fig. 3 Observed light curves (left) and radial velocity curves (right) of DW Car with best-fitting models shown using solid lines. Radial velocity observations for the primary and secondary stars are indicated using circles and squares, respectively, and rejected observations are shown using open symbols

there are only four He I spectral lines of reasonable strength. These lines have been analysed individually using cross-correlation, the TODCOR algorithm (Zucker and Mazeh, 1994), Gaussian fitting and spectral disentangling (Simon and Sturm, 1994). Good results have been obtained for disentangling and Gaussian fitting, whilst cross-correlation is significantly affected by line blending. Circular orbits were fitted using SBOP, and the orbit from fitting the He I λ4471 line with a double Gaussian is shown in Fig. 3.

The resulting masses and radii of DW Car are $M_A = 11.4 \pm 0.2\,M_\odot$, $M_B = 10.7 \pm 0.2\,M_\odot$, $R_A = 4.52 \pm 0.07\,R_\odot$ and $R_B = 4.39 \pm 0.07\,R_\odot$. Strömgren index calibrations and the flux ratio from the light curves give $T_{\text{eff}A} = 27\,500 \pm 1000$ K and $T_{\text{eff}B} = 26\,750 \pm 1250$ K. These parameters are acceptably fitted by the predictions of the Cambridge models using the $Z = 0.03$ ZAMS (Fig. 4), but no conclusions can be drawn from this until definitive values for the radii are obtained.

DW Car shows a double-peaked emission line at Hα with a sharp central absorption characteristic of a Be star. The line profile does not change with the orbital motion so must come from circumbinary rather than circumstellar matter, as expected given the closeness of the stars to each other. DW Car is a very young system and the rotational velocities are 'only' about 170 km s^{-1}. These two facts are very unusual for the Be phenomenon, which is thought to increase slightly with age and only be present in stars which are rotating at above 70–80% of their critical velocities (Porter and Rivinius, 2003).

4. Where next?

The study of dEBs in open clusters has been shown to be an excellent way to determine the parameters of clusters by comparison with theoretical stellar models (Southworth et al., 2004b), but the goal of simultaneously fitting models to

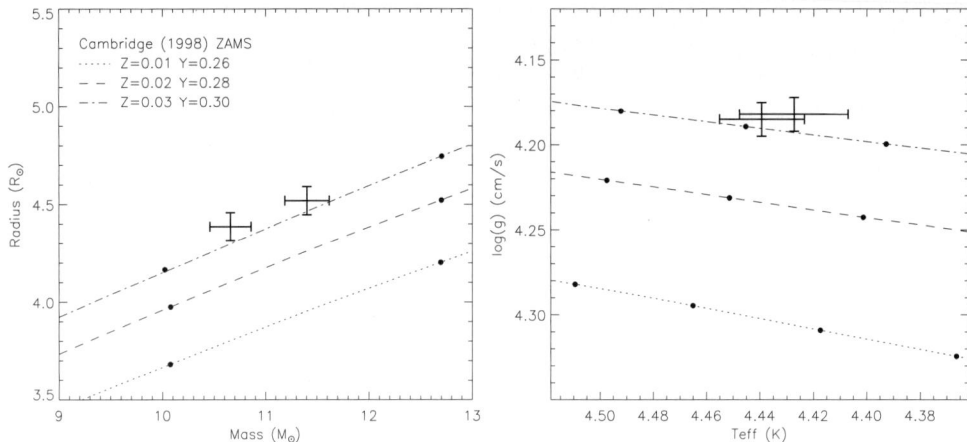

Fig. 4 Mass–radius and $T_{\rm eff}$–$\log g$ comparison plots between the properties of the components of DW Car and the predictions of the Cambridge theoretical models

both the cluster and the two stars in the dEB remains elusive. The main problems are that definitive observations of both the cluster and the dEB requires extensive telescope time and that the clusters are often too sparsely populated to be useful.

Clusters containing several dEBs are excellent targets for study, because accurate fundamental parameters can be found for four or more stars with the same age, chemical composition and distance, and using the same photometric CCD observations. Several Galactic open clusters, in both the Northern and Southern hemispheres, are known to contain at least four high-mass dEBs, and full studies of these should provide excellent tests of theoretical models.

References

Andersen, J.: Accurate masses and radii of normal stars. A&ARv **3**, 91 (1991)
Andersen, J.: Spectroscopic observations of eclipsing binaries. III. Definitive orbits and effects of line blending in CV Velorum. A&A **44**, 355–362 (1975)
Chaboyer, B.: Absolute ages of globular clusters and the age of the universe. ApJ **444**, L9–12 (1995)
Clausen, J.V.: Eclipsing binaries as precise standard candles. New Astronomy Reviews **48**, 679–685 (2004)
Etzel, P.B.: A photometric analysis of WW Aurigae Master's Thesis, San Diego State University (1975)
González, J.F., Lapasset, E.: Spectroscopic binaries and kinematic membership in the open cluster NGC 3532. AJ **123**, 3318–3324 (2002)
Pols, O.R., Schröder, K.-P., Hurley, J.R., Tout, C.A., Eggleton, P.P.: Stellar evolution models for Z = 0.0001 to 0.03. MNRAS **298**, 525–536 (1998)
Popper, D.M., Etzel, P.B.: Photometric orbits of seven detached eclipsing binaries. AJ **86**, 102–120 (1981)
Porter, J.M., Rivinius, T.: Classical Be stars. PASP **115**, 1153–1170 (2003)
Simon, K.P., Sturm, E.: Disentangling composite spectra. A&A **281**, 286–291 (1994)
Southworth, J., Maxted, P. F. L., Smalley, B.: Eclipsing binaries in open clusters. I. V615 Per and V618 Per in h Persei. MNRAS **349**, 547–559 (2004a)
Southworth, J., Maxted, P.F.L., Smalley, B.: Eclipsing binaries in open clusters. II. V453 Cygni in NGC 6871. MNRAS **351**, 1277–1289 (2004b)
Southworth, J., Zucker, S., Maxted, P.F.L., Smalley, B.: Eclipsing binaries in open clusters. III. V621 Persei in χ Persei. MNRAS **355**, 986–994 (2004c)
Southworth, J., Maxted, P.F.L., Smalley, B.: Eclipsing binaries as standard candles: HD 23642 and the distance to the Pleiades. A&A **429**, 645–655 (2005a)
Southworth, J., Smalley, B., Maxted, P.F.L., Claret, A., Etzel, P. B.: Absolute dimensions of detached eclipsing binaries. I. The metallic-lined system WW Aurigae. MNRAS **363**, 529–542 (2005b)
Thompson, I.B., et al.: Cluster AgeS Experiment: The age and distance of the globular cluster ω Centauri determined from observations of the eclipsing binary OGLEGC 17. AJ **121**, 3089–3099 (2001)
Wilson, R.E., Devinney, E.J.: Realization of accurate close binary light curves: application to MR Cygni. ApJ **166**, 605–619 (1971)
Zucker, S., Mazeh, T.: Study of spectroscopic binaries with TODCOR. I. A new two-dimensional correlation algorithm to derive the radial velocities of the two components. ApJ **420**, 806–810 (1994)

Eclipsing Binaries in Local Group Galaxies

R.W. Hilditch

Received: 1 November 2005 / Accepted: 1 February 2006
© Springer Science + Business Media B.V. 2006

Abstract A review is presented of the progress that has been made in the last 3 years towards quantifying the properties of high-mass detached and semi-detached eclipsing binaries in Local Group galaxies. Comparisons between these observational results on masses, radii, temperatures and luminosities for stars in detached binaries and evolution models for single stars at the appropriate metallicity are found to be very good. New evolution models for interacting binaries passing through case A mass exchange are being calculated, and indicate a requirement for some mass loss to find agreement with the observational data. The observational data on such semi-detached systems show similar properties to those in the Milky Way galaxy. The directly-determined distances to all these eclipsing binaries are proving to be most valuable for strengthening the distance scale amongst the Local Group galaxies.

Keywords Stars: binaries: eclipsing · Stars: evolution · Abundances · Techniques: photometric

1. Introduction

There are three main reasons for studying eclipsing binaries in Local Group galaxies: to establish the properties and evolutionary states of close binaries in galaxies with different chemical abundances and star formation histories from the Milky Way Galaxy, as tests of single-star and binary-star evolution theory; to determine distances to these binaries directly, independently of all other techniques, and thereby establish mean distances to these galaxies together with their three-dimensional structure; and lastly, to confirm the values of these galaxies in providing a strengthening of the cosmic distance scale – $cf.$, for example, that the Andromeda galaxy (M31) is a fundamental marker for the Tully-Fisher relationship for spiral galaxies.

All the various methods of distance determination that contribute to the cosmic distance ladder sensibly overlap either over kiloparsec scales within the Milky Way, or over megaparsec scales between clusters of galaxies. The link between such stellar methods and extragalactic methods necessarily has been reliant on the few techniques involving intrinsically luminous stellar sources, such as the use of classical Cepheid variables and their relationship between pulsation period and luminosity, and the use of stars at the tip of the red-giant branch. To date, the observational results with these methods show discrepancies at the 10% level in distances to Local Group galaxies, which is dissatisfying to most astronomers. The eclipsing binary method is an independent technique, requiring significant allocations of telescope time for both photometry and spectroscopy in order to provide the necessary data to establish complete astrophysical parameters for both stars in each eclipsing binary, and therefore the intrinsic luminosity of each binary. The photometry provides the apparent brightnesses and the basic inverse-square law for propagation of light then establishes the distance to each binary since the luminosity is determined in absolute units from the combined solution of the orbital velocity data and the eclipsing binary light curve.

The spectroscopic observations provide measures of the radial velocities of both stars in each binary due to their orbital motion about their common centre of mass, and hence the sizes of their orbits projected into the observer's line of sight as well as the minimum masses ($m_{1,2} sin^3 i$) of the two stars. They also provide spectral energy distributions and absorption-line profiles to determine effective temperatures,

R.W. Hilditch
Scottish Universities Physics Alliance, School of Physics and Astronomy, University of St Andrews, Scotland, U.K.

or at least spectral types and flux ratios between components. The photometric observations provide complete light curves, preferably in several passbands, and their analysis establishes the orbital inclination, the sizes of the two stars relative to their separation, and the flux ratio, as well as other factors. The combination of both sets of data provides a complete astrophysical specification of each binary system – masses, radii, temperatures, luminosities in absolute units of both stars independently of the distance to each binary. Of course, corrections for any interstellar extinction must be made in order to determine a direct distance, but such extinction does not influence the determinations of the astrophysical parameters, except for the temperatures and only then if the continuum fluxes are used directly.

The observational/technological requirements are substantial for these eclipsing binaries. Whilst the photometric images can be obtained reasonably straightforwardly via CCD cameras on modest telescopes with apertures of 1–2 m, the extraction of unbiased photometry from fairly crowded star fields is a substantial demand on programming and computer time. The necessary spectroscopy, at a sufficient spectral resolution and time resolution, requires telescopes with apertures of 4m for the brighter eclipsing binaries in the nearby Magellanic Clouds, and telescopes with apertures of 8–10 m for those in M31 and M33. Fibre-fed multi-object spectrographs ensure efficient use of such large telescopes, and the subsequent analysis of the secured spectra requires the latest spectral matching and cross-correlation techniques such as TODCOR(Zucker and Mazeh, 1994), and/or the spectral disentangling techniques (Simon and Sturm, 1994; Hadrava, 1995; Harries et al., 2003), in order to be successful.

The nearest galaxies of the Local Group are sufficiently close that current technology with ground-based telescopes allows us to secure observations on points of light that are angularly resolved from other such points. Many of those points of light have been discovered in recent years to show periodic variations in apparent brightness that are characteristic of normal eclipsing binary stars. For the Magellanic Clouds, these major advances have occurred as by-products of the searches, by the EROS,MACHO, OGLE, and MOA teams, for gravitational micro-lensing of the LMC and SMC stars by massive compact objects in the halo of the Milky Way Galaxy. In addition, there have been detailed photometric studies of individual eclipsing binaries discovered via earlier photographic surveys. For more distant galaxies, such as M31, the crowding of stellar images is a potentially serious problem, which must be addressed for all methods of distance determination involving stars(cf.,e.g., Mochejska et al., 2000). The DIRECT project has been gathering photometric data on fields in M31 and M33, and there are other research teams starting studies of some of the dwarf galaxies of the Local Group.

At the IAU General Assembly in Sydney, Australia, in July 2003, the Joint Discussion No.13 on *Extragalactic Binaries* provided a very clear account of the historical development of the subject, and its major recent advances up to 2003.(cf. the proceedings edited by Ribas and Gimenez (2004)). This review will concentrate on the results published from 2003 to the present. The allocated review time, and this Journal space, did not allow a review of the many interesting developments on high-mass x-ray binaries in Local Group galaxies that have been published recently, much due to the XMM-Newton mission. (cf.,e.g., Pietsch et al., 2004a,b, 2005).

2. The Magellanic Clouds

The photometric studies by the OGLE team of the Large and Small Magellanic Clouds have produced the largest and most consistent set of light curves for eclipsing binaries (EBs) in both galaxies. The original OGLE-II survey (Udalski et al., 1998) was extended over 4 observing seasons, and analysed via the more precise difference-image-analysis (DIA) photometry to yield better determined orbital periods and more precise I-band light curves for 1350 EBs (Wyrzykowski et al., 2004). A complementary DIA study of the LMC in the OGLE-II survey has provided a total of 2580 EBs with well-determined orbital periods and I-band light curves (Wyrzykowski et al., 2003). These light curves are formed from typically 300 observations, with rms scatter in the range 0.005–0.025 mag. for stars of maximum apparent magnitudes in the range 12–19 mag respectively. Udalski (2004) announced that the OGLE-III surveys of both galaxies (with much larger fields of view) were started in 2001, and $\simeq$ 150 observations per field had been obtained by 2003. This programme is ongoing. It is unfortunate that the OGLE surveys are conducted only in the I-band, rather than having multi-colour data, but this deficiency is offset by the quality and quantity of the data, and by the OGLE team publishing BVI colour indices for all EBs at maximum brightness. The issues surrounding the definition of the magnitude or flux scale derived from DIA photometry in crowded fields (cf., Michalska and Pigulski 2005) have also been mentioned by Wyrzykowski et al. (2004), who note that only a few sources in their SMC sample show small (up to $0^m.05$) systematic differences between the reference levels of the DIA and the PSF light curves.

In addition to such discovery surveys, detailed studies of individual EBs have continued via dedicated CCD photometry at Mount John Observatory, New Zealand, and at the European Southern Observatory. Bayne et al. (2004) published precise $uV_J I_C$ light curves of one EB discovered by the MACHO survey in the LMC, and two EBs discovered by the MOA survey in the SMC. Model fits to these light curves were obtained but no spectroscopic observations of these systems have yet been obtained. Clausen et al. (2003) also published $uvby$ photometry for 4 EBs in the Magellanic Clouds, establishing very good light curves and solutions. Whilst spectroscopic data was only available for one of these four systems, this programme is ongoing with high-dispersion spectra from the UVES spectrograph on the ESO VLT being secured. There is also another ongoing programme involving spectroscopy with the VLT and its FLAMES multi-object spectrograph being conducted by P.North on B-type eclipsing binaries in the LMC (private communication).

Fitzpatrick et al. (2003) continued their detailed studies of individual EBs in the LMC, combining BV light curves, high-resolution spectra, and HST UV and optical spectrophotometry to establish complete system parameters and distances to each binary. Again this is an ongoing programme, with much more data yet to be completely analysed and published. The distances for 4 LMC EBs published by this group lie in the range 43.2 to 50.2 kpc, showing that the three-dimensional structure of the LMC is an important component in determining a mean distance. Allowing for the known tilt of the LMC, with the Eastern side of the LMC being nearer to the Milky Way, the average distance modulus for these 4 binaries is 18.38 ± 0.11 (47 kpc) (Ribas, 2004), whereas the HST Key Project on Cepheids adopted a canonical value of 18.50 (50 kpc).

Comparisons of the derived astrophysical parameters for unevolved detached systems by the above authors, and most recently by González et al. (2005) for another 8 LMC EBs, with the theoretical models for single stars at the appropriate metallicity, show excellent overall agreement. The two components of each binary lie sensibly on single isochrone lines in the mass-radius and HR diagrams, and the observed and theoretical mass-luminosity relationships are in very good agreement.

From a large-scale survey of 50 EBs in the SMC, Harries et al. (2003) and Hilditch et al. (2005) have drawn similarly positive conclusions in comparisons of detached systems with published evolution models for single stars. They found that the average difference between the evolution mass, inferred from the HR diagram and evolution tracks, and the measured dynamical mass for 42 stars in 21 detached EBs was $-0.9 \pm 2.2 \, M_\odot (s.d.)$. Interestingly, they also found that 28 systems were in semi-detached states, the slow phase of case A mass exchange, with properties that are probably consistent with new models (not yet published) for close binary evolution at $Z = 0.004$ (cf. Wellstein and Langer, 1999; Wellstein et al., 2001; Nelson and Eggleton, 2001 for $Z = 0.020$). The observed secondaries are clearly very evolved with low surface gravities, whilst the observed primaries are on the main sequence as expected, and their evolution masses agree broadly with their measured dynamical masses, the average difference being $-0.1 \pm 4.4 \, M_\odot (s.d.)$. Such agreement suggests rejuvenation of the stellar cores during the mass transfer event, but it is unclear why there is such a large dispersion about the mean. Only one binary in this sample of 50 was found to be a true contact binary. Further work on modelling the case A evolution of close binaries with SMC metallicity is being conducted by de Mink and Pols (2005, private communication), and direct comparisons are being made with the results from the above survey of 50 SMC eclipsing binaries. These suggest that some mass loss is required during the rapid phase of case A mass transfer in order to explain the observed properties of these semi-detached systems.

The average distance modulus to the SMC derived from the 50 EBs in this sample is $18.91 \pm 0.03 (s.e., internal) \pm 0.1 (s.e., external)$, corresponding to a distance of $60.6 \pm 1.0 \pm 2.8$ kpc. This value is one of the most precise available determinations of the distance to the SMC. Groenewegen (2000) found a line-of-sight depth of $\simeq 0.07$ in distance modulus ($\simeq$ 2kpc) over the $\simeq 2°$ spatial extent of this EB survey, from studies of Cepheid variables, and these EB data are consistent with this finding. The difference in distance modulus between the SMC and the LMC is found to be $+0.50 \pm 0.02$ (Groenewegen 2000) or $+0.44 \pm 0.05$ (Cioni et al., 2000), which implies a distance modulus for the LMC in the range $18.41 - 18.47$, consistent with that determined by Ribas (2004) of 18.38 ± 0.11 from 4 eclipsing systems when corrected to the centre of the known line-of-sight depth of the LMC. Clearly we need many more observations of many more LMC and SMC EBs to resolve the apparent discrepancies between distances from binaries and distances from Cepheids, stars at the tip of the red-giant branch, or star clusters.

3. M31 and M33

The photometric observations that provide the basis for the DIRECT project to determine direct distances to the nearby spiral galaxies M31 and M33 were all obtained in the interval 1996–2000, and most of the final data had been published by 2003. Macri (2004) provides a useful summary, and reference list, of the status of this project and includes commentaries on the follow-up observations that have been conducted on eclipsing binaries and Cepheid variables. It would appear from such summaries that securing the necessary spectroscopic observations to establish the orbital parameters of some of these eclipsing binaries is proving to be very difficult. Such is to be expected, since the most luminous binaries in M31 composed of two O-type stars would have apparent magnitudes of $V \simeq 19^m$ at maximum brightness, and securing spectra at a sufficient spectral resolution in a time interval that is short compared to the orbital period on such stars necessitates 8–10m telescopes.

Bonanos et al. (2005), and this conference proceedings, has achieved the first successes in securing sufficient spectra to provide an orbital solution for one eclipsing binary system in M33, together with determinations of the effective

temperatures of both stars from model-atmosphere fits to the O7-type spectra. The light curve data are taken from the DIRECT project database, together with additional B, V photometry from the 2.1m telescope at Kitt Peak. The spectroscopy was obtained with ESI on the Keck 10m telescope and with GMOS on the Gemini North 8.2m telescope. The complete solution indicates a normal detached eclipsing binary composed of two massive O7-type stars (33 $M_\odot$ and 30 $M_\odot$) with significant stellar winds, as expected for such early-type stars. The distance to the binary is found to be 942 ± 73 kpc, the first direct distance to an eclipsing binary in a nearby spiral galaxy.

Vilardell et al. (2005) have used the 2.5m Isaac Newton telescope with its wide-field camera (INT+WFC) to study the North-East quadrant of M31 to establish global stellar properties of the region, and to discover many eclipsing binaries, or conduct more detailed follow-up observations of binaries discovered by the DIRECT project. They have secured good quality B, V light curves of many systems, with typical uncertainties of $0^m.01$ per observation at $V \simeq 19 - 20$ mag from data obtained over four years. Complementary spectroscopic observations have been obtained with the 8.2m Gemini North telescope and its GMOS spectrograph on 2 systems so far, and the first results on one system have been determined (Ribas et al., 2005). This system is found to be a post-mass-transfer semi-detached system, composed of a detached O9 primary of 23 $M_\odot$ with a Roche-lobe-filling B1 secondary of 15 $M_\odot$, at a distance of 772 ± 44 kpc. This derived distance is in very good agreement with recent determinations of the distance to M31 via other techniques (McConnachie et al., 2005, and references therein).

Todd et al. (2005) have made use of the archived INT+WFC photometry of the North East quadrant of M31 (a subset of that used by Ribas et al. noted above) to search for eclipsing binaries and other variable stars, finding 280 EBs and determining their orbital periods via a matched filter technique. Unfortunately, their published light curves are given in instrumental units of ADUs per second rather than being converted to fluxes via the CCD gain values. They warn that many stellar images are blends of several stars, and therefore conversion to correct fluxes in such crowded regions will be difficult (cf., Michalska and Pigulski, 2005).

4. Summary

Progress on studies of eclipsing binaries in the Magellanic Clouds is rapid, and already we have quantitative information about more high-mass eclipsing binaries in these Clouds than we do about those in the Milky Way galaxy. For the spiral galaxies, M31 and M33, a start has been made on establishing the astrophysical parameters and distances of their individual eclipsing binaries. It is early days, but subject to sufficient allocations of observing time on 8–10m telescopes for spectroscopy, this subject could progress very rapidly in the next 3–5 years. Photometric studies of EBs in dwarf galaxies of the Local Group are also underway.

References

Bayne, G.P., Tobin, W., Pritchard, J.D., Pollard, K.R., Albrow, M.D.: MNRAS **349**, 833 (2004)
Bonanos, A.Z., Stanek, K.Z., Kudritski, R., Macri, L., Sasselov, D.D., Kaluzny, J., Bersier, D., Bresolin, F., Matheson, T., Mochejska, B., Przybilla, N., Szentgyorgyi, A., Tonry, J., Torres, G.: in preparation (2005)
Cioni, M.-R.L., van der Marel, R.P., Loup, C., Habing, H.J.: A&A **359**, 601 (2000)
Clausen, J.V., Storm, J., Larsen, S.S., Giménez, A.: A&A **402**, 509 (2003)
Fitzpatrick, E.L., Ribas, I., Guinan, E.F., Maloney, F.P., Claret, A.: ApJ **587**, 685 (2003)
González, J.F., Ostrov, P., Morrell, N., Minniti, D.: ApJ **624**, 946 (2005)
Groenewegen, M.A.T.: A&A **363**, 901 (2000)
Hadrava, P.: A&AS **114**, 393 (1995)
Harries, T.J, Hilditch, R.W., Howarth, I.D.: MNRAS **339**, 157 (2003)
Hilditch, R.W., Howarth, I.D., Harries, T.J: MNRAS **357**, 304 (2005)
Macri, L.M.: In *Variable Stars in the Local Group*, IAU coll. No. 193, eds., Kurtz,D.W., Pollard,K,R., ASP Conf.Ser., Vol. **310**, p.33 (2004)
McConnachie, A.W., Irwin, M.J., Ferguson, A.M.N., Ibata, R.A., Lewis, G.F., Tanvir, N., MNRAS **356**, 979 (2005)
Michalska, G., Pigulski, A.: A&A **434**, 89 (2005)
Mochejska, B.J., Macri, L.M., Sasselov, D.D., Stanek, K.Z.: AJ **120**, 810 (2000)
Nelson, C.A., Eggleton, P.P.: ApJ **552**, 664 (2001)
Pietsch, W., Mochejska, B.J., Misanovic, Z., Haberl, F., Ehle, M., Trinchieri, G.: A&A **413**, 879 (2004a)
Pietsch, W., Misanovic, Z., Haberl, F., Hatzidimitriou, D., Ehle E., Trinchieri, G.: A&A **426**, 11 (2004b)
Pietsch, W., Freyberg, M., Haberl, F.: A&A **434** , 483 (2005)
Ribas, I.: New Astron. Rev. **48**, 731 (2004)
Ribas, I., Giménez, A., eds.: New Astron. Rev. **48**, 647–762 (2004)
Ribas, I., Jordi, C., Vilardell, F., Fitzpatrick, E.L., Hilditch, R.W., Guinan, E.F.: ApJ **635**, L37–L40 (2005)
Simon, K.P., Sturm, E.: A&A **281**, 286 (1994)
Todd, I., Pollacco, D., Skillen, I., Bramich, D.M., Bell, S., Augusteijn, T.: MNRAS in press (2005)
Udalski, A.: New Astron. Rev. **48**, 667 (2004)
Udalski, A., Soszyński, I., Szymański, M., Kubiak, M., Pietrzyński, G., Woźniak, P., Żebruń, K.: Acta Astron. **48**, 563 (1998)
Vilardell, F., Ribas, I., Jordi, C.: in preparation (2005)
Wellstein, S., Langer, N.: A&A **350**, 148 (1999)
Wellstein, S., Langer, N., Braun, H.: A&A **369**, 939 (2001)
Wyrzykowski, L., Udalski, A., Kubiak, M., Szymański, M., Żebruń, K., Soszyński, I., Woźniak, P.R., Pietrzyński, G., Szewczyk, O.: Acta Astron. **53**, 1 (2003)
Wyrzykowski, L., Udalski, A., Kubiak, M., Szymański, M.K., Żebruń, K., Soszyński, I., Woźniak, P.R., Pietrzyński, G., Szewczyk, O.: Acta Astron. **54**, 1 (2004)
Zucker, S., Mazeh, T.: ApJ **420**, 806 (1994)

The First DIRECT Distance to a Detached Eclipsing Binary in M33

A. Z. Bonanos · K. Z. Stanek · R. P. Kudritzki ·
L. Macri · D. D. Sasselov · J. Kaluzny · D. Bersier ·
F. Bresolin · T. Matheson · B. J. Mochejska ·
N. Przybilla · A. H. Szentgyorgyi · J. Tonry · G. Torres

Received: 1 November 2005 / Accepted: 1 February 2006
© Springer Science + Business Media B.V. 2006

Abstract We present the first direct distance determination to a detached eclipsing binary in M33, which was found by the DIRECT Project. Located in the OB 66 association, it was one of the most suitable detached eclipsing binaries found by DIRECT for distance determination, given its 4.8938 day period. We obtained follow-up BV photometry and spectroscopy from which we determined the parameters of the system. It contains two O7 main sequence stars with masses of $33.4 \pm 3.5\ M_\odot$ and $30.0 \pm 3.3\ M_\odot$ and radii of $12.3 \pm 0.4\ R_\odot$ and $8.8 \pm 0.3\ R_\odot$, respectively. We derive temperatures of 37000 ± 1500 K and 36000 ± 1500 K and determine the reddening $E(B-V) = 0.14 \pm 0.03$. Using HST photometry for flux calibration in the V band, we obtain a preliminary distance modulus of 24.87 ± 0.16 mag (942 ± 73 kpc). The photometry and thus distance is subject to revision in the final paper.

A. Z. Bonanos (✉) · K. Z. Stanek · D. D. Sasselov · A. H. Szentgyorgyi · G. Torres
Harvard-Smithsonian CfA

R. P. Kudritzki · F. Bresolin · J. Tonry
IfA

L. Macri · T. Matheson
NOAO

J. Kaluzny
Copernicus Astronomical Center

D. Bersier
STScI

B. J. Mochejska
Purdue University

N. Przybilla
University of Erlangen-Nuremberg

1. Introduction

Starting in 1996 we undertook a long term project, DIRECT (i.e. "direct distances"), to obtain the distances to two important galaxies in the cosmological distance ladder, M31 and M33. These "direct" distances are obtained by measuring the absolute distance to detached eclipsing binaries (DEBs), which have the potential to achieve an unprecedented accuracy of 5%. DEBs (for reviews and history of method see Andersen, 1991; Paczynski, 1997; Kruszewski and Semeniuk, 1999) offer a single step distance determination to nearby galaxies and may therefore provide an accurate zero point calibration of various distance indicators – a major step towards very accurate determination of the Hubble constant. In the last few years, eclipsing binaries have been used to obtain accurate distance estimates to the Large Magellanic Cloud (e.g. Guinan et al., 1998; Fitzpatrick et al., 2003) and the Small Magellanic Cloud (Harries et al., 2003; Hilditch et al., 2005). Distances to individual DEBs in these papers are claimed to be accurate to better than 5%.

DEBs have yet to be used as distance indicators to M31 and M33. The DIRECT project has initiated a search for DEBs and new Cepheids in the M31 and M33 galaxies. We have analyzed five 11 arcmin ×11 arcmin fields in M31, A–D and F (Kaluzny et al., 1998; Stanek et al., 1998, 1999; Kaluzny et al., 1999; Mochejska et al., 1999) and one 22 arcmin ×22 arcmin field, Y (Bonanos et al., 2003). A total of 674 variables, mostly new, were found: 89 eclipsing binaries, 332 Cepheids and 253 other periodic, possible long-period or non-periodic variables. We have analyzed two fields in M33, A and B (hereafter Paper VI; Macri et al., 2001) and found 544 variables: 47 eclipsing binaries, 251 Cepheids and 246 other variables. Follow up observations of fields M33A and M33B produced 280 and 612 new variables, respectively

(Mochejska et al., 2001a,b), including 101 new eclipsing binaries.

In this paper, we present the preliminary distance we obtained to a DEB discovered in M33A (Paper VI).

2. Observations

The DIRECT Project discovered the detached eclipsing binary in field M33A (Paper VI), using the F.L. Whipple Observatory 1.2 meter telescope between 1996 September and 1997 October. In 1999 and 2001 we obtained followup BV photometry with the KPNO 2.1 m telescope, from which we constructed the light curves used for modeling. We also retrieved UBV archival Hubble Space Telescope data taken with WFPC2 outside of eclipse, which we used for flux calibration. We extracted PSF photometry for the DEB and our preliminary results are $U = 18.44 \pm 0.01$, $B = 19.47 \pm 0.01$, $V = 19.58 \pm 0.01$.

We obtained spectra of the DEB over 7 epochs, in the fall of 2002, 2003 and 2004 with the Echellette Spectrograph and Imager (or ESI, Sheinis et al., 2002) on the 10-meter Keck-II telescope on Mauna Kea. In the fall of 2004, we also obtained 4 additional epochs at quadrature with the Gemini Multi Object Spectrograph (GMOS, Hook et al., 2003) on the Gemini-North 8-meter telescope on Mauna Kea. The signal to noise (S/N) of the spectra ranges from 15–40 and we extracted them with IRAF[1] and IDL routines. Absorption lines from both stars are clearly resolved in the spectrum.

3. Analysis

We determined the light curve parameters by simultaneously fitting our BV light curves with a model of a detached binary with the Wilson-Devinney (WD) program (Wilson and Devinney, 1971; Wilson, 1979, 1990; van Hamme and Wilson, 2003). Figure 1 presents the fit to the V band light curve. We found the following best fit parameters: eccentricity $e = 0.17$, $\omega = 251.8$ deg, inclination $i = 86.9$ deg, light ratio $L_2/L_1 = 0.4912$ in V and 0.4923 in B and flux ratio $F_1/F_2 = 1.039$ in V and 1.036 in B, in agreement with the definition of the primary being the hotter star.

We used the method of two dimensional cross correlation or TODCOR, developed by Zucker and Mazeh (1994), to measure radial velocities of the stars in the DEB. We initially ran TODCOR with ATLAS9 template spectra (Kurucz, 1993) and found preliminary values for the semi-amplitudes, mass ratio and thus the semi-major axis. We thus obtained estimates for log(g) and the masses from which we computed

---

[1] IRAF is distributed by the National Optical Astronomy Observatories, which are operated by the Association of Universities for Research in Astronomy, Inc., under cooperative agreement with the NSF.

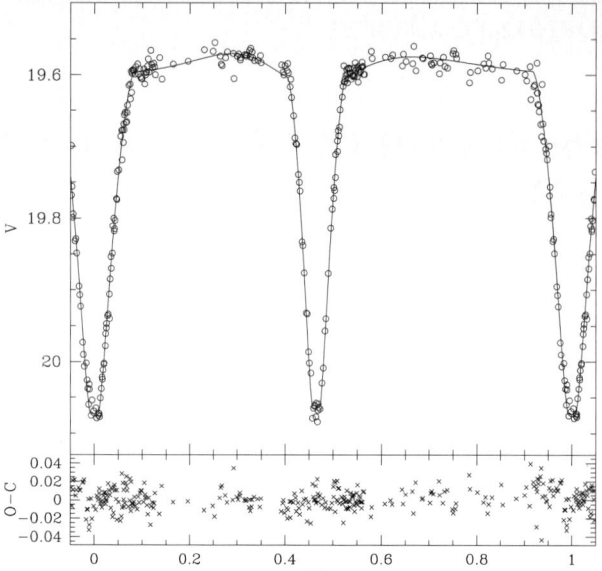

Fig. 1 V-band light curve of the DEB with model fit from the Wilson-Devinney program. The rms is 0.01 mag

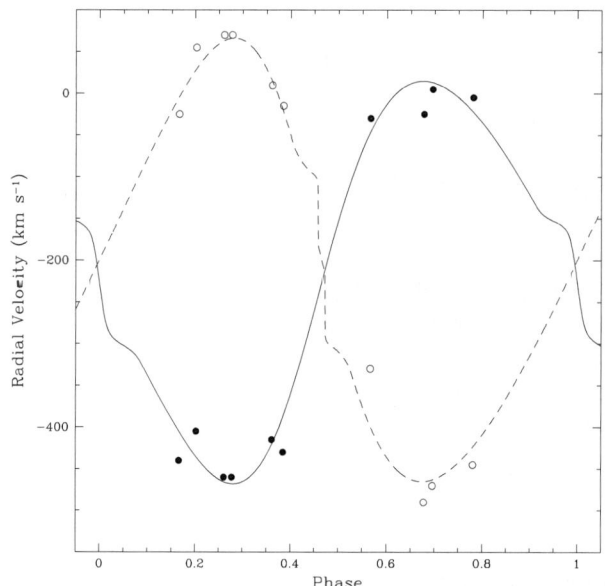

Fig. 2 Radial velocities for the DEB measured by TODCOR with FASTWIND synthetic spectra. Model fit is from Wilson-Devinney program, including the Rossiter effect. The rms is 26.0 km s^{-1} for the primary (filled circles) and 28.0 km s^{-1} for the secondary (open circles)

a grid of NLTE (non-local thermodynamic equilibrium) spectra with FASTWIND (Santolaya-Rey et al., 1997; Puls et al., 2005). We found the best fit spectra to have effective temperatures $T_{\text{eff1}} = 37000$ K and $T_{\text{eff2}} = 36000$ K.

Finally, a simultaneous WD fit to the BV light curves and radial velocities (see Fig. 2) yielded masses $M_1 = 33.4 \pm 3.5\ M_\odot$, $M_2 = 30.0 \pm 3.3\ M_\odot$, radii $R_1 = 12.3 \pm 0.4\ R_\odot$, $R_2 = 8.8 \pm 0.3\ R_\odot$, and values for log($g$) of 3.78 for the primary and 4.03 for the secondary.

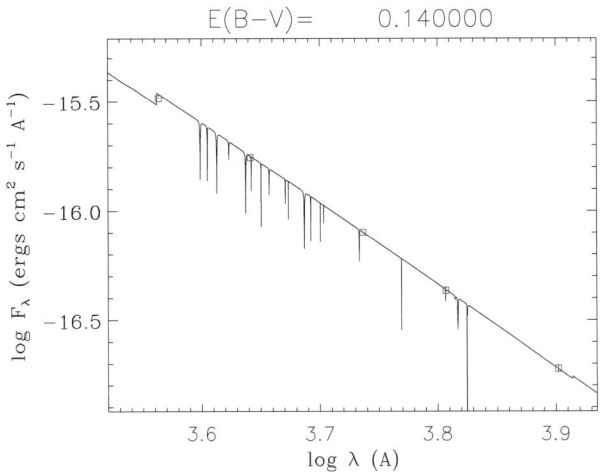

Fig. 3 Fit of the DEB model spectrum to the UBV HST photometry and RI spectrophotometry from the ESI spectrum of 20041011. Fixing the reddening to $E(B-V) = 0.14$, yields a distance modulus of 24.87 ± 0.16 mag (942 ± 73 kpc)

4. Distance determination

Having measured the temperatures of the stars from the spectra, we calculated the surface fluxes and absolute magnitude for each star and for the binary in each band. Taking the difference between the absolute color predicted from the models $(B - V)_0 = -0.25$ and the observed color $B - V = -0.11$ from the HST photometry, we derive $E(B - V) = 0.14$. The total extinction A_V follows by multiplication with $R = 3.1$ (Cardelli et al., 1989) as prescribed in (Schlegel et al., 1998). Armed with the extinction, radii and fluxes, we determine a distance modulus to the DEB and thus M33 of 24.87 ± 0.16 mag (942 ± 73 kpc). The fit of the model spectrum to the V-band HST photometry and the comparison with UB HST photometry and RI spectrophotometry is shown in Fig. 3. The error is computed by adding in quadrature the individual errors: 0.05 mag in the V-band flux calibration, 4% in the radii which translates to 0.085 mag in the distance modulus, 0.093 in A_V, assuming a conservative value for $E(B - V) = 0.14 \pm 0.03$, and 4% in $T_{\text{eff},1}$ which corresponds to 0.06 mag.

5. Discussion

We present the first distance to a detached eclipsing binary in a galaxy farther than the SMC. This distance determination is a significant step towards replacing the current anchor galaxy of the extragalactic distance scale, the LMC, with galaxies more similar to those in the HST Key Project (Freedman et al., 2001), such as M33 and M31. We have chosen a detached eclipsing binary to simplify the modeling and derived a distance modulus of 24.87 ± 0.16 mag. This result is preliminary and subject to revision in the final paper.

Previous distance determinations to M33 fall into two categories: "short" distances ∼24.6 mag, and "long" distances ∼24.8 mag. Our measurement, although completely independent of the others, agrees with the long distance to M33. Taken at face value, our DIRECT distance determination combined with the HST Key Project distance to M33 yields a LMC distance modulus of 18.75 ± 0.22 mag. The error is obtained by adding in quadrature the individual errors in the two distance measurements. We note that the Key Project reddening value is significantly larger than the reddening we have measured for the DEB. Furthermore, our LMC distance implies a 12% decrease in the Hubble constant to $H_0 = 63$ km s^{-1} Mpc^{-1}. This demonstrates the importance of accurately calibrating the distance scale and thus H_0, which are both vital for constraining the dark energy equation of state (Hu, 2005) and complementing the cosmic microwave background measurements from the Wilkinson Microwave Anisotropy Probe (Spergel et al., 2003).

References

Andersen, J.: A&A Rev. **3**, 91 (1991)
Bonanos, A.Z., Stanek, K.Z., Sasselov, D.D., et al.: AJ **126**, 175 (2003)
Cardelli, J.A., Clayton, G.C., Mathis, J.S.: ApJ **345**, 245 (1989)
Fitzpatrick, E.L., Ribas, I., Guinan, E.F., et al.: ApJ **587**, 685 (2003)
Freedman, W.L., Madore, B.F., Gibson, B.K., et al.: ApJ **553**, 47 (2001)
Guinan, E.F., Fitzpatrick, E.L., Dewarf, L.E., et al.: ApJ **509**, L21 (1998)
Harries, T.J., Hilditch, R.W., Howarth, I.D.: MNRAS **339**, 157 (2003)
Hilditch, R.W., Howarth, I.D., Harries, T.J.: MNRAS **357**, 304 (2005)
Hook, I., Allington-Smith, J.R., Beard, S.M., et al.: Proceedings of the SPIE, Volume 4841 (2003)
Hu, W.: ASP Conf. Ser. 339: Observing Dark Energy (2005)
Kaluzny, J., Stanek, K.Z., Krockenberger, M., et al.: AJ **115**, 1016 (1998)
Kaluzny, J., Mochejska, B.J., Stanek, K.Z., et al.: AJ **118**, 346 (1999)
Kruszewski, A., Semeniuk, I.: Acta Astronomica **49**, 561 (1999)
Kurucz, R.: ATLAS9 Stellar Atmosphere Programs and 2 km/s grid. Kurucz CD-ROM No. 13. Cambridge, Mass.: Smithsonian Astrophysical Observatory **1993**, 13 (1993)
Macri, L.M., Stanek, K.Z., Sasselov, D.D., et al.: AJ **121**, 870 (2001)
Mochejska, B.J., Kaluzny, J., Stanek, K.Z., et al.: AJ **118**, 2211 (1999)
Mochejska, B.J., Kaluzny, J., Stanek, K.Z., et al.: AJ **121**, 2032 (2001a)
Mochejska, B.J., Kaluzny, J., Stanek, K.Z., et al.: AJ **122**, 2477 (2001b)
Paczynski, B.: in The Extragalactic Distance Scale 273–280 (1997)
Puls, J., Urbaneja, M.A., Venero, R., et al.: A&A **435**, 669 (2005)
Santolaya-Rey, A.E., Puls, J., Herrero, A.: A&A **323**, 488 (1997)
Schlegel, D.J., Finkbeiner, D.P., Davis, M.: ApJ **500**, 525 (1998)
Sheinis, A.I., Bolte, M., Epps, H., et al.: PASP **114**, 851 (2002)
Spergel, D.N., Verde, L., Peiris, H.V., et al.: ApJS **148**, 175 (2003)
Stanek, K.Z., Kaluzny, J., Krockenberger, M., et al.: AJ **115**, 1894 (1998)
Stanek, K.Z., Kaluzny, J., Krockenberger, M., et al.: AJ **117**, 2810 (1999)
van Hamme, W., Wilson, R.E.: in ASP Conf. Ser. 298: GAIA Spectroscopy: Science and Technology 323 (2003)
Wilson, R.E.: ApJ **234**, 1054 (1979)
Wilson, R.E.: ApJ **356**, 613 (1990)
Wilson, R.E., Devinney, E.J.: ApJ **166**, 605 (1971)
Zucker, S., Mazeh, T.: ApJ **420**, 806 (1994)

Absolute Parameters of Early-Type Close Binaries in the LMC

S. Nesslinger · H. Drechsel

Received: 1 November 2005 / Accepted: 1 February 2006
© Springer Science + Business Media B.V. 2006

Abstract A project is presented which aims at high-precision determination of absolute parameters of close early-type eclipsing binaries in the LMC. We will use multi-object spectrographs (MOS) to measure RV curves of a large number of program stars selected from the MACHO archive. Spectroscopic mass ratios will be used as input for our light curve analysis code MORO. Application of the Simplex-based algorithm FITSB2 will achieve spectrum disentangling and fitting of NLTE model (TLUSTY) atmospheres, yielding orbital and atmospheric parameters. The method was extensively tested by application to time series of synthetic binary spectra for the expected range of S/N and MOS instrumental resolution. $\log g$ and $T_{\rm eff}$ were reestablished with a precision of better than 5%, radial velocity amplitudes with errors of <3%. An important by-product of our project will be the improvement of the distance modulus of the LMC, a topic which is still being intensely discussed.

Keywords Close binaries · Early-type binaries · LMC · Synthetic spectra

1. Program overview

The data base of well-studied OB-type binary stars with absolute parameters is quite scarce at present. Deriving precise absolute parameters for this type of star is of crucial importance to calibrate stellar evolution theories for early-type single and binary stars. The need is especially serious for systems in low-metallicity environments like the LMC, as only

S. Nesslinger (✉) · H. Drechsel
Dr. Remeis Observatory Bamberg, Astronomical Institute of the University of Erlangen-Nürnberg, Germany
e-mail: nesslinger@sternwarte.uni-erlangen.de

very few of these systems have been thoroughly analyzed so far. Hence it is most fortunate that during recent years a considerable number of high-quality light curves of extragalactic eclipsing binaries has been collected as by-products of several microlensing surveys like OGLE (Udalski et al., 1998; Wyrzykowski et al., 2003) EROS (Grison et al., 1995) and MACHO (Alcock et al., 1997). We intend to exploit the most promising of these vast sources to derive absolute parameters of close OB-type binaries in the LMC. In order to achieve this goal we need to gather spectroscopic observations. Therefore we intend to make use of state-of-the-art multi-object spectrographs, which are ideally suited for observing our a large number of our program stars.

As a by-product, our research will improve the value of the distance to the LMC, which is still under discussion, and no agreement has been found yet on a long or short distance scale.

The feasibility of such a project has been recently demonstrated for the SMC by Harries et al. (2003) and Hilditch et al. (2005). The focus of these studies lay on distance determination and on absolute parameters to provide tests of binary star evolution models. This project aims to increase the quantity and quality of the necessary data for such tests.

2. Photometric data

In spite of the vast number of eclipsing binary light curves in the OGLE archives we will primarily use the MACHO data base. This archive does not offer the multitude of light curves that OGLE does, but other than in the case of OGLE, all light curves benefit from complete and dense phase coverage in two (R and V) bands (see Fig. 1).

Of the 611 identified eclipsing binaries (Alcock et al., 1997) we selected a subset suitable for our project by

Fig. 1 Two examples of typical MACHO light curves – a close, but detached system (left) and an overcontact binary (right)

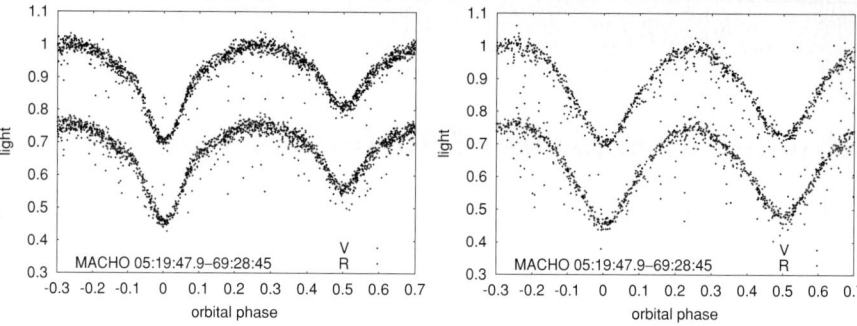

application of the following criteria: orbital periods should be shorter than two days, ensuring ease of spectroscopic observations at any phase and enhancing the probability of finding total eclipses. The $V - R$ color index is selected to be smaller than 0, ensuring our focus on early-type stars. Application of these criteria results in 227 remaining systems, from which our target objects will be chosen according to further selection criteria.

3. Planned observations

We will conduct our observations with powerful multi-object spectrographs (MOS) which perfectly meet our specific needs. Since our objects are distributed over a relatively small area of sky, we can observe many binaries simultaneously with reasonably long exposure times. This very time-efficient observing strategy requires only four or five nights (which do not even have to be consecutive) to collect a sufficient number of spectra at good S/N of most targets. Typical MOS instruments allow between 50 and 150 of our program stars to be observed simultaneously during one run (see Fig. 2). The field sizes of various instruments contain from five to as much as 50 program stars. If there should be unused instrument capacity, it will be assigned to early-type field stars, further enhancing the scientific value of our project by abundance analysis of even more objects.

4. Analysis of spectroscopic data

We will analyze the spectra of our program stars by a powerful new software called FITSB2 (Napiwotzki et al., 2004). This program can fit model spectra from any grid of synthetic spectra to the observed data. Different synthetic spectra can be convolved to account for double-lined binary stars while individual rotational velocities can be adjusted. When observation times and orbital periods are given, the program constrains the solution accordingly, constructing a radial velocity curve and thereby yielding the spectroscopic mass ratio. Furthermore, $\log g$ and T_{eff} of both components can be derived accurately. The minimization procedure employs the Simplex algorithm, thereby ensuring convergence of solutions and high numerical stability.

For our analysis of OB-type stars we will use the OSTAR2002 grid of synthetic spectra (Lanz and Hubeny, 2003) which are based on NLTE model atmospheres calculated with TLUSTY (Hubeny and Lanz, 1995). This grid is available for several metallicities and includes effective temperatures from 27500 K up to 55000 K. Therefore it is ideally suited for our targets.

5. Assessment of the method

In order to check the practicability and reliability of the fitting procedure in advance, we constructed a synthetic eclipsing binary time series. Based on a fictitious O8+B3 system, we synthesized six observations near the quadrature phases for different S/N and instrumental resolutions. These parameters were based on the performance of real MOS instruments.

These synthetic observations were analyzed by FITSB2. Several runs were conducted with a wide range of start parameters to emulate the uncertainty one has to cope with when dealing with real observations. Practically all of these runs were highly successful in that they not only converged to plausible solutions, but also recovered the original input parameters to a very satisfying degree of precision. Moreover, in the majority of cases the recovery could be achieved simultaneously, i. e. all relevant parameters were adjusted at the same time (see Table 1 for parameter values and Figs. 3 and 4 for line fits).

Table 1 Parameters derived by FITSB2 during simultaneous fitting of all lines shown in Figs. 3 and 4 together with statistical errors and true values

Parameter	Derived value	True value
$T_{\text{eff},1}$	36470 ± 179 K	35800 K
$T_{\text{eff},2}$	30195 ± 1484 K	30000 K
$\log g_1$	3.96 ± 0.016	3.97
$\log g_2$	4.05 ± 0.080	3.97
k_1	270.4 ± 0.8 km/s	264.8 km/s
k_2	351.5 ± 1.8 km/s	348.0 km/s
$v_{\text{rot},1}$	232.0 ± 1.2 km/s	253.1 km/s
$k_{\text{rot},2}$	240.4 ± 14.5 km/s	220.3 km/s

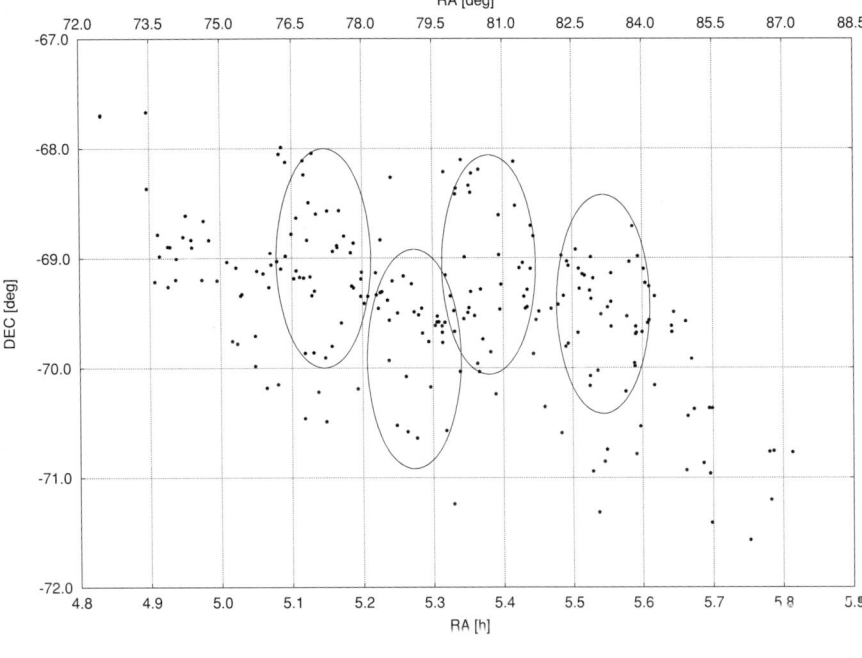

Fig. 2 Spacial distribution of selected program stars vs. sizes of four circular MOS fields. Setup corresponds to 2dF instrument of AAO, Australia; field coordinates have been optimized for maximum target number coverage

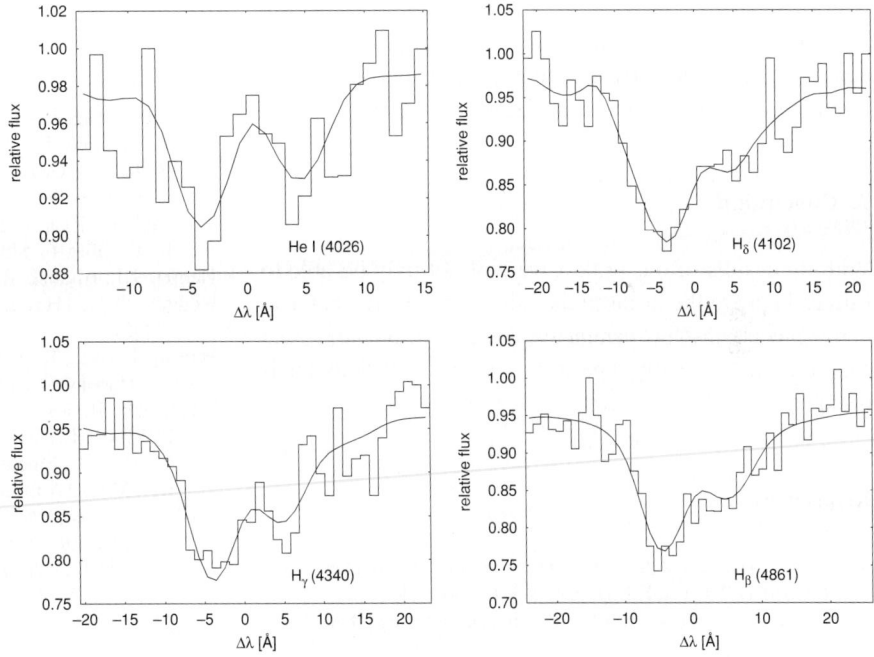

Fig. 3 Synthetic double-lined binary spectra of selected H and He lines with $S/N = 30$ at orbital phase 0.225 shown together with best fits simultaneously achieved by FITSB2

6. Light curve analysis

The primary goal of the analysis of the MACHO light curves in our project is the determination of the orbital inclination in order to get correct masses of the stars. Of course, other parameters like radii and temperature ratio are also derived and checked against the values extracted from the spectra.

We will use the software MORO (MOdified ROche model, denoting the use of a Roche model which takes into account radiation pressure effects in a physically and geometrically plausible way), which also employs the Simplex algorithm as parameter optimization scheme (Drechsel et al., 1995). In order to find good light curve solutions, it is crucial that suitable start parameters are chosen, especially for detached systems where the parameter space is very complex. Due to this need, we have constructed a synthetic light curve grid with the most important parameters (i, T_{eff}, Ω, q) containing 6×10^6 light curves for detached and semi-detached / overcontact configurations. The density of parameter space coverage is dependent on local sensitivity of parameter adjustment on the

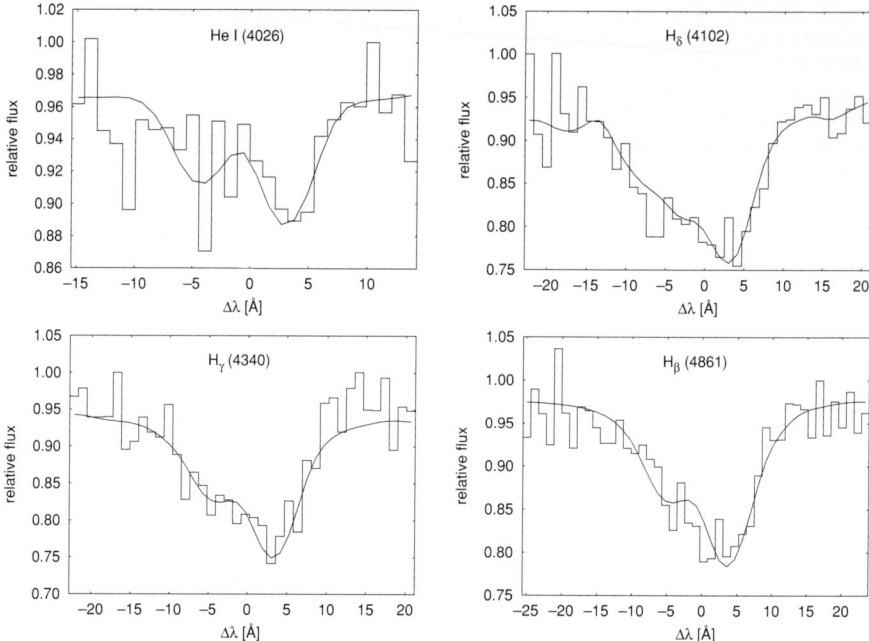

Fig. 4 Same as Fig. 3 for orbital phase 0.678

light curve shape. The observed light curves will be checked against the entire grid or a selected part of our library (depending on how much the solutions are already constrained) to find the most promising sets of start parameters.

7. Conclusion

With the results of our project we will not only be able to further improve the distance modulus of the LMC, but also derive precise absolute parameters and chemical abundances for a multitude of early-type stars in a low-metallicity environment.

References

Alcock, C., Allsman, R.A., Alves, D., Axelrod, T.S., Becker, A.C., Bennett, D.P., Cook, K.H., Freeman, K.C., Griest, K., Lacy, C.H.S., Lehner, M.J., Marshall, S.L., Minniti, D., Peterson, B.A., Pratt, M.R., Quinn, P.J., Rodgers, A.W., Stubbs, C.W., Sutherland, W., Welch, D.L.: AJ **114**, 326 (1997)

Drechsel, H., Haas, S., Lorenz, R., Gayler, S.: A&A **294**, 723 (1995)

Grison, P., Beaulieu, J.-P., Pritchard, J.D., Tobin, W., Ferlet, R., Vidal-Madjar, A., Guibert, J., Alard, C., Moreau, O., Tajahmady, F., Maurice, E., Prevot, L., Gry, C., Aubourg, E., Bareyre, P., Brehin, S., Gros, M., Lachieze-Rey, M., Laurent, B., Lesquoy, E., Magneville, C., Milsztajn, A., Moscoso, L., Queinnec, F., Renault, C., Rich, J., Spiro, M., Vigroux, L., Zylberajch, S., Ansari, R., Cavalier, F., Moniez, M.: A&AS **109**, 447 (1995)

Harries, T.J., Hilditch, R.W., Howarth, I.D.: MNRAS **339**, 157 (2003)

Hilditch, R.W., Howarth, I.D., Harries, T.J.: MNRAS **357**, 304 (2005)

Hubeny, I., Lanz, T.: ApJ **439**, 875 (1995)

Lanz, T., Hubeny, I.: ApJS **146** (2003)

Napiwotzki, R., Karl, C., Nelemans, G., Yungelson, L., Christlieb, N., Drechsel, H., Heber, U., Homeier, D., Leibundgut, B., Koester, D., Marsh, T.R., Moehler, S., Pauli, E.-M., Reimers, D., Renzini, A.: Rev. Mex. AA Conf. Ser. **20**, 113 (2004)

Udalski, A., Soszyński, I., Szymański, M., Kubiak, M., Pietrzynski, G., Woźniak, P.R., Żebruń, K.: AcA **48**, 563 (1998)

Wyrzykowski, Ł., Udalski, A., Kubiak, M., Szymański, M., Żebruń, K., Soszyński, I., Woźniak, P.R., Pietrzyński, G., Szewczyk, O.: AcA **53**, 1 (2003)

Searching for Planetary Eclipses in Open Clusters: NGC 7086

Russell Robb · John Vincent · Joanne Rosvick

Received: 1 November 2005 / Accepted: 1 February 2006
© Springer Science + Business Media B.V. 2006

Abstract In an attempt to discover planetary eclipses, we observed the open cluster NGC 7086. On one night we observed the cluster and standard stars through the B and V filters, enabling us to make a colour-magnitude diagram. Plots of the differential magnitudes were inspected for variability, but no planetary eclipses were found. New variable stars were discovered and their light curves show some of them to be eclipsing binary stars. The location on the colour-magnitude diagram of some of the variable stars is consistent with cluster membership.

Keywords Planets · Open clusters

1. Introduction

To discover a planet orbiting a solar type star, we searched for planetary transits in the populous and relatively unstudied open cluster NGC 7086. We modelled our project on that of Kaluzny and Shara, 1988 and that of Gilliland et al. (2000), who searched for planetary transits in the globular cluster 47 Tucanae. Was their null result due to the low metallicity of the cluster or was it a consequence of the star's crowded environment? We chose to observe an open rather than a globular cluster of stars, since the stars would have a higher metallicity and they would have been less likely to have experienced a disruptive gravitational interaction.

Eleven nights of nearly consecutive observations were made using the 1.8 meter telescope, R filter and SITE-5 CCD camera of the Herzberg Institute of Astrophysics. On one night we observed the cluster and standard stars (Landolt, 1992; Stetson, 1988) through the Johnson B and V filters, enabling us to make a colour-magnitude diagram for NGC 7086. The same telescope was used for two nights in spectrographic mode to observe standard and cluster stars. These classification spectra had a resolution of about 4Å, a dispersion of 0.72Å per pixel and wavelength coverage from 3771 to 5035Å.

2. The planet transit search

Approximately 1000 R observations were made of 939 stars ranging in brightness from $V=14.5$–19.0. The brightest (not overexposed), isolated star, of intermediate colour, which seemed most constant was S478. It was used as a comparison star for the calculation of the differential magnitudes. Plots of the 11 nightly light curves of the 939 stars were inspected for variability. No planetary transits were seen. For bright stars the standard deviations of the observations made during each night ranged from 0.004 to 0.006 magnitudes. The nightly means had a standard deviation of 0.002 magnitudes. The expected depth of a "Hot Jupiter" planet transiting a solar type star would be approximately 0.017 magnitudes, so such an eclipse would have been obvious for this signal to noise ratio.

3. Colour-magnitude diagram

The instrumental magnitudes B and V were measured and transformed to the standard system in the usual manner. The new CMD supersedes the earlier one from Hassan (1967) since the use of a CCD detector and modern data reduction software has improved the accuracy and precision of

R. Robb (✉) · J. Vincent
University of Victoria

J. Rosvick
Thompson Rivers University

Table 1 Variable stars discovered in the field of NGC 7086

Star	RA(J2000)	Dec(J2000)	V	(B−V)	Type	Period in days
S199	21^{h}30^{m}38.70^s	51°33′13.9″	16.29(1)	1.12(1)	γDor	0.46, 0.32
S276	21^{h}30^{m}44.46^s	51°33′54.7″	15.36(1)	1.10(1)	EA	0.9013(18)
S361	21^{h}30^{m}44.60^s	51°34′39.5″	14.63(1)	1.11(1)	EA	1.9487(16)
S505	21^{h}30^{m}41.19^s	51°36′06.8″	14.43(1)	0.75(1)	EA	12.54?
S654	21^{h}30^{m}10.87^s	51°37′34.3″	17.30(2)	1.64(2)	EW?	0.189?

Table 2 Times of minima of eclipsing binary stars

Star	HJD	Uncertainty	Star	HJD	Uncertainty
S276	2452850.9281	0.0005	S361	2452851.8655	0.0005
S276	2452851.8297	0.0005	S361	2452853.8169	0.0011
S276	2452855.8859	0.0005	S361	2452855.7629	0.0008
S276	2452856.7878	0.0005	S505	2453210.8980	0.0006
S276	2452859.9406	0.0005			

the magnitudes and has vastly increased the number of stars observed. In addition the magnitude limit of the new observations is close to $V = 20$, nearly 5 magnitudes fainter than in previous studies.

To determine the reddening of the cluster, six stars were observed and classified according to spectral type. For example from the hydrogen lines, the lack of Calcium K line and the existence of the He I 5017 and 4922Å lines, we classify the star GSC 3602–0101 as B7V ±1. Neglecting GSC 3602-0874, we find from the remaining stars' spectral classes and observed $(B − V)$'s, a reddening of $0.83 \pm .02$.

In Fig. 1. the cluster V and $(B − V)$ data have been shifted by a reddening of 0.83 and a distance modulus of 13.4 (∼ 1500pc), matching isochrones from Girardi, et al. (2002) for a $z = 0.008$ metallicity for ages of 100 (dashed) and 200 (dotted) Megayears. The lack of a well developed giant branch precludes a much older, closer and less reddened cluster.

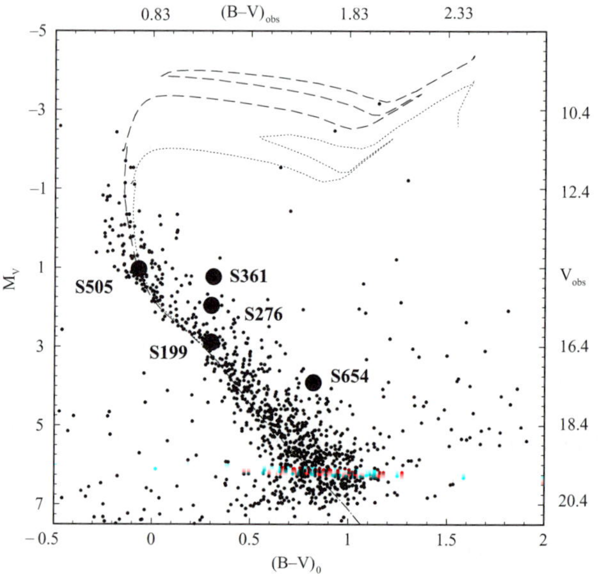

Fig. 1 Color Magnitude Diagram for the open cluster NGC 7086 with isochrones of 100 and 200 Megayears (Girardi et al., 2002)

4. The variable stars

The star S199 varied by up to ∼ 0.1 magnitudes during some of the nights. From a nine term fit found using Period04 (Lenz and Breger, 2005) the periods with the largest amplitudes were found to be 0.318 and 0.459 days. The most likely classification would be one of the newly defined Gamma Doradus stars (Handler and Shobbrook, 2002). These stars are similar to the Delta Scuti stars, but have a period of 0.4–3 days and amplitudes of 0.1 magnitudes in V. They have a spectral classification of A7-F5 V which is consistent with S199's $(B-V)_o = 0.30$ (=F0V) assuming $E(B − V) = 0.83$. The physical interpretation of the variability is due to g-mode oscillations, so multiple periods are to be expected.

The serendipitously discovered eclipsing binary stars are listed in Table 1. The times of minimum light found for the eclipsing binaries using the method of Kwee and van Woerden (1956) are given in Table 2. From these times of minima and the light curves we found the orbital periods given in Table 1.

5. Conclusion

We did not find any planetary transits in the open cluster NGC 7086. The total time spent observing the cluster was approximately 2.75 days, which is a large fraction of the typical orbital period of a "Hot Jupiter". The reddening of the cluster was so large that most of our well exposed stars (brighter than $V \sim 17.5$) have too large a radius for a planetary transit to be seen. We expected roughly 1 transiting planet out of 1000 F, G, K main sequence stars (eg. Gilliland et al., 2000), which is consistent with our null result.

We did however find some interesting variable stars. Our new colour-magnitude diagram gives a vast improvement in the determination of the age, reddening and distance to NGC 7086. Further observations of other clusters will be necessary to ascertain whether the cluster environment precludes the existence of "Hot Jupiters".

References

Gilliland, R. et al.: ApJ **545**, L47 (2000)
Girardi, L. et al.: A&A **195**, 391 (2002)
Handler, G. Shobbrook, R.: MNRAS **330**, 57 (2002)
Hassan, S.M.: Z. Astrophys **66**, 6 (1967)
Kaluzny, J., Shara, M.: AJ **95**, 785 (1988)
Kwee, K.K., van Woerden, H.: Bull. Astr. Inst. Neth. **12**, 327 (1956)
Landolt, A.U.: AJ **104**, 340 (1982)
Lenz, P., Breger, M.: CoAst **146**, 5 (2005)
Stetson, P.B., Harris, W.E.: PASP **102**, 932 (1988)

Binaries with Total Eclipses in the LMC: Potential Targets for Spectroscopy

Pierre North

Received: 1 October 2005 / Accepted: 18 November 2005
© Springer Science + Business Media B.V. 2006

Abstract 35 Eclipsing binaries presenting unambiguous total eclipses were selected from a subsample of the list of Wyrzykowski et al. (2003). The photometric elements are given for the I curve in DiA photometry, as well as approximate T_{eff} and masses of the components. The interest of these systems is stressed in view of future spectroscopic observations.

Keywords LMC · Stars: fundamental parameters

1. Introduction

As recalled by Wyithe and Wilson (2001), "systems where the stars are completely eclipsed are particularly important because they can provide robust measurements of the ratio of radii", k. Accurate determination of k is indeed a well known problem in partially eclipsed systems. An impressive demonstration of this is provided by Fig. 3 of Gonzáles et al. (2005). Here, I draw attention to a few tens of totally eclipsing systems in the LMC, which would deserve spectroscopic observations for an orbit determination, and possibly additional photometric ones for a more accurate determination of radii and surface brightness ratio. Such systems will be useful, not only for distance determination, but also for comparison with stellar structure models.

2. Sample, lightcurve solution and stellar parameters

From the sample of Wyrzykowski et al. (2003) based on OGLE photometry, we have selected a subsample of 510

P. North
Laboratoire d'astrophysique, Ecole Polytechnique Fédérale de Lausanne (EPFL), Observatoire, CH–1290 Sauverny, Switzerland

binaries with $I_{max} \leq 18.0$, a depth of the secondary minimum ≥ 0.20 mag and an EA type. This is the same sample as that used by North and Zahn (2004). All lightcurves were solved interactively with the EBOP code, assuming a linear limb-darkening coefficient $u_p = u_s = 0.18$ except in the few cases of clearly cool components. Out of this sample, we selected visually 35 systems with clearly total eclipses. The fundamental stellar parameters were determined through interpolation in the evolutionary tracks of Schaerer et al. (1993) computed for a metal-content $Z = 0.008$ typical of the LMC. This was done as in North and Zahn (2004), but without the hypothesis of identical components and assuming $(m - M)_0 = 18.5$. The relative radii, orbital period and surface brightness ratio J_s were used to constrain the solution. J_s was calibrated in terms of T_{eff} ratio through the models of Kurucz (1979). In addition, $E(B - V)$ and $A_V = 3.1\, E(B - V)$ were determined simultaneously, using the measured $B - V$ index, following North (2004). The condition that both components lie on the same isochrone was not implemented, because of the presence of some post-mass exchange systems; for the latter, the masses given are just those of single stars with same T_{eff} and luminosity, and therefore may be wrong. For main sequence systems lacking a $B - V$ index, stellar parameters were determined in a cruder way, assuming both components lie on a $\log t = 7.0$ isochrone. The T_{eff} of cool giants in a few systems were derived from the $B - V$ or the $V - I$ index assuming $E(B - V) = 0.143$, while the masses were assumed identical to those of stars with same M_V on the isochrone. Some parameters of the I DIA lightcurve as well as T_{eff} and mass of both components are given in Table 1, a more complete version of which is available at the site http://obswww.unige.ch/~north/DEB/tot_param.

The errors are the formal ones given by the EBOP code and give an idea of the quality of the fit, though one has to

Table 1 Totally eclipsing systems in the LMC. The errors on r_p, k and $e\cos\omega$ are given as the last digit(s) of the corresponding values. The null uncertainties stand for parameters which were fixed during the fit, though they were generally adjusted in a previous iteration. Eleven stars in the list are common to Michalska and Pigulski (2005): W03 No 114, 362, 1148, 1291, 1520, 1551, 1748, 1880, 1996, 2279 and 2462. The T_{eff} of cool stars were estimated from $B-V$ with the calibration of Hauck and Künzli (1996) or from $V-I$ with the calibration of Kenyon and Hartmann (1995) (their Table A5)

No	I_{quad}	P_{orb}	r_p	k	$e\cos\omega$	$L_p(I)$	$T_{\text{eff }p}$	M_p	$T_{\text{eff }s}$	M_s
114	15.125	2.98958	0.258 2	0.537 04	0.0704 00	0.842	34768	23.2	26930	11.1
362	15.266	3.37025	0.306 1	0.733 03	−0.0093 03	0.682	21999	9.6	20959	7.8
398	16.862	3.33782	0.176 4	1.666 47	0.0031 00	0.563	23133	7.6	8256	2.7
406	18.037	13.16448	0.315 5	0.427 09	0.0000 00	0.965	6387	6.1?	4056	1.6?
416	16.736	5.16691	0.175 3	0.473 05	−0.2392 09	0.829	19066	6.6	17758	4.6
460	16.618	2.68062	0.278 4	0.423 04	0.0005 00	0.866	19924	7.2	18539	4.8
670	16.604	4.14459	0.178 4	1.632 35	0.0058 00	0.561	21405	7.0	8226	2.9
810	16.743	3.31516	0.342 4	0.357 04	0.0061 00	0.896	14032	4.9	13591	3.1
841	17.861	5.40797	0.158 3	0.542 09	−0.0012 09	0.798	12686	3.5	11453	2.5
1148	17.224	2.16054	0.262 2	0.653 05	−0.0341 09	0.741	18293	5.7	16106	4.1
1208	17.602	2.58542	0.258 3	0.476 06	0.0192 18	0.867	16308	5.0	12647	2.8
1212	18.034	2.05110	0.245 7	0.497 09	0.0535 23	0.858	7348	5.9?	6440	2.7?
1263	17.076	6.12966	0.175 2	0.556 05	0.1714 06	0.762	14506	4.6	14514	3.6
1278	17.567	7.42814	0.168 3	0.413 07	0.0011 00	0.852	11135	3.3	11313	2.4
1291	16.898	5.23194	0.184 3	0.504 05	0.1656 08	0.816	17385	5.9	15840	4.0
1344	17.710	4.14494	0.233 5	0.505 08	0.0009 17	0.812	11088	3.2	10558	2.3
1381	16.860	79.17160	0.126 2	0.526 07	−0.0016 10	0.819	5552	8.6?	5214	4.7?
1450	15.439	2.72713	0.330 1	0.605 02	0.0001 05	0.765	31983?	14.6?	23557?	8.2?
1469	17.334	1.56400	0.316 3	0.482 04	0.0006 00	0.889	19480	6.1	13044	2.8
1520	16.082	1.33832	0.354 1	0.694 04	0.0000 00	0.720	27348?	10.8?	21970?	7.2?
1551	16.393	1.53805	0.349 2	0.778 05	0.0000 00	0.716	22260	7.9	17536	5.2
1566	17.566	3.23738	0.223 6	0.450 09	−0.0256 13	0.864	15696	4.7	13351	2.9
1675	17.574	3.39105	0.295 8	0.364 09	0.0088 00	0.886	11436	3.5	11451	2.4
1748	15.544	5.45728	0.248 2	0.535 03	0.0254 06	0.800	19860	8.4	18469	5.8
1880	17.202	1.34524	0.333 3	0.668 07	−0.0001 00	0.753	18153	5.4	15305	3.8
1996	17.045	1.82795	0.258 3	0.535 05	−0.0037 12	0.850	23050	8.1	16899	4.2
2009	17.570	3.22935	0.204 4	0.447 09	−0.1751 23	0.869	18001	5.5	13873	3.0
2073	17.007	2.89578	0.303 5	0.386 06	0.0159 16	0.883	14864	4.8	14108	3.1
2279	16.289	3.31752	0.257 2	0.691 07	−0.1347 09	0.680	18293	6.5	18112	5.5
2289	17.789	3.80450	0.196 4	0.568 10	0.0001 15	0.769	13681	3.9	13130	3.0
2380	17.374	2.83324	0.275 4	0.447 07	0.0939 20	0.852	14296	4.4	13320	3.0
2462	16.767	4.26120	0.189 2	0.617 06	0.0993 10	0.766	23378?	8.1?	17646?	4.9?
2482	16.272	8.07316	0.177 4	1.547 39	0.0000 00	0.616	8800?	9.6?	7050?	8.0?
2533	17.748	3.27882	0.269 6	0.640 11	−0.0394 19	0.736	10035	2.8	9311	2.2
2583	17.717	2.07166	0.266 4	0.567 09	−0.0346 20	0.788	14909	4.2	13389	3.0

keep in mind that some parameters were kept fixed in the fit, so that the errors displayed are rather lower limits to the real uncertainties.

The HR diagrams in Figure 1 show that all but four binaries host main sequence components.

3. Discussion

Many systems with total eclipses have also relatively shallow minima. Although this might be due to 3rd light in some cases, this cannot explain all of them. Many such systems are certainly pairs of main sequence stars with a relatively small ratio of masses and radii. Among them, those for which $J_s \sim 1$ are especially interesting: they are composed of two stars close to the turn-off, with an evolved primary and an unevolved secondary. The ratio of radii is often near 0.5. Such systems allow to probe efficiently the global metallicity and helium content of each component – as far as the stellar structure models can be trusted – according to the method of Ribas et al. (2000). Since the metallicity of the LMC is less than half that of the Sun, this can potentially improve the $\Delta Y/\Delta Z$ relation obtained by Ribas et al. on the basis of Galactic systems. Adding totally eclipsing systems of the SMC will further improve the determination of this relation, even providing an independent estimate of the primordial He abundance through extrapolation to zero metallicity.

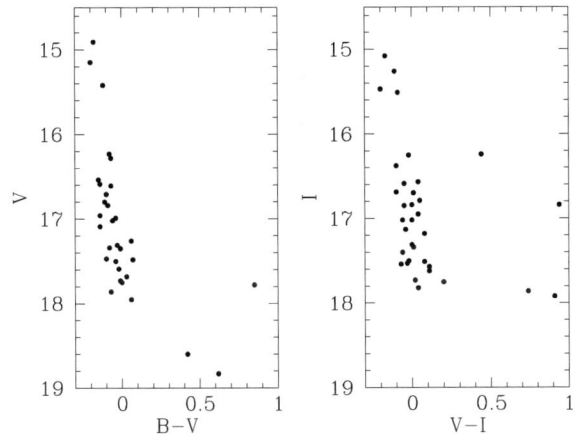

Fig. 1 HR diagram of the totally eclipsing systems in the LMC. Notice the three systems with giant components (a fourth one, in the I vs V-I diagram, is an Algol-type system)

However, the great interest of these systems has a price as regard to spectroscopic observations: the small k implies a small luminosity ratio – especially in the blue (λ domain of choice because of the number of spectral lines) if the secondary is cooler than the primary – so that high S/N spectra will be needed. The UVES instrument on the VLT, used with a wide slit, may be appropriate. Another possibility would be the use of FLAMES-GIRAFFE in the IFU "low" resolution mode ($R \sim 10000$), which would allow to observe a few systems simultaneously.

References

Gonzáles, J.F., Ostrov, P., Morrell, N., Minniti, D.: ApJ **624**, 946 (2005)
Hauck, B., Künzli, M.: Baltic Astronomy **5**, 303 (1996)
Kenyon, S.J., Hartmann, L.: ApJS **101**, 117 (1995)
Kurucz, R.L.: ApJS **40**, 1 (1979)
Michalska, G., Pigulski, A.: A&A **434**, 89 (2005)
North, P.: ASP Conf. Series **318**, 273 (2004)
North, P., Zahn, J.-P.: New Astron. Rev. **48**, 741 (2004)
Ribas, I., Jordi, C., Torra, J., Giménez, Á.: MNRAS **313**, 99 (2000)
Schaerer, D., Meynet, G., Maeder, A., Schaller, G.: A&AS **98**, 523 (1993)
Wyithe, J.S.B., Wilson, R.E.: ApJ **559**, 260 (2001)
Wyrzykowski, L., Udalski, A., Kubiak, M. et al.: AcA **53**, 1 (W03) (2003)

Eclipsing Binary Stars in Nearby Galaxies

Ian Todd · Don Pollacco · Ian Skillen · D. M. Bramich · Steve Bell · Thomas Augusteijn

Received: 1 November 2005 / Accepted: 1 February 2006
© Springer Science + Business Media B.V. 2006

Abstract We briefly discuss the survey programme we are conducting to detect eclipsing binaries in local group galaxies. Some lightcurves from studies of M31, IC 1613 and NGC 6822 are presented along with details of future work.

Keywords Stars · Local group · Eclipsing binaries · Distance scale

1. Introduction

Local Group Galaxies serve as a natural laboratory for the study of astrophysical phenomenon over a wide range of environments. The determination of their distance allows the calibration of standard candles such as Cepheids, enabling the distance scale to be accurately extended beyond the Local Group. Ultimately, one goal is to provide a better determination of the Hubble Constant, which is known to a precision of around 10% (Altavilla et al., 2004). There are two major factors that contribute to this uncertainty: the absolute distance to the Large Magellanic Cloud and the possible dependance of the Cepheid Period-Luminosity relationship on metallicity.

Eclipsing Binaries (EBs) however allow the possibility of distance determination to better than 5%, as it is possible to determine the *absolute* properties of the components with great accuracy (Andersen, 1991; Clausen, 2004). Traditionally, detached eclipsing binaries are thought to be more appropriate for distance determination, but even semi-detached eclipsing binary systems can be used, as they contain more information throughout their lightcurves and are subject to more constraints (Wilson, 2004). Modern software can fit these eclipsing binary lightcurves using the latest physical models.

The use of eclipsing binaries for distance determination is not a new concept. Gaspshkin (1968) first suggested their use in the Magellanic clouds. The distance to the SMC has also been derived using eclipsing binaries (Hilditch et al., 2005), and also Bonanos et al. (2006) report on the first distance to M33 based on an eclipsing binary.

I. Todd (✉) · D. Pollacco
Dept of Physics and Astronomy, Queen's University Belfast, Antrim, BT7 1NN, UK
e-mail: i.todd@qub.ac.uk

I. Skillen
Isaac Newton Group of Telescopes, Apartado 321, E-38700 Santa Cruz de la Palma, Tenerife, Spain

D. M. Bramich
Dept. of Physics and Astronomy, University of St. Andrews, Fife, KY16 9SS, UK

S. Bell
HM Nautical Almanac Office, CCLRC Rutherford Appleton Laboratory, Oxon, OX11 0QX, UK

T. Augusteijn
Nordic Optical Telescope, Apartado 474, E-38700 Santa Cruz de la Palma, Tenerife, Spain

2. Data reduction

The data were reduced using the difference imaging software of Bramich et al. (2005). This method has several advantages over aperture photometry and profile fitting in crowded extragalatic fields. Difference Image Analysis (DIA) matches the point–spread–function (PSF) between frames in a time series by generating a best–seeing reference frame and then degrading that frame to the seeing of the other frames via the generation of a kernel solution. The kernel solution models

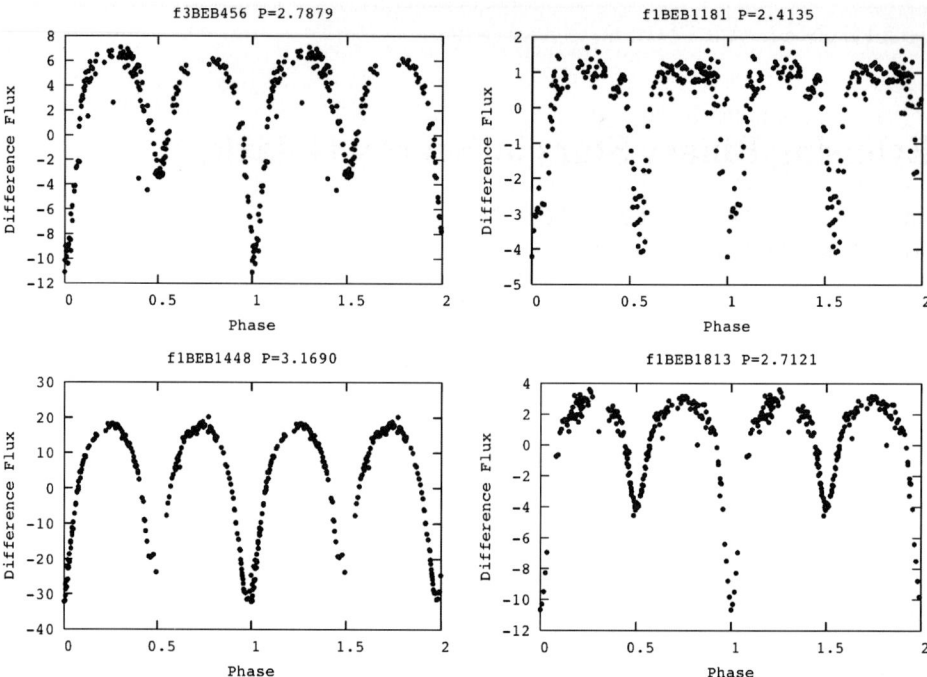

Fig. 1 A selection of eclipsing binary lightcurves from M31

the changes in the PSF from frame–to–frame as well as within a frame. In this way, a difference flux lightcurve is generated for each star found on the reference, and converted to magnitudes by measuring the brightness of each of the stars on the reference. In crowded fields however, problems with blending may cause star magnitudes to be measured inaccurately, resulting in erronous amplitudes of lightcurves. Thus, it is better to *clearly* state that this will be the case or give lightcurves in units of differential flux. A more detailed account of these issues can be found in Todd et al. (2005).

3. Observations and results

The data for M31 were extracted from the Isaac Newton Group Data Archive[1] over a four–year baseline. All relevant data over this period were extracted from the archive.[2] The Wide Field Camera (WFC) attached to the Issac Newton Telescope (INT) on La Palma was used to collect all M31 data. Data for IC 1613 and NGC 6822 were collected using MOSCA on the Nordic Optical Telescope (NOT).

3.1. M31

M31 is an important Local Group Galaxy for stellar population studies and distance determination. Its proximity to the Galaxy means that its bright stellar population can be resolved relatively easily. It has been shown to have many eclipsing binaries and Cepheids. As it is a spiral galaxy, geometric effects are relatively easy to model (unlike the irregular LMC and SMC), and it is the first step of the Tully-Fisher relation for spiral galaxies. It is therefore important that its distance is known accurately and systematics are identified.

We have used DIA to extract lightcurves for variable objects in the Eastern Spiral Arm of M31 over a four year baseline. Using a matched filter approach to detect binaries and Phase Dispersion Minimization (PDM) to determine periods (Stellingwerf, 1978), we have discovered around 280 binary systems in our data, with 98 newly discovered systems (Todd et al., 2005) and 128 have well determined periods. The data also contains many Cepheids, another primary distance indicator. A selection of binary lightcurves is shown in Fig. 1.

3.2. IC 1613 and NGC 6822

We have also been surveying IC 1613 and NGC 6822 for variability. Previously, Baade surveyed IC 1613 for variable content (Sandage, 1971). Antonello (2002) have discovered at least nine eclipsing binaries in IC 1613 and NGC 6822. So far, we have discovered around one hundred eclipsing binary lightcurves in these galaxies and are currently collecting more data to increase our sampling. We expect to have a catalogue of bright eclipsing binaries in these galaxies in the future.

---

[1] ING Archive at http://archive.ast.cam.ac.uk.

[2] see Acknowledgements.

4. Other work

Currently we are conducting surveys in other galaxies which are mostly beyond the Local Group. It is important that in the study of eclipsing binaries and particularly in the case of Cepheids, we survey environments with different metallicities to investigate its effect on the Period-Luminosity relationship. Galaxies that are currently under investigation include NGC 2403, M81, M83, Leo A, NGC 3109 and Sex A. We intend to characterise the Cepheid population of these targets.

Acknowledgements We would like to thank the ING Data Archive at Cambridge for maintaining and making accessible the data that made this study possible, and the multitude of observers and their associates who took time at the telescope through the various surveys to collect this data on the INT (T. Augusteijn, P. Boyce, M. Bremer, P. Dunclark, R. Corrandi, P. Dobbie, A. Ferguson, S. Hodgkin, C. Jordi, Kenynon, P. Lacerda. J. Mendez, S. Rawlings, V. Reyes, I. Ribas, S. Rix, S. Smartt, W. Sutherland, P. Sorenson, N. Walton). We would also like to thank the Nordic Optical Telescope for allowing our group time to make our own observations of IC 1613 and NGC 6822.

References

Altavilla, G., Fiorentino, G., Marconi, M., Musella, I., Cappellaro, E., Barbon, R. et al.: MNRAS **349**, 1344 (2004)
Andersen, J.: A&AR **3**, 91 (1991)
Antonello, E., Fugazza, D., Mantegazza, L., Stefanon, M., Covino, S.: A&A **386**, 860 (2002)
Clausen, J.V.: NewAR **48**, 679 (2004)
Bonanos, A.Z., Stanek, K.Z., Kudritzki, R.P. et al.: DOI: 10.1007/s10509-006-9112-1 (2006)
Bramich, D.M. et al.: MNRAS **359**, 1096 (2005)
Gasposhkin, S.: PASP **80**, 556 (1968)
Hilditch, R.W., Howarth, I.D., Harries, T.J.: MNRAS **357**, 304 (2005)
Sandage, A.: ApJ **166**, 13 (1971)
Stellingwerf, R.F.: ApJ **224**, 953 (1978)
Todd, I., Pollacco, D., Skillen, I., Bramich, D.M., Bell S.A., Augusteijn, T.: MNRAS **362**, 1006 (2005)
Wilson, R.E.: NewAR **48**, 695 (2004)

EROS-II Variable Stars: Eclipsing Binaries in the EROS Microlensing Surveys*

J. Pritchard · J. B. Marquette · Patrick Tisserand · J. P. Beaulieu · E. Lesquoy · A. Milsztajn

Received: 8 June 2006 / Accepted: 14 June 2006
© Springer Science + Business Media B.V. 2006

Abstract More than 10 years of microlensing survey observations by the EROS Collaboration have monitored several million stars, amongst them several thousand eclipsing binary stars. In this poster we present some of the difficulties and rewards of the study of this immense database.

Keywords Eclipsing binaries · Pre-main sequence stars · Eros1061

1. Introduction

Pre-Main-Sequence evolution is an important phase of evolution, yet despite much progress on the theoretical models in recent years, significant descrepancies between the predictions of different sets of models remain. Accurate stellar parameters of pre-main sequence stars are therefore needed to allow rigorous tests of the models. The eclipsing binary EROS1061 in the Large Magellanic Cloud (LMC) is suspected to consist of two pre-main-sequence components, and could thus provide much needed data for massive, low-metallicity PMS stars.

---

*Based on observations made at ESO by the EROS collaboration

J. Pritchard (✉)
European Southern Observatory, Casilla 19001, Vitacura 19, Santiago, Chile
e-mail: j.pritchard@eso.org

J. B. Marquette · P. Tisserand · J. P. Beaulieu · E. Lesquoy
Institut d'Astrophysique de Paris, 98bis Boulevard Arago, 75014 Paris, France

P. Tisserand · E. Lesquoy · A. Milsztajn
CEA, DSM/DAPNIA, Centre d'Études de Saclay, 91191 Gif-sur-Yvette, France

The *Expérience de Recherches d'Objets Sombres* (EROS) was begun in 1990 with the search for microlensing events caused by MAssive Compact Halo Objects (MACHOs). Five years (1990–1994) of ESO-Schmidt telescope photographic

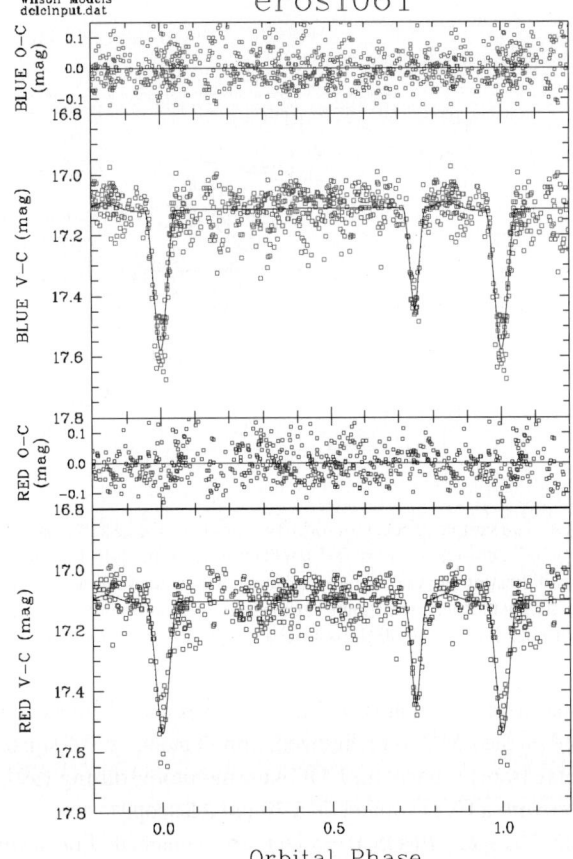

Fig. 1 EROS *B* and *R* light curves phased with a period of 4.5380 days. Wilson & Devinney fits and the residuals to the fit for each light curve from Pritchard (1997) are shown

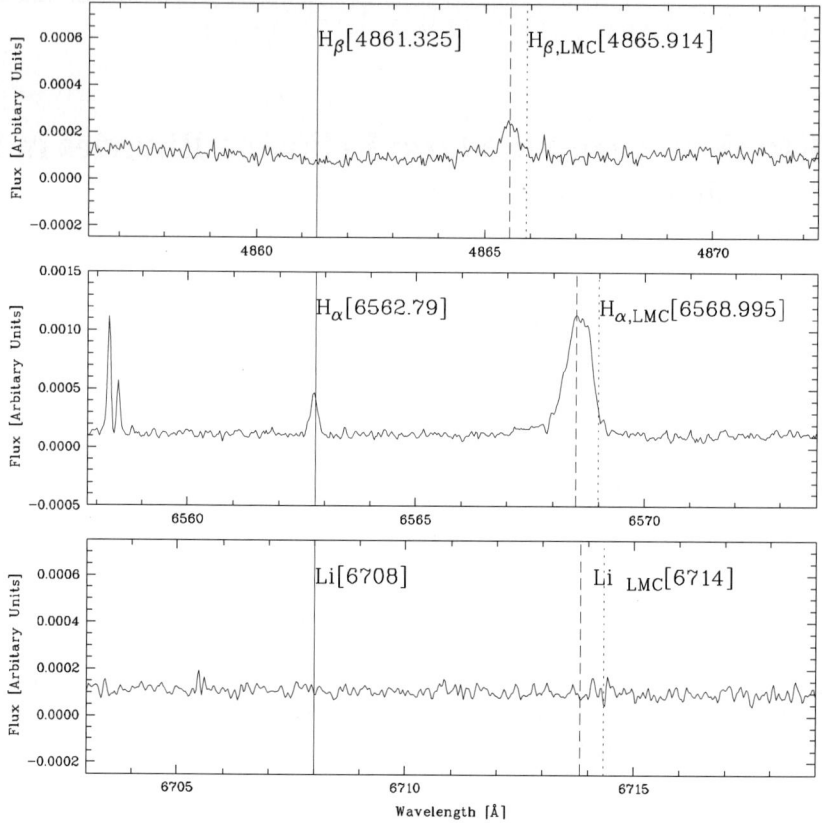

Fig. 2 FEROS spectra in the regions of H$_\beta$ (top), H$_\alpha$ (middle) and Li[6708Å] (bottom). The solid vertical line indicates the rest wavelengths of each spectra feature, the dotted line the redshifted wavelength for the radial velocity of the LMC. The radial velocity of eros1061 is estimated from the H$_\alpha$ emission line (dashed line) and indicated in the H$_\beta$ and Li[6708Å] plots (dashed lines)

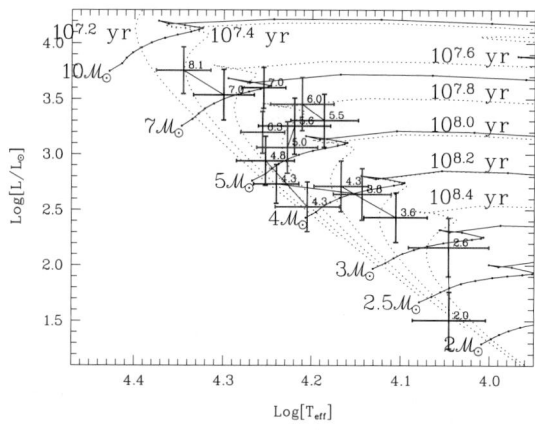
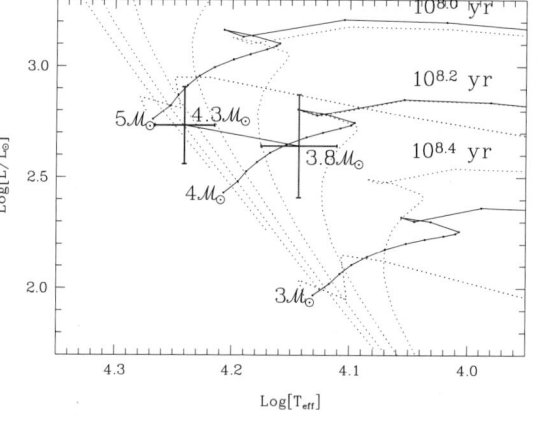

Fig. 3 The sample of eccentric-orbit binaries from the EROS catalogue (left) and a closeup on eros1061 (right) plotted in the HR diagram. A cross indicating the uncertainties is drawn for each component with a line joining the two components of each binary. Evolutionary tracks and isochrones for $Z = [0.008]$ from Schaerer et al. (1993) are shown. In the closeup of eros1061, PMS tracks from Iben (1965) are also shown. All the EROS eccentric-orbit systems appear to lie in the main sequence band and are consistent with the standard assumption of coeval creation, except eros1061, which is however broadly consistent with coeval PMS stars

plates covering an area of 5.25 × 5.25 square degrees centered on the LMC were digitized. Simultaneously, 0.5 square degree from the bar of the LMC were monitored during 1991–1994 with a CCD camera on a 40-cm telescope.

In 1996 the EROS-II project was launched. For seven years, 55 square degrees centered on the LMC and five square degrees centered on the SMC were monitored with a CCD camera on a 1-m telecscope. Other lines of sight in the direction of the spiral arms Gamma Normae and Theta Muscae and the Galactic bulge were also monitored. The fields were monitored on a semi-regular basis with a frequency between 2 and 14 visits per week.

Light curves are available from both the EROS (Fig. 1) and EROS-II surveys. Both sets of light curves consist of two colour photometry obtained over several years. Only standard EROS DoPhot based pipeline results are so far available for the EROS-II data. Re-reduction using difference imaging techniques is underway.

Four spectra totalling 160 min of exposure time have been obtained with FEROS at the MPG/ESO-2.20 m telescope, ESO La Silla (Fig. 2.)

2. Discussion

EROS1061 was discovered in the EROS data (Grison et al., 1995). Assuming a distance for the LMC the calibrated photometric light curves have been analysed to find the two stars simultaneously consistent with the light curves and theoretical models (Schaerer et al., 1993). Of the eight eccentric systems analysed this way, EROS 1061 is the only one whose solution does not give two components on the same Main Sequence star isochrones (Fig. 3) (Pritchard, 1997).

The solution proposed by Pritchard (1997) is that EROS1061 is a binary consisting of two Pre-Main Sequence components. The two components of the solution are consistent with Pre Main Sequence evolutionary tracks of Iben (1965) (Fig. 3) and the FEROS spectrum (Fig. 2) clearly indicates H_α emission, a typical indicator of PMS stars. H_β emission is also present; however, there is no indication of any Lithium absorption at 6708Å.

Further spectra are required to confirm or rebut the proposed evolutionary status of EROS1061.

Acknowledgements We are very grateful to the observatories that support our science (European Southern Observatory) via the generous allocations of time that make this work possible. This research has made use of the SIM-BAD database, Aladin and Vizier catalogue operation tools, from CDS, Strasbourg, France and the Atomic Line List at http://www.pa.uky.edu/~peter/atomic/.

References

Grison, P., Beaulieu, J.-P., Pritchard, J., Tobin, W., Ferlet, R., Vidal-Madjar, A., Guibert, J., Alard, C., Moreau, O., Tajahmady, F., Maurice, E., Prévot, L., Gry, C., Aubourg, E., Bareyre, P., Brehin, S., Gros, M., Lachiéze-Rey, M., Laurent, B., Lesquoy, E., Magneville, C., Milsztajn, A., Moscoso, L., Quiennec, F., Rich, J. Spiro, M., Vigroux, L., Zylbrajch, S. Ansari, R., Cavalier, F., Moniez, M.: EROS catalogue of eclipsing binary stars in the bar of the Large Magellanic Cloud. A&AS **109**, 447 (1995)

Iben, Jr., I.: Stellar Evolution. I. The approach to the main sequence. AJ **141**, 993 (1965)

Pritchard, J.: CCD Photometry of Eclipsing Binary Star Systems in the Large and Small Magellanic Clouds'. Ph.D. Thesis, University of Canterbury, New Zealand (1997)

Schaerer, D., Meynet, G., Maeder, A., Schaller, G.: Grids of stellar models. II. From 0.8 to 120M. at Z = 0.008. A&AS **98**, 523 (1993)

First Results from ROTES: The ROtse Telescope Eclipsing-binary Survey

I. Ribas · J.C. Morales · C. Allende Prieto · C. Jordi · D. H. Bradstreet · S. J. Sanders

Received: 1 November 2005 / Accepted: 1 February 2006
© Springer Science + Business Media B.V. 2006

Abstract Detached eclipsing binaries (EBs) provide a unique opportunity to carry out stringent tests on stellar evolution models. The value of EBs is enhanced by their membership in open clusters, but the number of known systems is still very scarce. We have started a systematic search for late-type EBs in the nearest open clusters with the fully robotic ROTSE3b telescope at McDonald Observatory in West Texas. On our first campaigns on the Hyades and Collinder 359, we have identified a number of previously unknown eclipsing binary candidates. Some of these stars have been selected for spectroscopic and photometric follow-up. Here we present details of the observing and reduction strategy as well as the first results of this ongoing survey.

I. Ribas (✉)
Institut de Ciències de l'Espai – CSIC, Campus UAB, Facultat de Ciències, Torre C5 – parell – 2a planta, 08193 Bellaterra, Spain;
Institut d'Estudis Espacials de Catalunya (IEEC), Edif. Nexus, C/ Gran Capità 2-4, 08034 Barcelona, Spain

J. C. Morales
Institut d'Estudis Espacials de Catalunya (IEEC), Edif. Nexus, C/ Gran Capità, 2-4, 08034 Barcelona, Spain

C. Allende Prieto
McDonald Observatory and Dept. of Astronomy, University of Texas, Austin, TX 78712, USA

C. Jordi
Institut d'Estudis Espacials de Catalunya (IEEC), Edif. Nexus, C/ Gran Capità 2-4, 08034 Barcelona, Spain; Dept. d'Astronomia i Meteorologia, Universitat de Barcelona, Av. Diagonal 647, 08028 Barcelona, Spain

D. H. Bradstreet · S. J. Sanders
Dept. of Physical Science, Eastern College, St. Davids, PA 19087, USA

Keywords Binaries:eclipsing · Binaries:spectroscopic · Stars:fundamental parameters · Techniques:photometric

1. Introduction

During the last years, research on the properties of low-mass stars in eclipsing binaries (EBs) has uncovered discrepancies between models and observations. While for stars with masses between 1 and 10 $M_\odot$ models reproduce the measured radii and effective temperatures to within 1-σ errors of 6% and 2%, respectively (Allende Prieto and Lambert, 1999), for stars less massive than the Sun observations give radii ~10% larger and temperatures ~5% lower than theory (e.g., Torres and Ribas, 2002; Ribas, 2003). This effect is probably related to the high magnetic activity level of the stars studied, which are components of close binary systems with fast rotation rates (e.g., Torres et al., 2006; Chabrier et al., 2005). Unfortunately interferometric measurements (Ségransan et al., 2003) of stars with lower magnetic activity do not yet have sufficient accuracy to check such hypothesis.

Double-lined EB stars are well suited to test models because the radii, masses, temperatures and luminosities of their components can be determined accurately (~1–2%) from analyses of their light and radial velocity curves and photometric measurements. EBs in open clusters are especially valuable because the age and metallicity of the binary system components can be estimated from the cluster membership. These are two parameters that can be fixed when comparing with stellar models. There is still a small number of EBs with well-determined masses and radii spanning a wide range of orbital periods, which would be appropriate to study the effects of magnetic activity on stellar properties.

To solve this paucity of known systems, since 2003 we have been conducting a systematic search for late-type EBs

Table 1 EBs in the Collinder 359 (c prefix) and Hyades (h prefix) regions

Star	P (days)	V_{max}	$(V-K)_0$	Star	P (days)	V_{max}	$(V-K)_0$
cstar268.26082	0.31	13.13	1.77	cstar268.47580	3.47	12.05	3.40
cstar271.42918	0.50	13.78	1.07				
hstar61.44196	0.29	13.93	2.50	hstar61.90892	1.76	13.31	3.09
hstar62.27151	1.56	13.42	2.35	hstar62.31986	0.59	12.53	1.13
hstar62.81043	2.21	13.23	2.47	hstar63.68436	1.37	12.81	2.24
hstar64.40810	0.38	14.39	1.81	hstar64.52607	4.09	14.90	3.41
hstar65.02993	0.44	14.20	1.60	hstar65.39291	0.63	14.35	1.79
hstar66.23733	0.61	14.92	2.55	hstar66.40923	0.62	13.33	1.63
hstar66.93768	0.53	14.65	2.04	hstar67.31705	0.41	10.95	1.66
hstar67.62950	0.85	11.61	2.52	hstar68.35001	0.37	14.65	2.03
hstar68.38767	0.27	14.81	3.00	hstar68.55581	0.67	15.14	3.25
hstar69.05671	2.07	13.79	2.46	hstar69.12546	1.33	13.64	1.82
hstar69.46529	1.14	13.22	1.86	hstar69.72262	3.80	12.16	3.13
hstar69.79089	1.09	13.81	1.88	hstar69.83558	0.36	14.20	2.65

in nearby extended clusters using the wide-field and robotic capabilities of the ROTSE3b telescope. The observations of the so-called ROTES project have been mostly focused on the Hyades and Collinder 359 clusters in the first years. Other nearby open clusters are being observed at the moment.

2. Survey characteristics

ROTSE3b is a fully-robotic 0.450 m f/1.9 telescope equipped with a 2 k × 2 k Marconi CCD (Akerlof et al., 2003). The telescope was designed to provide white-light wide-field imaging (1.85° × 1.85°) with its main scientific goal being the detection and follow-up of GRB optical transients. Since such observations only take up a fraction of the available time, the telescope is also used for other projects, mostly time-series photometry of variables sources, supernovae searches, and sky surveys. Our EB survey is one of such projects. The exposure time is set to 5 s, providing white-light photometry accurate to better than 0.01–0.02 mag down to V ∼ 14 mag.

Observations of open clusters are routinely taken every clear night. Several fields (17 for Hyades and 5 for Collinder 359) are used to cover a large fraction of the area of the clusters. Depending on various constraints, each field is visited a few times per night to attain optimal phase coverage in the resulting light curves. The images are flat-fielded and calibrated astrometically by the reduction ROTSE pipeline. We have designed a custom photometric routine within IRAF that performs aperture photometry of all stars. With the full image sequence for each cluster, a set of UNIX shell scripts and FORTRAN programs is used to carry out a refinement of the astrometry, cross-matching of the stars, determination of a photometric zero-point and finally to construct time-series photometry files for each star.

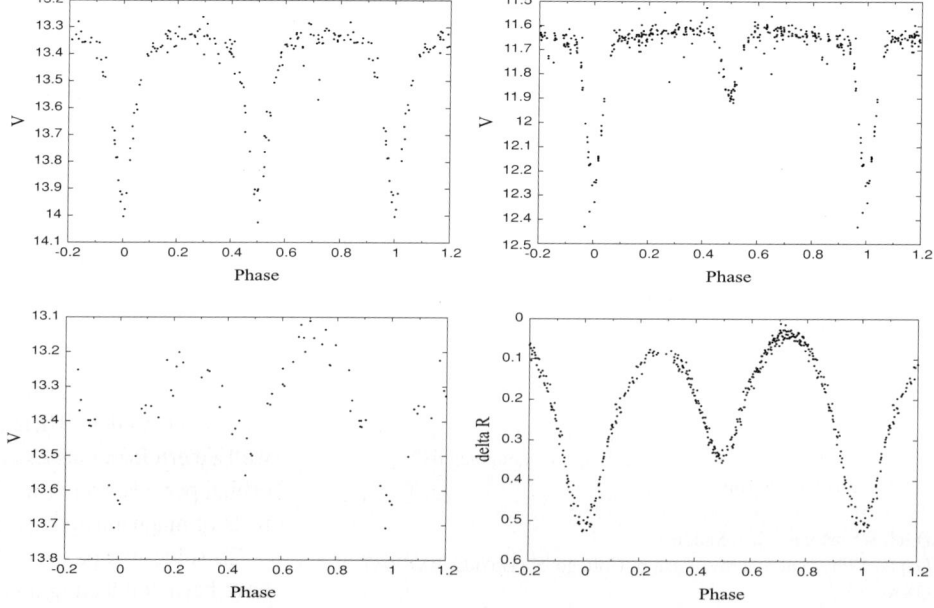

Fig. 1 Top: ROTES light curves of the two EBs hstar66.40923 (left) and hstar67.62950 (right). Bottom: Light curves of cstar268.26082 from the ROTES survey (left) and follow-up (right) using a 40-cm telescope equipped with a CCD

3. Eclipsing binaries in the Hyades and Collinder 359

To design an EB selection method, we first simulated observations of the well-studied CU Cnc and YY Gem systems. The result was that 2 to 7 observations out of 50 randomly distributed in time and 6 to 24 observations out of 130 would occur during the eclipses. Thus, the strategy we use with our data is to evaluate the number of observations with a magnitude 2-σ below the average and to select as EB candidates those with a certain number of such occurrences. Such scheme favours the detection of EBs because they have small values of σ since most of the measurements are out of the eclipses. Periods were computed for the selected variable stars using the analysis of variance algorithm (Schwarzenberg-Czerny, 1996). Finally, the light curves were used to identify the best EB candidates for further study.

The selected EB candidates were cross-matched with the 2MASS, USNO2, UCAC2 and Tycho2 catalogues. B_T and V_T magnitudes from Tycho2 of the brightest stars in the survey were used to calibrate the ROTSE photometry as a function of (J-K) transformed to the Johnson system. An estimation of T_{eff} was made from the $(V-K)_\circ$ index assuming single main-sequence stars with the reddening of the cluster. We derived the distance using a calibration of M_K as a function of $(V-K)_\circ$. Combining this with UCAC2 proper motions an evaluation of cluster membership probability can be performed.

The Hyades data from our survey have an average of 130 observations (October'04–March'05) so we have selected stars with over 6 measurements 2-σ below the mean magnitude. We have found 24 suitable EB candidates with periods smaller than 4 days and temperatures ranging from 3900 to 6500 K. Such detection rate represents a 0.1% of the surveyed stars. For Collinder 359 we only have about 50 observations per star (July–October'04), so we have selected those with over 2 measurements 1.5-σ below the mean magnitude. Although we have over 50,000 stars measured in this cluster region, only three have been found to be good EB candidates. They have periods of 0.3–3.5 days and temperatures of 3900–6600 K. More observations on this cluster are underway for a more accurate census of EBs. Table 1 lists some properties of our EB candidates. Figure 1 shows light curves from the ROTES survey and the accurate photometric follow-up of one of our EB candidates proving the efficiency of the selection method.

References

Akerlof, C.W., Kehoe, R.L., McKay, T.A. et al.: PASP **115**, 132 (2003)
Allende Prieto, C., Lambert, D.L.: A&A **352**, 555 (1999)
Chabrier, G., Baraffe, I., Allard, F., Hauschildt, P.H.: in Resolved Stellar Populations, ASP. Conf. Ser. Valls-Gabaud D. and Chavez M. (eds.), in press (astro-ph/0509798) (2005)
Ribas, I.: A&A **398**, 1195 (2003)
Schwarzenberg-Czerny, A.: ApJ **460**, L107 (1996)
Ségransan et al.: A&A **397**, L5 (2003)
Torres, G., Ribas, I.: ApJ **567**, 1140 (2002)
Torres, G., Lacy, C.S., Marschall, L.A., Sheets, H.A., Mader, J.A.: ApJ **640**, 1018 (2006)

Planets and Planet-Sized Binaries from the OGLE Transit Survey

Frédéric Pont · François Bouchy

Received: 10 October 2005 / Accepted: 1 February 2006
© Springer Science + Business Media B.V. 2006

Abstract The OGLE survey for transiting planets has identified 177 transit candidates. Subsequent radial velocity follow-up of these candidates has allowed the detection of five transiting planets, as well as several dozen eclipsing binaries.

Some of these systems consist of solar-type stars transited by small M dwarf companion, including the smallest stellar companions yet measured by transit. As a result, the OGLE transit survey has yielded a wealth of data on the mass-radius relation of planets and low-mass stars. In particular, two planet-sized stars were found, an empirical proof of the model predictions on Jupiter-sized main-sequence stars.

Keywords Planets · Close binaries

1. The OGLE planetary transit survey

Using the 1.3-m Warsaw University Telescope at Las Campanas Observatory (Chile), the OGLE-III survey (Optical Gravitational Lensing Experiment) has realized an extensive photometric search for planetary and low-luminosity object transits. In two seasons, about 200,000 stars were monitored with photometric accuracy better than 1.5% and analyzed for periodic eclipse signals with depth from a few per cent down to slightly below one per cent. Altogether 177 stars with multi-transiting low-luminosity objects were detected and announced (Udalski et al., 2002a,b,c 2003, 2004).

F. Bouchy
LAM, Marseille, France

F. Pont (✉)
Observatoire de Genève, Switzerland

The estimated radii of these objects range from 0.5 Jupiter radius to 0.5 solar radius and their orbital periods from 0.8 to 8 days. The smallest objects could be suspected to be extrasolar giant planets, but the radius estimated from the photometric signal is not sufficient to conclude on the planetary nature of the objects. They could as well be brown dwarfs or low-mass stars, since in the low mass regime ($M < 0.1\ M_\odot$) the radius becomes practically independent of the mass. Some configurations of grazing binary eclipses and of eclipsing binaries in multipe systems can also mimic a planetary transit signal.

2. Doppler follow-up of OGLE candidates

Radial velocity follow-up of transiting candidates is the only way to confirm the nature of the companions and to discriminate true central transits by a planet, brown dwarf or small star from other cases such as, for example, grazing eclipsing binaries, blended systems and stellar activity. The spectroscopy of the central star, which is a by-product of the radial velocity measurement, allows to constrain the radius of the star and hence the real size of the transiting companion. The measurement of the true mass of the companion by the radial velocity orbit, coupled with the measurement of its radius, leads to a direct measurement of its mean density, an essential parameter for the study of the internal structure of extrasolar giant planets, brown dwarfs and low-mass stars.

The difficulty of Doppler follow-up of OGLE candidates comes from the faintness of the stars (with V magnitudes in the range 15–18) located in very crowed fields. To characterize a Hot Jupiter, one needs radial velocity precision better than 100 m s^{-1} and the capability to distinguish whether the system is blended by a third star.

We have observed the 60 most interesting objects among the OGLE transit candidates from the first two seasons

(OGLE-TR-1 to TR-137) with the FLAMES facility on the VLT. FLAMES is a multi-fiber link which feeds the UVES echelle spectrograph with up to 7 targets in a field-of-view of 25 arcmin diameter, in addition to a simultaneous Thorium calibration. The fiber link allows a stable illumination at the entrance of the spectrograph, and the simultaneous Thorium calibration is used to track instrumental drift. As a result the systematics in the radial velocity measurements are reduced to less than 35 $m\,s^{-1}$. Combined with the photon noise uncertainty, typical precisions of 40–60 $m\,s^{-1}$ are reached on each individual Doppler measurement.

3. Four families of impostors

The spectroscopic follow-up of OGLE transit candidates presented a complete panorama of the configurations that can mimic the photometric signal of a planetary transit. These fall into four categories:

Grazing eclipsing binaries. Two large stars, when eclipsing at a high angle, can produce shallow transit-like dips in the light curve. These cases are the easiest to discriminate. Several hints are usually present in the light curve itself, such as a V-shaped transit curve, or ellipsoidal modulations in the light curve due to tidal effects and reflected light. Nevertheless, at low signal-to-noise such systems can also be mistaken for good planetary transit candidates. They are easy to resolve with spectroscopic observations, thanks to the presence of two sets of lines in the spectra with large velocity variations.

M-dwarf transiting companions. A small M-dwarf transiting a larger star can produce a photometric signal closely similar to a planetary transit. If the companion is not larger than a Hot Jupiter, and the orbital distance is too large for tidal and reflection effects to be detectable in the light curve, then the photometric signal is strictly identical to that of a planetary transit. In both cases, an opaque, Jupiter-size object transits the target star. These cases can only be resolved by Doppler observations, the amplitude of the reflex motion of the star revealing the mass of the transiting companion. Two nice example of planet-size transiting stellar companions were found among the OGLE candidates, OGLE-TR-122 and OGLE-TR-123 (see below).

Multiple systems. An eclipsing binary can produce shallow transit-like signals if the – deep – eclipse is diluted by the light of a third star. There are many possible configurations for such systems, and as a result they can be very difficult to disentangle, even with Doppler informations. In some cases, the conjuration of the parameters can be so good as to mimic not only the light curve of a planetary transit, but also induce planet-like variations of the inferred radial velocity, produced by the blending of several sets of lines in the spectra. OGLE-TR-33 (Konacki et al. 2004) is such a case.

False detections. Stellar variability, and systematic trends in the photometry, can produce fluctuations in the light curve interpreted as a possible transit signal, especially as one tries to detect shallower signals near the detection threshold. Several cases of low signal-to-noise detections showed no significant radial velocity variations, making them suspected false positives of the photometric transit detection procedure. Further photometric observations at the epoch of the detected transit are needed in these cases to distinguish bona fide transits from false positives.

4. The mass-radius diagram from stars to planets

Our Doppler follow-up of the 60 most promising OGLE candidates led to the characterisation of:

- 24 low-mass-star transiting companions,
- 5 grazing eclipsing binaries,
- 12 low-mass-star transiting companions in triple or quadruple systems,
- 7 false positives of the transit detection,
- 5 exoplanets.

Seven cases were left unsolved, because of the early type or faintness of the primary (none of them promising planet transit candidates).

The main outcome of our FLAMES campaigns (Bouchy et al., 2005; Pont et al., 2005b) is the characterization of 5 true planetary transits (see Table 1 and Fig. 1). However, thanks to the multiplexing capacity of FLAMES, we could also the measurement of the mass of many transiting low-mass stellar companions initially tagged as possible planetary transit (see Fig. 2 and Table 2). Small M-dwarf transiting companions were detected with masses down to the brown dwarf limit, providing the first direct radius measurement for such planet-size stars.

5. Implication for planet properties

Three of the five Hot Jupiters detected by the OGLE survey have periods shorter than 2 days. This is in stark contrast with the pile-up of periods above 3 days for the Hot Jupiters detected by radial velocity surveys. Taking into account the strong selection biases that favour the detection of very short-period objects in transit surveys, these "very Hot Jupiters" are thought to represent a few percent of all Hot Jupiters.

These three planets are also heavier than other Hot Jupiters. They revealed the presence of a clear relation

Table 1 Parameters of the five OGLE exoplanets. The discrepancy on the radius of OGLE-TR-10b comes from the disagreement on the stellar parameters. New spectroscopic measurements provided by Santos et al. (2005) indicate an intermediate value. For OGLE-TR-56b and OGLE-TR-113b, an improvement of the mass determination is given taking into account the RV measurements from both teams

Name	Period [days]	Mass [M_{Jup}]	Radius [R_{Jup}]	Reference
OGLE-TR-10b	3.101	0.66 ± 0.21	1.52 ± 0.12	Bouchy et al. (2005)
	3.101	0.57 ± 0.12	1.24 ± 0.09	Konacki et al. (2005)
	3.101	0.63 ± 0.13	1.43 ± 0.10	Santos et al. (2005)
OGLE-TR-56b	1.212	1.45 ± 0.23	1.23 ± 0.16	Torres et al. (2004)
	1.212	1.18 ± 0.13	1.25 ± 0.09	Bouchy et al. (2005)
	1.212	1.34 ± 0.13	1.25 ± 0.09	Combined RVs
OGLE-TR-111b	4.016	0.53 ± 0.11	$1.00^{+0.13}_{-0.06}$	Pont et al. (2004)
OGLE-TR-113b	1.433	1.35 ± 0.22	$1.08^{+0.07}_{-0.05}$	Bouchy et al. (2004)
	1.433	1.08 ± 0.28	1.09 ± 0.10	Konacki et al. (2004)
	1.433	1.29 ± 0.17	$1.08^{+0.07}_{-0.05}$	Combined RVs
OGLE-TR-132b	1.690	1.01 ± 0.31	$1.15^{+0.8}_{-0.13}$	Bouchy et al. (2004)
	1.690	1.19 ± 0.13	1.13 ± 0.08	Moutou et al. (2004)

Table 2 Parameters of the small transiting stars from our FLAMES follow-up of the OGLE survey. References: (a) Bouchy et al. (2005), (b) Pont et al. (2005b), (c) Pont et al. (2005a), (d) Pont et al. (2005c)

Name	Period [days]	Mass [$M_\odot$]	Radius [$R_\odot$]	Reference
OGLE-TR-5b	0.81	0.263 ± 0.012	0.271 ± 0.035	a
OGLE-TR-6b	4.55	0.393 ± 0.018	0.359 ± 0.026	a
OGLE-TR-7b	2.72	0.282 ± 0.013	0.281 ± 0.029	a
OGLE-TR-18b	2.22	0.390 ± 0.040	0.387 ± 0.049	a
OGLE-TR-34b	8.58	0.435 ± 0.033	0.509 ± 0.038	a
OGLE-TR-78b	5.32	0.240 ± 0.013	0.243 ± 0.015	b
OGLE-TR-106b	2.54	0.181 ± 0.013	0.116 ± 0.021	b
OGLE-TR-114a	3.42	0.73 ± 0.09	0.82 ± 0.06	b
OGLE-TR-114b	3.42	0.72 ± 0.09	0.82 ± 0.08	b
OGLE-TR-120b	9.17	0.40 ± 0.02	0.47 ± 0.04	b
OGLE-TR-122b	7.27	$0.120^{+0.024}_{-0.013}$	0.092 ± 0.009	c
OGLE-TR-123b	1.80	0.133 ± 0.009	0.085 ± 0.011	d
OGLE-TR-125b	5.30	0.211 ± 0.027	0.209 ± 0.033	b

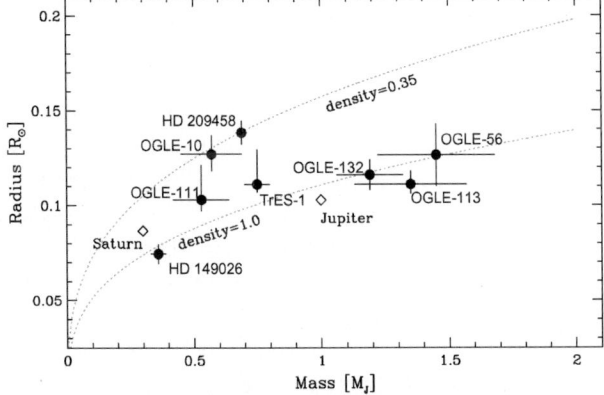

Fig. 1 The mass-radius diagram of transiting exoplanets. The values for HD209458b and Tres-1 are from Laughlin et al. (2005). The values for HD149026b are from Sato et al. (2005). The three heaviest OGLE planets (56b, 113b and 132b) have the shortest orbital periods and reveal the class of "very hot Jupiters"

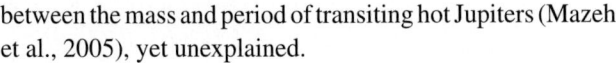

between the mass and period of transiting hot Jupiters (Mazeh et al., 2005), yet unexplained.

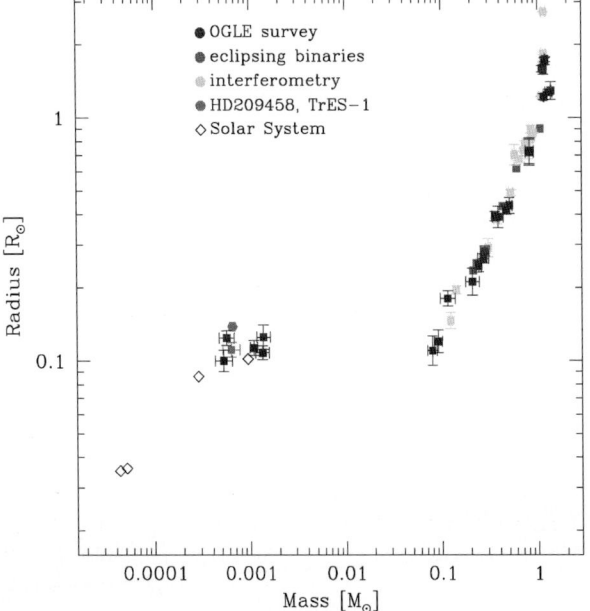

Fig. 2 The mass-radius diagram from stars to planets. The two smallest stars OGLE-TR-122b (Pont et al., 2005a) and OGLE-TR-123b (Pont et al., 2005c) have a size comparable to exoplanets

The radii of the gas giants detected among the OGLE candidates also define a mass-radius relation that is only slightly above that traced by Saturn and Jupiter, and markedly below the position of HD209458b. Therefore, the OGLE planets have shown that HD209458b had an exceptionally large radius, even for its close orbit.

6. Two planet-sized stars

Two eclipsing systems are particularly noteworthy: OGLE-TR-122 (Pont et al., 2005a) and OGLE-TR-123 (Pont et al., 2005c). In both systems, a solar-type primary is eclipsed by a M dwarf near the Hydrogen-burning limit, with a mass smaller than 0.1 $M_\odot$ and a size comparable to that of Hot Jupiters. These objects provide the first direct radius measurements in an unexplored part of the mass-radius diagram, that of stars in the degenerate electron-pressure regime. In this regime, the radius is expected to depend only very weakly on mass, thus leading to the expectation of a more or less constant radius from M dwarfs through brown dwarfs to gas giant planets. This is what we observe.

References

Bouchy, F., Pont, F., Santos, N.C. et al.: A&A **421**, L13 (2004)
Bouchy, F., Pont, F., Melo, C. et al.: A&A **431**, 1105 (2005)
Konacki, M., Torres, G., Sasselov, D. et al.: ApJ **609**, L37 (2004)
Konacki, M., Torres, G., Sasselov, D. et al.: ApJ **624**, 372 (2005)
Laughlin, G., Wolf, A., Vanmunster, T. et al.: ApJ **621**, 1072 (2005)
Mazeh, T., Zucker, S., Pont, F.: MNRAS **356**, 955 (2005)
Moutou, C., Pont, F., Bouchy, F. et al.: A&A **424**, L31 (2004)
Pont, F., Bouchy, F., Queloz, D. et al.: A&A **426**, L15 (2004)
Pont, F., Melo, C., Bouchy, F. et al.: A&A **433**, L21 (2005a)
Pont, F., Bouchy, F., Melo, C. et al.: A&A **438**, 1123 (2005b)
Pont, F., Bouchy, F., Behrend, R. et al.: A&A **447**, 1035–1039 (2005c)
Santos, N.C., Pont, F., Bouchy, F. et al.: In: Arnold, L., Bouchy, F., Moutou, C. (eds.), Tenth Anniversary of 51 Peg-b : Status and Prospect of hot Jupiter studies, colloquium held in OHP, France, August 22–25, 2005. Publ. Platypus Press (2005)
Sato, B., Fischer, D.A., Henry, G.W. et al.: ApJ **633**, 465 (2005)
Torres, G., Konacki, M., Sasselov, D.: ApJ **609**, 1071 (2004)
Torres, G., Konacki, M., Sasselov, D.: ApJ **641**, 979 (2004)
Udalski, A., Paczynski, B., Zebrun, K. et al.: Acta Astronomica **52**, 1 (2002a)
Udalski, A., Zebrun, K. et al.: Acta Astronomica **52**, 115 (2002b)
Udalski, A., Zebrun, K. et al.: Acta Astronomica **52**, 317 (2002c)
Udalski, A., Pietrzynski, G. et al.: Acta Astronomica **53**, 133 (2003)
Udalski, A., Szymanski, M.K., Kubiak, M. et al.: Acta Astronomica **54**, 313 (2003)

Eccentricities of Planets in Binary Systems

Genya Takeda · Frederic A. Rasio

Received: 1 November 2005 / Accepted: 1 February 2006
© Springer Science + Business Media B.V. 2006

Abstract The most puzzling property of the extrasolar planets discovered by recent radial velocity surveys is their high orbital eccentricities, which are very difficult to explain within our current theoretical paradigm for planet formation. Current data reveal that at least 25% of these planets, including some with particularly high eccentricities, are orbiting a component of a binary star system. The presence of a distant companion can cause significant secular perturbations in the orbit of a planet. At high relative inclinations, large-amplitude, periodic eccentricity perturbations can occur. These are known as "Kozai cycles" and their amplitude is purely dependent on the relative orbital inclination. Assuming that every planet host star also has a (possibly unseen, e.g., substellar) distant companion, with reasonable distributions of orbital parameters and masses, we determine the resulting eccentricity distribution of planets and compare it to observations? We find that perturbations from a binary companion always appear to produce an excess of planets with both very high ($\gtrsim 0.6$) and very low ($e \lesssim 0.1$) eccentricities. The paucity of near-circular orbits in the observed sample implies that at least one additional mechanism must be increasing eccentricities. On the other hand, the overproduction of very high eccentricities observed in our models could be combined with plausible circularization mechanisms (e.g., friction from residual gas) to create more planets with intermediate eccentricities ($e \simeq 0.1$–0.6).

Keywords Binaries: general · Celestial mechanics · Stellar dynamics · Planetary systems · Stars: low-mass · Brown dwarfs

G. Takeda (✉) · F. A. Rasio
Department of Physics and Astronomy, Northwestern University, 2145 Sheridan Road, Evanston, IL 60208, USA

1. Introduction

As of September 2005, more than 160 extrasolar planets have been discovered by radial-velocity surveys.[1] At least ~10% are orbiting a component of a wide stellar binary system (Eggenberger et al., 2004). In contrast to the planets in our own solar system, one of the most remarkable properties of these extrasolar planets is their high orbital eccentricities. These high orbital eccentricities are probably not significantly affected by observational selection effects (Fischer and Marcy, 1992). Thus, if we assume that planets initially have circular orbits when they are formed in a disk, there must be mechanisms that later increase the orbital eccentricity. A variety of such mechanisms have been proposed (Tremaine and Zakamska, 2004). Of particular importance is the Kozai mechanism, a secular interaction between a planet and a wide binary companion in a hierarchical triple system with high relative inclination (Kozai, 1962; Holman et al., 1997; Ford et al., 2000). When the relative inclination angle i_0 between the orbital planes is greater than the critical angle $i_{crit} = 39.2°$ and the semimajor-axes ratio is sufficiently large (to be in a small-perturbation regime), long-term, cyclic angular momentum exchange occurs between the planet and the distant companion, and long-period oscillations of the eccentricity and relative inclination ensue. To lowest order, the maximum of the eccentricity oscillation ($e_{1,max}$) is given by a simple analytic expression:

$$e_{max} \simeq \sqrt{1 - (5/3)\cos^2 i_0} \tag{1}$$

---

[1] For an up-to-date catalog of extrasolar planets, see exoplanets.org or www.obspm.fr/encycl/encycl.html.

(Innanen et al., 1997; Holman et al., 1997). Note that e_{max} depends just on i_0. Other orbital parameters, such as masses and semimajor axes of the planet and the companion, affect only the period of the Kozai cycles. Thus, a binary companion as small as a brown dwarf or even another Jupiter-size planet can in principle cause a significant eccentricity oscillation of the inner planet.

Our motivation in this study is to investigate the possible global effects of the Kozai mechanism on extrasolar planets, and its potential to reproduce the unique distribution of observed eccentricities. In practice, we run Monte Carlo simulations of hierarchical triple systems. We have tested many different plausible models and broadly explored the parameter space of such triple systems.

2. Methods and assumptions

The purpose of our study is to simulate the orbits of hierarchical triple systems and calculate the probability distribution of final eccentricities reached by the planet. For each model, 5000 sample hierarchical triple systems are generated, with initial orbital parameters based on various empirically and theoretically motivated distributions, described below. Our sample systems consist of a solar-type host star, a Jupiter-mass planet, and a distant F-, G- or K-type main-sequence dwarf (FGK dwarf) or brown dwarf companion. The possibility of another giant planet being the distant companion is excluded since it would likely be nearly coplanar with the inner planet, leading to very small eccentricity perturbations.

The initial orbital parameters of the triple systems are randomly generated using the model distributions described in Table 1. In this paper, we present six models, each with different initial conditions that are listed in Table 2.

For the calculation of the eccentricity oscillations, we integrated the octupole-order secular perturbation equations (OSPE) derived in Ford et al. (2000). These equations also include GR precession effects, which can suppress Kozai oscillations. As noted by Holman et al. (1997) and Ford et al. (2000), when the ratio of the Kozai period (P_{KOZ}), to the GR precession period (P_{GR}) exceeds unity, the Newtonian

Table 2 Initial conditions of the triple systems

Model	$a_{2,FGK}$ (AU)	$a_{2,BD}$[a] (AU)	e_2^b	BDs[c]
A	Using P_2, < 2000	100–2000	10^{-5}–0.99	5%
B	Using P_2, < 2000	100–2000	10^{-5}–0.99	10%
C	Using P_2, < 2000	100–2000	10^{-5}–0.99	20%
D	Using P_2, < 2000	100–2000	10^{-5}–0.99	30%
E	–	10–2000	0.75–0.99	100%
F	–	10–2000	0.75–0.99	5%

[a] Uniform in logarithm
[b] All from thermal distribution, $P(e_2) = 2e_2$
[c] The fraction of brown dwarfs in 5000 samples

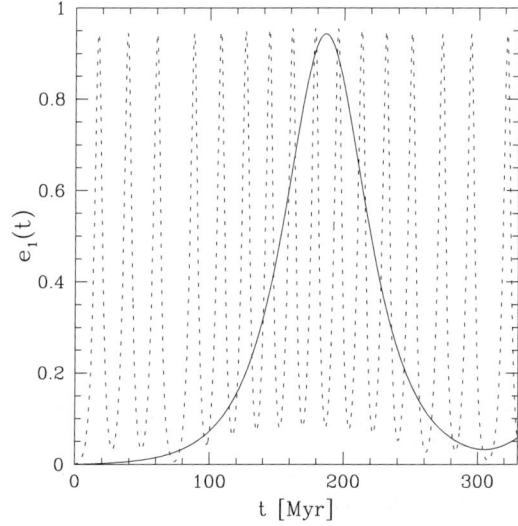

Fig. 1 Eccentricity oscillation of a planet caused by a distant brown dwarf companion ($M = 0.08 M_\odot$, solid line) and by a main-sequence dwarf companion ($M = 0.9 M_\odot$, dotted line). For both cases, the mass of the planet host star $m_0 = 1 M_\odot$, the planet mass $m_1 = 1 M_J$, the planet semimajor axis $a_1 = 2.5$ AU, the semimajor axis of the companion $a_2 = 750$ AU, the initial eccentricity of the companion $e_2 = 0.8$, and the initial relative inclination $i_0 = 75°$. Note that $e_{1,max}$ is the same in both cases, as it is dependent only on i_0, but the smaller mass of a brown dwarf companion results in a much longer oscillation period P_{KOZ}

secular perturbations are suppressed, and the inner planet does not experience significant oscillation.

Figure 1 shows typical eccentricity oscillations in two different triple systems. One contains a distant brown dwarf companion and the other a solar-mass stellar companion.

Table 1 Model parameter distributions

Parameter	Model distribution function	Ref.
Host-star mass m_0 ($M_\odot$)	Uniform in 0.9–1.3 $M_\odot$	
Planet mass m_1 (M_{Jup})	Uniform in $\log m_1$, 0.3–10M_{Jup}	[1]
Secondary mass m_2 ($M_\odot$)	$\xi(q \equiv m_2/m_1) \sim \exp\{\frac{-(q-0.23)^2}{0.35}\}$	[2]
Semimajor axis a_1 (AU) of planet	Uniform in $\log a_1$, 0.1–10 AU	[1, 3]
Binary period P_2 (days)	$f(\log P_2) \sim \exp\{\frac{-(\log P_2 - 4.8)^2}{10.6}\}$	[2]
Eccentricity of planet e_1	10^{-5}	
Age of the system τ_0	Uniform in 1–10 Gyr	[4]

[1] Zucker and Mazeh (2002),
[2] Duquennoy and Mayor (1991), [3] Ida and Lin (2004),
[4] Donahue (1998)

The two systems have the same initial orbital inclination ($i_0 = 75°$), and we see clearly that the amplitude of the eccentricity oscillation is about the same but with a much longer period P_{KOZ} for the lower mass companion.

To find the final orbital eccentricity distribution, each planetary orbit in our systems is integrated up to the assumed age of the system (τ_0), and then the final eccentricity (e_f) is recorded. The results for representative models are compared to the observed eccentricity distribution in §3. For more details, see Takeda and Rasio (2005).

3. Results for the eccentricity distribution

Figure 2 shows the final eccentricity in various models. Each model is compared with the distribution derived from all the observed single planets with $a_1 > 0.1$, from the California & Carnegie Planet Search Catalogue. The statistics of the final eccentricities for our models and for the observed sample are presented in Table 3. Models A–D represent planets in hierarchical triple systems with orbital parameters that are broadly compatible with current observational data and constraints on stellar and substellar binary companions. All the models produce a large excess of planets with $e_f < 0.1$ (more than 50%), compared to only 19% in the observed sample – excluding multiple-planet systems. An excess of planets which remained in low-eccentricity orbits was evident in most of the models we tested. Changing the binary parameters, such

Table 3 Statistics of eccentricity distributions

Model	Mean	First quartile	Median	Third quartile
Observed	0.319	0.150	0.310	0.430
A	0.213	0.000	0.087	0.348
B	0.215	0.000	0.091	0.341
C	0.201	0.000	0.070	0.322
D	0.203	0.000	0.066	0.327
E	0.245	0.000	0.141	0.416
F	0.341	0.071	0.270	0.559

as separations or frequency of brown dwarf companions, did not change this result.

A major difference between most of the simulated and observed eccentricity distributions in the low-eccentricity regime ($e < 0.1$) mainly arises from a large population of binary companions with low orbital inclination angle. For an isotropic distribution of i_0, about 23% of the systems have $i_0 < i_{crit}$, leading to negligible eccentricity evolution. For completeness, biased distributions of i_0 and e_2 are tested in model E and F, as an attempt to achieve the best possible agreement with the observations. With all the binary companions having sufficient inclination angles, model F shows a better agreement with the observed sample in the low-eccentricity regime. However, the number of planets remaining in nearly circular orbit ($e < 0.1$) is still larger than in the observed sample. Moreover, model F produces the largest excess of planets at very high eccentricities ($e > 0.6$). Note that these extreme models are clearly artificial, and our aim here is merely to quantify how large a bias would be needed to match the observations "at any cost."

4. Summary and discussion

In most of our simulations, too many planets remain at very low orbital eccentricities. The fraction of planets with $e < 0.1$ is at least 25% in our models, but only ~15% in the observed sample. There are several reasons for this overabundance of low eccentricities in our models. First, the assumption of an isotropic distribution of i_0 automatically implies that 23% of the systems have $i_0 < i_{crit}$, resulting in no Kozai oscillation. This fraction already exceeds the observed fraction of nearly circular orbits ($e_1 < 0.1$) which is ~15%. Systems with sufficient initial relative inclination angles still need to overcome other hurdles to achieve highly eccentric orbits. If many of the binary companions are substellar or in very wide orbits, Kozai periods become so long that the eccentricity oscillations are either suppressed by GR precession, or not completed within the age of the system (or both). This can result in an additional 15–40% of planets remaining in nearly circular orbits. Even when the orbits of the planets do undergo eccentricity oscillations, there remains

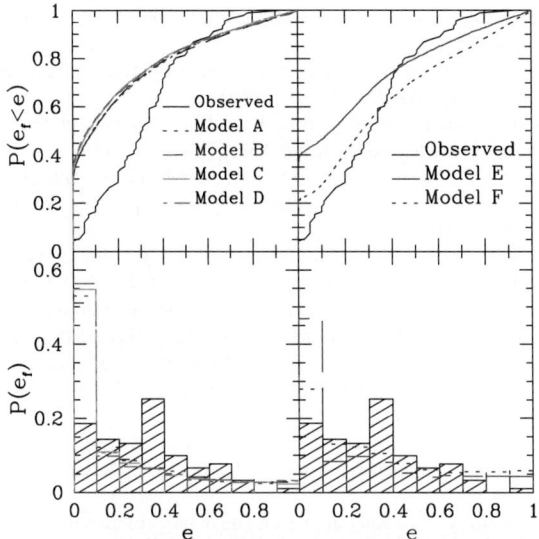

Fig. 2 Final cumulative eccentricity distributions (*top*) and normalized probability distributions in histogram (*bottom*). Four models with different fractions of brown dwarf and stellar companions. Initial inclinations (i_0) are distributed uniformly in $\cos i_0$ (*left*). Two extreme models where all the binary companions have orbits inclined by more than 40° (*right*)

still 8–14% that simply happen to be observed at low eccentricities. Thus, our results suggest that the observed sample has a remarkably small population of planets in nearly circular orbits, and other dynamical processes must clearly be invoked to perturb their orbits. Among the most likely mechanisms is planet–planet scattering in multi-planet systems, which can easily perturb eccentricities to modest values in the intermediate range ∼0.2–0.6 (Rasio and Ford, 1996; Weidenschilling and Marzari, 1996; Marzari and Weidenschilling, 2002). Clear evidence that planet–planet scattering must have occurred in the υ Andromedae system has been presented by Ford (2005). Even in most of the systems in which only one giant planet has been detected so far, the second planet could have been ejected as a result of the scattering, or it could have been retained in a much wider, eccentric orbit, making it hard to detect by Doppler spectroscopy.

In the high-eccentricity region, where $e_1 \gtrsim 0.6$, our models show much better agreement with the observed distribution. The Kozai mechanism can even produce a small excess of systems at the highest eccentricities ($e_1 > 0.7$), although it should be noted that the observed eccentricity distribution in this range is not yet well constrained. It is evident that the observed planets are rather abundant in intermediate values of eccentricity. The Kozai mechanism tends to populate somewhat higher eccentricities, since during the eccentricity oscillation planets spend more time around $e_{1,\max}$ than at intermediate values. However, this slight excess of highly eccentric orbits could easily be eliminated by invoking various circularization processes. For example, some residual gas may be present in the system, leading to circularization by gas drag (Adams and Laughlin, 2003). In another scenario, decreased periastron distances can consequently remove the orbital energy of the planet by tidal dissipation. This mechanism, referred to as "Kozai migration", was proposed by Wu and Murray (2003) to explain the orbit of HD80606 b. Kozai migration can also circularize the planetary orbit. It is worth to note that the only three massive hot Jupiters in the observed sample, τBoo b, GJ86 b and HD195019 b ($M \sin i > 2M_{\text{Jup}}$, $P < 40$ days) are all in wide binary systems (Zucker and Mazeh, 2002). Their tight orbits with low eccentricity can be a consequence of wider orbits with small periastron distances, initially invoked by the Kozai mechanism.

Clearly, even by stretching our assumptions, it is not possible to explain the observed eccentricity distribution of extrasolar planets solely by invoking the presence of binary companions, even if these companions are largely undetected or unconstrained by observations. However, our models suggest that Kozai-type perturbations could play an important role in shaping the eccentricity distribution of extrasolar planets, especially at the high end. In addition, they predict what the eccentricity distribution for planets observed around stars in wide binary systems should be. The frequency of planets in binary systems is still very uncertain, but the search for new wide binaries among exoplanet host stars has been quite successful in the past few years (e.g., Mugrauer et al., 2005).

Acknowledgements We thank Eric B. Ford for many useful discussions. This work was supported by NSF grants AST-0206182 and AST-0507727.

References

Adams, F.C., Laughlin, G.: Migration and dynamical relaxation in crowded systems of giant planets. Icarus **163**, 290–306 (2003)

Donahue, R.A.: Stellar Ages Using the Chromospheric Activity of Field Binary Stars. In: ASP Conf. Ser. 154: Cool Stars, Stellar Systems, and the Sun. p. 1235 (1998)

Duquennoy, A., Mayor, M.: Multiplicity among solar-type stars in the solar neighbourhood. II – Distribution of the orbital elements in an unbiased sample. A&A **248**, 485–524 (1991)

Eggenberger, A., Udry, S., Mayor, M.: Statistical properties of exoplanets. III. Planet properties and stellar multiplicity. A&A **417**, 353–360 (2004)

Fischer, D.A., Marcy, G.W.: Multiplicity among M dwarfs. ApJ **396**, 178–194 (1992)

Ford, E.B., Kozinsky, B., Rasio, F.A.: Secular evolution of hierarchical triple star systems. ApJ **535**, 385–401 (2000)

Holman, M., Touma, J., Tremaine, S.: Chaotic variations in the eccentricity of the planet orbiting 16 CYG B. Nature **386**, 254–256 (1997)

Ida, S., Lin, D.N.C.: Toward a deterministic model of planetary formation. I. A desert in the mass and semimajor axis distributions of extrasolar planets. ApJ **604**, 388–413 (2004)

Innanen, K.A., Zheng, J.Q., Mikkola, S., Valtonen, M.J.: The Kozai mechanism and the stability of planetary orbits in binary star systems. AJ **113**, p. 1915 (1997)

Kozai, Y.: Secular perturbations of asteroids with high inclination and eccentricity. AJ **67**, 591–598 (1962)

Marzari, F, Weidenschilling, S.J.: Eccentric extrasolar planets: The jumping jupiter model. Icarus **156**, 570–579 (2002)

Mugrauer, M., Neuhäuser, R., Seifahrt, A., Mazeh, T., Guenther E.: Four new wide binaries among exoplanet host stars. A&A **440**, 1051–1060 (2005)

Rasio, F.A, Ford, E.B.: Dynamical instabilities and the formation of extrasolar planetary systems. Science **274**, 954–956 (1996)

Takeda, G, Rasio F.A.: High orbital eccentricities of extrasolar planets induced by the Kozai mechanism. ApJ **627**, 1001–1010 (2005)

Tremaine, S, Zakamska N.L.: Extrasolar Planet Orbits and Eccentricities. In: AIP Conf. Proc. 713: the Search for Other Worlds, pp. 243–252 (2004)

Weidenschilling, S.J., Marzari, F.: Gravitational scattering as a possible origin for giant planets at small stellar distances. Nature **384**, 619–621 (1996)

Wu, Y., Murray, N.: Planet migration and binary companions: the case of HD 80606b. ApJ **589**, 605–614 (2003)

Zucker, S., Mazeh, T.: On the mass-period correlation of the extrasolar planets. ApJ **568**, L113–L116 (2002)

A Three Body Solution for the DI Her System

Keith Hsuan · Rosemary A. Mardling

Received: 1 November 2005 / Accepted: 1 February 2006
© Springer Science + Business Media B.V. 2006

Abstract The DI Herculis system has been extensively studied over the past few decades because its observed rate of apsidal advance is less than a quarter of that which is expected from its physical and orbital properties. Work by Khaliullin et al. (1991) proposed that this slow rate of apsidal advance is a result of the presence of a third (stellar mass) body orbiting the system, however, observations by Guinan et al. (1994) severely restrict the orbital properties of such a solution. We show that a planetary mass object in a highly inclined orbit relative to the binary is capable of producing the observed apsidal motion, while remaining within the bounds of the most recent set of observations. A wide range of stable solutions are possible.

Keywords Planetary systems · Stars: binaries: eclipsing · Celestial mechanics

1. Introduction

The DI Herculis system is an 8th magnitude eclipsing binary composed of two B-type stars with masses $4.52 M_\odot$ and $5.15 M_\odot$ in a tight eccentric orbit ($a = 0.496 AU$, $e = 0.2$) (Guinan and Maloney, 1985). The main sequence radii of B-type stars are relatively small ($R/R_\odot \simeq (M/M_\odot)^{0.8}$) so that while close enough to produce significant relativistic apsidal advance ($\dot\varpi_{\rm rel}^{\rm th} = 2.34°$ per century), binaries containing such stars are nonetheless uncircularized (their relative youth naturally contributes to this as well). The apsidal advance produced by the spin-induced quadrupole moments of the stars, assuming they are aligned with the orbit nor-

K. Hsuan (✉) · R. A. Mardling
Centre for Stellar and Planetary Astrophysics, Monash University, Clayton, Australia

mal, is $\dot\varpi_{\rm quad}^{\rm th} = 1.93°$ per century, giving a total apsidal motion of $\dot\varpi_{\rm tot}^{\rm th} = 4.27°$ per century. The observed rate is $\dot\varpi_{\rm tot}^{\rm obs} = 1.04°$ per century (Guinan et al., 1994): this discrepancy is the subject of the present paper.

A number of solutions have been proposed to account for this including new theories of gravity (Guinan and Maloney, 1985; Moffat, 1984), misaligned stellar spins (Company et al., 1988; Reisenberger and Guinan, 1989) and rapid circularisation of the binary orbit (Guinan and Maloney, 1985). Observations by Guinan and Maloney (1989) indicate that the rotational axes of the stars are most likely parallel, while rapid circularization requires unrealistic Q-values (Guinan and Maloney, 1985).

In this paper we propose that the anomolous apsidal advance is a result of the presence of a low-mass body, possibly a planet, in a near-polar orbit around the binary. While the third-body hypothesis is not new (see Guinan and Maloney, 1985; Khaliullin et al., 1989), previous studies have involved a stellar mass body whose presence would be detectable via its influence on the inclination of the massive binary to our line of site. Observations over a thirty year period constrain the rate of change of inclination to be no more than $0.0002°$ yr^{-1}.

2. Apsidal advance

In Newtonian dynamics, the orientation of a binary pair consisting of two *spherically symmetric* (or point mass) objects will remain fixed in space (there is no preferred direction except that defining the initial orientation), and this is expressed mathematically via conservation of the Runge-Lenz vector (see, for example, Goldstein, 1959). In general relativity, however, the finite speed of light breaks this symmetry and introduces a length scale so that the orbit rotates in the

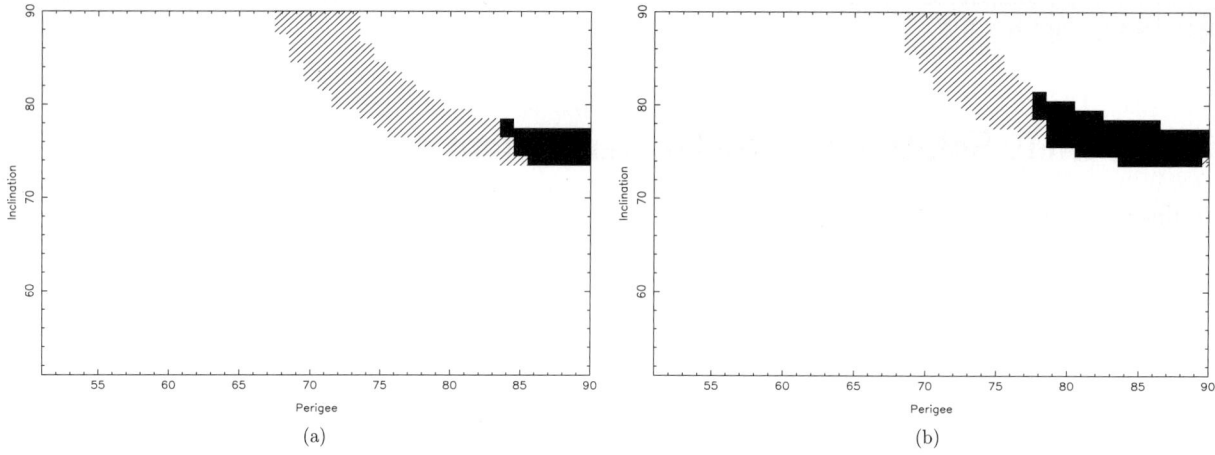

Fig. 1 Parameter space plots showing orbital orientations which cause the observed apsidal motion; (a) $1 M_{\rm JUP}$, $a = 0.80$ AU, (b) $2 M_{\rm JUP}$, $a = 1.00$ AU, (c) $5 M_{\rm JUP}$, $a = 1.40$ AU, (d) $10 M_{\rm JUP}$, $a = 1.80$ AU

orbital plane at a rate proportional to $(v/c)^2$, where v is the orbital speed and c is the speed of light. Similarly, if one or both of the bodies rotates, producing equatorial bulges and hence spherical asymmetry, the line of apsides rotates in response, as it does if the bodies are close enough to raise tides. The presence of a third body will also break the symmetry of a point mass pair, producing apsidal advance in a direction which depends on the inclination of the third body's orbit to that of the binary.

The contributions to the apsidal advance from these effects depend on various parameters in the system such as (in the case of spin and tidal bulges) the ratio of the semimajor axis to the radius of the star(s), or in the case of a third body, the ratio of its semimajor axis to the binary orbit. Mass ratios also naturally play a role as do the apsidal motion constants of the bodies. Expressions for these may be found in Mardling and Lin (2002) (in which is described the numerical code used in this study) and Innanen et al. (1997).

3. Three-body dynamics

Not only does the presence of a third body produce apsidal motion in a binary, if its orbit is not coplanar it will produce a change in the inclination of the binary to our line of sight. This constrains the possible mass, eccentricity and semimajor axis of such a body, as well as the orientation of its orbit with respect to that of the binary. In the absence of other effects (ie, for the point-mass case), highly inclined orbits produce large variations in the eccentricity and inclination (the so-called Kozai effect: Innanen et al., 1997). However the amplitude of these variations can be severely moderated when other effects such as stellar quadrupole moments are present, and this is certainly the case for DI Her. For certain ranges of orbital orientation the apsidal motion due to the presence of a third body is in the opposite direction to that produced

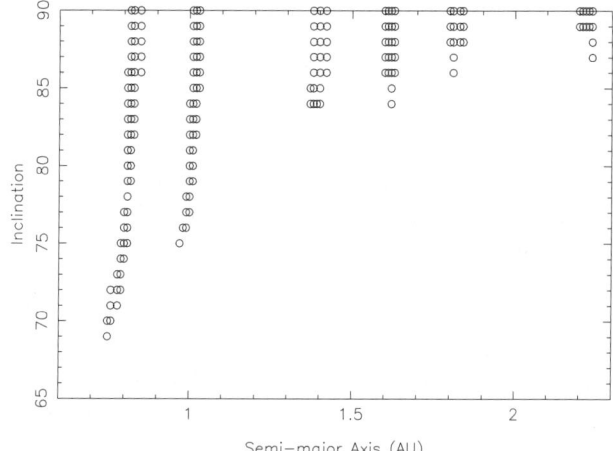

Fig. 2 Summary of inclinations and semi-major axes for third bodies which meet both apsidal motion and di/dt criteria. The points to the far left correspond to $1 M_{\rm JUP}$, then $2 M_{\rm JUP}$, $5 M_{\rm JUP}$, $7 M_{\rm JUP}$, $10 M_{\rm JUP}$ and $20 M_{\rm JUP}$ to the far right

by (aligned) stellar spin and the relativistic potentials of the stars. Thus the problem becomes determining suitable ranges of masses and orbital parameters for a hypothetical third body which *retard* the apsidal advance at the same time as limiting the rate of change of inclination of the binary to our line of sight. The following section describes a numerical study of the problem; a mathematical analysis of the problem will be presented elsewhere.

4. Results

Using the numerical integration code described in Mardling and Lin (2002), we performed a parameter space search for orbits which satisfy the current observations; that is, the body must cause an overall apsidal motion of $1.04 \pm 0.12°$ per century, while simultaneously causing the inclination of the

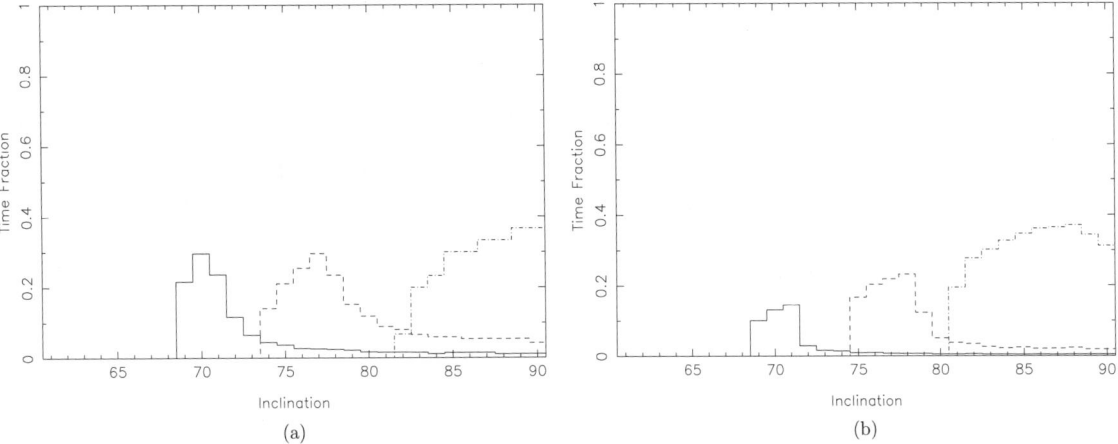

Fig. 3 Time fraction against inclination for (a) $2\,M_{JUP}$ at $1.00\,AU$ and (b) $5\,M_{JUP}$ at $1.40\,AU$

observed binary to vary by less than $0.006°$ in a 35 year span. The parameter space search is performed across the orbital inclination of the third body, and the inital value of the line of apsides. Each search is performed for a series of third body masses between $1 - 20 M_{JUP}$, and for a range of semi-major axes.

The results for two such parameter space searchs are presented in Fig. 1. In each case, the unshaded region in the upper right-hand corner of the graph represents third body orbits which have too large an effect on the binary's apsidal motion, while the area below represents orbits which have too little effect on the binary, creating an apsidal motion higher than that observed. The region of solid shading represents orbital configurations which meet both the apsidal motion and di/dt criteria, while the cross hatched area represent configurations which only meet the apsial motion criterion.

From the graphs in Fig. 1, we find that the initial value of the line of apsides of the third body has a significant effect on the apsidal motion induced in the binary orbit; lower inclinations can cause the required effect on the binary's apsidal motion, provided that the initial line of apsides is sufficiently high. As all simulations began with the stars of the inner binary aligned along the line of sight, such a relationship indicates that the largest effect occurs when the third body has its closest approach to the binary between the component stars, rather than when it is closer to one star than the other.

Figure 2 shows the inclinations and semi-major axes of third body orbits in which both conditions have been met for the range of masses studied; 1, 2, 5, 7, 10 and $20 M_{JUP}$. As the mass of the third body increases, we see that its orbit must increase in size to satisfy the criteria; however, the range of inclinations possible decreases markedly, such that the plane of the third body orbit must be perpendicular to that of the binary. For low mass companions, there is a larger range of inclinations that can cause the observed effect. While the apparent trend in the Fig. 2 indicates that masses lower than $1 M_{JUP}$ will allow for even lower inclination orbits, such orbits would be exceptionally close to the binary, and would likely be unstable. A $1 M_{JUP}$ third body is likely to be the lower limit for a companion for this reason, as the ratio between the semi-major axes of the inner and outer orbit would be 1:4.

We find that the possible range of orbits which obey both the apsidal motion and di/dt requirements to be quite narrow; however, we can potentially expand this range by requiring that the orbits need only to cause a $di < 0.006°$ over any 35 year period, corresponding with the period of time over which we have reliable observations of the DI Herculis system. As the change in inclination of the binary orbit is sinusoidal, it is possible that the current observations have been made in a period of minimum inclination change. By calculating the fraction of time which the effect of the third body induces a $di < 0.006°$, we can determine a maximum range of orbital configurations which satisfy the observations. Figure 3 shows the fraction of time that the third body causes a $di < 0.006°$ in the binary for two third body masses, each orbiting at several different semi-major axes; the first is for a $2 M_{JUP}$ mass, while the second is for a $5 M_{JUP}$ body. The abcissa in each graph represents the initial inclination of the third body orbit, while the ordinate shows the fraction of time that such a body causes a $di < 0.006°$.

The two graphs in Fig. 3 show similar features; larger orbits will shift the required inclination for the correct apsidal motion towards $90°$, and that within each set of results, lower inclinations allow for a larger time fraction below the di/dt threshold. In each case however, we find there is a region where the time fraction is considerable, providing a reasonable window in which the current observations can be made.

5. Discussion and conclusions

The results presented here are a relatively limited set of the total number of three body solutions available which can explain the apsidal motion of DI Herculis, while still causing a small enough change in the inclination to remain undetected. It is possible to show that a range of possible orbital configurations exist for third bodies ranging in mass from $1 M_{\rm JUP}$ to $10 M_{\rm JUP}$. For larger masses, the parameter space in which both these conditions are met are greatly reduced, requiring the third body's orbit to be perpendicular to that of the inner binary.

Further observation of this system will be useful to extend the time over which reliable eclipse timings are available; the most recent observations were presented some 15 years ago. These new measurements will extend the historical observation time to 50 years, and can be used to establish the nature of any third body present. A more accurate value of di/dt will aid in either supporting the presented results, or further constraining parameters of a third body.

References

Claret, A.: Some notes on the relativistic apsidal motion of DI Herculis. A&A **330**, 533–540 (1998)

Company, R., Portilla, M., Gimenez A.: On the apsidal motion of DI Herculis. ApJ **335**, 962–964 (1988)

Eggleton, P.P., Kiseleva, L.G., Hut, P.: The equilibrium tide model for tidal friction. ApJ **499**, 853–870 (1998)

Goldstein, H.: Classical Mechanics. Addison-Wesley Publishing Company, Inc (1959)

Guinan, E.F., Maloney F.P.: The apsidal motion of the eccentric eclipsing binary DI Herculis – an apparent discrepancy with general relativity. AJ **90**, 1519–1528 (1985)

Guinan, E.F., Marshall, J.J., Maloney, F.P.: A new apsidal motion determination for DI Herculis'. IBVS **4101**, 1–+ (1994)

Hut, P.: Tidal evolution in close binary systems. A&A **99**, 126–140 (1981)

Innanen, K.A., Zheng, J.Q., Mikkola S., Valtonen, M.J.: The Kozai mechanism and the stability of planetary orbits in binary star systems. AJ **113**, 1915–+ (1997)

Mardling, R.A., Lin, D.N.C.: Calculating the tidal, spin, and dynamical evolution of extrasolar planetary systems. ApJ **573**, 829–844 (2002)

Moffat, J.W.: The orbital motion of DI Herculis as a test of a theory of gravitation. ApJL **287**, L77–L79 (1984)

Murray, C.D., Dermott, S.F.: Solar system dynamics. Cambridge University Press (1999)

Reisenberger, M.P., Guinan, E.F.: A possible rescue of general relativity in DI Herculis. AJ **97**, 216–221 (1989)

Common Proper Motion Search for Faint Companions Around Early-Type Field Stars – Progress Report

Valentin D. Ivanov · G. Chauvin · C. Foellmi ·
M. Hartung · N. Huélamo · C. Melo · D. Nürnberger ·
M. Sterzik

Received: 20 June 2006 / Accepted: 3 July 2006
© Springer Science + Business Media B.V. 2006

Abstract The multiplicity of early-type stars is still not well established. The derived binary fraction is different for individual star forming regions, suggesting a connection with the age and the environment conditions. The few studies that have investigated this connection do not provide conclusive results. To fill in this gap, we started the first detailed adaptive-optic-assisted imaging survey of early-type field stars to derive their multiplicity in a homogeneous way. The sample has been extracted from the Hipparcos Catalog and consists of 341 BA-type stars within ∼300 pc from the Sun. We report the current status of the survey and describe a Monte-Carlo simulation that estimates the completeness of our companion detection.

Keywords Stars: binaries:general · Stars:binaries: visual · Stars: early-type

1. Introduction

The multiplicity of pre-main sequence (PMS) and main sequence (MS) late-type stars have been extensively studied: Duquennoy and Mayor (1991), Reipurth and Zinnecker (1993), Prosser et al. (1994), Brandner et al. (1996). PMS late-type stars in low-density clouds like Taurus show higher binary fractions than PMS late-type stars formed in massive and dense star forming regions (SFR) like Orion. This difference was explained with an environmental dependence of the binary fraction: low-mass stars born in dense SFRs show a higher probability of dynamic interactions with more massive members, so that they are ejected resulting in a low binary fraction, as shown by Sterzik and Durisen (1995). On the other hand, stars in low-density SFRs undergo low rate of stellar encounters, resulting in a higher binary fraction.

It is unclear if this result remains valid for early-type stars, mostly because of the difficult detection of faint companions near bright O- and B-type stars. The works of Petr et al. (1998), Preibisch et al. (1999) and Shatsky and Tokovinin (2002) indicate that the binary fraction of early-type stars varies from one SFR to another but current statistical basis is still not solid enough. The age of the SFRs can also affect the binary statistics: older systems are expected to have undergone through more dynamical interaction, reducing their binarity fraction in comparison with the younger ones, as point by Tokovinin et al. (1999).

2. The survey

The cluster multiplicity studies can cover only a limited range of density and age. This prompted us to estimate the binarity fraction of a representative, volume-limited sample of early-type field stars. We designed a survey able to detect at ∼10σ level an M4-type companion at the mean distance of our sample (∼200 pc) down to 0.4 arcsec separation from 100 Myr old A-type primary. The companions around B-type stars will be younger (∼10 Myr), brighter, and easier to detect. Last but not least, the physical nature of the candidate components is verified by their common proper motion. Our goal is to compare the properties such as incidence and mass ratio of the multiple stars in the field and in different star-forming regions.

V. D. Ivanov (✉) · G. Chauvin · C. Foellmi · M. Hartung ·
N. Huélamo · C. Melo · D. Nürnberger · M. Sterzik
European Southern Observatory, Ave. Alonso de Cordova 3107,
Casilla 19, Santiago 19001, Chile

3. Sample selection

The sample stars were selected according to the following criteria:

- Apparent color $B - V \leq 0.2$ mag – conservative criterion met by all unreddened BA stars, and a few contaminating later-type stars. The spectral types for all stars were verified to be BA.
- The sample contains field stars only. Members of the OB-associations listed in de Zeeuw et al. (1999) were excluded.
- Distance $D \leq 300$ pc from the Sun (HIPPARCOS). At $D = 300$ pc the telescope's diffraction limit of 0.07 arcsec corresponds to 21 A.U (Shatsky and Tokovinin (2002) probed separations 45–900 A.U).
- Proper motions ≥ 27 mas/yr allowing to confirm/reject physical companion candidates taking two epoch separated by 1–2 yr.
- The apparent $V = 5 - 6$ mag, so the targets are suitable self-references for NACO even under poor weather conditions.
- The stars have DEC ≤ 0 deg, i.e. visible from the VLT.

The final sample contains 341 field B- and A-type stars. The average distance is 114 pc, the median distance is 104 pc.

4. Observations and current status

The observations were carried out with NAOS–CONICA (Nasmyth Adaptive Optics System – Near-Infrared Imager and Spectrograph) at the ESO VLT over the last two years. The pixel scale was 27.03 mas px^{-1}, giving 27.7×27.7 arcsec field of view. Each target was observed at 9 different position on the detector, collecting total of ~7.5 min of integration. The data reduction includes sky subtraction, flat-fielding, aligning and combination of the images into a single frame. Next, we perform PSF fitting/subtraction and search for faint companions.

As of Sept 2005 we have observed 196 objects from our sample. We have analyzed 152 of them: 81 appear single, and 71 show companion candidates whose nature will be tested with the second epoch observations with ≥ 2 yr baseline.

5. Analysis: Modeling the survey

To estimate the sensitivity and the completeness of the survey we have created a Monte-Carlo simulation that takes into account all available information for the survey stars (Fig. 1). The model input parameters are:

- The known distances, spectral types, absolute luminosities for all primaries;

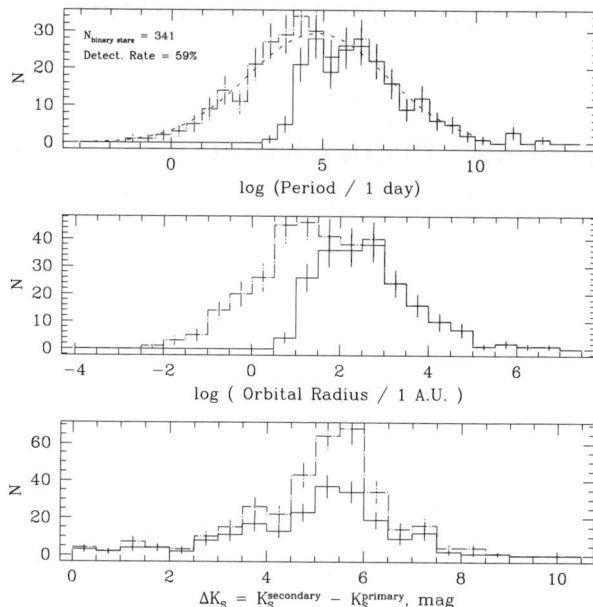

Fig. 1 Monte-Carlo simulation of the survey completeness: orbital period distribution (top panel), orbital radii distribution (middle), and primary-secondary K-band magnitude difference (bottom). The dot-dashed lines show the "true" adopted/generated distributions and the solid lines are the "observed" distributions after the observational detectability conditions have been applied. The theoretical period distribution of Duquennoy and Mayor (1991) is also shown (doted line on the top panel)

- Adopted binarity fraction;
- Secondary star mass randomly sampled from the Kroupa et al. (1993) IMF;
- Secondary star's spectral type and absolute magnitude calculated from the mass;
- Orbital periods – randomly generated from Duquennoy and Mayor (1991) distribution;
- Major axis calculated from the Kepler's low and the period;
- Random ellipticity, random orbital inclination;
- Visibility criterion based on the magnitude difference and the angular separation between the primary and the companion.

The model predicts: the distributions of periods, angular separations, magnitude differences and spectral types for the detected binaries. The simulations indicate that we will detect about 2/3 of the physical companions.

Acknowledgements We are grateful to our colleagues from the ESO-Paranal Science Operations Department who carried out these observation in Service Mode.

References

Brandner, W., Alcala, J.M., Moneti, A., Zinnecker, H.: A&A **307**, 121 (1996)

de Zeeuw, P.T., Hoogerwerf, R., de Bruijne, J.H.J., Brown, A.G.A., Blaauw, A.: AJ **117**, 354 (1999)
Duquennoy, A., Mayor, M.: A&A **248**, 485 (1991)
Kroupa, P., Tout, C.A., Gilmore, G.: MNRAS **262**, 545 (1993)
Petr, M.G., Coude Du Foresto, V., Beckwith, S.V.W., Richichi, A., McCaughren, M.J.: ApJ **500**, 825 (1998)
Preibisch, T., Balega, Y., Hofmann, K.-H., Weiglet, G., Zinnecker, H.: New Astr. **4**, 531 (1999)
Prosser, C.F., Stauffer, J.R., Hartmann, L., Soderblom, D.R., Jones, B.F., Werner, M.W., McCaughren, M.J.: ApJ **421**, 517 (1994)
Reipurth, B., Zinnecker, H.: A&A **278**, 81 (1993)
Shatsky, N., Tokovinin, A.: A&A **382**, 92 (2002)
Sterzik, M.F., Durisen, R.H.: A&A **304**, 9 (1995)
Tokovinin, A., Chalabaev, A., Shatsky, N., Beuzit, J.L.: A&A **346**, 481 (1999)

ns Sci (2006) 304:249
DOI 10.1007/s10509-006-9164-2

ORIGINAL ARTICLE

The Asiago Extrasolar Planet Transit Search

R. U. Claudi · M. Cancian · M. Barbieri · R. Gratton ·
S. Desidera · M. Montalto · G. P. Piotto · S. Scuderi

Received: 1 November 2005 / Accepted: 1 February 2006
© Springer Science + Business Media B.V. 2006

Abstract RATS is an Italian project devoted to Hot Jupiter search with the transit method. A planet transiting in front of a host star can be mimed by several, and well defined, astrophysical phenomena (Brown, 2003). In order to recognize these false alarms we can utilize a preventive strategy to limit false alarm rates and a spectroscopic follow up to refuse no transit candidates. As preventive strategy it is important to develop an accurate target field selection, with well defined requisites, in order to maximize the solar type star numbers and to minimize the risk of possible astrophysical false alarms.

Keywords Extrasolar planets · Transit method

1. Overview

RATS is a collaboration between several INAF Observatories (Padova, Catania, Napoli and Palermo), the Astronomy and Physic Departments of the Padova University and ESA in order to search for extra solar planets with the photometric transit technique and spectroscopic false alarms detection. The survey main goal is to find about 10 (20) Hot Jupiters in 5 years exploiting both the 67/92 Schmidt Telescope for photometric transit search and the 182 cm "COPERNICO" Telescope for radial velocities measurements and target characterization.

R. U. Claudi (✉) · M. Barbieri · R. Gratton · S. Desidera
INAF, Astronomical Observatory of Padova

M. Cancian · M. Montalto · G. P. Piotto
Dip. di Astronomia, Universita' di Padova

S. Scuderi
INAF, Astronomical Observatory of Catania

2. Observational constraints

The RATS strategy imposes the survey stellar magnitude range. The faintest magnitude is defined by the efficiency of the spectrograph and its radial velocity precision. The brightest magnitude will be a compromise between star counts and the possibility to observe the brighter stars without saturate the detector. In order to satisfy this requisite we defocalize the Schmidt telescope.

3. Selection of the fields and expected performances

The RATS stellar fields will be selected with the following requisites:

– Stellar Magnitude Range: $9 < V < 13(14?)$
– $\delta > 13°$: More than eight hours of visibility from $00^h46^m17.3^s$ E long, and $+45°50'36.2''$N Lat.
– $|b| < 40°$ (maximize the star counts)
– Higher star counts without Telescope PSF blend
– Higher FGK dwarf number

In Asiago we expect to be able to obtain around 50 clear nights/year resulting in few real transits against tens of spurious candidates every year.

References

Brown, T. M.: Expected detection and false alarm rates for transiting jovian planets. ApJ **593**, L125–128 (2003)

The WASP Project and SuperWASP Camera

D. Pollacco · I. Skillen · A. Cameron · D. Christian ·
J. Irwin · T. Lister · R. Street · R. West · W. Clarkson ·
N. Evans · A. Fitzsimmons · C. Haswell · C. Hellier ·
S. Hodgkin · K. Horne · B. Jones · S. Kane · F. Keenan ·
A. Norton · J. Osborne · R. Ryans · P. Wheatley

Received: 1 October 2005 / Accepted: 1 December 2005
© Springer Science + Business Media B.V. 2006

Abstract The WASP project and infrastructure supporting the SuperWASP Facility are described. As the instrument, reduction pipeline and archive system are now fully operative we expect the system to have a major impact in the discovery of bright exo-planet candidates as well in more general variable star projects.

Keywords Binaries · Eclipsing: extrasolar planets · Transits

S. Hodgkin
WFS Unit, Institute of Astronomy, Madingley Road, Cambridge, CB3 0HA, UK

I. Skillen · J. Irwin
Isaac Newton Group of Telescopes, Apartado de correos 321, E-38700 Santa Cruz de La Palma, Tenerife, Spain

N. Evans · C. Hellier · B. Jones
Astrophysics Group, School of Chemistry and Physics, Keele University, Staffs, ST5 5BG, UK

R. West · J. Osborne · P. Wheatley
Dept. of Physics and Astronomy, University of Leicester, Leicester, LE1 7RH, UK

W. Clarkson · C. Haswell · A. Norton
Dept. of Physics and Astronomy, The Open University, Walton Hall, Milton Keynes, MK7 6AA, UK

D. Pollacco (✉) · D. Christian · R. Street · A. Fitzsimmons ·
F. Keenan · R. Ryans
Dept. of Physics and Astronomy, Queen's University of Belfast, University Road, Belfast BT7 1NN, UK

A. Cameron · T. Lister · K. Horne · S. Kane
School of Physics and Astronomy, University of St Andrews, North Haugh, St Andrews, KY16 9SS, UK

1. Introduction

In recent years, interest has grown in relatively insensitive but inexpensive wide-field imaging systems, essentially composed of large CCDs mounted directly to high-quality wide-angle camera optics. The first prominent success of such an instrument was the spectacular discovery of the neutral sodium tail of comet Hale-Bopp (Cremonese et al., 1997). Since then, similar cameras have resulted in imaging of a gamma-ray burst (GRB) during the burst period (Akerlof et al., 1999), and the first detection of the transits of an extra-solar planet in front of its parent star, HD209458 (Charbonneau et al., 2000). These instruments are ideal for projects requiring photometry of bright but rare objects.

The search for extra-solar planetary systems is currently dominated by precise radial velocity work (e.g. Butler et al., 1998). These surveys have discovered Jupiter-sized objects in orbits out to 3 AU around 6% of the nearby Sun-like stars surveyed, with ∼30% parked in ∼4-day 0.05 AU orbits. About 10% of these *Hot Jupiters* in randomly inclined orbits will transit their host star. Hence, roughly 1 in every 1000 solar-type stars should exhibit short period transits. Since Charbonneau et al. (2000) pioneering observations several other groups have published transit detections, most notably candidates from the OGLE survey (Udalski et al., 2002a,b) and TrES-1 (Alonso et al., 2004). Horne (2003) lists some 23 projects either in operation or construction.

2. The WASP consortium and superWASP camera

The Wide Angle Search for Planets or WASP Consortium was established in 2000 by a group of primarily UK based astronomers. Our first instrument was WASP0, which is composed completely of commercial parts and utilised a

Fig. 1 The SuperWASP facility on the night of first light (2004 November 23)

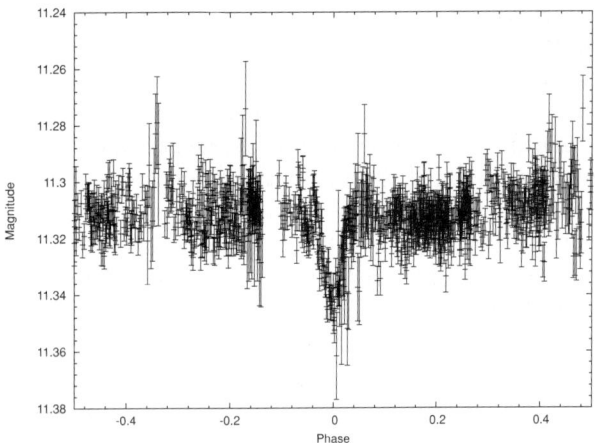

Fig. 2 An example of a low amplitude eclipsing system discovered with SuperWASP. This is composed of *many* nights of data

Nikon 200 mm f 2.8 telephoto lens coupled to an Apogee AP10 CCD detector. WASP0 was used in summer 2000 and shown able to detect the transit of HD209458 amongst other variables (Kane et al., 2004/5). On the strength of this the Consortium was able to raise sufficient funding for a more ambitious project – the multi-detector SuperWASP camera. The limited development required from this is reflected in the project timescale: funding approved in March 2002, first light achieved in November 2003 and first candidates August 2005.

2.1. The superWASP camera and data reduction system

The mounting is a traditional equatorial fork mount constructed by Optical Mechanics Inc. of Iowa, USA. The mount is easily capable of slewing at rate of 10 degrees per second. In place of a traditional telescope, multiple, short focus camera are held within a custom cradle. In keeping with other wide angle surveys use Canon 200 mm f 1.8 lenses (from ebay!). Due to the wide field nature of the instrument we use a enclosure based around a simple run-off roof design (but use the space under the rails as a computer room). Figure 1 shows the enclosure at La Palma.

The SuperWASP detectors were manufactured by Andor Technology (Belfast) and use a 5 stage peltier device to cool the e2v 2048×2048 pixel CCDs. Each has a controller card able to read the detector out with 16-bit accuracy in 4 seconds. Due to the high data rate each detector has a dedicated control computer, while the observatory itself is controlled via a further machine.

We use the USNO-B1 catalogue as the photometric input catalogue. We carry out aperture photometry at the positions of all stars in the catalogue brighter than a given limiting magnitude. This has two important advantages for subsequent data retrieval and analysis. All photometric measurements are associated with known objects from the outset, and the aperture for every object is always centred at a precisely-determined and consistent position on the CCD. After all known objects are removed we also search the image for "orphan" objects. The instrumental magnitudes are then subjected to trend removal (e.g. extinction) and various quality control procedures before being accepted in to the archive.

Internally to the archive both the raw images and the processed photometric data are held in the form of FITS files and indexed within a Relational Database Management System which allows the data relevant to a user query to be found in a fast and efficient manner.

The complete SuperWASP facility is described in greater detail by Pollacco et al. (2006, PASP in print).

3. Initial results

For the 2004 season, SuperWASP produced some 5Tb of data whilst monitoring some 2500 square degrees of sky. The data shows good stability with bright stars having rms errors of only a few millimags over a month or more. Exo-planet candidates have been identified from this dataset and are being further investigated. Needless to say the instrument is finding many previously unclassified variables (e.g. Fig. 2).

The WASP Consortium is currently upgrading Super-WASP to its full compliment of 8 cameras (field of view 481 square degrees) and constructing a clone facility at SAAO.

Acknowledgements The WASP Consortium consists of representatives from the University of Cambridge (Wide Field Astronomy Unit), Instituo de Astrofisìca de Canarias, Isaac Newton Group (La Palma), University of Keele, University of Leicester, The Open University, Queen's University Belfast, and the University of St. Andrews. The SuperWASP and WASP0 Cameras were constructed and operated with funds made available from Consortium Universities and PPARC.

References

Akerlof, C. et al.: Nature **398**, 400 (1999)
Alonso, R. et al.: ApJ **613**, L153 (2004)
Butler, R.P. et al.: PASP **110**, 1389 (1998)
Charbonneau, D. et al.: ApJ **529**, L45 (2000)
Cremonese, G. et al.: ApJ **490**, L199 (1997)
Horne, K.: In: Deming, D., Seager, S. (eds.), Scientific Frontiers in Research on Extrasolar Planets (2003)
Kane, S.R. et al.: MNRAS **353**, 689 (2004)
Kane, S.R. et al.: MNRAS **362**, 117 (2005)
Udalski, A. et al.: Acta Astron. **52**, 1 (2002a)
Udalski, A. et al.: Acta Astron. **52**, 115 (2002)

X-Ray Binaries in Nearby Galaxies

Vassiliki Kalogera

Received: 1 November 2005 / Accepted: 1 February 2006
© Springer Science + Business Media B.V. 2006

Abstract In the last few years *Chandra* observations have revolutionized the study of X-ray binaries by extending observed samples to extragalactic distances and stellar environments of varied star formation history. Here we summarize some of our results related to the interpretation of current observations of young star clusters and of elliptical galaxies.

Keywords X-rays · Binaries · Compact objects

1. Introduction

One of the most exciting and interesting developments made possible through *Chandra* observations is the abundance of extragalactic point sources, identified as X-ray binaries (XRBs). Current observed samples include XRBs identified in association with dense star clusters (young and old) and starburst, spiral, and ellitpical galaxies. Furthermore, a class of ultra-luminous X-ray sources has also been identified and are hypothesized to harbor intermediate-mass black holes. For a recent review of these developments (see Fabbiano and White, 2001). In my conference talk I presented recent results from theoretical studies of the X-ray luminosity function of an example starburst galaxy (Belczynski et al., 2004), of the ability of intermediate-mass black holes to acquire mass transferring binary companions and producing high X-ray luminosities (Blecha et al., 2005), of XRB formation in young dense star clusters (Sepinsky et al., 2005), and of the upper-end X-ray luminosity function of elliptical galaxies (Ivanova and Kalogera, 2005). In what follows we summarize the findings of the latter two studies.

V. Kalogera
Department of Physics and Astronomy, Northwestern University,
2145 Sheridan Road, Evanston, IL 60208, USA

2. X-Ray binaries formed in young star clusters

Optical and infrared observations of starburst galaxies show massive, young star-forming clusters, often referred to as super star clusters. They have been shown to range in mass from $\sim 10^4\,M_\odot$ to $\sim 10^7\,M_\odot$ (Smith and Gallagher, 2001; Harris et al., 2001), and in age from just ~ 1 Myr to ~ 50 Myr, with the majority falling at $\simeq 10-20$ Myr.

Recently, Kaaret et al. (2004); hereafter K04 have studied 3 nearby starburst galaxies (M82, NGC1569, and NGC5253) each containing a significant number of identified young star clusters and point X-ray sources. They find that XRBs brighter than $\sim 10^{36}$ erg s^{-1} are preferentially found within distances of $\sim 30-100$ pc from their nearest cluster, but there is a clear lack of XRBs found coincident with (i.e., within) the clusters. Similar results have been found in observations of the Antennae (Zezas et al., 2002).

In what follows we outline self-consistent models that address XRB formation in young clusters. We track the kinematic evolution of compact object binaries in the absence of dynamical interactions and we follow their X-ray luminosity (L_X) evolution. We focus on two specific, testable points of comparison between the published observations and our calculations: the average number of XRBs per cluster, and the median distance of XRBs from their parent cluster (or nearest cluster in the observations).

2.1. Theoretical modeling

To generate the necessary stellar populations for the modeled clusters, we use the population synthesis program *StarTrack* (developed by Belczynski et al., 2002, Belczynski et al., 2005). We generate and evolve a population of binaries under a given set of conditions, such as the initial mass function (IMF), supernova kick distribution, common envelope efficiency, etc. With the resultant evolutionary parameters of the binaries at the time of neutron star or black hole formation, we place them in a cluster potential and track their motion and X-ray luminosity as a function of time. We do not account for any stellar interactions in these young clusters, as our goal is to examine whether supernova kicks alone can account for the observed spatial distribution of XRBs with respect to their parent clusters. This may not be a well justified assumption in general. Nevertheless the clusters relevant to our study are very young (few Myrs to $\sim$10–20 Myr), so significant cluster evolution is not expected, except for possibly in the most massive and most compact (small half-mass radius) clusters.

For the orbital evolution we use the calculated post-core-collapse systemic velocities of XRB progenitors and combine them consistently with a Plummer model for the gravitational potential of the model cluster with a given half-mass radius. To determine the spatial distribution of the XRB progenitors right after compact-object formation, we assume that the number density of stars is proportional to the mass density. Initial systemic velocities consistent with the Plummer potential are also generated. We then apply the calculated post-core-collapse systemic velocities randomly oriented with respect to the initial cluster velocities. We follow the motion of the binary as a function of time and correlate position with the X-ray luminosity evolution calculated with the binary evolution code.

To determine the number of binaries to evolve for a given cluster mass, we express the total mass and total number as a function of the IMF and mass-ratio-distribution parameters and the binary fraction and derive a relation between the two (turns out to be a proportionality). We adopt a flat mass-ratio distribution and a binary fraction equal to unity in order to represent an upper limit on the number of binaries, and therefore on the total number of X-ray sources. We consider cluster masses in the range $10^4 M_\odot$–$10^6 M_\odot$, with IMF indices of 2.35 and 2.7. It is interesting to note that although orbits are calculated typically for few to several hundreds of Myr, binaries are X-ray sources for only a small part of the orbits and they are *bright* (i.e., Lx $\geq 5 \times 10^{35}$ erg s^{-1}) X-ray sources for even smaller parts.

We note that statistical effects play a significant role, especially in the low mass ($\sim 10^4 M_\odot$) clusters. Typically, no more than one XRB is bright enough to be seen in these clusters and the position can vary significantly across the cluster for each different simulation. Therefore we consider a large number of Monte Carlo realizations for each parameter set (cluster mass, half-mass radius, and IMF index). The lower mass clusters have the smallest number of initial binaries and hence require the most realizations. We chose the number of realizations for each cluster mass, so that our results averaged over the many realizations remained unaffected at the 1% level.

2.2. Results & discussion

Our calculations have yielded a wealth of results. Specifically, we focus on the statistical averages of the two quantites quoted observationally in K04: the median distance of XRBs from the nearest (parent in the models) cluster, and the mean number of bright XRBs per cluster (within 1000 pc).

In Fig. 1 we plot the model average number of XRBs per cluster as a function of distance from the parent clusters, for a variety of cluster ages and for two sets of clusters with mass of $5 \times 10^4 M_\odot$ (left panel) and $5 \times 10^5 M_\odot$ (right panel). It is evident that the XRB spatial distributions have a dramatic time dependence. For "young" clusters, the average XRB number per cluster rises to a maximum rapidly and very few XRBs are found at large distances. This is primarily because even the unbound XRBs have not had enough time to move away from their parent clusters. It is also evident that at certain ages in which XRBs are distinctly more numerous than at others. This peak age is dependent on the L_X cut-off. Fully exploring these dependencies could allow us to derive general conclusions about XRB populations dependent only on the averages ages of the young cluster population. The behavior is similar for all masses (we also examined $5 \times 10^6 M_\odot$ clusters), except that the average number of XRBs at a given radius scales with the mass of the cluster almost linerally. This is due to the direct relationship between the number of binary systems run and the total mass of the cluster.

Mean number of XRBs per cluster: We find the theoretical mean XRB number per cluster to vary significantly from $\sim$0.1 to $\sim$10, depending on the cluster mass. Therefore, it is possible to reproduce the results derived in K04 by taking contributions from a number of clusters of different masses. Two of the three clusters discussed in K04 (M82 and NGC 5253) have a mean number of observed XRBs of $\sim$1 per cluster, while a third (NGC 1569) seems to have an very small number of XRBs, with only $\sim$0.25 per cluster. This difference would point towards NGC 1569 having on average smaller-mass clusters, even though outliers at high masses can still exist. A difficulty in the comparison arises because the properties of the clusters in these galaxies are very difficult to determine orbservationally. Those with measured masses are skewed to higher masses ($> 1 \times 10^5 M_\odot$) and younger ages ($<$15 Myr) simply because these are the brightest, biggest, and easiest to measure (J. Gallagher, private communication). So, developing a proper theoretical cluster distribution for comparison is rather challenging without further observational studies of the cluster populations.

Median distance of XRBs from the cluster: Our results indicate a strong dependence of the median XRB distance on the age and moderate on the cluster mass. For clusters with a half-mass radius of 10 pc and masses $< 5 \times 10^5 M_\odot$, median distances reach values of 30–100 pc (similar to

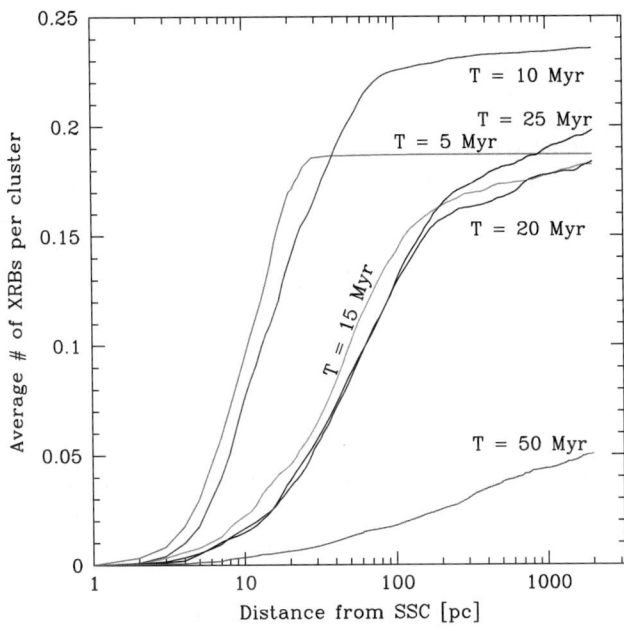

Fig. 1 Average number of XRBs per cluster within a given distance from its parent cluster. We use the average of 1000 realizations of a $5 \times 10^4 \, M_\odot$ cluster (left panel) and of 100 realizations of $5 \times 10^5 \, M_\odot$ cluster (left panel), using an initial mass function with a power law index of -2.35 and a limiting L_X of 5×10^{35} erg s^{-1}

those observed) at ages of 15 Myr and older. Only very massive clusters of $\sim 5 \times 10^6 \, M_\odot$ reach such distances later at ~ 50 Myr. This is consistent with the current observations, although massive and older clusters are also present in the photometrically selected clusters in K04 (Gallagher, 2004, private communication).

3. Black-hole X-ray binaries in elliptical galaxies

With *Chandra* observations, elliptical galaxies out to the Virgo cluster have now been studied. These observations have revealed a large number of point X-ray sources and (Kim and Fabbiano, 2004) have found that the combined sample of sources from all observed galaxies requires two power laws and a break at $5 \pm 1.6 \times 10^{38}$ ergs s^{-1} and the best-fit slope of the upper end to be $\alpha_d = 2.8 \pm 0.6$.

We analyze the upper-end of XLF assuming: (i) the XLF above the break at 5×10^{38} erg s^{-1} is populated by X-ray binaries (XRBs) with black hole (BH) accretors (?, Sarazin et al., 2000); (ii) the vast majority of these BH-XRBs are part of the galactic-field stellar population in ellipticals; (iii) donor masses are lower than $\simeq 1-1.5 \, M_\odot$ (in accordance to the current estimates for the ages of stellar populations in ellipticals) and (iv) we adopt the current understanding for the origin of transient behavior in XRBs (King, 2005) and we assume that during the outburst the X-ray luminosity is equal to the BH accretor Eddington luminosity.

Mass transfer in BH XRBs with low-mass main sequence (MS) donors is driven by angular momentum losses due to magnetic braking (MB) and gravitational radiation (GR). From detailed binary evolutionary calculations using the stellar evolution and MT code (Ivanova and Taam, 2004), we find that if MB follows the Skumanich law (Rappaport et al., 1983, taken as in Rappaport et al., 1983), then XRBs with BHs $\leq 10 M_\odot$ are persistent as long as the donor masses are $> 0.3 \, M_\odot$. The MT rates are about $0.01 - 0.25$ of the BH's Eddington rate. As a result, the persistent X-ray luminosity for these systems is $\leq 10^{38}$ erg s^{-1}. For $M_{BH} > 10 \, M_\odot$, the outburst luminosity is in excess of $\simeq 1.5 \times 10^{39}$ erg s^{-1}. This limit is comparable to the highest luminosity seen currently in XLFs of ellipticals (Kim and Fabbiano, 2004), and therefore these systems cannot contribute significantly to the observed XLFs. Outbursts are possible from transient BH-MS with $M_{BH} < 10 \, M_\odot$ and donors less massive than $\simeq 0.3 M_\odot$; these low mass donors are out of thermal equilibrium. In the case of the MB prescription derived for fast rotating systems, which are BH-MS binaries (Ivanova and Taam, 2003), IT, BH-MS systems are transient for all BHs masses $M_{BH} > 3 M_\odot$ and for all low-mass MS donors. If a donor is a low-mass subgiant or red giant (RG), it has been shown (King, 2005) that such XRBs are transient, regardless of the BH mass. The evolution of mass-transferring BH systems with a white dwarf (WD) very weakly depends on the BH mass. A typical life-time of such a system in the persistent stage is $\sim 10^7$ years and only during 10^6 years a BH-WD will have MT rates in excess of the Eddington limit. We conclude that all BH binaries that contribute to the current upper-end XLFs of ellipticals are expected to be transient sources.

3.1. Weighting factor for the BH mass spectrum and the duty cycle

The transient BH XRBs (those in outburst) contributing to the upper-end XLF is a sub-set of the true population of BH XRBs in ellipticals determined by the duty cycle of BH transient binaries. For the general case of a transient duty cycle that is dependent on the BH accretor mass, the differential XLF $n(L)_{\rm obs}$ and the underlying BH mass distribution in XRBs $n(m)_{\rm BH}$ are connected by:

$$n(L_X)_{\rm obs} = n(m_{\rm BH}) \times W(m_{\rm BH}), \qquad (1)$$

where $W(m_{\rm BH})$ is a weighting factor related to the dependence of the transient duty cycle on $m_{\rm BH}$. The observed slope of the differential upper-end XLF is $\alpha_{\rm d} = 2.8 \pm 0.6$: $n(L_X)_{\rm obs} \propto L_X^{-\alpha_{\rm d}}$. Assuming that $n(m_{\rm BH}) \propto m_{\rm BH}^{-\beta}$ and $W(m_{\rm BH}) \propto m_{\rm BH}^{-\gamma}$, the slope characterizing the underlying BH mass distribution in XRBs is $\beta = \alpha_{\rm d} - \gamma$.

At present there are no strong constraints on the duty cycles either from observations or from theoretical considerations. Among known Galactic X-ray transients, typical duty cycles of a few % is favored for hydrogen donors (Tanaka and Shibazaki, 1996) and there are no data on duty cycles for transients with a WD companion. For the standard assumption of a constant duty cycle, $\beta = \alpha_{\rm d} = 2.8 \pm 0.6$. However, according to binary population synthesis models for the Milky Way published so far, the number of formed BH-WD LMXBs exceeds the number of BH-MS LMXBs by a factor of 100 (Hurley et al., 2002). If the duty cycle were similar for both types of systems, we would observe a few hundreds BH-WD binaries; however none such binaries have been identified. We therefore investigate how plausible duty cycle assumptions affect the upper-end XLF shape, considering a variable (dependent on MT rates) duty cycle equal to $\eta = 0.1 \times \dot{M}_{\rm d}/\dot{M}_{\rm crit}$. This specific choice of the dependence is not solidly motivated, however, it implies a correlation of the duty cycle with how strong a transient the system is: the further away from the critical MT rate, the smaller the duty cycle.

The expression for η in the case of WD and RG donors can be found analytically as described in detail in (Ivanova and Kalogera, 2005). For RG we obtain $\gamma = 0$ and:

$$W(T) \simeq 0.03 \times T^{-0.5}. \qquad (2)$$

For WD donors we find:

$$W(T; m_{\rm BH}) = 1.6 \times 10^{-4} m_{\rm BH}^{-0.5} t_1^{-7/4} \frac{1 - (t_2/t_1)^{-3/4}}{1 - (t_2/t_1)} \qquad (3)$$

Here t_1 and t_2 are the time in Gyrs from the current moment to the start and the end of BH-WD binaries formation. Therefore for WD donors $\gamma = 0.5$ and W depends on both the BH mass and the galaxy age.

As discussed previously, for IT MB, a BH-MS system is transient throughout the MT phase. It also can be seen that the MT dependent duty cycle is about few %, consistent with the observations. Prolonged mass accretion onto the BHs affects their mass spectrum. Since this effect cannot be included analytically, we have examined it quantitatively using simple Monte Carlo simulations. We find that: (i) due to accretion the BH mass spectrum slope increases by about 0.2, i.e., $\beta = \beta_0 + 0.2$, where β_0 is the BH mass slope at MT onset; (ii) the slope of the BH mass spectrum at the beginning of mass transfer best reproduces the observations with $\beta_0 = 2.3 \pm 0.6$ (see Fig. 1). We also find that the relation between β and β_0 is not sensitive to the age of the elliptical.

3.2. Conclusions

We conclude that all BH binaries contributing to the upper-end XLF of ellipticals are transient. A constant transient duty cycle independent of the donor type can be excluded unless BH-WD transients have very weak outbursts and can not be detected. The upper-end XLF is formed by an underlying mass spectrum of accreting BHs. The BH X-ray transients have a dominant donor type and an accreting BH mass spectrum slope β that depends on the strength of MB angular momentum loss. In particular, in the case of Skumanich-type MB, the XLF is dominated by BH-RG binaries and $\beta = 2.8 \pm 0.6$. In the case of MB for fast-rotators, only BH-MS transients significantly contribute to the upper-end XLF and $\beta = 2.5 \pm 0.6$ ($\beta_0 = 2.3 \pm 0.6$). If the relative fraction of BH-RG transients in ellipticals is larger than the observed relative fraction in our Galaxy, we expect that the BH-RG binaries contribution will lead to a time-dependence of XLF slopes, where younger ellipticals will have a slope predicted for BH-RG binaries, and older ellipticals a steeper slope predicted for BH-MS binaries.

Acknowledgements We are grateful to J. Gallagher, T. Maccarone and P. Kaaret for useful discussions. This work was partially supported by a Packard Foundation fellowship, a NASA Chandra Award to V. Kalogera and a *Chandra* Theory Award to N. Ivanova.

References

Belczynski, K., Kalogera, V., Bulik, T.: ApJ **572**, 407 (2002)
Belczynski, K., Kalogera, V., Zezas, A. et al.: ApJ **601**, L147 (2004)
Belczynski et al.: ApJ, submitted (2005)
Blecha, L. et al.: ApJ, submitted [astro-ph/0508597] (2005)
Fabbiano, G., White, N.E.: In: Lewin, W.H.G., van der Klis, M. (eds.) Compact Stellar X-ray Sources, Cambridge University Press, in press [astro-ph/0307077] (2003)
Harris, J., Calzetti, D., Gallagher, J.S. III, Conselice, C.J., Smith, D.A.: ApJ **122**, 3046 (2001)
Hurley, J. R., Tout, C.A., Pols, O.R.: MNRAS **329**, 897 (2002)

Ivanova, N., Kalogera, V.: ApJ, in press [astro-ph/0506471] (2005)
Ivanova, N., Taam, R.E.: ApJ **599**, 516 (2003)
Ivanova, N., Taam, R.E.: ApJ **601**, 1058 (2004)
Kaaret, P., Alonso-Herrero, A., Gallagher, J.S. III, Fabbiano, G., Zezas, A., Rieke, M.J.: ApJ **348**, L28 (2004)
Kilgard, R.E., Kaaret, P., Krauss, M.I., Prestwish, A.H., Raley, M.T., Zezas, A.: ApJ **572**, 138 (2002)
King, A.: In: Lewin, W.H.G., van der Klis, M. (eds.) Compact Stellar X-ray Sources, Cambridge University Press (2005).
Kim, D., Fabbiano, G.: ApJ **611**, 846 (2004)
Rappaport, S., Verbunt, F., Joss, P.C.: ApJ **275**, 713 (1983)
Sepinsky, J., Kalogera, V., Belczynski, K.: ApJ **621**, L37 (2005)
Smith, L.J., Gallagher, J.S. III: MNRAS **326**, 1027 (2001)
Zezas, A., Fabbiano, G., Rots, A.H., Murray, S.S.: ApJ **577**, 710 (2002)

Synthesis of Line Profiles and Radial Velocity Curves for X-Ray Binary Systems

A. M. Cherepashchuk

Received: 1 November 2005 / Accepted: 1 February 2006
© Springer Science + Business Media B.V. 2006

Abstract Line profiles and radial velocity curves for optical stars in X-ray binaries are calculated taking into account ellipticity of the optical star and X-ray heating effect.

Keywords Line profiles · Synthesis · X-ray binary system · Masses · Relativistic objects

1. Introduction

Tidal deformation of the optical star in X-ray binaries and its heating by the companion's radiation lead to changes of its projected shape and temperature distribution on its surface with orbital phase. Therefore, absorption line profiles vary with orbital phase. This variability can be used for independent determination of the basic system parameters such as the mass ratio of the components $q = m_x/m_v$ (m_x and m_v are the masses of relativistic and optical components) and the inclination of the orbital plane i. First estimates of the ellipticity and X-ray heating effects on the radial velocity curves and line profiles of X-ray binary systems were done by Wilson and Sofia (1976) and Milgrom (1976). Modeling line profiles in close binaries was realised in the papers by Antokhina and Cherepashchuk (1994), Antokhina (1996), Mukherjee, et al. (1996). A new method for estimation of q and i in X-ray binaries from orbital variability of absorption line profiles was suggested by Antokhina and Cherepashchuk (1997) and Shahbaz (1998). A method of line profile synthesis for X-ray binaries taking into account ellipticity and X-ray heating effect was developed recently by Antokhina et al. (2003, 2005).

A. M. Cherepashchuk
Sternberg Astronomical Institute, Moscow University, Moscow, Russia

2. Calculations of local line profiles

The standard Roche model (Wison, Devinney, 1971) at the eccentric orbit accounting for asynchronous rotation was used. Radial velocity V_r for each elementary area is calculated by the formulas $\vec{V} = [\vec{\omega}_{rot} \vec{r}]$, $V_r = -(\vec{V} \vec{a}) + V_c^{star}$, where $\vec{V}$ is the velocity of an elementary area relative to the center of gravity of the optical star ($\vec{\omega}_{rot}$ is the angular rotation velocity of the star, $\vec{r}$ is the radius-vector of an elementary area), V_r is the radial velocity of an elementary area ($\vec{a}$ is a unity vector along the line of sight, V_c^{star} is the radial velocity of the center of gravity of the optical star).

For each elementary area the structure of the stellar atmosphere was calculated taking into account X-ray heating from a point-like X-ray source. LTE-approximation was used for the solution of radiation transfer equation together with radiative and hydrostatic equilibrium equations which were solved for various values of the heating parameter $K_x = \frac{L_x}{L_v}$ where L_x and L_v are bolometric luminosities of the relativistic and the optical star, respectively. The expression for the distribution of the source function over optical depth is given by the formula:

$$S(\tau_\nu) = S_0(\tau_\nu) + a_2 S_2(\tau_\nu) + a_3 S_3(\tau_\nu) + a_4 S_4(\tau_\nu)$$

where $S_0(\tau_\nu) = \sigma T^4$ is the distribution of the source function for a non-radiated atmosphere,

$$S_2 = \int_{\nu_1}^{\nu_2} \frac{dH^+}{d\tau_\nu} d\nu_x, \quad S_3 = \int_{\nu_1}^{\nu_2} H_\nu^+(\tau = 1) d\nu_x,$$

$$S_4 = \int_{\nu_1}^{\nu_2} \int_0^{\tau_\nu} [S_2(\tau_\nu') + H^+(\tau_\nu')] d\tau_\nu' d\nu_x,$$

Fig. 1 *Left*: Distribution of the electron temperature, T_e, in the local atmosphere of a star for different values of the heating factor K_x. At the bottom are the corresponding local line profiles. *Right*: Dependence of the integral line profile CaI λ 6439 Å on the orbital phase for $K_x = 10$, $i = 90°$.

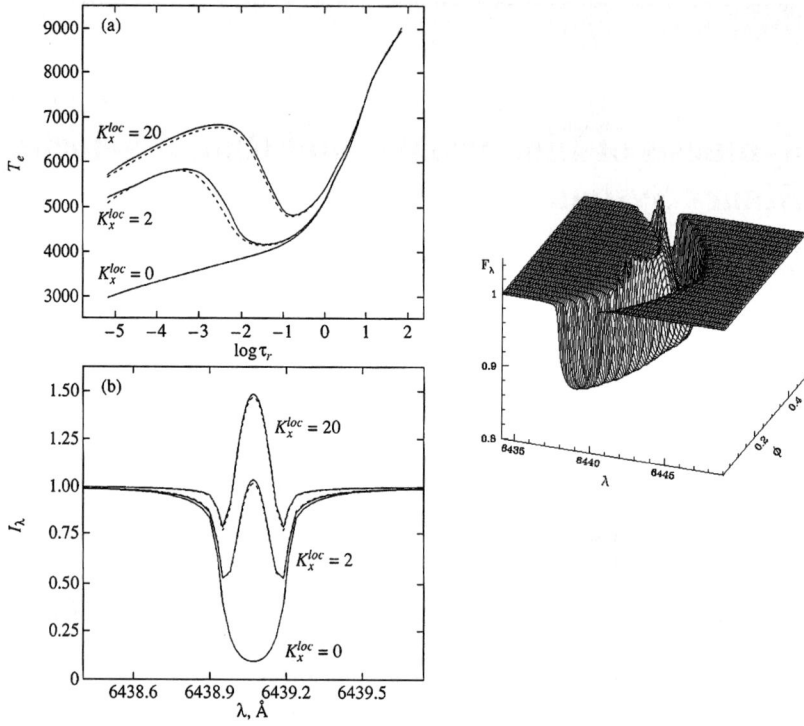

a_2, a_3, a_4 are the functions of α_x/α_v (H^+ is the X-ray flux, α_x, α_v are absorption coefficients for X-ray and optical radiation respectively). For more details see the papers by Antokhina et al. (2003, 2005).

In Fig. 1 the distributions of the electron temperature, T_e, in a local part of the atmosphere of the optical star ($T_{ef} = 4752K$, $\lg g = 2.63$) for various K_x are presented. Strong increase in T_e in outer parts of the atmosphere for $K_x > 1$ leads to formation of an emission component for the local atmospheric absorption line CaI λ6439 Å (lower part of Fig.1).

3. Calculations of integral line profiles

Local line profiles (Fig. 1) are integrated over visible surface of the optical star taking into account Doppler shifts caused by radial velocities of the elementary areas. The radial velocities of the optical star were derived by averaging the width of a line profile at three levels: one-third, one-half and two-thirds of the depth at the line center. In Fig. 1 three-dimensional dependence of the distribution of intensity in the CaI λ6439 Å line profile is presented ($K_x = 10$, $m_v = 1M_\odot$, $m_x = 10M_\odot$, Roche lobe filling factor $\mu = 1$, $T_{ef} = 5000K$, $i = 90°$). Orbital Doppler shifts of the line as well as variability of emission and absorption components with the phase of orbital period are clearly seen. High resolution and high signal-to-noise spectroscopy of X-ray binaries will allow one to estimate basic parameters of the binary system directly from variability of the line profile (Antokhina, Cherepashchuk, 1997; Shahbaz, 1998).

4. Effects of star Ellipticity and X-ray heating on line profiles

We selected input parameters that were close to the parameters of low-mass X-ray binaries (X-ray novae) to model the theoretical line profiles. The mass of $m_v = 1M_\odot$ and temperature $T = 5000K$ of the optical star were kept fixed. The gravitational-darkening coefficient was assigned $\beta = 0.08$. The stellar rotation was assumed to be synchronous with the orbital revolution, and the orbit to be circular. The orbital period was taken to be $p = 5$ days ($p = 12$ days for the computation of the radial velocity curves from line profiles). The spectrum of the compact object in the wavelength range (12–0.5)Å (photon energies 1–20 keV) was modelled by the power law $I_x(\nu) = I_x^0 \nu^{-0.6}$. We studied the behaviour of the line profiles of the optical star during the orbital motion for various relative powers of the incident X-ray flux K_x. The component mass ratio, $q = m_x/m_v$, Roche lobe filling factor of the optical star, μ_v, and orbital inclination, i, were varied as well. Theoretical profiles were computed for the CaI λ6439.075 Å absorption line. Fig. 2 shows the variations of the shape and intensity of the line profile for $K_x = 10$ and $i = 45°$. Orbital Doppler shifts are eliminated. We can clearly see the evolution of the summed line profile of the optical star as it turns the X-ray-heated side toward the observer. At

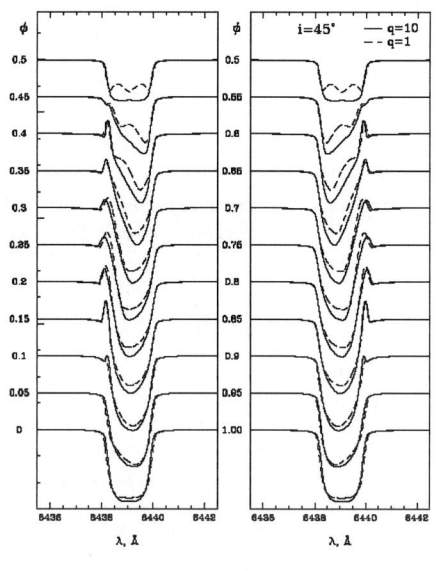
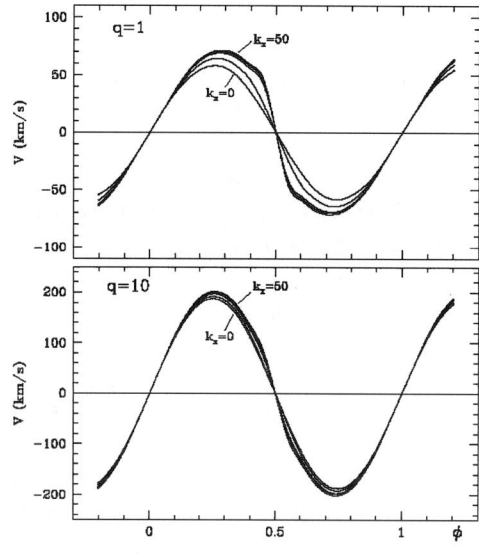

Fig. 2 *Left*: Orbital-phase variations of the CaI λ 6439 Å line profile in the spectrum of an X-ray irradiated star ($K_x = 10$, $i = 45°$). *Right*: Dependence of the radial velocity curves of the optical star on the K_x ($K_x = 0.1, 10, 20, 50$) for $q = 1$ (up) and $q = 10$ (bottom)

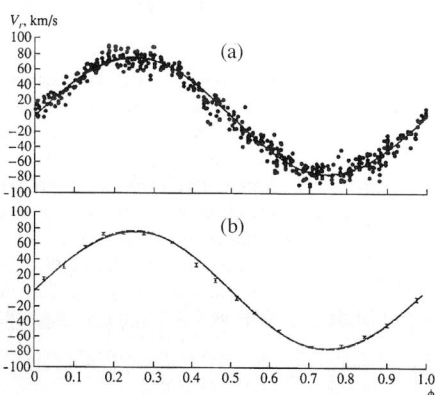
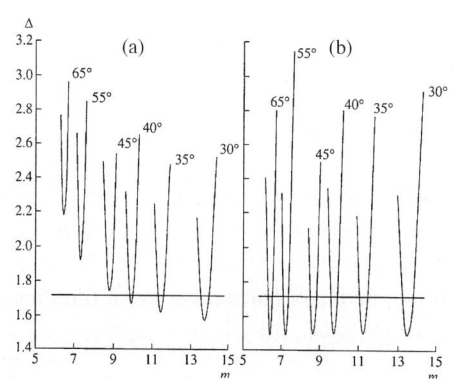

Fig. 3 *Left*: An observed high-precision radial velocity curve of the optical star in the X-ray binary Cyg X-1 combined 502 observational nights. Right: Residuals Δ obtained by fitting the mean high-precision radial velocity curves for Cyg X-1 for different i. (a) – the Roche model for the optical star; (b) – the point-like model.

orbital phases when both heated and unheated regions of the star moving with different velocities can be observed, the absorption profile becomes narrower due to the formation of the emission component. We emphasize that, in the case of a pure ellipticity effect for an optical star without X-ray heating, the width of the line increases at quadrature, when star is viewed from the side, since it has its maximum extent along the line connecting the component centers.

The radial velocity curves as functions of the X-ray heating power K_x are shown in Fig. 2. The amplitude of the radial velocity grows with K_x, this effect being more pronounced for $q = 1$ than for $q = 10$.

4.1. High-mass X-ray binary Cyg X-1

In the paper by Abubekerov et al. (2004b) the estimates of the orbital inclination and the mass of the black hole were obtained from a high-precision average radial velocity curve of Cyg X-1 including 502 observational nights (corresponding accuracy 3 percent), see Fig. 3. Dependence of residuals, Δ, between the observed and theoretical radial velocity curves, on the parameter i is shown in Fig. 3. In the case of the point-like model of optical star minimum residuals are not dependent on i, therefore, the classical statement is confirmed: an iclination of the orbital plane, i, cannot be determined from a radial velocity curve. But in the case of a tidally deformed optical star, clear dependence of minimal residuals on i is observed which allows us to constrain the value of i: $i < 44°$. The lower limit of i can be obtained from optical ellipsoidal photometric variability of Cyg X-1 and information about the distance from this system: $i > 31°$. A corresponding estimate of the black hole mass is $13.6 M_\odot > m_x > 8.5 M_\odot$.

4.2. Low-mass X-ray binary 2S 0921-63

In the paper by Jonker et al. (2005) a high-quality radial velocity curve of 2S 0921-63 was obtained ($P = 9.006$ days, $e = 0$, $K_v \simeq 99$ km/s). Strong X-ray heating effect is observed in this system, $K_x \simeq 10$. Taking into account X-ray heating allows us to decrease the mass of the relativistic companion by 0.5-1 solar masses. Therefore, we can suggest the presence of a heavy neutron star in the system 2S

0921-63 rather than a low-mass black hole (for more details see Abubekerov et al. 2005).

A strong emission component of the line CaI λ6439 appears to become visible in the quadrature for this system due to heating effect implying considerable increase of the amplitude of the radial velocity curve. Taking into account this effect allows us to obtain a correct estimate of the mass of the relativistic companion.

4.3. X-ray pulsars in massive X-ray Binaries

It is shown in the paper by Abubekerov, Antokhina, Cherepashchuk (2004a) that the amplitude of the radial velocity curve of the optical star in the case of a realistic Roche model of this star is less comparing with the point-like model of this star. Therefore, previous mass values obtained for X-ray pulsars seem to be underestimated by 5–10 percent. Application of our method when interpreting spectroscopic observations of X-ray binaries allowed us to obtain more correct mass estimates for X-ray pulsars.

5. Conclusion

Our method for computing line profiles for X-ray binary systems, taking into account the ellipticity effect and the heating of the stellar atmosphere due to external radiation, can be applied to detailed analyses of high-resolution and high-signal-to-noise spectra ($R \simeq 30000 - 50000$, $S/N > 50$). We hope that applying our method to the interpretation of such spectra will make it possible to obtain more precise estimates of the masses of relativistic objects and reduce the errors in these masses.

Acknowledgements The author thanks E.A. Antokhina, M.K. Abubekerov and V.V. Shimanskii for valuable discussions.

References

Abubekerov, M.K., Antokhina, E.A., Cherepashchuk, A.M.: Astronomy Reports **48**, 89 (2004a)
Abubekerov, M.K., Antokhina, E.A., Cherepashchuk, A.M.: Astronomy Reports **48**, 550 (2004b)
Abubekerov, M.K., Antokhina, E.A., Cherepashchuk, A.M., Shimanskii, V.V.: Astronomy Reports **49**(10), 801 (2005)
Antokhina, E.A.: Astronomy Reports **40**, 483 (1996)
Antokhina, E.A., Cherepashchuk, A.M.: Astronomy Reports **38**, 367 (1994)
Antokhina, E.A., Cherepashchuk, A.M.: Astron. Lett. **23**, 773 (1997)
Antokhina, E.A., Cherepashchuk, A.M., Shimanskii, V.V.: Izvestia of Russian Acad. Sci. Phys. Ser. **67**, 293 (2003)
Antokhina, E.A., Cherepashchuk, A.M., Shimanskii, V.V.: Astronomy Reports **49**, 109 (2005)
Jonker, P.G. et al.: MNRAS **356**, 621 (2005)
Milgrom, M.: ApJ **206**, 869 (1976)
Mukherjec, J.D., Peters, G.J., Wilson, R.E.: MNRAS **283**, 613 (1996)
Shahbaz, T.: MNRAS **298**, 153 (1998)
Wilson, R.E., Devinney, E.J.: ApJ **166**, 605 (1971)
Wilson, R.E., Sofia, S.: ApJ **203**, 182 (1976)

Hot Subdwarfs in Precataclysmic Binaries

Vladislav Pustynski · Izold Pustylnik

Received: 1 November 2005 / Accepted: 1 February 2006
© Springer Science + Business Media B.V. 2006

Abstract We present a brief review of the recent results in modeling physical processes in strongly irradiated atmospheres of unevolved companions in precataclysmic binary systems (PCB) and their light curves. Constraints on physical parameters of the hot sdws primaries, thermal instability in upper irradiated atmosphere, monochromatic albedos, the deficit of the total emergent flux compared to the incoming flux from hot sdw primary are briefly discussed.

Keywords Binaries: close · Subdwarfs · Radiative transfer · Instability

1. Introduction: Modeling of strongly irradiated atmospheres of unevolved companions of PCBs

PCBs are a small group of detached close binary systems, many of which are observed at the centers of PN. They consist of a hot white dwarf or sdw and an essentially unevolved low mass cool companion with $M_2 < 1 M_\odot$. PCB orbital periods are so short (typically less than two days) that they can only have been formed via a common envelope evolutionary stage (e.g. Bond et al., 1992; Ritter, 1986; see also Pustylnik and Pustynski, 1999, 2001; Pustylnik et al., 2001). The substantial proximity of the components and high temperatures of the hot primary lead to extraordinarily high amplitude of reflection effect in these systems. Our studies of the irradiation effects in PCBs were motivated by the following considerations.

(a) Conventional treatment of the reflection effect may not be adequate for PCBs due to small separations and very high primary temperatures (about $T_1 \sim 3 \cdot 10^4 \div 10^5$ K). In that case substantial portion of the incident flux is concentrated in the Lyman continuum (L_c). Radiation from a hot source alters considerably the structure of corona and chromosphere of the cool companion and may be a cause for instability of various types. (b) Modeling physical conditions in atmospheres may help us to determine energy distribution in the impinging flux and thus estimate independently primary temperature. (c) Pulsations have been discovered in EC14026-type stars (rapidly oscillating sdB stars). Such pulsations may be caused by instabilities in the secondary atmosphere.

In the course of our studies we have elaborated an original computer code; adopting a set of PCB system orbital and physical parameters, it calculates two-layer model atmospheres at different regions of the secondary disc, intensities and angular distributions of the emergent radiation and constructs continuum light curves with due account of both annular eclipses and occultations or grazing eclipses.

Our model is based on the following assumptions. (a) The whole L_c flux is supposed to be absorbed in the upper atmosphere which is assumed to be transparent for optical quanta. The validity of this simplification was proved to be well-founded post-factum. (b) The free path of a photon is assumed negligible compared to the layer thickness. (c) The recombinational re-emission mechanism of L_c absorbed flux is adopted. The diffusion approximation is applied in the deeper atmosphere. The models of upper and deeper layers are coordinated via boundary variables. (d) The impinging flux is found by dividing the secondary disc into circular zones symmetrical with respect to the sub-stellar point; the flux is calculated taking into account possible occultation of

V. Pustynski (✉)
Tartu Observatory, Tõravere, 61602 Estonia and Tallinn Technical University, Ehitajate tee 5, 19086, Tallinn, Estonia
e-mail: vladislav@aai.ee

I. Pustylnik
Tartu Observatory, Tõravere, 61602 Estonia
e-mail: izold@aai.ee

the primary disc by the local horizon, for each zone an independent model is built and emergent radiation intensity is computed. Thereafter intensities from different zones are integrated to yield the total luminosity, and inputs of possibly eclipsed portions of the both stellar discs are consequently subtracted. A more detailed description of our model and the results obtained with it may be found in the following our works: Pustylnik and Pustynski, 1999, 2001; Pustylnik et al., 2001; Pustylnik and Pustynski, 2004, 2005a,b,c.

2. Constraints on physical parameters of the hot sdws in PCBs

We have applied our model to two well-studied eclipsing PCBs, UU Sge and V477Lyr.

The first of them is a central star of the planetary nebula Abell 63, an eclipsing binary with the orbital period of 0.465 days. According to Pollaco and Bell (1993), different models based on different assumptions have been proposed for this system, and various methods have been applied to obtain its parameters. Masses of the components are known to be $M_1 = 0.63 \pm 0.06\, M_\odot$, $M_2 = 0.29 \pm 0.04\, M_\odot$, but large uncertainty exists in respect of the temperatures of the components and the secondary radius. Using V observations provided in the above-mentioned article, we were able to find two sets of parameters that fit well the observational light curve. The first set is the following: $T_1 = 80\,000$K, $R_1 = 0.33\, R_\odot$, $T_2 = 5500$K, $R_2 = 0.54\, R_\odot$ and $i = 88°$. The second set is the following: $T_1 = 85\,000$K, $R_1 = 0.34\, R_\odot$, $T_2 = 5600$K, $R_2 = 0.54\, R_\odot$ and $i = 88°$. This data constrains the system parameters range established by other authors, especially in respect of the primary temperature, for which different sources give values in the range of $T_1 \sim (35\,000 \div 117\,500)$K. More details see in our article Pustylnik and Pustynski, 2005c.

V477 Lyr is an eclipsing central star of Abell 46 planetary nebula. According to Pollaco and Bell (1994), the masses of the components are estimated as $M_1 = 0.51\, M_\odot$, $M_2 = 0.15\, M_\odot$ and $T_1 = 60\,000\, K$ with the uncertainty about 20%. We have found four sets of parameters that fit well the observational light curve. These are: $T_1 = 74\,000 - 102\,000$K, $R_1 = 0.23\, R_\odot, T_2 = 4000$K, $R_2 = 0.58 - 0.64\, R_\odot, i \approx 80°$. As one can see, the range of the primary temperature remains broad, so additional data is necessary to constrain the system parameters. At the same time it should be emphasized that since we deal with the direct modeling of irradiated photospheres, our formalism does not involve physically unfounded parameters of the limb brightening x_2 for the irradiated component nor the heating efficiency α, the latter being dependent on T_1 and T_2 values (Pollaco and Bell, 1994 assumed $x_2 = -1.0$ and $\alpha_2 = 1.5$ for V477 Lyr).

We superposed our data upon the evolutionary tracks calculated by Iben and Tutukov, (1993) where one can see the cooling tracks for white dwarfs of different masses and position of several PCBs. Even for eclipsing binaries like UU Sge and V477 Lyr with a good coverage of observational points of full light curves one cannot reliably discriminate between the cases of WD with helium core or COWD (see Fig. 1 in Pustylnik and Pustynski, 2005a).

3. Some recent model results for strongly irradiated atmospheres of PCB-s

We briefly report here some recent results of modeling strongly irradiated atmospheres of the unevolved

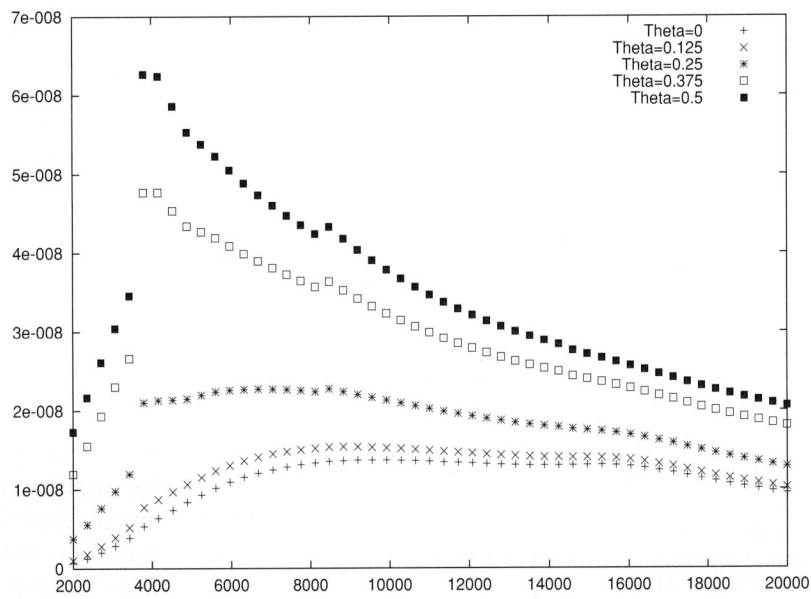

Fig. 1 Spectral distribution of emergent flux from irradiated atmosphere for different orbital phases

companions which can be helpful in better understanding of the observed peculiarities of PCBs.

(a) *Changes of spectral distribution with phase.* The plot denoted Fig. 1 illustrates systematic variations of spectral distribution for the irradiated component with orbital phase. One can see systematic progressive shift from a smooth shallow maximum in spectral energy distribution characteristic for intrinsic energy distribution for non-irradiated hemisphere to shorter wave-lengths and a peaked distribution for orbital phase where the "hot spot" becomes visible. Contribution from recombinational continuum is visible in form of Balmer and Pashen jumps. Using these data and calculating respective effective temperature variations of irradiated atmosphere with orbital phase one can get a good estimate of the effective temperature of the hot sdw primary from the multi-color light curves of non-eclipsing PCBs.

(b) *Deficit of the total emergent flux.* The plot denoted as Fig. 2 indicates the ratio of the emergent flux to the incoming flux for different angle distance from substellar point. The deficit of the flux is in evidence (high values close to the limb is an artefact, the contribution to the total emergent radiation for zones close to the limb is negligible). The deficit is in a good accord with our earlier result – namely, our models indicated that roughly 50% up to 75% of incoming flux is spent on heating of electronic gas and the re-structuring of the upper layers of the irradiated atmosphere (Pustylnik and Pustynski, 2005c).

(c) *Monochromatic albedos.* The plot denoted as Fig. 3 illustrates the run of monochromatic albedos (the ratio of

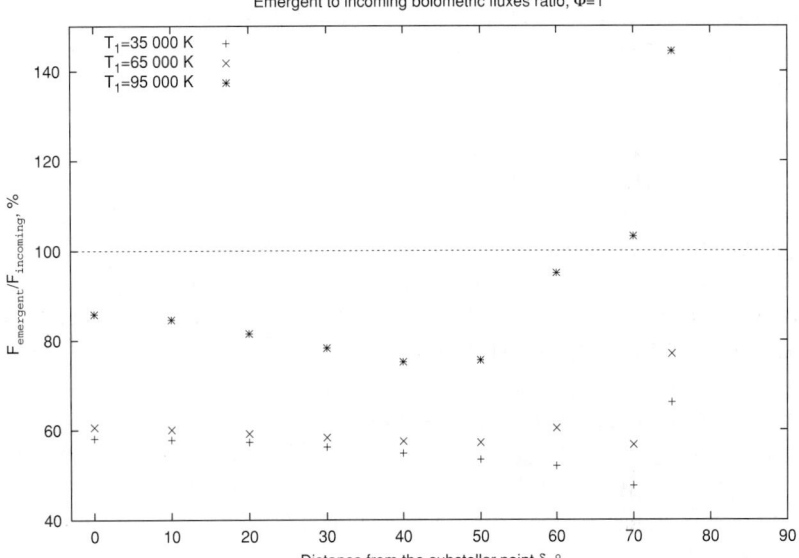

Fig. 2 Ratio of emergent and incoming total flux for irradiated atmosphere of unevolved companion in PCB

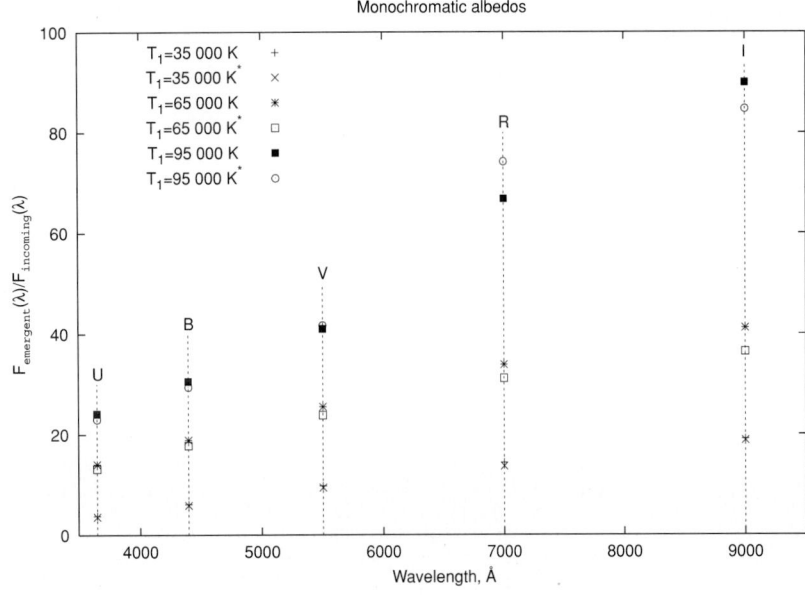

Fig. 3 Monochromatic albedoes for irradiated atmosphere of unevolved companion in PCB

emergent monochromatic flux to the incident flux in the same wave length) for the central wave length of UBVRI bands for 2 cases: (a) full account of recombinational radiation; and (b) without it for typical parameters of PCBs (albedos being systematically lower for the latter case). High values of albedos in all colors testify to the effectiveness of reprocessing of Lyman continuum to the visible and near IR region: from each UV photon several less energetic visible photons are produced more effectively at higher values of the hot sdw effective temperature. This circumstance explains why one finds high albedo values (2–5) from the analysis of the UBV light curves of PCBs.

(d) *Thermal instability.* From our models of strongly irradiated atmospheres we have found recently the evidence on thermal instability of the whole layer where L_c flux is effectively absorbed (Pustylnik and Pustynski, 2005c). The thermal time-scale for the layer where the L_c flux is absorbed is comparable to the orbital period of a typical PCB (for column masses dm $\simeq 10^{-3}$ g/cm^{-2} and an effective thickness of the layer 10^7 cm). A similar value is obtained assuming that T_{equil} is the equilibrium temperature of neutral H, T_e being the temperature of the electronic gas, and estimating the cooling time scale $t_{th} \approx 1.8 \cdot 10^{11} N_e^{-1} T_e^{-1/2}$ s from Spitzer equation for the exchange of energy between neutral H atoms and free electrons (see Pustylnik and Pustynski, 2005c). For typical values of $T_e \sim 10^5\,K$ and $N_e \sim 10^{12}$ cm^{-3} we find $t_{th} \simeq 10^4$ s. Whether the onset of a thermal instability will cause pulsations, deserves a special investigation. As we see, the results of model calculations suggest that transient effects on a time-scale comparable to the orbital period of PCB-s are anticipated.

Acknowledgements One of the authors (I.P.) expresses his gratitude to Dr. Niarchos for hospitality and fruitful discussions. This investigation has been supported by Grant 5760 of the Estonian Science Foundation.

References

Bond, H.E., Ciardullo, R., Meakes, M.G.: In: Kondo, Y., Sistero, R.F., Polidan, R.S. (eds.) IAU Symp. No. 151, Evolutionary Processes in Interacting Binary Stars, 517 (1992)
Iben, I., Jr. Tutukov, A.V.: ApJ **418**, 343 (1993)
Pollaco, D.L., Bell, S.A.: MNRAS **262**, 377 (1993)
Pollaco, D.L., Bell, S.A.: MNRAS **267**, 452 (1994)
Pustylnik, I., Pustynski, V.V.: In: 11th European Workshop on White Dwarfs, ASP Conf. Series **169**, 289 (1999)
Pustylnik, I., Pustynski, V.V.: In: 12th European Workshop on White Dwarfs, ASP Conf. Series **226**, 253 (2001)
Pustylnik, I., Pustynski, V.-V., Kubát, J.: In Odessa Astron. Publ. 14, 87–90 **14**, 87 (2001)
Pustylnik, I., Pustynski, V.V.: Baltic Astronomy **13**, No.1, 122 (2004)
Pustylnik, I., Pustynski, V.V.: In: Zdeněk Kopal's Binary Star Legacy, ASP Conf. Series **296**, Nos. 1–4, 251 (2005a)
Pustylnik, I., Pustynski, V.V.: In: Ap&SS **299**, No. 2, 177 (2005b)
Pustylnik, I., Pustynski, V.V.: In: 14th European Workshop on White Dwarfs, ASP Conf. Series **334**, 639 (2005c)
Ritter, H.: A&A **169**, 139 (1986)

On the Binary Nature of SS 433

Linda Schmidtobreick · Katherine Blundell

Received: 1 November 2005 / Accepted: 1 February 2006
© Springer Science + Business Media B.V. 2006

Abstract We present time–resolved optical spectroscopy of the famous X-ray binary SS 433. We obtained 61 medium resolution spectra spread over three months and thus cover roughly five orbits and about half a precession phase. We used various emission lines, that we attribute to the accretion disc, to determine the radial velocities of the compact component. They are of course modulated with the orbital period but in addition show a variation of the system velocity on a longer time scale. With the present data it is not possible to determine whether this is a transient effect or a periodic variation, although we present various possible interpretations of this effect.

Keywords Microquasars · SS 433

1. Introduction

SS 433 has become famous as the first known relativistic jet source in the Galaxy. It is believed to be a high inclination binary system with an orbital period of 13.08 days and consists of a compact source accreting matter from a companion. The observed periodic variations in the velocities of the jet lines are generally explained by a kinematic model including precession and nodding of the disc (Margon, 1984; Eikenberry et al., 2001), while the stationary lines are supposed to either originate from the donor star (e.g. Fuchs et al., 2002)
or from the accretion disc around the compact object (e.g. Hillwig et al., 2004).

We here analyse the stationary lines only. We determine their basic properties in the high S/N average spectrum, determine the individual radial velocities and discuss the variations with the orbital phase and on longer time–scales.

2. The data

The time–resolved optical spectroscopy was obtained between August and November 2004 using EMMI on ESO's New Technology Telescope (NTT) at La Silla, Chile. Grating #6 together with a 0.5 arcsec slit has been used to cover the wavelength range from 5800 to 8700 Å with a spectral resolution of 2.2 Å/FWHM. The individual exposure time was 300 s, and we usually observed one spectrum per night, sometimes more. In total, 61 spectra were taken spread over three months and thus covering roughly five orbits and about half a precession phase.

IRAF was used for the basic data reduction including overscan subtraction, flat–fielding, and wavelength calibration. No flux calibration was performed. Instead, the spectra have been normalised for the continuum using the program SPLOT kindly written and provided by Jochen Liske. The moving Hα jet lines were individually fitted for all spectra and subtracted to get a clean data set of stationary lines. All further analysis have been done using MIDAS.

3. Results

In Fig. 1, the average spectrum of all the nights is displayed. It is dominated by the strong Hα emission, but also shows the

L. Schmidtobreick (✉)
European Southern Observatory, Casilla 19001, Santiago 19, Chile

K. Blundell
University of Oxford, Department of Physics, Keble Road, Oxford, OX1 3RH, UK

Fig. 1 The average of the 61 normalised spectra, from which the jet components had been subtracted, is plotted above. Below, a zoom on the weaker lines is presented. Apart from Hα and the Paschen series in the red, the spectrum is dominated by low excitation lines of He I, O I, and C II

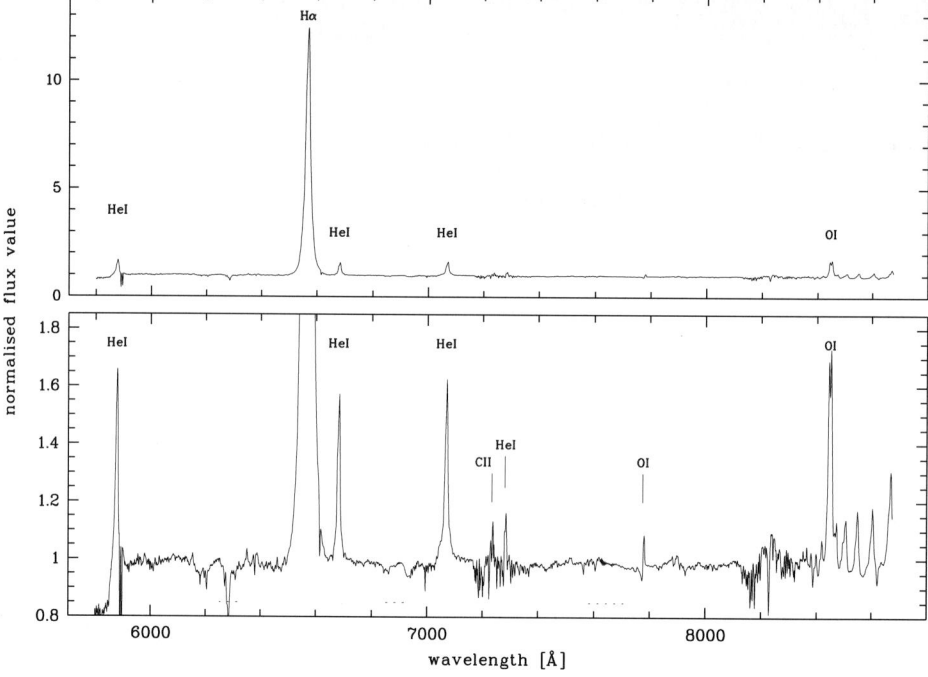

Paschen series at the red end and the presence of several low ionisation emission lines from He I, O I, and C II. Most of these lines show a clear double-peak profile in the individual spectra (see Fig. 2 for an example), which strengthens the theory that their origin is actually in a rotating disc. However, this profile gets washed out in the average spectrum.

Table 1 gives an overview on properties of the mayor emission lines. Except for Hα, the lines agree more or less on the FWHM and give an average projected rotation velocity of $V_{rot} = 540$ km/s. Hα instead has a rotation velocity of about 1000 km/s. The variation in time is illustrated in Fig. 3. It shows the different behaviour of the Hα line whose FWHM is continuously increasing after about HJD = 2453270. Instead, the other lines remain nearly constant, the FWHM even decreases a bit, and then rises around day HJD = 2453300. The low values of the Paschen lines at this time are actually due to a deep P-Cyg-profile, which appears around this day and cuts away part of the line and thus reduces its FWHM.

The radial velocities have been determined by fitting broad Gaussians to the emission lines. We have used the emission lines Hα, He I at 6678Å and 7065Å, and Paschen 11, 12 and 13. The resulting radial velocities are plotted against the heliocentric Julian date in Fig. 4.

They show the expected variation with the orbital period (13 d). From the first two orbits of the average velocities, we derive a radial velocity amplitude $K_1 = 110$ km s^{-1} and a system velocity $\gamma = 100$ km s^{-1}. However, both these values seem to change with increasing time. While the variation of K_1 seems to depend strongly on the line that is used for its

Table 1 The FWHM with the corresponding projected rotation velocity, and the equivalent width are given for the main emission lines

Transition	FWHM [Å]	V_{rot} [km/s]	$-W$ [Å]
Hα	22.12(6)	1011(3)	285.4(1)
He I λ6678	12.18(4)	547(2)	7.7(2)
He I λ7065	14.23(3)	604(2)	11.3(2)
O I λ8446	16.24(2)	577(1)	14.6(8)*
Paschen 11	13.5(3)	470(10)	4.2(1)
Paschen 12	11.9(1)	412(4)	3.0(1)
Paschen 13	15.2(1)	534(4)	2.6(1)

*O I includes also Paschen 14 and 15

Fig. 2 Some line profiles from the data taken on HJD = 2453274.48. The double–peak profile typical for rotating discs is clearly visible

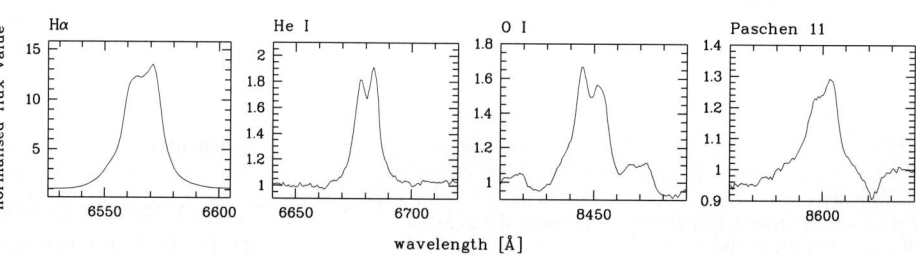

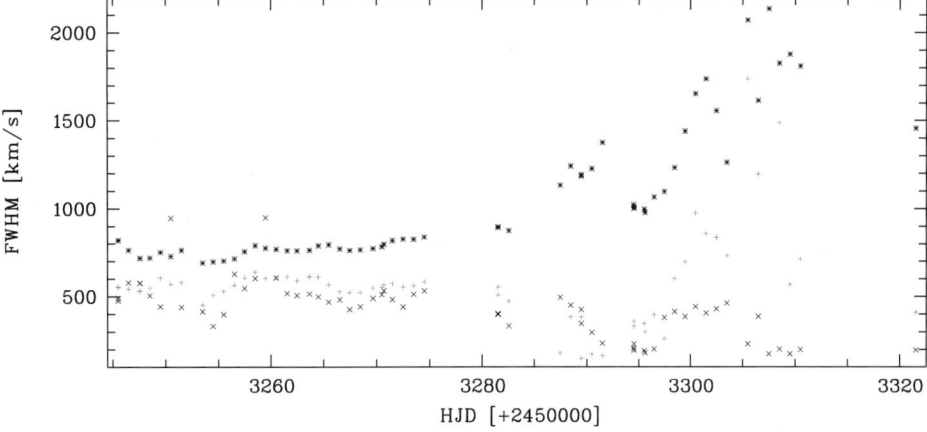

Fig. 3 The FWHM of the Hα (*), He I,7065 (+), and Paschen 11 (×) emission lines is plotted against the heliocentric Julian date

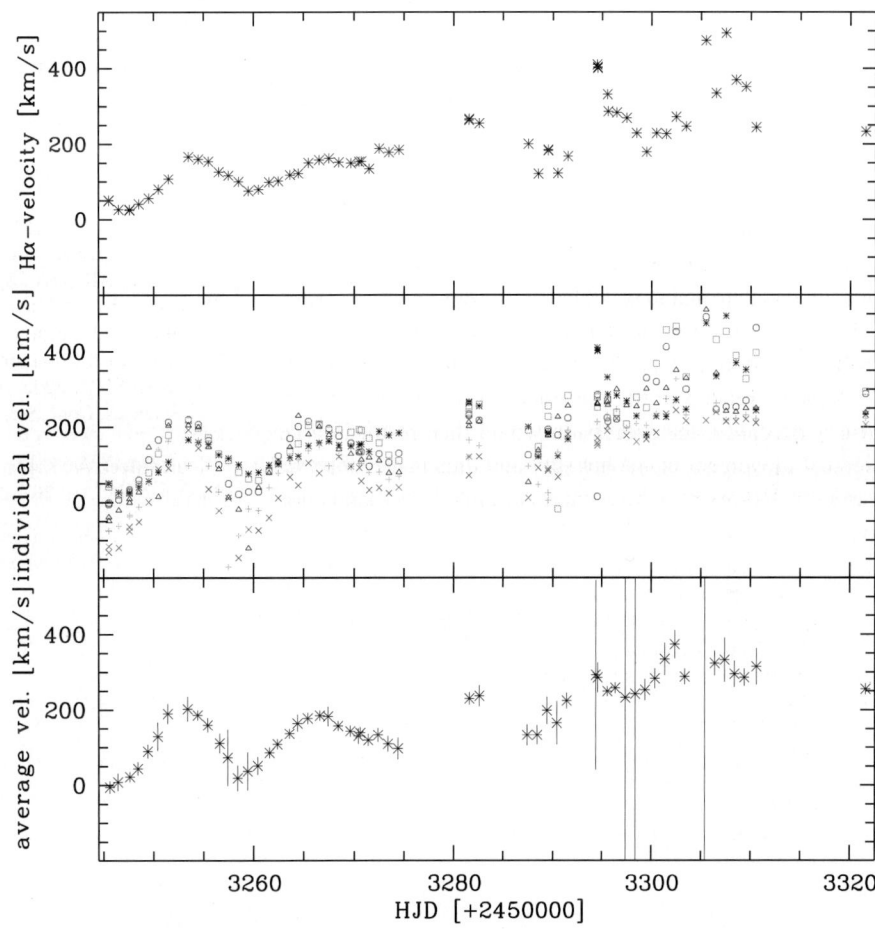

Fig. 4 The radial velocities are plotted against the heliocentric Julian date: Hα (upper plot), all the emission lines as indicated in the text (middle plot), and the average of these lines with the corresponding sigma (lower plot)

determination, the system velocity increases systematically for all analysed lines and reaches a value of $\gamma = 350$ km s^{-1} before reaching a plateau and even starts decreasing again.

We tried to fit this modulation with an additional trigonometric function. Unfortunately, the data span is too short to confirm a periodicity in this variation or, in any case, to get its period. Periods between 80 and 130 d yield equally good fits.

4. Discussion

There are three main questions to discuss: (1) Is the observed variation of the system velocity real? and if yes, then (2) Is it periodic and (3) What causes this variation?

Concerning the first question, the fact that all analysed emission lines show the same increase of system velocity is a strong argument in favour of the robustness of this variation. Another argument comes from the literature. Unfortunately, no similar study has been done before, or at least the radial velocities have been plotted against phase rather than actual date. However, we can use the variation of the individual data points in the phase–plots from Crampton and Hutchings (1981) or Gies et al., (2002). They all show a scatter of 150 to 300 km/s which is compatible with the variation of the system velocity that we found. Crampton et al. (1980) even give a table with the Hβ and Hα peak velocities and their Julian date. These values also show a variation of the system velocity with time, however the data points are too sparse to allow for a statement on its periodicity. This requires a large data set of continuous monitoring with a decent sampling of the orbital period and over at least one precession phase.

As to what causes the variation there are several possibilities. If it is not periodic it must be a transient effect which is then probably related to the radio flare which took place around HJD = 2453294 (see Blundell et al., 2006, for a discussion of this event and its effect on the optical spectra). One can see from Fig. 3 that this event coincides with the steep increase of the FWHM. This is mainly due to a broad component in the lines, which is most probably a wind feature. It increases and broadens dramatically around the time of the radio-flare and thus influences the emission line profile and possibly also the measurement of the radial velocity.

If the variation actually turns out to be a periodic one it might be related to the precession. E.g., if the accretion disc is eccentric, the precession would then result in a periodical movement of the hot spot and thus in the observed variation. However such a scenario is unlikely as a sinusoidal variation of 162 d period does not agree with the observed variation.

The most interesting possibility is of course the presence of a third body which forces the binary system on its orbit. Such a triple star scenario has been suggested before for SS 433 (Fabian et al., 1986), although on a rather smaller scale. Again, more data are needed to confirm any of these ideas.

Acknowledgements We wish to thank the La Silla team for continuously performing these observations in service mode. We acknowledge the intense use of the SIMBAD database operated at CDS, Strasbourg, France.

References

Blundell, K.M., Bowler, M.G., Schmidtobreick, L.: in preparation (2006)
Crampton, D., Cowley, A.P., Hutchings, J.B.: ApJ **235**, L131 (1980)
Crampton, D., Hutchings, J.B.: ApJ **251**, 604 (1981)
Fabian, A.C., Eggleton, P.P., Pringle, J.E., Hut, P.: ApJ **305**, 333 (1986)
Gies, D.R., McSwain, M.V., Riddle, R.L., Wang, Z., Wiita, P.J., Wingert, D.W.: ApJ **566**, 1069 (2002)
Margon, B.: ARA&A **22**, 507 (1984)
Eikenberry, S.S., Cameron, P.B., Fierce, B.W., Kull, D.M., Dror, D.H., Houck, J.R., Margon, B.: ApJ **561**, 1027 (2001)
Fuchs, Y., Koch-Miramond, L., Abraham, P.: In: Combes, F., Barret, D. (eds.) SF2A-2002: Semaine de l'Astrophysique Française, EdP-Sciences, Conf. Ser, p. 317 (2002)
Hillwig, T.C., Gies, D.R., Huang, W., McSwain, M.V., Stark, M.A., van der Meer, A., Kaper, L.: ApJ **615**, 422 (2004)
Shafter, A.W.: ApJ **267**, 222 (1983)

On Possible Nature of Pre-eclipse Dips in Light Curves of Semi-detached Systems with Steady-state Disks

D. V. Bisikalo · P. V. Kaygorodov · A. A. Boyarchuk · O. A. Kuznetsov

Received: 1 November 2005 / Accepted: 1 February 2006
© Springer Science + Business Media B.V. 2006

Abstract In order to explain pre-eclipse dips at phases ∼0.7 in light curves of cataclysmic variables, the hypothesis of the thickening at the outer edge of the accretion disk is usually used. Shock-free nature of the interaction between the stream and the steady-state disk in the stationary solution poses the question on the cause of the of matter presence at significant height above the accretion disk in systems of this type. The results of 3D modeling have shown that in the absence of direct collision between the stream and the disk the formation of thickening of the halo above the disk is also possible. In the framework of the hot line model the significant part of the matter gets the acceleration in vertical direction. Gas movement in the vertical direction together with its movement along the outer edge of the disk leads to the gradual increase of the near-disk halo width. Maximum of the calculated thickening located above the outer part of the disk corresponds to the ∼0.7 phase that is in agreement with the observed values for cataclysmic variables with steady-state disks. This fact confirms the hot line model suggested earlier for description of the flow structure in semi-detached binaries and gives new opportunities for the interpretation of the light curves of such systems.

Keywords Close binaries · Light-curves · Gasdynamic · Mass transfer · Accretion disk

D. V. Bisikalo (✉) · P. V. Kaygorodov · A. A. Boyarchuk · O. A. Kuznetsov
Institute of Astronomy, Pyatnitskaya str. 48, Moscow, 119017 Russia

Introduction

Observations of low-mass X-ray binaries (LMXBs) have revealed pre-eclipse dips in the X-ray light curves for several systems. Similar light-curve features in various wavelength ranges have also been recorded for a number of cataclysmic binaries in outburst, such as U Gem (Mason et al., 1988; Long et al., 1996), OY Car (Naylor et al., 1988 Billington et al., 1996), and Z Cha (Harlaftis et al., 1992). Further observations showed that lightcurve dips can also appear when a system is in a stationary state. Studies of the ultraviolet light curves of the eclipsing nova-like cataclysmic binaries UX UMa and RW Tri (Mason et al., 1997) confirmed this result, and suggested that this phenomenon was universal in semi-detached binaries with accretion disks. It is interesting that, in contrast to the cataclysmic systems, systems with stationary disks display pre-eclipse dips at much earlier phases, $0.6 \div 0.7$ (Mason et al., 1997; Froning et al., 2003).

This raises the question of what leads to the presence of matter at considerable heights above the accretion disk in the case of a stationary interaction between the stream and the disk. Gasdynamical studies demonstrate that the interaction between the stream and the disk is collisionless in this case (Bisikalo et al., 1998, 2003). Our aim here is to study possible ways of thickening the halo above a disk, which give rise to eclipses of the central source and the presence of dips in the light curves during the stationary flow of matter in semi-detached binaries.

Results of the computations

We used the numerical model described in paper by Bisikalo et al. (2003). We considered a semi-detached binary consisting of a donor with mass M_2 that fills its Roche lobe and an accretor with mass M_1. The following parameters were adopted for the system: $M_1 = 1.02 M_\odot$, $M_2 = 0.5 M_\odot$, and $A = 1.42 R_\odot$, corresponding to $P_{\rm orb} = 3.79^h$. The disk in the model had a temperature of $13600°$ K. For the specified rate of matter entering the system, the corresponding accretion rate in the model was $\approx 10^{-10} M_\odot/\text{yr}$.

The morphology of the matter flows in a semi-detached binary with a stationary, cool ($T = 13600°$ K) disk was described by Bisikalo et al. (2003). Let us briefly summarize the main features of the computed flow structure. The region of the interaction between the stream and circumdisk halo is shown in Fig. 1 (Figs. 7, 8 from Bisikalo et al., 2003). The left panel of Fig. 1 displays contours of constant density and velocity vectors, while the right panel of Fig. 1 is the so-called texture: a visualization of the velocity field by means of numerous tracks of test particles.

The results presented demonstrate that the interaction between the circumdisk halo and the stream possesses all the characteristic features of an oblique collision of two flows. The resulting structure of two shock waves with a tangential discontinuity between them is clearly visible in Fig. 1. The region of the shock interaction between the stream and halo has a complex shape. The parts of the halo far from the disk have low density, and the shock due to their interaction with the stream lies along the edge of the stream. The shock bends as the gas density in the halo increases, finally assuming a position along the edge of the disk. The solution has the following qualitative features: the interaction between the stream and the disk is collisionless, the region of enhanced energy release is due to the interaction between the gas in the circumdisk halo and in the stream and is located outside the disk, and the resulting system of shocks is extended and can be considered as a "hot line".

It follows from the above general features of the flow pattern that, at the interaction zone, the halo gas and stream gas pass through the shocks corresponding to their own flows, are mixed, and then move along the tangential discontinuity between the two shocks. Subsequently, the disk itself, halo, and inter-component envelope are formed of precisely this matter. The jump in the gas parameters after the passage of the shock leads to a the increase of density and temperature in the region between the shocks and, consequently, to the appearance of a pressure gradient along the z axis, perpendicular to the system's plane of rotation. As a result, the gas begins to expand vertically, increasing the z component of the velocity, until the pressure gradient is balanced by the gravitational force.

The vertical gas pressure due to the presence of the z component of the velocity, together with the motion of the gas along the tangential discontinuity at the disk's outer edge, lead to a gradual increase of the thickening of the circumdisk halo (along the z axis). This thickening of the halo along the outer edge of the accretion disk is clearly visible in the three-dimensional constant-density surface shown in Fig. 2.

The region of vertical acceleration is restricted to the hot line zone. After passing this region, the gas has a large vertical velocity component that makes it climb higher, until its store of kinetic energy is exhausted. The point where the upward motion ceases corresponds to the maximum height, which is reached at phase ~ 0.7, i.e., already substantially outside the hot line.

To quantitatively analyze the thickening of the halo above the disk, let us consider the behavior of the streamlines and the distributions of the gas parameters along them. In a cylindrical coordinate system with its origin coincident with the accretor ($x = A, y = 0.0, z = 0.0$) and the angle ϕ measured from the point L_1 opposite to the direction of rotation of the matter (coincident with the system's direction of rotation), each point of the streamline is described by the coordinates (r, ϕ, z). The $z(\phi)$ relations for four streamlines originating at points in the neighborhood of L_1 are presented in the left panel of Fig. 2. We can see that, after entering the hot line region (phase ~ 0.975), the streamlines begin to climb due to the increase of the velocity's z

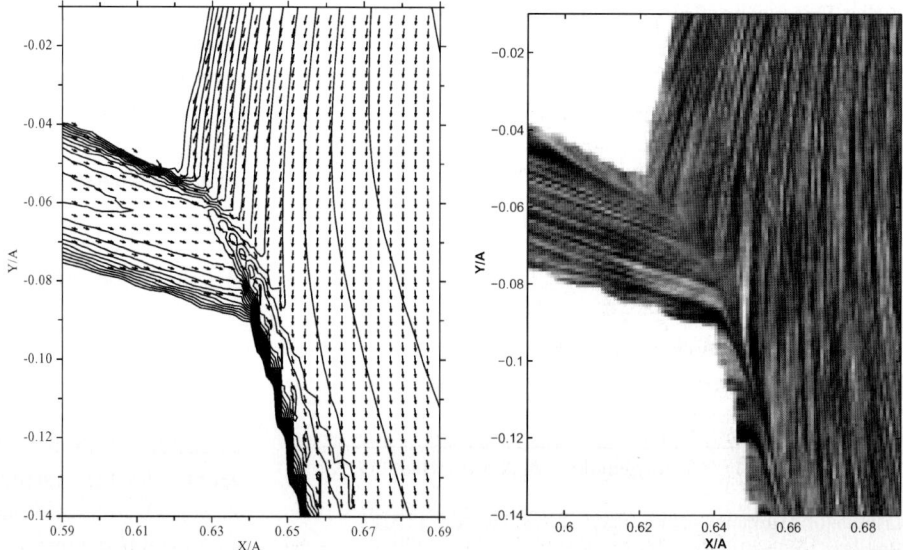

Fig. 1 Contours of constant density and velocity vectors (left panel) and a visualization of the velocity field (right panel) in the region of interaction between the stream and the circumdisk halo, in the system's equatorial plane. This Figure was first published in Bisikalo et al. 2003)

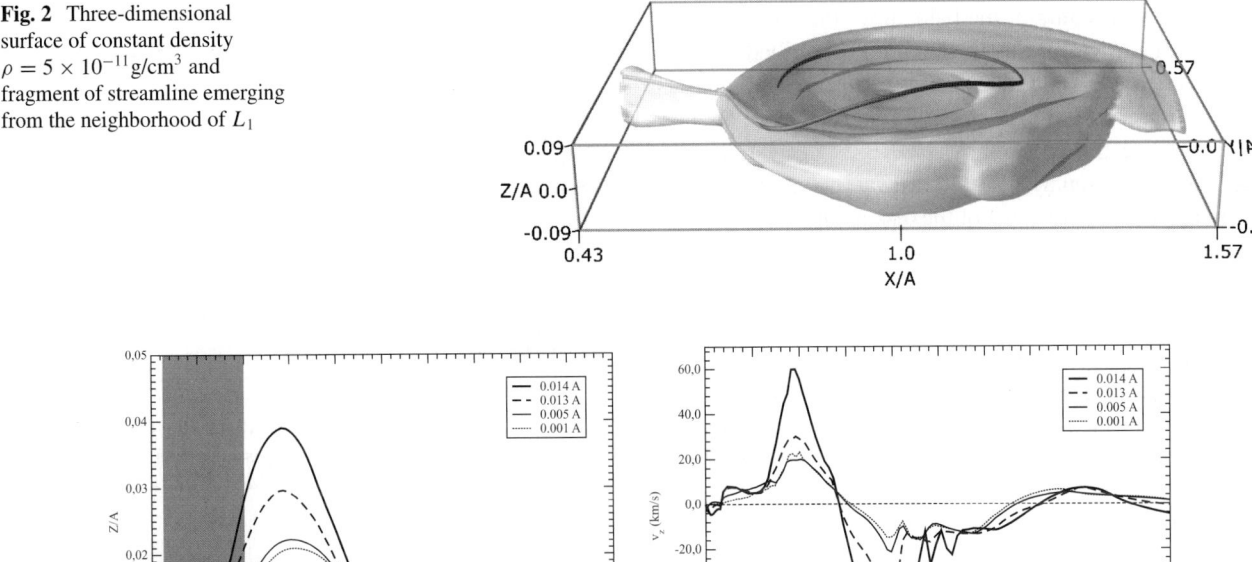

Fig. 2 Three-dimensional surface of constant density $\rho = 5 \times 10^{-11}\,\mathrm{g/cm}^3$ and fragment of streamline emerging from the neighborhood of L_1

Fig. 3 $z(\phi)$ and $V_z(\phi)$ relation for four streamlines. The streamlines originate in the vicinity of L_1 and have the coordinates $(x_{L_1}, 0, z_0)$. The z_0 values for each streamline are indicated in the plot. The shaded area corresponds to the hot line region where v_z increases

component. The phase relation of the vertical velocity for the same four streamlines is displayed in the right panel of Fig. 3. When the gas emerges from the hot line (phase ∼0.8), the force due to the pressure jump behind the shocks disappears, and the gas simply moves in the accretor's gravitational field. The vertical velocity of the gas becomes zero at phase ∼0.7, corresponding to the maximum ascent of the streamlines. The height of the halo at this position reaches ∼0.04A.

Conclusions

The results of our three-dimensional numerical modeling of the flow structure in semi-detached binaries with stationary disks confirm our earlier conclusions that the interaction between the stream and disk is collisionless, the region of increased energy release is due to the interaction between the gas in the circumdisk halo and the stream and is located outside the disk, and the system of shocks that forms is extended and can be considered as a "hot line." The interaction between the circumdisk halo and the stream possesses all the characteristic features of an oblique collision of two flows that results in the formation of a structure consisting of two shock waves and a tangential discontinuity.

In the hot line region, the halo and stream gases pass through the shocks corresponding to their own flows, are mixed, and then move along the tangential discontinuity between the two shocks. During the interaction between the stream and the circumdisk halo, a considerable fraction of the matter acquires a vertical acceleration. The vertical motion of the gas due to the z component of the velocity, together with its motion along the tangential discontinuity at the outer edge of the disk, results in a gradual growth of the thickness of the circumdisk halo. The region of vertical acceleration is restricted to the hot line zone, and its angular size does not exceed ∼65°. However, once it has passed this region, the gas has a sufficiently large vertical velocity component to rise until its store of kinetic energy is exhausted. The point where the upward motion ceases corresponds to the maximum height, which is reached at phase ∼0.7, already substantially outside the hot line. The thickening extends appreciably higher than the scale height of the disk, reaching values ∼0.04A (this corresponds to a ratio of the thickening's height to the distance to the accretor >0.1), and its angular size exceeds ∼130°. Our computations also show that, in addition to the primary maximum, the system also displays a height minimum at phase ∼0.3, when $z \sim (0.005 \div 0.006)A$, and a secondary maximum at phase ∼0.1, when the height of the halo is ∼0.01A.

Our analysis of these results leads us to conclude that the dips in the light curves of semi-detached binaries with stationary disks (i.e., in the absence of a collisional interaction between the stream and disk), can be explained by the formation of a thickening of the halo above the outer edge of the disk. The origin of this thickening is described well by the hot line model, and its quantitative features are consistent with observations. The proposed model can be applied to both semi-detached systems (LMXBs, cataclysmic variables) in their stationary state and to dwarf novae in outburst, provided that the outer parts of the disk are not strongly distorted in the outburst.

Acknowledgements This study was supported by the Russian Foundation for Basic Research (project nos. 05-02-16123, 05-02-17070, and 06-02-16097), the Program of Support for Leading Scientific Schools of Russia (grant no. NSh162.2003.2), and the basic-research programs "Mathematical Modeling and Intellectual Systems" and "Unstable Phenomena in Astronomy" of the Presidium of the Russian Academy of Sciences.

References

Billington, I., Marsh, T.R., Horne, K., Cheng, F.H., Thomas, G., Bruch, A., O'Donoghue, D., Eracleous, M.: MNRAS **279**, 1274–1288 (1996)

Bisikalo, D.V., Boyarchuk, A.A., Chechetkin, V.M., Kuznetsov, O.A., Molteni, D.: MNRAS **300**, 39–48 (1998)

Bisikalo, D.V., Boyarchuk, A.A., Kaygorodov, P.V., Kuznetsov, O.A.: Astron. Rep. **47**, 809–820 (2003)

Froning, C.S., Long, K.S., Knigge, C.: ApJ **584**, 433–447 (2003)

Harlaftis, E.T., Hassall, B.J.M., Naylor, T., Charles, P.A., Sonneborn, G.: MNRAS **257**, 607–619 (1992)

Long, K.S., Mauche, C.W., Raymond, J.C., Szkody, P., Mattei, J.A.: ApJ **469**, 841 (1996)

Mason, K.O., Cordova, F.A., Watson, M.G., King, A.R.: MNRAS **232**, 779–791 (1988)

Mason, K.O., Drew, J.E., Knigge, C.: MNRAS **290**, L23–L27 (1997)

Naylor, T., Bath, G.T., Charles, P.A., Hassal, B.J.M., Sonneborn, G., van der Woerd, H., van Paradijs, J.: MNRAS **231**, 237–255 (1988)

After the Supernova: Runaway Stars and Massive X-ray Binary Populations with Metallicity

L. M. Dray · J. E. Dale · M. E. Beer · A. R. King · R. Napiwotzki

Received: 1 November 2005 / Accepted: 1 February 2006
© Springer Science + Business Media B.V. 2006

Abstract One of the main pathways by which massive runaways are thought to be produced is by the disruption of a binary system after the supernova (SN) of one of its components. Under such a scenario, the populations of runaway stars in different phases should reflect the input binary population and its evolution. Conversely, if the system stays together after the SN, a High Mass X-Ray Binary (HMXB) may result. We present simulations exploring the behaviour of such runaway and HMXB populations with metallicity, and compare them to observations. As many as two-thirds of massive runaway stars may be produced by supernovae in binaries. Decreasing metallicity lowers the fraction of O stars which are runaway, but increases the Wolf-Rayet runaway fractions and the number of potential HMXBs.

Keywords stars: early-type · binaries: close · stars: kinematics

1. Introduction

Massive binary evolution may lead to many interesting phenomena. One notable feature of the most massive stars in particular is that around 10% of them have unusually high space velocities – greater than 30 kms^{-1} (Mason et al. 1998), or about three times as large as an 'average' O star velocity. This runaway fraction applies not only to O stars but

L. M. Dray (✉) · J. E. Dale · M. E. Beer · A.R. King
Theoretical Astrophysics Group, Department of Physics and Astronomy, University of Leicester, UK
e-mail: lynnette.dray@astro.le.ac.uk

R. Napiwotzki
Centre for Astrophysics Research, University of Hertfordshire, UK

also to Wolf-Rayet (WR) stars (Moffat et al. 1998) even though the O and WR phases are at opposite ends of a massive star's lifespan. However, less massive stars have smaller runaway fractions (e.g. Portegies Zwart, 2000). This implies that the runaway-producing mechanism results specifically from massive stellar evolution.

Two main mechanisms are thought to contribute to the observed population of runaway stars. Firstly, a proportion of runaways will result from binary-single, binary-binary and higher-order multiple system dynamical interactions (Poveda et al., 1967; Clarke and Pringle, 1992). Systems which form as unstable triples and subsequently break up may also eject stars with runaway velocity. The most probable dynamical ejection scenario is that in which two binaries interact in a young dense cluster, ejecting the two least massive stars at high velocity and forming a tight binary from the remaining two. Such interactions leave a clear observational signature: a binary and two single runaways with the same point of origin.

Secondly, stars may be made runaway by the explosion of their binary companion in a supernova (SN) explosion (Blaauw, 1961). For the binary to be split by this (most runaways being single stars, e.g. Mason et al., 1998), either half the system mass has to be lost or there has to be a suitable SN kick (Brandt and Podsiadlowski, 1995). If all runaway stars arise from this process then, assuming stars in binaries to be formed coevally, no runaways can be younger than the shortest possible time-to-SN of their companions (a few 10^6 years, Fig. 1). The presence of a runaway star, possibly displaying surface abundances which have been affected by binary mass transfer, which can be traced back to the same point of origin as a runaway neutron star (NS) is an observational signature of this process.

Hoogerwerf et al. (2001), following the paths of runaways back to their point of origin, find that the proportion of runaways which result from SNe in binaries is around

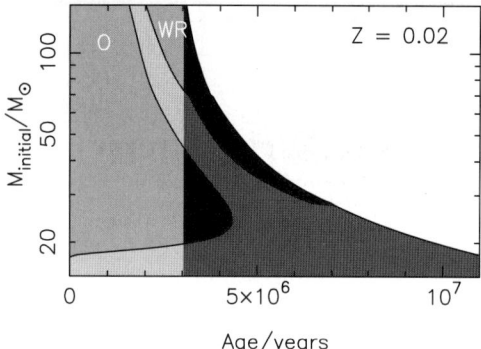

Fig. 1 Lifetimes in different evolutionary stages of massive single stars at solar metallicity, with mass. Light grey regions show portions of the stellar life which are incapable of becoming runaway stars in the binary supernova scenario, assuming both stars in a binary are born coevally. Black regions show regions in which O or WR stars may potentially be runaway. It is apparent that a much larger fraction of WR stars have the possibility of being made runaway in this simple picture

two-thirds of the total. However, in the case that binary SN runaways dominate, one might expect the runaway fraction amongst stages which are later in the star's lifetime to be much greater. Early stages in the stellar lifetime are strongly inhibited against attaining runaway status by the need for their companion to have already evolved to the SN stage (Fig. 1). Instead, the runaway fractions amongst O and WR stars (representing respectively the earliest and the last phase in the life of a massive star) are roughly similar (Moffat et al., 1998; Mason et al., 1998). This suggests the evolution of runaways is more complex than sketched above. In particular, runaway stars are strongly affected by the potentially rejuvenatory effect of mass transfer between binary components (Vanbeveren et al., 1998). As massive primaries tend to have massive companions (Garmany et al., 1980) and the primary's initial mass affects its evolution and hence how much mass is lost in the SN, which in turn affects the velocity imparted to its companion, it is also apparent that the secondary's O and/or WR evolution and whether it has high enough velocity to be classed as a runaway or not are not independent qualities. We simulate massive binaries to get an idea of the resulting runaway properties.

2. Simulations

We use Monte Carlo simulations assuming an initial population of massive binaries which is flat in mass ratio (M_2/M_1) and $\log P$, and for which the primary components obey a Salpeter IMF. The initial binary fraction is assumed close to 100% for O stars. To get stellar lifetimes and evolutionary tracks in the different phases we use the grid of interacting binary models of Dray and Tout (2005), with the single star models of Dray and Tout (2003) used for non-interacting binaries. Mass lost in the SN is calculated as in Portinari

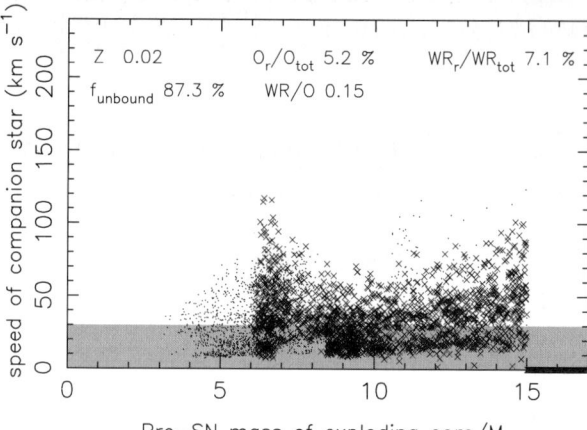

Fig. 2 Velocities of simulated secondary stars from systems with non-conservative mass transfer at solar metallicity, against the mass of the exploding core. Dots and crosses indicate stars which go through potentially-runaway O phase and/or WR phases. Resultant runaway fractions for O and WR stars are given at the top of each panel. The limit for a star to be classed a runaway here is 30 km s^{-1}

et al. (1998) and a Maxwellian distribution of kicks with mean 450 km s^{-1} (Lyne and Lorimer, 1994) is assumed. For the effect of the kick on the binary we use the formulae of Tauris and Takens (1998). Further details of the simulations are given in Dray et al. (2005). In general, adjusting the input

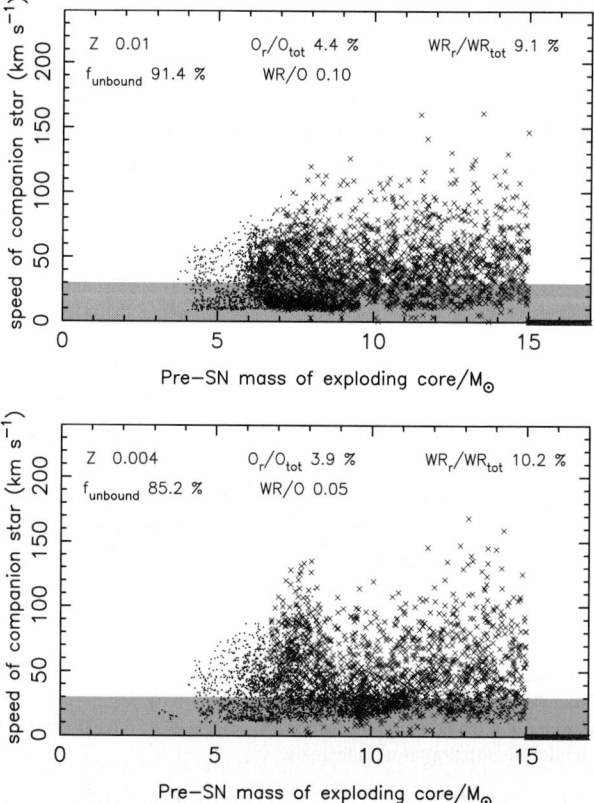

Fig. 3 As Figure 1, but showing the runaway population change with decreasing metallicity, for non-conservative mass transfer

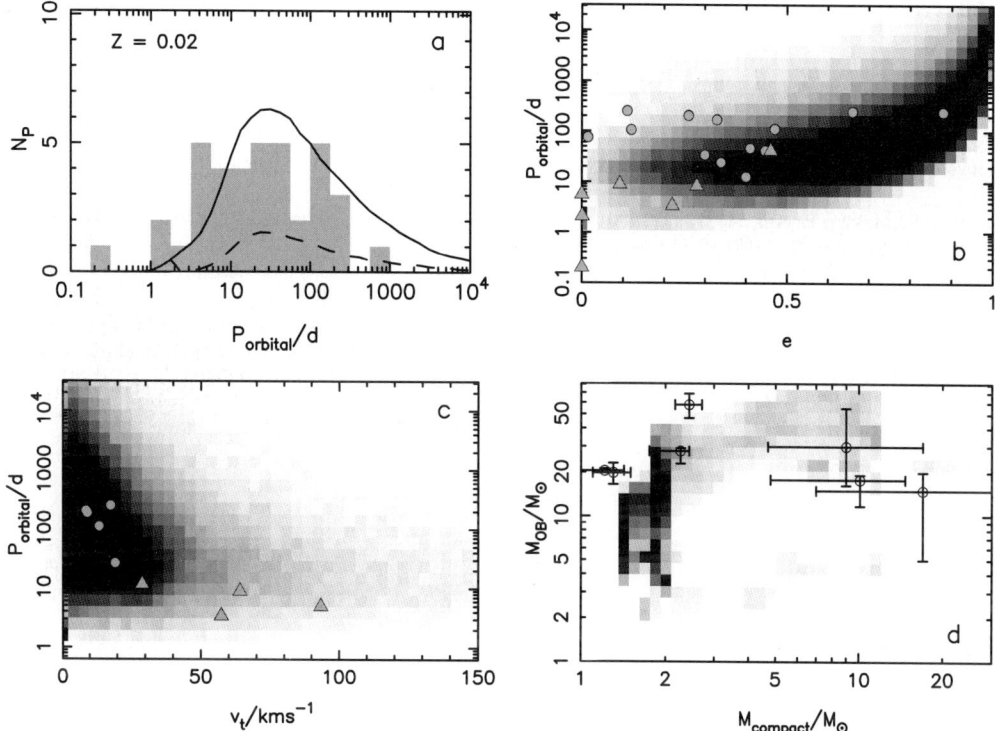

Fig. 4 Properties of non-separated non-conservative mass transfer binaries against observations of HMXBs. Panel a shows the observed orbital period histogram (grey) against model NS binaries (solid line) and BH binaries (dashed line). Panels b and c show the eccentricity and transverse velocity distributions with period against observations (grey circles for Be X-ray binaries, grey triangles for supergiant X-ray binaries). Panel d shows the mass distribution of binary components. Observations are taken from the catalogues of Liu et al. (2000) and Raguzova and Popov (2005) and references therein, with runaway velocities from Chevalier and Ilovaisky (1998)

parameters over reasonable values has relatively little effect on the resultant population. Exceptions are kicks for short-period systems (small kicks for initial period less than 10 days reduce the runaway fractions to around 1% in both cases) and the initial binary fraction, which needs to be large.

With these basic ingredients we run sets of 50000 binaries over a range of initial parameters. For solar metallicity we find non-conservative is marginally better than conservative mass transfer at fitting observations. Both produce runaway fractions which are consistent with 10% of O and WR stars being runaway, in the case that two-thirds of runaways arise from SNe (Fig. 2). For metallicities lower than solar, there are indications that the WR runaway fraction may not change significantly (Foellmi et al., 2003). We expect a small increase in WR runaway fraction and a small decrease in O star runaway fraction (Fig. 3).

3. HMXB progenitors

Not all systems are split by the first SN. Those systems which survive are the likely progenitors of HMXBs and may also have runaway velocities. In fact, most supergiant X-ray binaries are runaways, whereas most X-ray binaries with Be star companions have relatively normal space velocities Chevalier and Ilovaisky 1998), probably as a result of differing evolutionary paths van den (Heuvel et al., 2000). In our models we obtain a population of unseparated binaries whose initial properties appear to match well to the Galactic HMXB population (Fig. 4), although it should be noted that not all of them are likely to become X-ray bright. Since it has been suggested that HMXB populations could be used as a star formation tracer (Grimm et al., 2000), it is useful to know the metallicity dependence. We find that, if the HMXBs behave similarly to their progenitor populations, one would expect a number increase of around a factor of 3 as metallicity decreases between solar and SMC[1] values, and a decrease in the peak period.

References

Blaauw, A.: Bull. Astron. Inst. Netherlands **15**, 265 (1961)
Brandt, N., Podsiadlowski, P.: MNRAS **274**, 461 (1995)
Chevalier, C., Ilovaisky, S.A.: A&A **330**, 201 (1998)
Clarke, C.J., Pringle, J.E.: MNRAS **255**, 423 (1992)

---

[1] It should be noted that the SMC population itself is strongly affected by recent starbursts.

Dray, L.M., Dale, J.E., Beer, M.E., Napiwotzki, R., King, A.R.: MNRAS **364**, 59 (2005)

Dray, L.M., Tout, C.A.: In: Moffat, A.F.J., St.-Louis, N. (eds.), Massive Stars in Interacting Binaries. ASP Conf. Ser., in press (2005)

Dray, L.M., Tout, C.A.: MNRAS, **341**, 299 (2003)

Foellmi, C., Moffat, A.F.J., Guerrero, M.A.: MNRAS **338**, 1025 (2003)

Garmany, C.D., Conti, P.S., Massey, P.: ApJ **242**, 1063 (1980)

Grimm, H.-J., Gilfanov, M., Sunyaev, R.: MNRAS **339**, 793 (2003)

Hoogerwerf, R., de Bruijne, J.H.J., de Zeeuw, P.T.: A&A **365**, 49 (2001)

Liu, Q.Z., van Paradijs, J., van den Heuvel, E.P.J.: A&AS **147**, 25 (2000)

Lyne, A.G. Lorimer, D.R.: Nature **369**, 127 (1994)

Mason, B.D., Gies, D.R., Hartkopf, W.I., Bagnuolo, W.G. Jr., ten Brummelaar, T., McAlister, H.A.: AJ **115**, 821 (1998)

Moffat, A.F.J., Marchenko, S.V., Seggewiss, W., van der Hucht, K.A., Schrijver, H., Stenholm, B., Lundstrom, I., Gunawan, D.Y.A. Setia, Sutantyo, W., van den Heuvel, E.P.J., de Cuyper, J.-P., Gomez, A.E.: A&A **331**, 949 (1998)

Portegies Zwart, S.F.: ApJ **544**, 437 (2000)

Portinari, L., Chiosi, C., Bressan, A.: A&A **334**, 505 (1998)

Poveda, A., Ruiz, J., Allen, C.: Bol. Obs. Tonantzinla Tacubaya **28**, 86 (1967)

Raguzova, N.V., Popov, S.V.: MNRAS accepted (astro-ph/0505275) (2005)

Tauris, T.M., Takens, R.J.: A&A **330**, 1047 (1998)

Vanbeveren, D., de Donder, E., van Bever, J., van Rensbergen, W., de Loore, C.: New Astronomy **3**, 443 (1998)

van den Heuvel, E.P.J., Portegies Zwart, S.F., Bhattacharya, D., Kaper, L.: A&A **364**, 563 (2000)

X-rays from the Symbiotic Star RX Pup

Gerardo J. M. Luna · J. L. Sokoloski · Roberto D. D. Costa

Received: 1 November 2005 / Accepted: 1 February 2006
© Springer Science + Business Media B.V. 2006

Abstract We describe X-ray and optical observations of the symbiotic star RX Pup. From low resolution optical spectra, we obtain a reddening for RX Pup of $E(B-V) = 0.79$. We use the neutral column density corresponding to this reddening as a lower limit for the X-ray spectra fits. The X-ray spectra can be fitted with either a two-temperature thermal plasma model or a single-temperature plasma plus a narrow line at ≈ 0.55 keV, each modified by interstellar absorption. The RX Pup X-ray flux is not variable within the observation exposure time, suggesting that unlike in most CVs, an accretion disk boundary layer does not contribute significantly to the X-ray flux. Instead, the X-ray emission may come from shock-heated gas further away from the compact object.

Keywords Stars-binaries: Symbiotics · X-ray · Optical

1. Introduction

Symbiotic stars are binary systems with a red giant, a compact object (in most cases a white dwarf) and a nebula surrounding the binary. The nebulae are ionized by UV radiation from the compact object, and present different morphologies when observed with sufficient spatial resolution. A region of colliding winds may exist in these systems, as well as accretion disk and jet-like structures. Up to now, only two symbiotics have shown jet-like structures in X-rays, CH Cyg (Galloway and Sokoloski, 2004) and R Aqr (Kellog et al., 2001). X-ray emission of $\approx 10^6$ K is expected in the type of colliding wind environments expected to be present in some symbiotic stars, in which a *cold, slow* wind from the red giant component collides with a *hot, fast* wind from the compact source. In particular, a very detailed model for colliding winds in symbiotic stars was developed by Willson et al. (1984) in a series of papers discussing X-ray observations of HM Sge. Recently, a more detailed model has been published by Kenny and Taylor (2005). However, they only present the radio emission map expected in such environments.

2. Observations and reduction

The XMM-Newton observation of RX Pup (ObsId: 0153350101) was analyzed using the latest version of SAS (Science Analysis Software v6.5). We fit the EPIC spectra of RX Pup using XSPEC v.12 (Arnaud, 1996). SAS was used to construct response matrices and effective area curves for the specific source spectral extraction regions. The source region was centered on the position $\alpha = 08\ 14\ 12.239$, $\delta = -41\ 42\ 29.17$.

The nominal exposure time was 17.4 ks. The good time interval for the observation, after subtracting data contaminated by solar flares (3σ cleaning) were: 13936.59 s (pn), 16024.62 s (MOS 1) and 16133.79 s (MOS 2).

3. Results

In Fig. 1 (left panel), we show the spectra from the three EPIC cameras for RX Pup. The spectra can be fit with a thermal

G. J. M. Luna
Harvard-Smithsonian Center for Astrophysics, USA & Instituto de Astronomia, Geofisica e Ciencias Atmosfericas/USP, Brazil

J. L. Sokoloski
Harvard-Smithsonian Center for Astrophysics

R. D. D. Costa
Instituto de Astronomia, Geofisica e Ciencias Atmosfericas/USP, Brazil

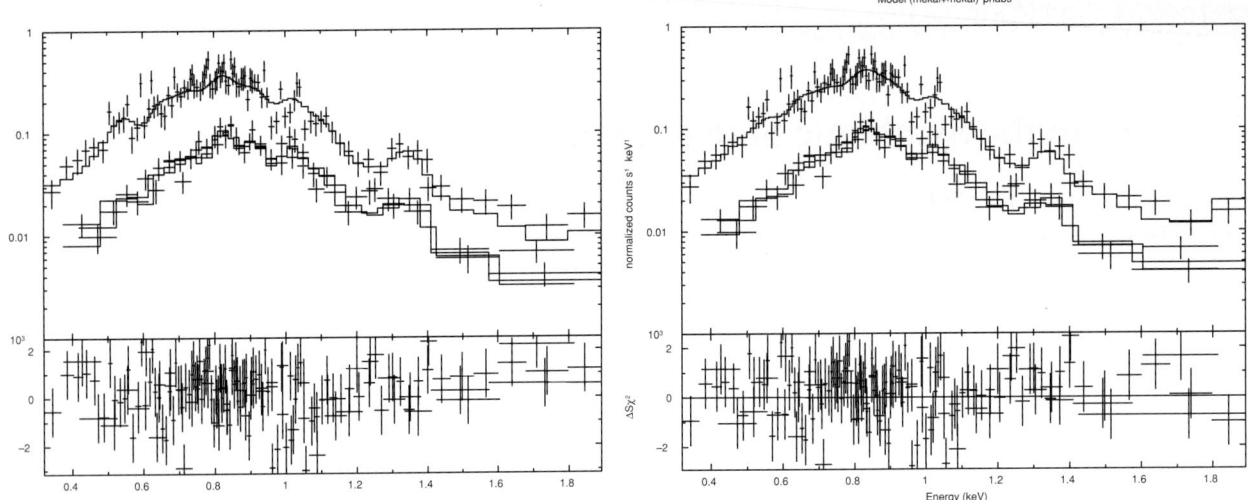

Fig. 1 RX Pup EPIC pn, MOS 1 and MOS 2 background-subtracted spectra (top panel) and χ^2 values for the fit

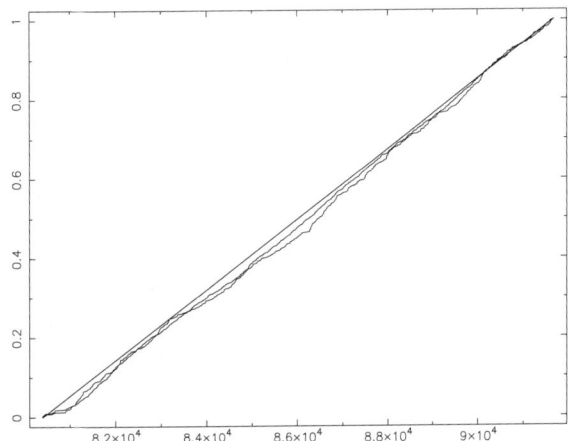

Fig. 2 Cumulative distribution for the source (traced line), background (full line) and constant model (dotted) for the pn camera in the range of 0.25 to 2.1 keV

plasma (*mekal*), as suggested by the presence of He-like Si XIII and Mg XI lines and a gaussian line accounting for the soft X-ray excess in $\approx$0.55 keV(possibly OVII). From low-resolution optical spectra, we determined the reddening to be $E(B - V) = 0.79 \pm 0.1$, which was estimated using the method described in Luna and Costa (2005). This reddening corresponds to a column density of $N_H = 3.0 \times 10^{21}$cm^{-2} (Groenewegen and Lamers, 1989), which was used as the starting value for the X-ray spectral fitting. The best-fit model ($\chi^2 \infty \nu = 1.22$) has kT $= 0.268^{0.291}_{0.249}$ keV ($3.11^{3.37}_{2.88} \times 10^6$ K), $N_H = 5.5^{6.3}_{5.2} \times 10^{21}$ cm^2 and line energy $0.55^{0.57}_{0.53}$ keV. Taking a distance of 1.8 kpc (Muerset et al., 1997), the unabsorbed 0.25–2.1 keV luminosity is $L_X = 1.35^{1.56}_{1.13} \times 10^{33}$ erg/s.

The spectra also can be fit with two thermal plasmas (Fig. 1 right panel). The best fit parameters ($\chi^2 \infty \nu = 1.17$) are: $kT_1 = 0.181^{0.199}_{0.156}$ keV ($2.10^{2.31}_{1.81} \times 10^6$ K); $kT_2 = 0.521^{0.566}_{0.480}$ keV ($6.04^{6.56}_{5.57} \times 10^6$ K); $N_H = 3.89^{4.53}_{3.33} \times 10^{21}$cm^{-3}, $L_{X_1} = 5.65^{6.55}_{4.75} \times 10^{32}$ erg/s, $L_{X_2} = 4.47^{5.18}_{3.75} \times 10^{32}$ erg/s

From ROSAT observations, Muerset et al. (1997) derived a plasma temperature of $T_X = 7.4^{10.47}_{5.25} \times 10^6$ K. They constrined the N_H value to $1.9^{2.66}_{1.52} \times 10^{21}$ atoms/cm^2 using $E(B - V)$ measurements. The luminosity derived was $L_X = 4.69^{8.15}_{2.7} \times 10^{32}$ erg/s.

We performed a timing analysis on the data to search for rapid variability. We extracted the arrival times of the source and background counts and applied a Kolmogorov-Smirnov test (KS test). The KS test shows that the cumulative distributions of the source, background and constant model are not significantly different. We therefore do not find any evidence for rapid variability. In Fig. 2 we show the cumulative distributions of source, background, and a constant model. The KS value is 0.03, and the null hypothesis probability

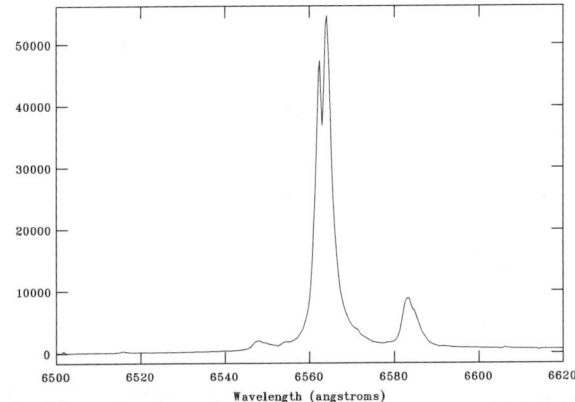

Fig. 3 High-resolution optical spectrum from CASLEO (Argentina) showing Hα and [NII] emission lines

is 0.02 (i.e., the probability that the two distributions are different is very small) when we compare source and background.

4. Discussion

The RX Pup X-ray spectra can be fitted using thermal plasma emission models. As pointed out by Muerset et al. (1997), when the wind from the hot compact source crashes into the cool star or its wind, the gas will be heated to several times 10^6 K and produce X-ray emission as shown by RX Pup. The values of column density (N_H) and plasma temperature (T_X) for RX Pup are similar to the values derived in other symbiotic stars (Muerset et al., 1997). The high-resolution optical spectrum (Fig. 3) shows asymmetrical profiles in the emission lines, which could be produced by colliding winds.

Acknowledgements This work was supported by CNPq and CAPES, Brazil and NSF grant number 0302055.

References

Galloway, D.K., Sokoloski, J.L.: ApJ **613L**, 61 (2004)
Girard, T., Willson, L.A.: A&A **183**, 247 (1987)
Groenewegen, M.A.T., Lamers, H.J.G.L.: A& AS **79**, 359 (1989)
Kellogg, E., Pedelti, J.A., Lyon, R.G.: ApJ **563**, L151 (2001)
Kenny, H.T., Taylor, A.R.: ApJ **619**, 527 (2005)
Luna, G.J.M., Costa, R.D.D.: A&A **435**, 1087 (2005)
Muerset, U., Wolff, B., Jordan, S.: A&A **319**, 201 (1997)
Whitelock, P., Barba, R., Garcia, L., Marang, F.: MNRAS **305**, 190 (1999)
Wallerstein, G., Willson, L.A., Salzer, J., Brugel, E.: A&A **133**, 127 (1984)
Willson, L.A., Wallerstein, G., Brugel, E.W., Stencel, R.E.: A&A **133**, 154 (1984)

Results of Optical Monitoring of Cataclysmic Variable GK Per in 2004

Irina Voloshina · Valerian Sementsov

Received: 1 November 2005 / Accepted: 1 February 2006
© Springer Science + Business Media B.V. 2006

Abstract The preliminary results of ground-based UBV photometric and CCD observations of cataclysmic variable GK Per during the outburst in autumn of 2004 are presented.

Keywords Close binaries · Cataclysmic variables · Nova · Dwarf nova · Intermediate polar · UBV photometry · CCD observations

Cataclysmic variable GK Per ($M_V = 14^m$) belongs to classical nova systems. It erupted in 1901 with an amplitude about 14 magnitude. The orbital period of GK Per is the longest known orbital period for a classical nova, $-1^d.996803$. From 1948 it began to show dwarf nova outbursts with recurrence time 880÷1240 days (Kim et al., 1992). During outbursts (which last 70 days on average) the brightness of GK Per is about $10^m.5$, the brightness in quiescence $- 13^m.5$. The system consists from a white dwarf as a primary component (with mass $M_1 \sim (0.7 - 0.8)M_\odot$) and an evolved K2 subgiant (with mass $M_2 \sim 0.25M_\odot$) as a secondary. The radii of the components are $0.01R_\odot$ and $1.92R_\odot$ accordingly, $i \sim 70°$ (parameters were taken from Catalog of Cherepashchuk et al., 1996). The secondary late-type star fills its Roche lobe, the gas stream flows from the secondary to the primary white dwarf. The mass loss is estimated as $(7 - 8.2)M_\odot yr^{-1}$. GK Per has a weak magnetic field (B $\simeq 10^5 - 10^6$ G), which does not prevent the formation of the accretion disk around white dwarf in this system.

The intermediate polars have two main periods: an orbital period and spin period of the white dwarf, which in the case of GK Per is equal 351 s (Crampton et al., 1986) seen in X-rays and in the optical range. The rotation of the components is asynchronous and degree of asynchronism P_{sp}/P_{orb} is 0.002.

GK Per displays complicated variations during the outburst cycle. There are cyclic variations of brightness with the full amplitude of $0^m.1 - 0^m.3$ on the timescale of 80 min (Patterson, 1994). In quiescence it shows variations with an amplitude about $0^m.3$ on the timescale of few days and with amplitude up to 1.5^m on the timescale of dozens days (Bianchini et al., 1990). There is also a flickering about $0^m.1$ on the timescale of a few minutes (Bruch, 1992). Photometric behavior of this object during the different periods of GK Per history was investigated by different authors: Hudec (1981), Sabbadin and Bianchini (1983), Bruch (1992). GK Per was detected as an X-ray source (Norton et al., 1988). Taking all this into account, GK Per was chosen as a good target for observations with INTEGRAL Space Observatory in autumn of 2004. These observations were made on October 1, 2004.

1. Observations

Our monitoring of GK Per in frame of announced international campaign has started after the beginning of the outburst, on October 13 and lasted up to the end of the outburst (JD 2453441). Optical observations were made with the help of UBV photometer on the 60-cm telescope of Sternberg Astronomical Institute in Crimea. The star SAO 38899 with magnitudes $V = 9^m.09$, $B = 9^m.38$, $U = 9^m.54$ from the nearest vicinity of GK Per served as the local standard. More than 200 photoelectric measurements of GK Per were obtained in 3 bands during the period of our observations. The measurement errors are 1% in the V and B bands and 2% in the U band.

CCD observations of this system during few nights in October were also performed with the goal to study short

I. Voloshina (✉) · V. Sementsov
Sternberg Astronomical Institute, Moscow, Russia
e-mail: vib@sai.msu.ru

Table 1 Journal of CCD observations of GK Per in 2004

Date	Start–end, JD	Band	Expos.	N
13 Oct. 2004	2453291.7488–2453291.9814	R	10 s	907
14 Oct. 2004	2453292.6011–2453292.9539	V	10 s	825
16 Oct. 2004	2453294.7923–2453294.8166	R	10 s	48
17 Oct. 2004	2453295.7575–2453295.8956	V	30 s	309
19 Oct. 2004	2453297.7611–2453297.8268	R	30 s	152

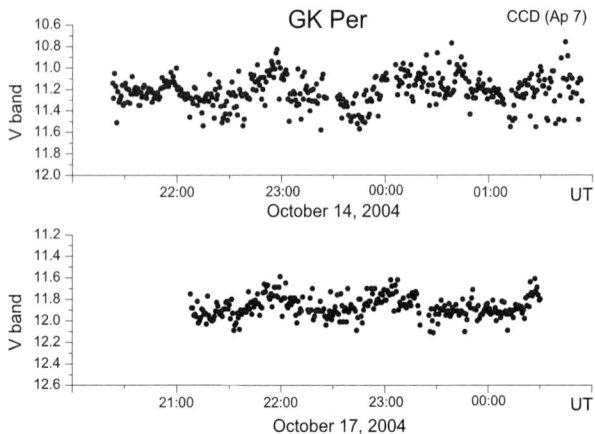

Fig. 1 The observational light curves of GK Per

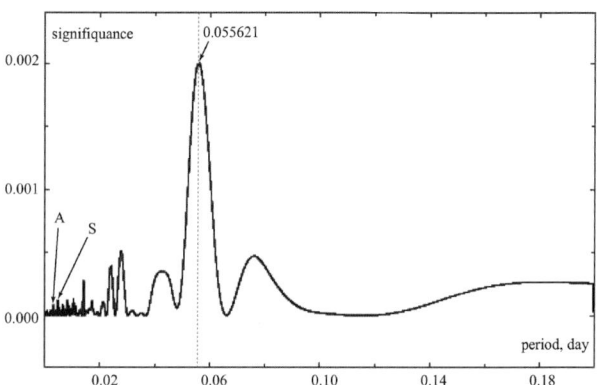

Fig. 2 The Scargle periodogram for October 14. Marked harmonics: (A) – atmospheric jitter, and (S) – second sub-harmonic of (A), or white dwarf spin

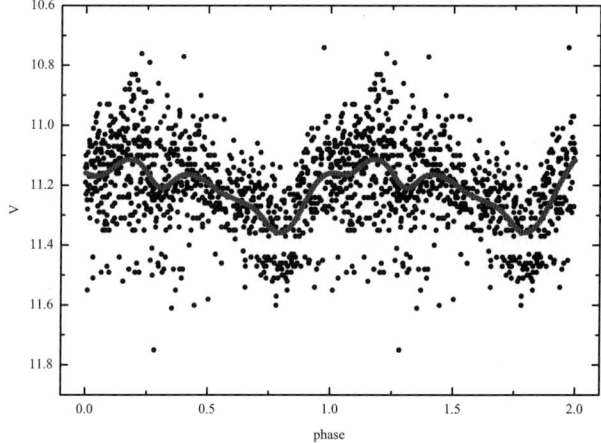

Fig. 3 Light curve of GK Per for October 14 folded with the period 80.094 min. Solid curve is 80-points FFT-smoothed data

variability of GK Per. Observations were made on the 60-cm telescope with CCD device Apogee 7p in the V, R and I bands (512 × 512 pixel, 1 pixel 24 μm). A brief description of our CCD observations is given in Table 1. The duration of individual observational sets about 4–5 hours (except JD2453294).

CCD frames were reduced using quadratic polynomials. Coordinates and photometric data for standard stars were taken from the well-known catalogue USNO UCAC-2 (Zacharias et al., 2004). The number of reference stars varies from 10 to 20. The magnitudes of these stars lies within the range of $12^m - 16^m$ (V).

2. Results

The different duration of observational sets and the wide range of suspected periods of GK Per led to the choice of particular method for light curve analysis. The Lomb–Scargle algorithm (Scargle, 1982) was used to search the periodicities in CCD data of GK Per in outburst. The highest peak on periodogram for data in the V band corresponds to 80 min period. This confirm the existence of 80-min periodicity in GK Per optical light curve which nature isn't quite clear yet. We also tried to find shorter periods on the time scale of 5 min, most likely related to the spin period of the white dwarf, but the interference of atmospheric jitter at 2 min and operating cycle of CCD-camera (approx. 40 s) cause strong artifacts on this frequency. The Scargle periodogram and phase curve for the most significant period 80.094^m determined from the data of the best observational night of October 14 are shown in Figs. 2 and 3.

From our analysis of GK Per observations during the outburst we could conclude:

- 80 min period variations exist in the V band with full amplitude about 0.6^m in the outburst.
- The color indices of GK Per are: $B - V = 0.444$, $U - B = -0.744$ (outburst), $B - V = 0.872$, $U - B = -0.404$ (quiescence)

The flickering at the timescale of few minutes masks the short periodicity of GK Per. So CCD-observations with 10 s exposures are not suitable for the search of 5.85 min variability. The orbital period must be excluded more precisely.

We thank N. Metlova for assistance in observations. This work was supported through the grant of Leading Scientific Schools of Russia NSh-388.2003.2 and RFBR grant 02-05-26818.

References

Bianchini, A., Margoni, R., Spogli, C.: IAUC, Circ. No. 5031 (1990)
Bruch, A.: A&A **266**, 237 (1992)
Cherepashchuk, A.M., Katysheva, N.A., Khruzina, T.S., Shugarov, S.Yu.: Catalog of Highly Evolved Close Binary Stars. Part 1. In: Cherepashchuk, A.M. (ed.) Brussel: Gordon and Breach Sci. Publ., p. 356 (1996)
Crampton, D., Fisher, W.A., Cowley, A.P.: Ap. J. **300**, 788 (1986)
Hudec, R.: Bull. Astron. Inst. of Czechoslovakia **32**, 93 (1981)
Kim, S.-W., Wheller, J., Mineshige, S.: Ap. J. **384**, 269 (1992)
Patterson, J.: PASP **106**, 209 (1994)
Scargle, J.D.: ApJ **263**, 835 (1982)
Zacharias, N., Urban, S.E., Wycoff, G.L. et al.: The Second US Naval Observatory CCD Astrograph Catalog (UCAC2). A.J. **127**, 3043 (2004)

Tomographic Simulations of Accretion Disks in Cataclysmic Variables – Flickering and Wind

Fabíola Mariana Aguiar Ribeiro · Marcos Diaz

Received: 30 September 2005 / Accepted: 30 October 2005
© Springer Science + Business Media B.V. 2006

Abstract Cataclysmic Variables (CVs) are close binary systems where mass is transferred from a red dwarf star to a white dwarf star via an accretion disk. The flickering is observed as stochastic variations in the emitted radiation both in the continuum and in the emission line profiles.

The main goal of our simulations is to compare synthetic Doppler maps with observed ones, aiming to constrain the flickering properties and wind parameters.

A code was developed which generates synthetic emission line profiles of a geometrically thin and optically thick accretion disk. The simulation allows us to include flares in a particular disk region. The emission line flares may be integrated over arbitrary "exposure" times, producing the synthetic line profiles. Flickering Doppler maps are created using such synthetic time series. The presence of a wind inside the Roche lobe was also implemented. Radiative transfer effects in the lines where taken into account in order to reproduce the single peaked line profiles frequently seen in nova-like CVs.

Keywords Accretion disks · Cataclysmic variables · Doppler tomography

1. Introduction

The flickering, or rapid variability, is observed as stochastic fluctuations in the emitted radiation. The typical timescales range from seconds to tenths of minutes, and the amplitudes vary from cents of magnitude to more than one magnitude. The flickering is a defining characteristic of cataclysmic variables, being frequently used to classify an object as a CV. Flickering is also observed in other classes of objects, as in some symbiotic stars (Mikolajewski et al., 1990). The first CV where flickering was observed is UX UMa (Linell, 1949). Since then, many photometric studies were made aiming to locate the flickering source region in many systems. Diaz (2001) proposed a tomographic method to map the flickering source regions using line profile data. The flickering tomography was applied to V442 Oph system, and an isolated flickering source region could not be identified. The objective of our simulations is to generate flickering tomograms from synthetic line profiles and compare these tomograms with the observed maps, aiming to constrain some parameters of the flickering and locate its forming region.

The presence of winds in cataclysmic variables is noticed by strong wind driven lines in UV, i.e. C IV, and the occurrence of P-Cygni profiles. As some tomograms present emission at low velocities, we have implemented the presence of a optically thick wind, aiming to reproduce this behavior. The comparison of model tomograms with observations may help us to constrain the main wind parameters.

In this proceeding we present the main physical concepts and parameterizations contained in the simulations and some preliminary results that arise from them.

2. Simulations

The first part of the simulation is the calculation of the line profiles from a Keplerian accretion disk. The disk steady line emissivity is described as a radial power law. In this disk, regions of enhanced emission could be marked to represent the hot spot and/or the boundary layer. The flickering and wind components are simply added to the underlying disk emission.

F. M. A. Ribeiro (✉) · M. Diaz
Universidade de São Paulo – Instituto de Astronomia, Geofísica e Ciências Atmosféricas – Brazil
e-mail: fabiola@astro.iag.usp.br

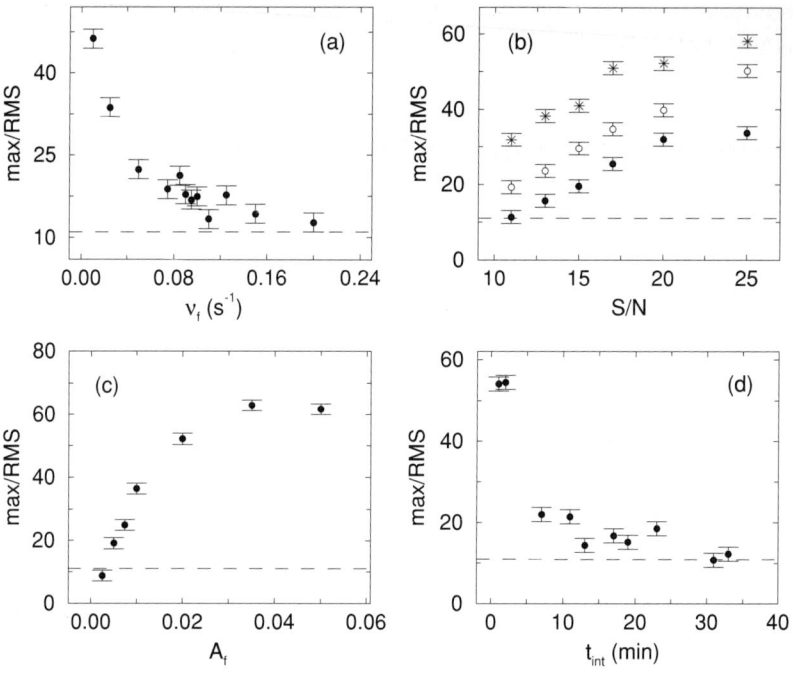

Fig. 1 Behavior of the quality factor (see text) with (a) flickering frequency, (b) S/N ratio, (c) flickering amplitude and (d) integration time. In (b) the "●" correspond to simulations with 1000 spectra and flickering amplitude 0.5%, the "○" were found with 500 spectra and amplitude 0.5% of the total line flux, and the "∗" correspond to simulations with 1000 spectra and amplitude of 1%. The dashed lines indicate max/RMS equal 11

2.1. Flickering

The flickering is implemented as a set of random flares in a predefined region of the accretion disk. In the simulations presented here the flickering was set to occur inside the hot spot region. The flares have instantaneous rise and exponential decay. They are integrated over an "exposure time" to generate the flickering contribution to each synthetic spectra.

To quantify our ability of detecting flickering spots, a quality factor was defined as being the ratio between the spot intensity in the flickering tomogram and the RMS over a steady region. The simulations suggest a detection criterion corresponding to quality factors greater than 11. The quality factor behavior was verified by varying the flickering parameters one by one (Fig. 1). As expected, one verifies that the detection is better for high S/N data and a large number of spectra. We also verified that the greater the flickering amplitude, the better is the detection. The information contained in the low amplitude flickering is lost first by dilution in the noise. The detection is also better for low frequency flickering and short integration times. Furthermore, we verified that the information contained in the high frequency flickering is lost first. As expected, high frequency structures are lost when long integration times are used.

2.2. Wind

The wind was implemented as coming from the accretion disk with a bipolar geometry, limited by the white dwarf and the disk outer rim. The velocity field of the wind is a composition of a Keplerian component, due to the disk angular moment conservation, and a poloidal component. The poloidal component follows a Castor and Abbot velocity field (Long and Knigge, 2002). The Keplerian wind component is dominant over the poloidal one at small radii while both components are comparable at the outer disk.

Radiative recombination is considered as the main Balmer and He line production mechanism in the wind. Radiative transfer in the wind must be included to reproduce the observed Doppler tomograms filled at low velocities. Was found that an optical depth $\tau_{line} > 10$ is needed to reproduce single peaked line profiles. Scattering is not included in the radiative transfer. The emissivity is calculated for each wind element, then the correspondent absorption inside the primary Roche lobe and along the line of sight is taken into account.

The response of the line profiles to each wind parameter was checked out. For instance, lower values of the wind acceleration coefficient produce a deficit in the red line wing, which can be interpreted as the effect of the poloidal wind component at the opposite side of the Roche lobe being more attenuated due to self absorption. As the wind get less collimated, the line profiles become wider. Other parameters, like the terminal velocity, effective acceleration scale and wind initial velocity, basically change the total line intensity.

3. Conclusions

Simulations of synthetic accretion disc line profiles including flickering and wind were performed. From the simulations one can see that high S/N and high time-resolution spectra are needed in order to obtain information from flickering tomograms. The information contained in the low amplitude and high frequency flickering is lost first. The flickering information is also lost if long integration times are used. From the wind simulations we conclude that a line optical depth greater than 10 is needed to obtain single peaked wind line profiles.

As incoming work this code will be used to generate synthetic tomograms with flickering, which will be compared to observed flickering tomograms of V3885 Sgr, aiming to locate the sites of flickering production and constrain the flickering parameters in this system.

Acknowledgements F.M.A.R is grateful from support from FAPESP fellowship 01/07078-8. MD acknowledges the support by CNPq under grant #301029.

References

Diaz, M.P.: ApJ **553**, L177 (2001)
Linnell, A.P.: S&T **8**, 166 (1949)
Long, K.S., Knigge, C.: ApJ **579**, 725 (2002)
Mikolajewski, M., Mikolajewska, J., Khudiakova, T.N.: A&A **235**, 219 (1990)

X1908+075: A Late O-Type Supergiant with a Neutron Star Companion

Thierry Morel · Yves Grosdidier

Received: 1 November 2005 / Accepted: 1 February 2006
© Springer Science + Business Media B.V. 2006

Abstract X1908 + 075 is a highly-absorbed Galactic X-ray source likely made up of a pulsar accreting wind material from a massive companion. We have used near-IR photometric data complemented by follow-up spectroscopy to identify the likely counterpart to this X-ray source and to assign a spectral type O7.5–9.5 If to the primary. Further details can be found in Morel and Grosdidier (2005).

Keywords Stars: early-type · Infrared: stars · X-rays: binaries · X-rays: individual: X1908+075

1. X1908 + 075 seen in X-rays

Because of their short lifetimes ($\sim$15,000 yr), OB supergiants harbouring a compact object (a neutron star or more rarely a black hole) constitute a rare, yet important and poorly understood evolutionary phase of massive binaries. The highly absorbed Galactic X-ray source X1908 + 075 has been recently found to belong to this class of objects and is made up of a neutron star accreting wind material (without RLOF) from the massive primary (Wen et al., 2000). X1908+075 displays energy-dependent, sinusoidal modulations of the X-ray power ($<L_X$ [5–100 keV]$>\sim 5 \times 10^{36}$ ergs s^{-1}) with a period of about 4.4 days. Pulsed emission with $\mathcal{P} = 605$ s has been detected. The following physical parameters for the primary star have been estimated based on RXTE data: $M = 9 - 31 M_\odot$, $R <22$ R$_\odot$ and a wind mass-loss rate greater than $1.3 \times 10^{-6} M_\odot$ yr^{-1} (Levine et al., 2004). This led to the suggestion that the mass donor could be a Wolf-Rayet star.

2. Identification of the potential counterparts

JHK_s-band images of the field were acquired on service observing mode in 2001 with the IR camera CAIN-II mounted on the TCS (Teide observatory, Canary Islands). Several sources are detected in the science frames going down to $J = 17.8$, $H = 16.5$ and $K_s = 16.5$ mag (Fig. 1). Note that these observations are of higher quality than the 2MASS data.

The colour-colour and colour-magnitude diagrams for the stars detected in the X-ray error boxes are shown in Fig. 2. The potential counterparts were pre-selected by considering the expected positions of two representative early-type stars (O5 V and O9 I) lying at $d = 7 \pm 3$ kpc and suffering $A_V = 15 \pm 5$ mag of extinction, as estimated from the X-ray data (Wen et al., 2000). Three candidates (A, B and C; see also Fig. 1 for their position) have near-IR properties consistent with the assumed distance and amount of visual extinction. However, candidate C lies slightly outside the intersection of the two X-ray error boxes.

3. Spectroscopic follow-up

Medium-resolution ($R\sim 800$) HK-band spectroscopy of the two prime candidates A and B was obtained on service observing mode with CGS4 at UKIRT in 2002. The observations were serendipitously carried out at superior conjunction of the neutron star ($\phi\sim 0.07$) according to the ephemeris for a circular orbit (Levine et al., 2004).

T. Morel (✉)
Katholieke Universiteit Leuven, Departement Natuurkunde en Sterrenkunde, Instituut voor Sterrenkunde, Celestijnenlaan 200B, B-3001 Leuven, Belgium
e-mail: thierry@ster.kuleuven.be

Y. Grosdidier
Department of Physics, McGill University, 3600 University St., Montréal, Québec, Canada, H3A 2T8

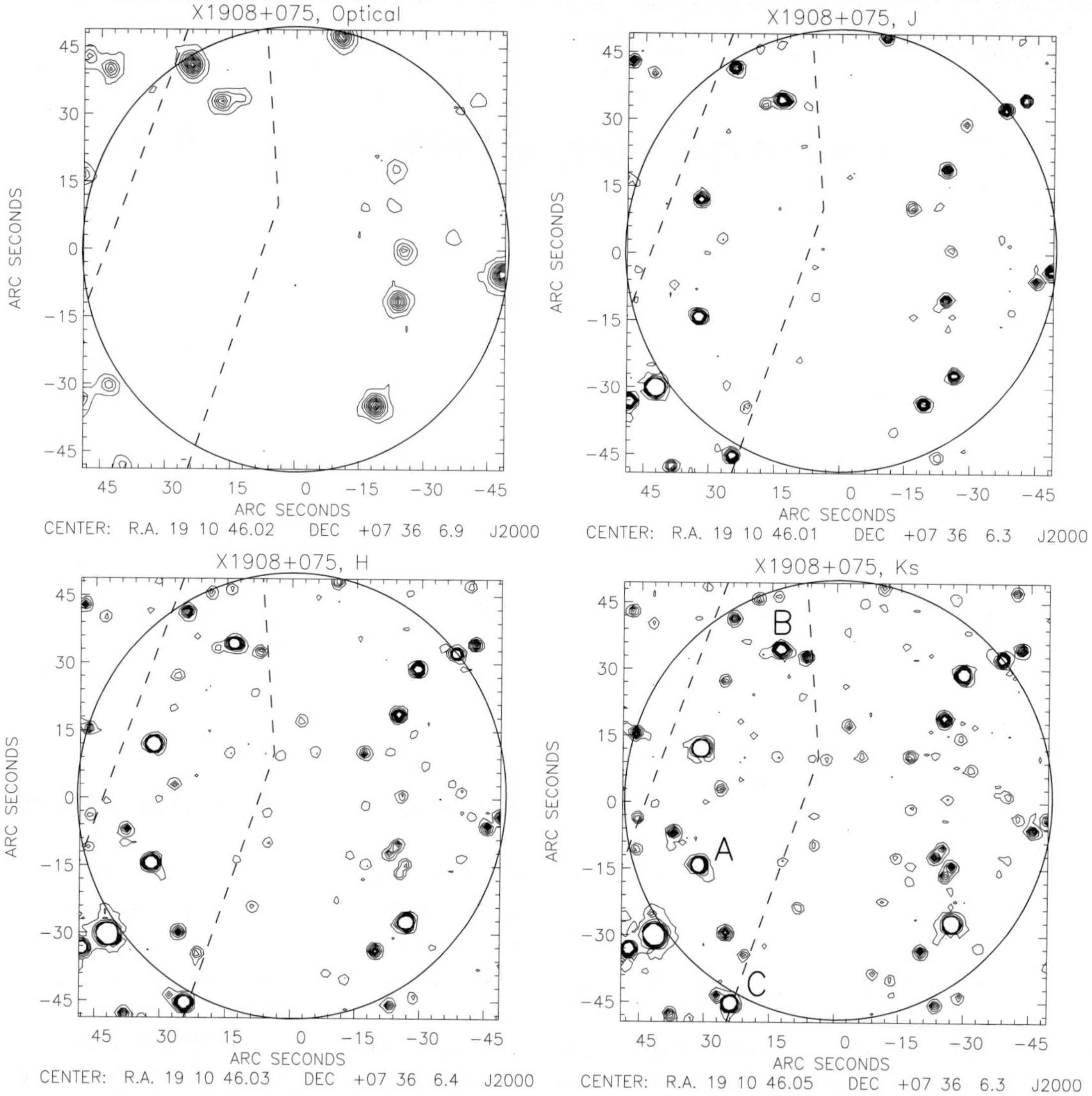

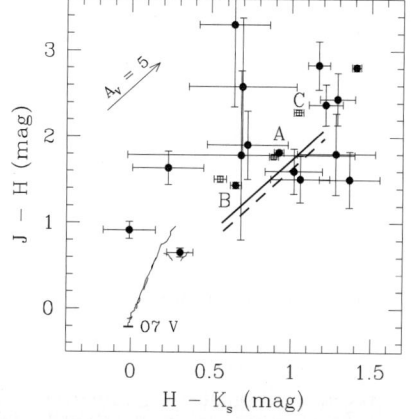

Fig. 1 Contour maps of the field in the optical (POSS-II red plate), J, H and K_s bands. The 50″ error circle of the *Einstein* satellite (*solid line*) and the diamond-like error box of the *HEAO 1* satellite (*dashed line*) are overlayed. The position of the potential candidates (A, B and C) is indicated in the lower, right-hand panel

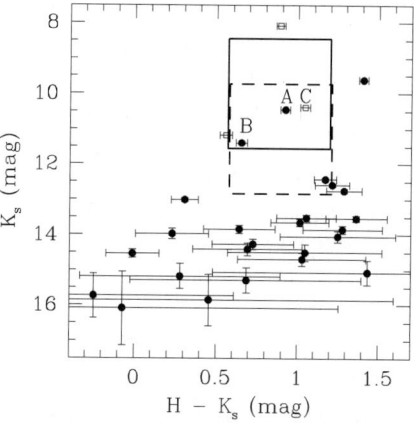

Fig. 2 Colour-colour and colour-magnitude diagrams for the stars in common between the *Einstein* and *HEAO 1* error boxes (*filled circles*). The 3 brightest sources to the South-East of the field lying within the *HEAO 1* error box, but just outside the *Einstein* error circle are also plotted (*open circles*). The position of candidates A, B and C is indicated. The diagonal lines in the left-hand panel and the boxes in the right-hand panel show the expected loci of the counterpart to X1908+075, assuming an O5 V (*dashed line*) or an O9 I (*solid line*) star

Fig. 3 HK-band spectra of candidates A (*top*) and B (*bottom*). The most prominent spectral features are labelled. The ⊕ symbols mark the position of the strong telluric lines. Some spurious features arising from an imperfect telluric subtraction can be seen at these locations

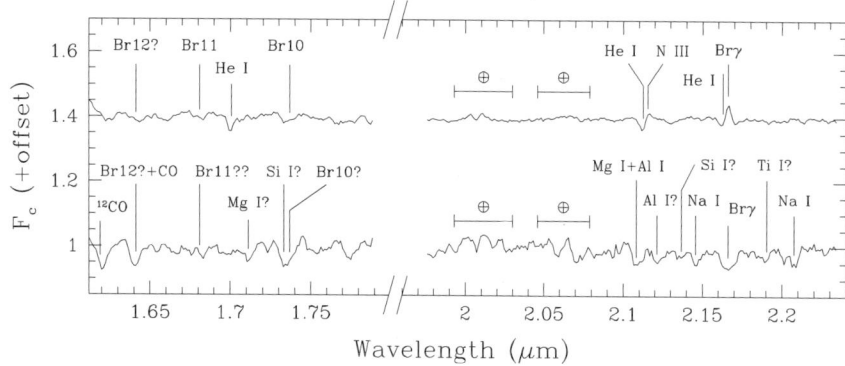

The spectrum of candidate B points to a cool star, most likely of luminosity class I–III (see Fig. 3). However, the measured strength of the diagnostic H, He and N lines all converge to a classification of candidate A as a late O-type star, while Brγ in emission strongly suggests a supergiant (see, e.g. Hanson et al., 1998). This leads us to propose a O7.5–O9.5 If classification.

4. Summary

- We have identified the likely counterpart to X1908 + 075 as a late O-type supergiant. This classification is compatible with the stellar parameters independently derived from X-ray data. Less than 10 binary systems of this type have been optically identified in the Galaxy.
- We obtain the following parameters for the counterpart: $\alpha =$ 19h 10m 48.204s and $\delta = +07°35'52.32''$ (J2000). $J = 13.199 \pm 0.018$, $H = 11.380 \pm 0.012$ and $K_s = 10.457 \pm 0.018$ mag. $d \sim 7$ kpc and $A_V \sim 16.5$ mag (the latter value is roughly consistent with the hydrogen column density derived from modelling of the X-ray spectrum).
- Our data do not support the primary being a Wolf-Rayet star and firmly rule out a Be/X-ray binary.
- The lack of evidence for an embedded young cluster in the line of sight to X1908 + 075 (Bica et al., 2003, and references therein) strengthens the case for candidate A being the counterpart. However, further X-ray observations with higher spatial resolution, as well as near-IR spectroscopic data for other possible candidates in the field, are needed for an unambiguous identification. We are currently attempting to detect in photometric data the orbitally modulated changes which would confirm our results.

References

Bica, E., Dutra, C.M., Soares, J., Barbuy, B.: A&A **404**, 223 (2003)
Hanson, M.M., Rieke, G.H., Luhman, K.L.: AJ **116**, 1915 (1998)
Levine, A.M., Rappaport, S., Remillard, R., Savcheva, A.: ApJ **617**, 1284 (2004)
Morel, T., Grosdidier, Y.: MNRAS **356**, 665 (2005)
Wen, L., Remillard, R.A., Bradt, H.V.: ApJ **532**, 1119 (2000)

Spectroscopy of the Candidate Pre-CV LTT 560

Claus Tappert · Boris T. Gänsicke ·
Ronald E. Mennickent · Linda Schmidtobreick

Received: 1 November 2005 / Accepted: 1 February 2006
© Springer Science + Business Media B.V. 2006

Abstract We present preliminary results on spectroscopic data of the candidate pre-cataclysmic variable LTT 560. A fit to the flux-calibrated spectrum reveals the temperature of the white-dwarf primary to be $T_{\rm eff} = 7000$–7500 K, and confirms the result of previous studies on the detection of an M5V secondary star. The analysis of radial velocity data from spectral features attributed to the primary and the secondary star show evidence for low-level accretion.

Keywords Spectroscopy · Cataclysmic variables · Evolution

1. Introduction

LTT 560 is a high proper motion system from the Luyten (1957) catalogue. A first photometry has been presented by Eggen (1968), who marked the object as a possible white dwarf (WD). It was later included as a nova-like variable in Vogt (1989), and appeared as such in the Downes et al. (1997) catalogue of cataclysmic variables (CVs) with the designation Scl2. A first spectrum was published by Hoard and Wachter (1998), who found narrow Hα emission and a red continuum that was fitted well by an M5 dwarf. Their data did not cover the blue part of the spectrum, so that they could not identify any contribution from the WD. However, a variable doubling of the Hα emission in their spectra let them deduce the binary nature of the object. Light curves presented by Tappert et al. (2004) showed ellipsoidal variation with a period $P_{\rm ph} = 3.54$ h, thus revealing LTT 560 as a potential pre-CV.

2. The flux-calibrated spectrum

In order to investigate the different stellar components, an optical spectrum was taken on 2003-07-19 with EMMI at the NTT, ESO La Silla. The data is shown in Fig. 1 together with the best fit, which comprises of a $T_{\rm eff} = 7000$ K, $\log g = 8.0$ WD and an M5.5 dwarf. Fits with $T_{\rm eff} = 7500$ K, $\log g = 8.5$ work equally well, as does an M5V star as secondary component. This confirms the earlier result by Hoard and Wachter (1998) on the secondary, and establishes LTT 560 as containing the coldest WD in a pre-CV besides RR Cae ($P_{\rm orb} = 7.3$ h), which has $T_{\rm eff} = 7000$ K, and with an M6V star also contains a secondary of similar spectral type (Bragaglia et al., 1995; Bruch and Diaz, 1998; see also Schreiber and Gänsicke, 2003, for a discussion in the context of CV evolution).

3. Radial velocities and line profiles

Time-resolved spectroscopy was taken on 2004-11-29 with the 4 m telescope at CTIO, Chile. A spectral range of 3500–7300 Å was covered at a FWHM resolution of 4.2 Å. Radial

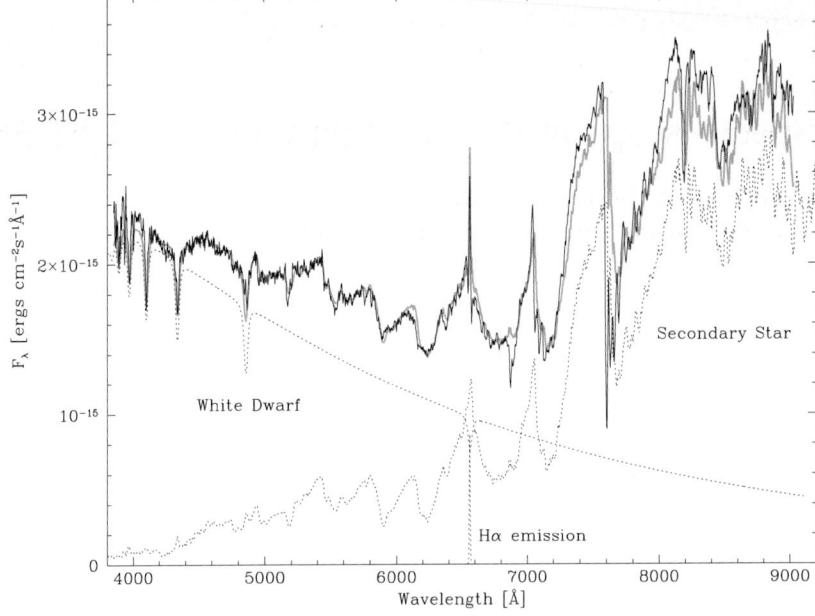

Fig. 1 Flux-calibrated spectrum of LTT 560 and the different components of the fit as indicated in the plot. The solid black line corresponds to the original spectrum, and the thick grey line gives the resulting fit

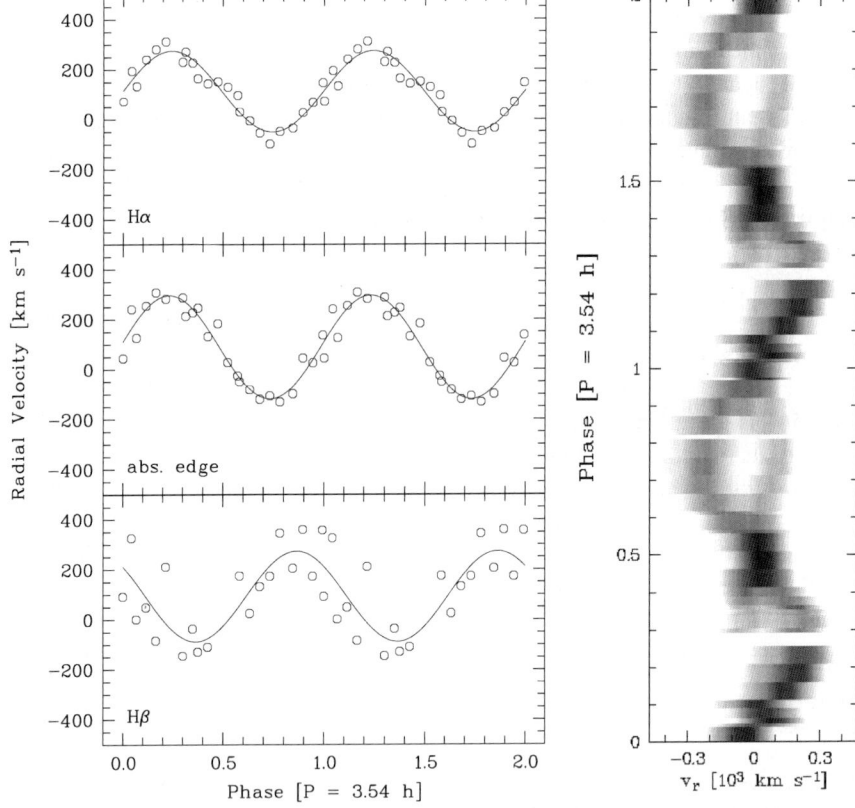

Fig. 2 Left: Radial velocities of the Hα emission (top), a TiO absorption edge near λ7040 Å (middle), and the Hβ absorption line (bottom). The data have been folded on the 3.54 h period. Zero phase has been set to the first data point of the spectroscopy. Right: Trailed spectrogram of the region around Hα

velocities were measured by fitting single Gaussians to several spectral features (Fig. 2). With this preliminary simple fit, especially the curve of the WD absorption (Hβ) therefore probably does not reflect well the true white dwarf radial velocity variation. This could explain why the parameters of the corresponding radial velocity curves yield mass ratios and phase offsets that are not conform with a pre-CV configuration. On the other hand, also the amplitudes for the two features attributed to the red dwarf show significant discrepancies ($K_{H\alpha} = 164(06)$ km s^{-1}, $K_{TiO} = 207(07)$ km s^{-1}).

A closer inspection of the Hα line profile shows the doubling at certain phases that had already been detected by

Hoard and Wachter (1998). It is obviously caused by two counterphased components of different strength, with the weaker one having a lower amplitude and disappearing at phases 0–0.3.

4. Conclusions

The general composition of LTT 560 appears clear. It is a binary star consisting of an M5-6V secondary star and a cold ($T_{\text{eff}} \approx 7000$ K) WD primary. The current orbital period amounts to 3.54 h, and, if the pre-CV configuration applies, the system should thus evolve into a semi-detached CV within Hubble time.

The derivation of other system parameters still needs a more thorough examination of the present data. The velocities of the WD absorption line are not yet reliable due to an insufficiently precise measurement of the WD absorption line. They will therefore be revised using a more sophisticated method. Furthermore, the Hα velocities are probably distorted due to the presence of an additional component.

Emission line doubling has been detected, e.g. also in the candidate pre-CV HS 2237+8154 (Gänsicke et al., 2004), but there it is due to a component that appears to be stationary at the system velocity. In LTT 560, this is clearly not the case. Since it is roughly in counterphase with the other component and the TiO band, it has to origin in some region beyond the centre of mass as seen from the secondary star, i.e. in the vicinity of the WD. A rough estimate of its semi-amplitude yields $K \approx 100$ km s^{-1}, thus, with respect to TiO, leading to much more plausible system parameters, e.g. a mass ratio $q \approx 0.5$. Still, this has to be confirmed by a more detailed analysis.

We conclude that our data shows evidence for emission from, or close to, the WD. It might be caused by sporadic accretion, e.g. due to wind from the secondary star. Accretion due to Roche lobe overflow appears less likely, since the late spectral type of the secondary implies that a semi-detached configuration is not expected for periods longwards of 2 h (e.g. Beuermann et al., 1998).

Acknowledgements This work was partly supported by Fondecyt grant 1051078. We gratefully acknowledge intensive use of the SIMBAD database, operated at CDS, Strasbourg, France.

References

Beuermann, K., Baraffe, L., Kolb, U., Weichhold, M.: A&A **359**, 518 (1998)
Bragaglia, A., Renzini, A., Bergeron, P.: ApJ **443**, 735 (1995)
Bruch, A., Diaz, M.P.: AJ **116**, 908 (1998)
Downes, R.A., Webbink, R.F., Shara, M.M.: PASP **109**, 345 (1997)
Eggen, O.J.: ApJ **153**, 195 (1968)
Gänsicke, B.T., Araujo-Betancor, S., Hagen, H.-J., et al.: A&A **418**, 265 (2004)
Hoard, D.W., Wachter, S.: PASP **110**, 906 (1998)
Luyten, W.J.: Catalogue of 9867 Stars in the Southern Hemisphere with Proper Motion Exceeding 0".2 Annually, University of Minnesota, Lund Press, Minneapolis (1957)
Schreiber, M.R., Gänsicke, B.T.: A&A **406**, 305 (2003)
Tappert, C., Gänsicke, B.T., Mennickent, R.E.: In: Tovmassian, G., Sion, E. (eds.) Compact Binaries in the Galaxy and Beyond, Rev. Mex. Astron. Astrof. (Ser. Conf.) **20**, 245 (2004)
Vogt, N.: In: Bode, M.F., Evans, A. (eds.) Classical Novae, Wiley, Chichester, 225–247 (1989)

The Enigmatic Behaviour of the Old Nova RR Pic

Linda Schmidtobreick · Claus Tappert ·
Alessandro Ederoclite · Ivo Saviane · Elena Mason

Received: 1 November 2005 / Accepted: 1 February 2006
© Springer Science + Business Media B.V. 2006

Abstract We report preliminary results of an observing run at the 1–m telescope at CTIO during which we followed RR Pic over several nights in February 2005. The resulting light curves show no sign of an eclipse. Apart from the 3.48 h period, which is usually interpreted as the orbital one, we find an additional period at $P = 3.78$ h, which we interpret as the superhump period of the system.

We furthermore present high–resolution Doppler maps in Hα, Hβ and He II, which we derived from observations at the NTT, La Silla in February 2004. They show strong variations in the emission distribution from one day to the next. While Hα and Hβ emission clearly show the accretion disc with some additional isolated sources, the He II emission is confined to an elongated region at low velocities.

Keywords Novae · Cataclysmic variables · Accretion discs

1. Introduction

RR Pic is an old nova which erupted in 1925 (Spencer Jones, 1931) and hence is well in its quiescence state by now. The binary is partly eclipsing and its orbital period of 0.145025 days places it just above the gap of the period distribution of cataclysmic variables. Recent time–resolved spectroscopy taken with the B&C at the 1.52–m ESO telescope on La Silla confirmed the presence of an additional emission source on the leading side of the disc (Schmidtobreick et al., 2003). These

L. Schmidtobreick (✉) · A. Ederoclite · I. Saviane · E. Mason
European Southern Observatory, Casilla 19001, Santiago 19, Chile

C. Tappert
Departamento de Astronomía y Astrofísica, Pontificia Universidad Católica, Casilla 306, Santiago 22, Chile

authors also found high velocity line wings moving with the orbital period but shifted by 0.4 phases. However, there is an offset of $\Delta\phi = 0.53$ between the zero-phase measured from radial velocities and the extrapolated eclipse phase from 1970 and 1984. A later analysis (Schmidtobreick et al., 2005a) revealed that this offset is varying in time, which could hint at a change of the orbital period. However, since the radial velocities are most likely influenced by the isolated emission sources in the disc, a variable emission structure in the disc of RR Pic might also explain these variations.

2. The light curves

Time resolved V-photometry was done using the 1–m telescope at CTIO, Chile. The data presented here, were taken in 2005 between Feb 07 and 14 and cover about 12.5 cycles. They are reduced following standard procedures (see, Schmidtobreick et al., 2005b, for details). Differential photometry has been performed using 5 stable comparison stars.

Apart from the hump–like variability, the light curves show the strong flickering and random variation typical for RR Pic. We do not find any convincing evidence for an eclipse in any of the light curves. The phase diagram is plotted in Fig. 1. The extrapolated eclipse phase coincides with a broad minimum, but comparing the overall shape of the light curve suggests that the eclipse should actually be located around phase 0.2, where a small minimum is present in the average curve. This phase offset points towards a change in the orbital period. The Scargle periodogram (Scargle, 1982) yields an orbital period $P = 0.1449(5)$ d, which is not yet precise enough to measure the small change of 10^{-4} needed to explain the phase shift. Some data taken later in 2005 will help to increase the accuracy to decide if a change of orbital period took place in RR Pic during the last decades (Schmidtobreick et al., 2005b).

Fig. 1 The averaged differential V–magnitude of RR Pic against orbital phase. The zero-phase was chosen to correspond to the eclipse–ephemeris of Warner (1986) and Kubiak (1984). On the first cycle, the sigma of the variation is overplotted

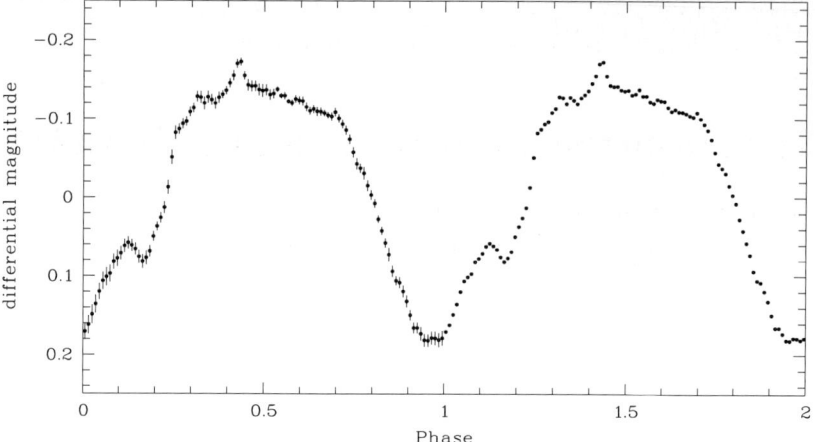

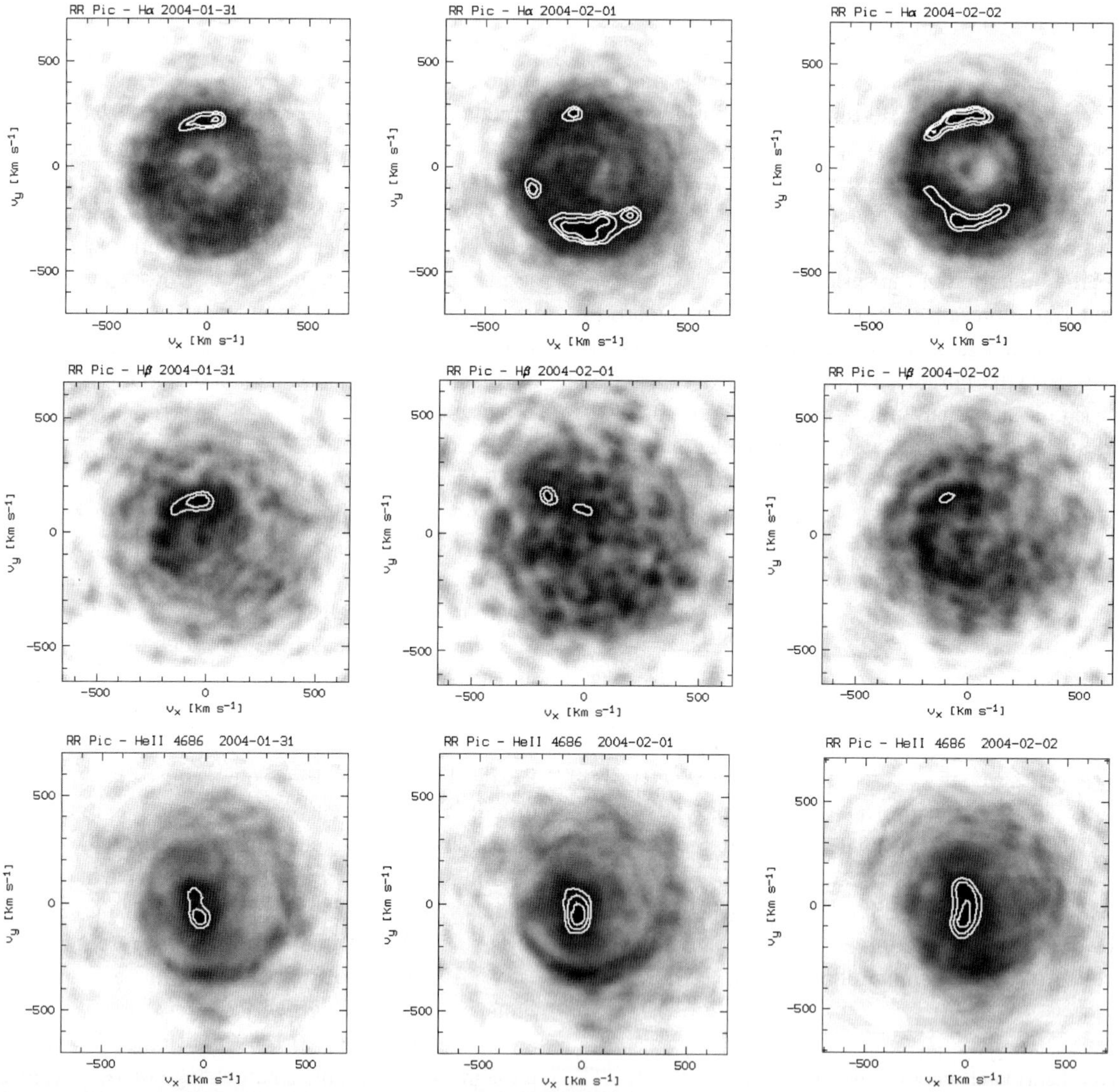

Fig. 2 The Doppler maps in Hα (upper row), Hβ (middle row) and HeII (lower row) on the three following nights (from left to right). The contours plotted refer to 90%, 85% and 80% of the maximum value, which has been normalised for all maps

To search for superhumps, we have subtracted the average light curve from the individual data points. A clear signal is present in the Scargle periodogram of the resulting differential light curve at a period of $P = 0.1577(6)$ d. Interpreted as a beat period between orbital and precession period, it yields a precession period of 1.8 d which is also present in the periodogram. The differential light curve averaged over the 0.1577 d period shows a clear hump, which we thus interpret as a superhump. Using the empirical formula $\epsilon = 0.22 \cdot q$ with $\epsilon = \frac{P_{SH} - P_{orb}}{P_{orb}} = 0.0862$ we find a mass ratio $q = 0.4$. Although this value is rather high, it is not unlikely for a high–mass transfer novalike.

3. Spectroscopy

Time resolved spectroscopy was obtained with EMMI on the NTT at ESO/La Silla on Jan 31 – Feb 02, 2004. The echelle mode with grating #9 centred on 5200Å, yielded a resolution of 0.7Å and allowed to use the strong emission lines Hα, Hβ, and He II (4686Å) for Doppler tomography techniques. We here use the code of Spruit (1998) with a MIDAS interface replacing the original IDL routines (Tappert et al., 2003).

Doppler maps $I(v_x, v_y)$ have thus been computed for each of the three nights and are given in Fig. 2. They display the flux emitted by gas moving with the velocity (v_x, v_y) and thus show the emission distribution in velocity coordinates with the centre of mass at (0, 0).

In general, we note that the emission distribution is rather unstable. The maps in Hα show a ring–like feature which is commonly interpreted as emission from the accretion disc. However the ring is patchy. Apart from the always present emission source around phase 0.9, which we attribute to a hot spot, additional emission sources appear around 0.75 and 0.5. The map of the old data (Schmidtobreick et al., 2003) resembles the emission distribution on Feb 01. The maps in Hβ more or less follow the behaviour in Hα but are more noisy due to the weaker emission line. The He II maps instead show a different structure. They are clearly dominated by an elongated emission region close to zero velocity which points towards a weak emission region at phase $\phi_r \approx 0.5$, the latter coinciding with the emission source present in Hα. The hot spot emission is only marginally present in He II. We do not find any evidence for disc emission in this line.

It is therefore well possible that this variation of the emission distribution distorts the line profile in such a way that the phase offset between radial velocities and photometry varies in the order of $\delta\phi \approx 0.3$. Although other effects cannot be excluded, this could already cause the observed mismatch.

References

Kubiak, M.: AcA **34**, 331 (1984)
Scargle, J.D.: ApJ **263**, 835 (1982)
Schmidtobreick, L., Tappert, C., Saviane, I.: MNRAS **342**, 145 (2003)
Schmidtobreick, L., Galli, L., Saviane, I. et al.: ASP Conf. Ser. **335**, 333 (2005a)
Schmidtobreick, L., Tappert, C., Ederoclite, A., Mason, E.: MNRAS, In preparation (2005b)
Spencer Jones, H.: Cape Obs. Ann. 10, Part 9 (1931)
Spruit, H.: preprint, astro–ph/9806141 (1998)
Tappert, C., Mennickent, R., Arenas, J. et al.: A&A **408**, 651 (2003)
Warner, B.: MNRAS **219**, 751 (1986)

3. Results and discussion

3.1. JHKL colours

A change in colour behaviour around JD 2446600, at the beginning of the first obscuration event, was observed. The colours became redder after that date, probably as a result of increased dust absorption. The colours of RR Tel during the observed period are compared with the mean values of normal Miras by use of data for Miras with mainly thin dust shells (Whitelock et al., 2000) and for IRAS selected Miras with relatively thick dust shells (Whitelock et al., 1994) (Fig. 1). It is evident that the observed RR Tel infrared colours are significantly shifted out of the range shown by normal Miras.

The colours of RR Tel can be explained by reddening which dominates over a small amount of dust emission. This interpretation was discussed by Whitelock (1987) who presented a sample of symbiotic binaries in a two-colour diagram, modeling their colours with a 2500K star plus an 800K dust shell (represented as a locus on her Fig. 3). The further up the locus, the thicker is the dust shell. RR Tel follows the same type of trend as the other objects showing obscuration events in that its colours move towards the upper right of the locus. At around 800K the dust is hotter than in typical solitary Miras and we might speculate that the extra heating is provided by the hot component of the symbiotic.

3.2. Distance estimate

For the distance estimate we use the mean values of the K magnitude and of the J-K colour index equal to 4.16 and 1.69, respectively, obtained from observations at the epoch preceding the first obscuration event, when the circumstellar absorption was at its minimum.

From the value of 387 days for the period of Mira pulsations and using the correlation between the Mira period and absolute magnitude (Feast et al., 1989) (assuming a distance modulus for the LMC of 18.5 mag), we obtain an absolute magnitude in K, $M_K = -8.0$. The unabsorbed $J - K$ was obtained by applying the period-colour relation for Mira variables in the solar neighbourhood from Whitelock et al. (2000), $(J - K)_0 = -0.39 + 0.71 \log P$. This leads to $(J - K)_0 = 1.45$ and thus $E_{J-K} = 0.24$. Using the reddening law specified by Glass (1999) we find $A_K = 0.13$. The interstellar extinction towards RR Tel is $E(B - V) = 0.10 - 0.11$ (Young et al., 2005) or $A_K \sim 0.03$ mag. Thus most of the measured extinction must be circumstellar rather than interstellar. This procedure outlined above assumes that the circumstellar extinction has the same reddening characteristics as does interstellar extinction and there are reasons for believing this to be a reasonable approximation (see Kotnik-Karuza et al. in preparation).

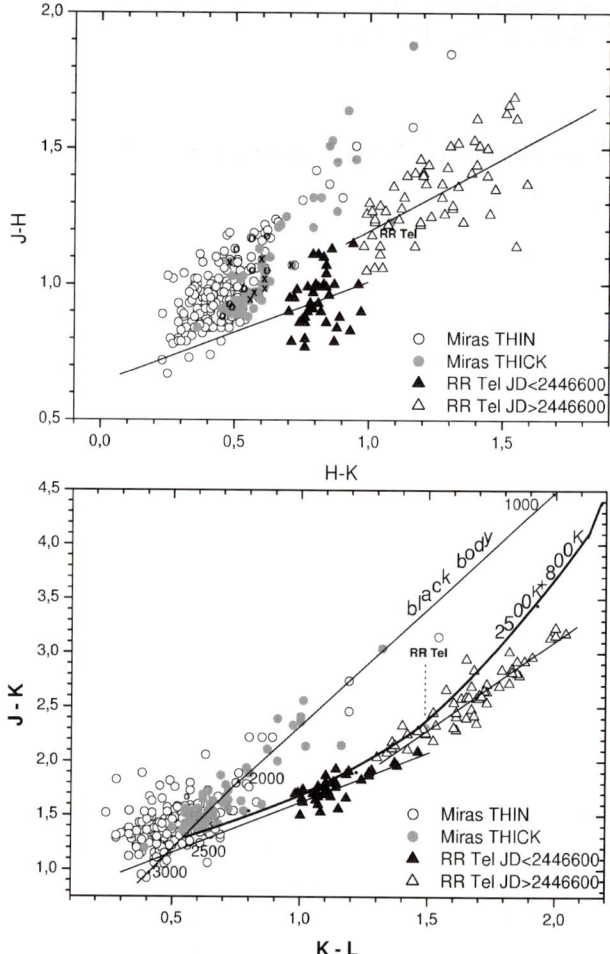

Fig. 1 Two colour diagrams of RR Tel before (full triangles) and after (empty triangles) JD 2446600 as well as of normal Miras with thin dust shells (empty circles) from Whitelock et al. (2000) and with thick dust shells (full circles) from Whitelock et al. (1994). The Miras with similar periods of pulsation to that of RR Tel are distinguished by crosses for thick dust shells and by bold circles for thin dust shells. Straight lines are least squares fits for RR Tel before and after JD 2446600. The black body curve and and the curve representing a 2500 K Mira + 800 K silicate dust shell of variable optical depth intersect at 2500 K blackbody temperature

The distance derived for RR Tel is therefore 2.5 kpc, which is very close to the value of 2.6 kpc found by Whitelock (1988).

References

Feast, M.W., Glass, I.S., Whitelock, P.A., Catchpole, R.M.: A period-luminosity-colour relation for Mira variables. MNRAS **241**, 375–392 (1989)

Glass, I.S.: Handbook of Infrared Astronomy, CUP, Cambridge (1999)

Whitelock, P.A.: Symbiotic miras. PASP **99**, 573–591 (1987)

Whitelock, P.A.: Infrared Observations of Symbiotic Miras. In: Mikolajewska, J., Friedjung, M., Kenyon, S.J., Viotti, R. (eds.) The symbiotic phenomenon, Kluwer, pp. 47–56 (1988)

Whitelock, P.A., Menzies, J., Feast, M.W., Marang, F., Carter B.S., Roberts, G., Catchpole, R.M., Chapman, J.: High-mass-loss AGB stars in the South Galactic Cap. MNRAS **267**, 711–742 (1994)

Whitelock, P., Marang, F., Feast, M.W.: Infrared colours for Mira-like long-period variables found in the Hipparcos Catalogue. MNRAS **319**, 728–758 (2000)

Whitelock, P., Feast, M.W.: Hipparcos parallaxes for Mira-like long-period variables. MNRAS **319**, 759–770 (2000)

Whitelock, P.: A comparison of symbiotic and normal miras. In Corradi, R.L.M., Mikolajewska, J., Mahoney, T.J. (eds.) Symbiotic stars probing stellar evolution. ASP Conf. Ser. **303**, 41–56 (2003)

Young, P.R., Berrington, K.A., Lobel, A.: Fe VII lines in the spectrum of RR Telescopii. A&A **432**, 665–670 (2005)

The Structure and the Partial Opacity of the Spiral Shocks on the Novalike Cataclysmic Variables

Christos Papadimitriou · Emilios Harlaftis · Danny Steeghs

Received: 1 November 2005 / Accepted: 1 February 2006
© Springer Science + Business Media B.V. 2006

Abstract V347 Pup is the first novalike system with clear spiral arms in its accretion disc, as evidenced from its HeI Doppler maps. Combining the Doppler maps of the various lines of the system V347 Pup a more complex structure is revealed on the first arm. On the Doppler maps of HeI $\lambda5875$ emission line the first arm splits to two smaller and thinner structures. The Doppler map of HeI $\lambda6678$ emission line shows that this line dominates the region of thin structure with higher velocities that we met on the Doppler map of HeI $\lambda5875$ emission line. On the contrary the Hα emission line dominates the region of thin structure with lower velocities. We therefore observe, that the Hα emission line dominates on the exterior of the first arm, and the emission line of HeI on the interior. Most of NaI is probably emanated from the white dwarf or from the interior of the disc, and then being absorbed by the spiral shocks.

Keywords Accretion · Accretion discs · Shock waves · Stars: Individual, V347 Pup · Novalike · Cataclysmic variables

C. Papadimitriou (✉)
Department of Physics, University of Athens, Institute of Astronomy and Astrophysics, National Observatory of Athens, Greece
e-mail: cpap@astro.noa.gr

E. Harlaftis
Institute of Space Application and Remote Sensing, National Observatory of Athens, Greece

D. Steeghs
Harvard-Smithsonian Center for Astrophysics, USA

1. The novalike V347 Pup

V347 Pup is a deeply eclipsing, 13th magnitude novalike cataclysmic variable. The radial velocity and eclipse analysis indicate an orbital inclination of 87° (Buckley et al., 1990). The novalike systems have never exhibited an outburst although they spectroscopically resemble dwarf novae in outburst. They are believed to be in a state of high mass-transfer rate, and are probably stars before or after the phase of a nova event. Previous work (Still et al., 1998; Papadimitriou et al., 2005) provides evidences for spiral waves in this system.

V347 Pup was observed the 25th December 1999, using the 3.5 m NTT (New Technology Telescope). During the night, 203 spectra were obtained, with 60 sec exposure time covering over 1 orbital cycle at a velocity resolution of 31 km/sec. All the CCD images were reduced and the spectra extracted using the optimal extraction technique (Marsh, 1989). The Doppler maps are computed using the maximum entropy method (Marsh and Horne, 1988).

2. Results

The structure of the trailed spectra, and in particular the variation of the double peak separation with phase, Fig. 1 present strong evidence for spiral waves in the disc (see for example Harlaftis et al., 1999; Steeghs et al., 1997). From the spectrophotometric lightcurves of Hα, HeI ($\lambda5875$) and HeI ($\lambda6678$) emission lines and from the absorption line of NaI (Fig. 2), two minima presented for the NaI, do not only relate to the secondary star (no cross-correlation exists on phases 0 or 0.5), but also to the spiral arms of the accretion disc.

Mapping of the Disc Structure of the Neutron Star X-ray Binary X1822-371

O. Giannakis · E. T. Harlaftis · P. G. Niarchos · S. Kitsionas · H. Barwig · M. Still · R. G. M. Rutten

Received: 1 November 2005 / Accepted: 1 February 2006
© Springer Science + Business Media B.V. 2006

Abstract We report the results of simultaneous optical/X-ray observations of X1822-371 at 0.1 second time resolution. The preliminary analysis finds no correlation between the optical/X-ray light curves. We aim to constrain the vertical structure and radius of the accretion disc.

Keywords Low-mass X-ray binaries · X1822-371

1. Introduction

The low-mass X-ray binary (LMXB) X1822-371 (V691 CrA), is viewed at high inclination ($85° \pm 3°$) with persistent X-ray emission (found also to be pulsed at 0.59 sec; Jonker and van der Klis, 2001). In this system we observe an accretion disc corona (ADC) which is partially obscured by the asymmetrically-thick accretion disc (Hellier and Mason, 1989).

2. Observations and analysis

We used simultaneous observations obtained on 2003 August 31, September 1 and September 2 with the proportional counter array (PCA; Jahoda et al., 1996) on-board the *Rossi X-ray Timing Explorer* (RXTE) satellite (Bradt et al., 1993) and with the ESO 2.2 m telescope at La Silla, Chile. X-ray observations yielded data with a sub-second time resolution in 255 energy channels covering the effective 2.0–60 keV energy range of RXTE. Optical observations were obtained with the MCCP fast photometer (Barwig et al., 1987) which yielded mean count rates of 160 cts s^{-1} in U-band, 1230 cts s^{-1} in B-band, 1030 cts s^{-1} in V-band and 330 cts s^{-1} in R-band at a time resolution of 0.1 secs. The three channels of the MCCP instrument can observe simultaneously the target star, a comparison star and the sky, providing very accurate photometry.

3. The light curves

In Fig. 1 (top panel) we present the BVRI light curves on 2 September 2003 which show the characteristic wide eclipse and flickering. A simple comparison of the B-band and I-band light curves in Fig. 1 (Top) shows that the optical flickering is extremely blue. A rise in intensity with complex structure at orbital phase $\sim$0.76 ($\sim$HJD 2452885.51) lasts for $\sim$2600 secs. We also present in Fig. 1 (Bottom) the X-ray light curve and the B-band light curve obtained on 2 September 2003.

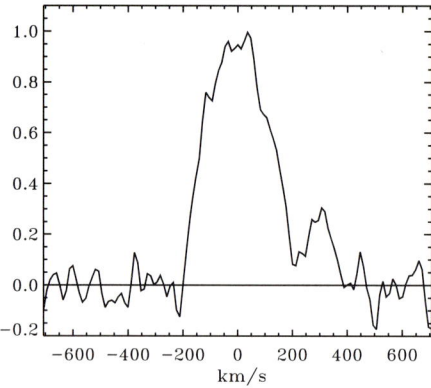

Fig. 3 The BF for SX Crv at the orbital phase of 0.25. Note that the secondary, at $q = 0.066$, is easily detectable in spite of its very low luminosity. Paradoxically, for for very small q, it is the heavily-broadened primary peak in the BF which must be properly measured for RV shifts as it carries most of information on $q = K_1/K_2$

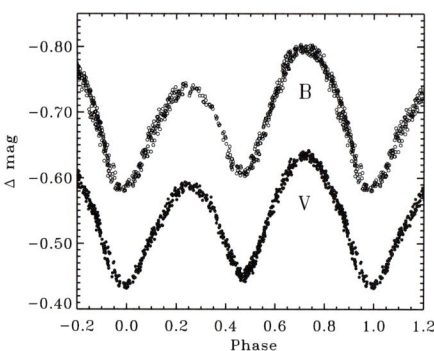

Fig. 4 The light curve of SX Crv from Zola et al. (2001) does not contain any indication that the binary has a very small mass ratio. Instead, a large asymmetry in light maxima is clearly visible. Without spectroscopic information, this system would be basically "unsolvable"

results (notwithstanding the fact that they cannot reproduce asymmetries in the BFs).

We do not see any preference for specific mass ratios in contact binaries; the distribution of q does not seem to peak at any specific value. The smallest q discovered by us for SX Crv is $q = 0.066$ (Paper V, see Fig. 3) while the largest is $q = 0.970$ for V753 Mon (Paper III). The case of SX Crv is of course particularly interesting as a second system (in addition to AW UMa) below the theoretically predicted low limit on the mass ratio Rasio (1995). The light curves of SX Crv (an example is shown in Fig. 4, Zola et al., 2004) do not carry any information on the mass ratio.

As mentioned before, we see many triple systems among our short-period binaries. This has prompted two recent investigations, both to be published soon: (1) Theo Pribulla and I have looked at the current numbers of triples among contact binaries, simply as they are, without analysis of biases, selection effects or types of companions which could not be discovered. Depending how we look at the data and what technique (astrometric, spectroscopic, etc.) we use, the lower limit to the frequency of companions comes out at the level 50% to 65%. (2) Caroline D'Angelo, Marten van Kerkwijk and I re-analyzed the spectra of the current DDO program by stacking them together, to average out signatures of the rapidly moving close binary and to strengthen the signature of the tertiary; this led to discovery of several faint companions. Both studies are consistent with the possibility that all contact binaries originate in multiple systems.

We estimate that we still have some 40 bright (<10 mag) binaries with periods shorter than one day at $\delta > -15$ to analyze within our program. Then, we can explore longer periods or observe fainter, very short period (<0.3 day) binaries in an attempt to bring our sample closer to a volume limited one.

Acknowledgements This paper is written partly to acknowledge contribution of, and express thanks to, many individuals: In addition to the participation of Hilmar Duerbeck and Wen Lu at the important initial stages, the program enjoyed enthusiastic support of several observers from the Department of Astronomy and Astrophysics of the University of Toronto (Mel Blake, Chris Capobianco, George Conidis, Heide DeBond, Stefan Mochnacki, Wojtek Pych, Jim Thomson), from Greece (Kosmas Gazeas), Poland (Piotr Ligeza, Waldek Ogloza, Piotr Rogoziecki, Greg Stachowski) and Slovakia (Theo Pribulla). I would like to thank all of them for their excellent work and contribution to one of the main legacies of the David Dunlap Observatory.

References

Lu, W., Rucinski, S.M.: Radial velocity studies of close binary stars. I. Astron. J. **118**, 515–526 (1999)

Lu, W., Rucinski, S.M., Ogloza, W.: Radial velocity studies of close binary stars. IV. Astron. J. **122**, 402–412 (2001)

Pych, W., Rucinski, S.M., DeBond, H., Thomson, J.R., Capobianco, C.C., Blake, R.M., Ogloza, W., Stachowski, G., Rogoziecki, P., Ligeza, P., Gazeas, K.: Radial velocity studies of close binary stars. IX. Astron. J. **127**, 1712–1719 (2004)

Rasio, F.A.: The minimum mass ratio of w ursae majoris binaries. Astrophys. J. **444**, L41–L43 (1995)

Rucinski, S.M.: Spectral-line broadening functions of w uma-type binaries. I. AW UMa. Astron. J. **104**, 1968–1981 (1992)

Rucinski, S.M.: Radial velocity studies of close binary stars. VII. Astron. J. **124**, 1746–1756 (2002)

Rucinski, S.M., Lu, W.: Spectral-line broadening functions of w uma-type binaries. II. AH Vir. Astron. J. **106**, 361–371 (1993)

Rucinski, S.M., Lu, W.: Radial velocity studies of close binary stars. II. Astron. J. **118**, 2451–2459 (1999)

Rucinski, S.M., Lu, W.: W Crv: The shortest-period algol with non-degenerate components? MNRAS **315**, 587–594 (2000)

Rucinski, S.M., Lu, W., Shi, J.: Spectral-line broadening functions of w uma-type binaries. III. W UMa. Astron. J. **106**, 1174–1180 (1993)

Rucinski, S.M., Lu, W., Mochnacki, S.W.: Radial velocity studies of close binary stars. III. Astron. J. **120**, 1133–1139 (2000)

Rucinski, S.M., Lu, W., Mochnacki, S.W., Ogloza, W., Stachowski, G.: Radial velocity studies of close binary stars. V. Astron. J. **122**, 1974–1980 (2001)

Rucinski, S.M., Lu, W., Capobianco, C.C., Mochnacki, S.W., Blake, R.M., Thomson, J.R., Ogloza, W., Stachowski, G.: Radial velocity studies of close binary stars. VI. Astron. J. **124**, 1738–1745 (2002)

Rucinski, S.M., Capobianco, C.C., Lu, W., DeBond, H., Thomson, J.R., Mochnacki, S.W., Blake, R.M., Ogloza, W., Stachowski, G., Rogoziecki, P.: Radial velocity studies of close binary stars. VIII. Astron. J. **125**, 3258–3264 (2003)

Rucinski, S.M., Pych, W., Ogloza, W., DeBond, H., Thomson, J.R., Mochnacki, S.W., Capobianco, C.C., Conidis, G., Rogoziecki, P.: Radial velocity studies of close binary stars. X. Astron. J. **130**, 767–775 (2005)

Zola, S., Rucinski, S.M., Baran, A., Ogloza, W., Pych, W., Kreiner, J.M., Stachowski, G., Gazeas, K., Niarchos, P., Siwak, M.: Physical parameters of components in close binary systems: III. Acta Astron. **54**, 299–312 (2004)

Abundances from Disentangled Component Spectra of Close Binary Stars: An Observational Test of an Early Mixing in High-Mass Stars

K. Pavlovski · D. E. Holmgren · P. Koubský · J. Southworth · S. Yang

Received: 30 September 2005 / Accepted: 1 February 2006
© Springer Science + Business Media B.V. 2006

Abstract Recent theoretical calculations of stellar evolutionary tracks for rotating high-mass stars suggests that the chemical composition of the surface layers changes even whilst the star is evolving on the Main Sequence. The abundance analysis of binary components with precisely known fundamental stellar quantities allows a powerful comparison with theory. The observed spectra of close binary stars can be separated into the individual spectra of the component stars using the method of spectral disentangling on a time-series of spectra taken over the orbital cycle. Recently, Pavlovski and Hensberge (2005, A&A, 439, 309) have shown that, even with moderately high line-broadening, metal abundances can be derived from disentangled spectra with a precision of 0.1 dex. In a continuation of this project we have undertaken a detailed abundance analysis of the components of another two high-mass binaries, V453 Cyg, and V380 Cyg. Both binaries are well-studied systems with modern solutions. The components are close to the TAMS and therefore very suitable for an observational test of early mixing in high-mass stars.

K. Pavlovski (✉)
Department of Physics, University of Zagreb, Zagreb, Croatia

D. E. Holmgren
SMART Technologies, Inc., Calgary, Canada

P. Koubský
Astronomcal Institute of the Academy of Sciences, Ondřejov, Czech Republic

J. Southworth
Niels Bohr Institute, Copenhagen University, Denmark

S. Yang
Department of Physics & Astronomy, University of Victoria, Canada

Keywords Stars: Abundances · Stars: Binaries: Eclipsing · Stars: Binaries: Close · Stars: Binaries: Spectroscopic

1. Introduction

New stellar evolution models which include the effects of rotationally induced mixing (Heger and Langer, 2000; Meynet and Maeder, 2000) have considerably changed our understanding of the evolution of high-mass stars, particularly during the early phases of core hydrogen burning. Rotation is now recognized as an important physical effect which substantially changes the lifetimes, chemical yields and stellar evolution. Theoretical predictions can be observationally tested, and some attempts at this have already been made (c.f. Venn et al., 2002).

Chemical analysis of the components of binary stars with precisely known fundamental stellar parameters allows a powerful comparison with theory. However, the precision of empirical abundances from double-lined binaries is hampered by increased line blending and by dilution of the spectral lines in the composite spectra. The techniques of spectral disentangling (Simon and Sturm, 1994; Hadrava, 1995) and Doppler tomography (Bagnuolo and Gies, 1991) overcome these difficulties by separating the spectra of the individual components contained in a time-series of composite spectra taken over the orbital cycle.

Pavlovski and Hensberge (2005) have performed a detailed spectral line analysis of disentangled component spectra of the eclipsing early-B binary V578 Mon in the open cluster NGC 2244, which is embedded in the Rosette Nebula. It is based on the disentangled spectra obtained by Hensberge et al. (2000) when deriving the orbit and the fundamental stellar parameters of this eclipsing, detached, double-lined system. V578 Mon consists of very young

Table 1 Fundamental parameters for the stars in V453 Cyg and V380 Cyg

Qnty/Comp	V453 Cyg A[1]	V453 Cyg B[1]	V380 Cyg A[2]	V380 Cyg B[2]
M [$M_\odot$]	14.36 ± 0.20	11.11 ± 0.13	11.1 ± 0.5	6.95 ± 0.25
$\log g$ [cgs]	3.731 ± 0.012	4.005 ± 0.015	3.148 ± 0.023	4.133 ± 0.023
T_{eff} [K]	$26\,600 \pm 500$	$25\,500 \pm 800$	$21\,350 \pm 400$	$20\,500 \pm 500$
$v \sin i$	107 ± 9	97 ± 20	98 ± 4	32 ± 6
ϵ_{He}[3]	0.13 ± 0.01	0.09 ± 0.01	0.14 ± 0.01	–

Notes. (1) Southworth et al. (2004); (2) Guinan et al. (2000); (3) This work

$(2.3 \pm 0.2 \times 10^6$ yr) high-mass stars, $M_A = 14.54 \pm 0.08$ $M_\odot$ and $M_B = 10.29 \pm 0.06$ $M_\odot$. The stars rotate moderately fast ($v \sin i \sim 100$ km s^{-1}). By comparison with spectra of single stars in the same open cluster (Vrancken et al., 1997), temperature-dependent, faint spectral features are shown to reproduce well in the disentangled spectra, which validates a detailed quantitative analysis of these component spectra. An abundance analysis differential to a sharp-lined single star, as applied earlier in this cluster to single stars rotating faster than the components of V578 Mon, revealed abundances in agreement with the cluster stars studied by Vrancken et al. (1997) and the large inner-disk sample of Daflon et al. (2004). Pavlovski and Hensberge (2005) have concluded that methods applicable to observed single star spectra perform well on disentangled spectra, given that the latter are carefully normalised to their intrinsic continua.

Since the fundamental stellar and atmospheric parameters of eclipsing, double-lined spectroscopic binaries are known with much better accuracy than in the case of single stars, the comparison with evolutionary models can be more direct and precise. The present work is a continuation of an observational project to test rotationally induced mixing in high-mass stars from disentangled component spectra of close binary stars.

We will now present preliminary results on two interesting early-B type systems, V453 Cyg and V380 Cyg. Both systems are detached, eclipsing, double-lined spectroscopic binaries and have reliable modern absolute dimensions, published by Southworth et al. (2004) for V453 Cyg and Guinan et al. (2000) for V380 Cyg (Table 1).

2. Spectroscopy and method

Several different sets of spectra were obtained for both binaries. We will briefly describe these observations.

V453 Cyg: This binary was observed in 1991 and 1992 with the 2.2-m telescope at German-Spanish Astronomical Center on Calar Alto, Spain. Four spectral windows were observed with the coudé spectrograph. A total of 28 spectra were collected. These spectra were kindly put at our disposal by Dr. Klaus Simon. Further description can be found in Simon and Sturm (1994). Another similar set, in two spectral windows, was secured by one of the authors (JS) in 2001 with the 2.5-m Isaac Newton Telescope at La Palma (Southworth et al., 2004). A total of 41 spectra were obtained. An additional set of six spectra in the red region centred on Hα were secured by DH on the 1.2-m telescope at the DAO in 2001.

V380 Cyg: Eight spectra centred on Hγ were obtained by PK and KP at the coudé spectrograph on the 2-m telescope in Ondřejov in 2004. An additional two spectra in the same region were obtained by PK on the 1.2-m telescope at the DAO, Victoria, also in 2004. An additional set of eight red spectra centred on Hα, from the same telescope, were obtained by SY in 2002 and are also used here.

To isolate the individual spectra of both components in V453 Cyg we have made use of the spectral disentangling technique (Simon and Sturm, 1994, Hadrava, 1995). The computer codes FDBINARY (Ilijić et al., 2004) and CRES (Ilijić, 2004), which rely on the Fourier transform technique (Hadrava, 1995), and the SVD technique in wavelength space (Simon and Sturm, 1994), respectively, were used. Spectral disentangling is a powerful method which has found a variety of applications in close binary research (c.f. Holmgren et al., 1998; Hensberge et al., 2000; Harries et al., 2003; Harmanec et al., 2004).

The non-LTE line-formation calculations are performed using DETAIL and SURFACE (Butler and Giddings, 1985). However, hydrostatic, plane-parallel, line-blanketed LTE model atmospheres calculated with the ATLAS9 code (Kurucz, 1983) have also been used. This hybrid approach has been compared with the state-of-the-art non-LTE model atmosphere calculations and excellent agreement has been found for the hydrogen and helium lines (Przybilla, 2005).

3. Results and conclusion

In the observed spectral ranges the helium abundance can be derived only from the lines centred at 4378, 4471, 4718 and 6678 Å. As discussed by Lyubimkov et al. (2004), calculations for He I 4378 Å are less reliable using DETAIL since only transitions up to level $n = 4$ are considered explicitly. Since level populations can be affected by the microturbulent

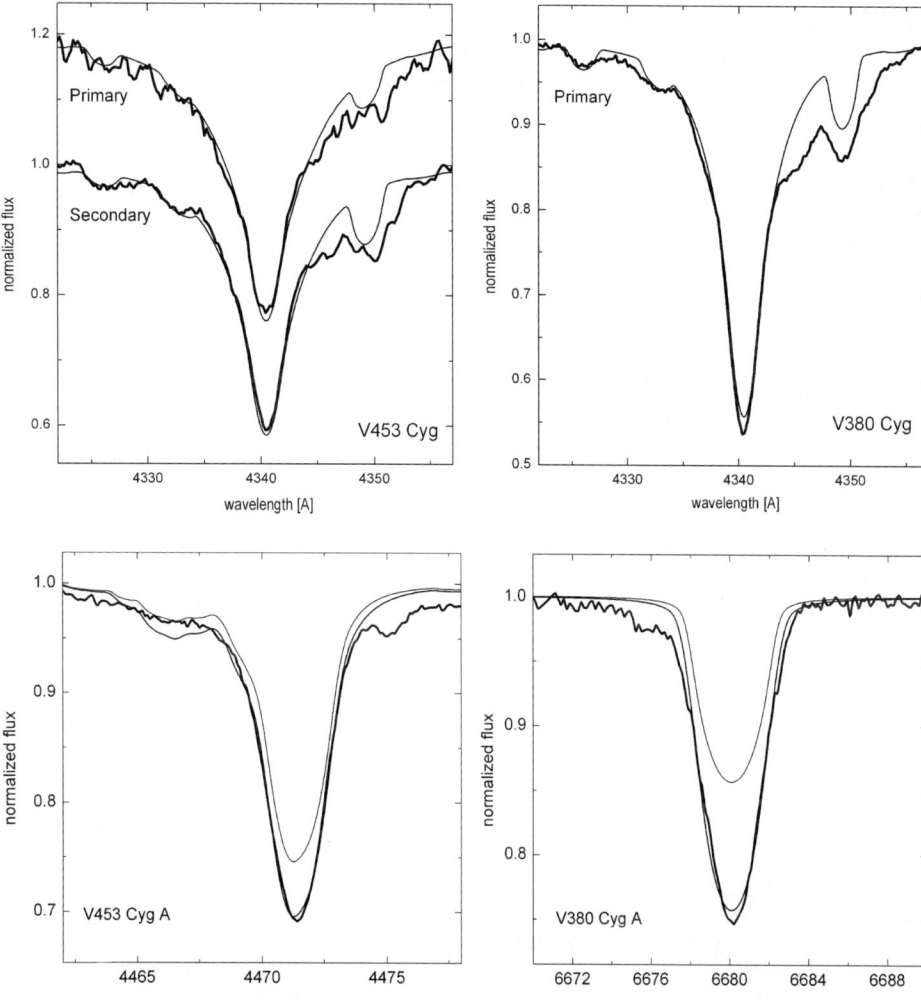

Fig. 1 The best fit of the calculated profiles (thin black line) of the Hγ line compared to the observed profiles (thick gray) for the components of V453 Cyg, left panel, and the primary component of V380 Cyg, right panel

Fig. 2 The best fitting calculated profiles (thin black) of the He I 4471 Å and 6678 Å lines compared to the observed profiles (gray) in the primary component of V453 Cyg (left panel), and the primary component of V380 Cyg (right panel). Light gray lines represent profiles for the solar helium abundance

parameter V_{turb} it should be also included in calculations and adjusted to the observed line profiles.

First, a check and slight adjustment of the effective temperature to the individual component spectra for V453 Cyg, and the primary of V380 Cyg, has been made. As an example, fitting of the calculated to the observed line profiles of the Hγ line is shown in Fig. 1. A simultaneous fit of the helium abundance ϵ_{He}, and microturbulent velocity V_{turb} has then been performed from the grid of the calculated spectra, while T_{eff} and log g have been kept fixed (Fig. 2).

The helium enrichment has been found for the primary component in the system V380 Cyg by Lyubimkov et al. (1996). The helium abundance they derived, $\epsilon_{\text{He}} = 0.19 \pm 0.05$, is considerably larger than the value derived in the present work. The complete analysis and discussion of possible sources of the discrepancy will be published elsewhere.

Recently, Lyubimkov et al. (2004) have derived the helium abundances in a large sample of early-B type stars. Their results are plotted in Fig. 3 as open symbols. This confirmed their finding that helium is becoming enriched in high-mass stars already on the main sequence. However, due to large

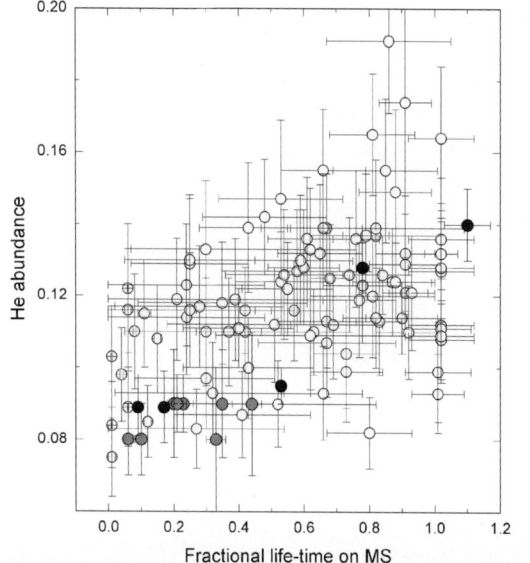

Fig. 3 Abundances of helium in the components of the close binaries (filled symbols; dark symbols show the results of this work) overplotted on the results for large sample of single early-B type stars (open symbols) of Lyubimkov et al. (2004)

errors in deriving the fundamental parameters for the single stars, there is considerable scatter in their diagram. Overplotted by filled symbols are results for the components of eclipsing, double-lined spectroscopic binaries, in light-gray (c.f. Pavlovski, 2004). In dark-gray are presented the results of this work. The general finding that in the later phases on the main sequence helium is enriched is confirmed, but there is disagreement for early phases for which results from the close binaries are very consistent and are giving a helium abundance close to the solar value. However, the sample is still rather limited and more work is needed to have complete picture on the helium enrichment on the MS for high-mass stars.

References

Butler, K., Giddings, J.R.: Newsletter on analysis of astronomical spectra, No. 9 (1985)
Daflon, S., Cunha, K., Becker, S.: ApJ **604**, 362 (2004)
Hadrava, P.: A&AS **114**, 393 (1995)
Guinan, E.F., Ribas, I., Fitzpatrick, E.L. et al.: ApJ **544**, 409 (2000)
Harmanec, P., Uytterhoeven, K., Aerts, C.: A&A **422**, 1013 (2004)
Harries, T.J., Hilditch, R.W., Howarth, I.D.: MNRAS **339**, 257 (2004)
Heger, A., Langer, N.: ApJ **544**, 1016 (2000)
Hensberge, H., Pavlovski, K., Verschueren, W.: A&A **358**, 553 (2000)
Holmgren, D., Hadrava, P., Harmanec, P. et al.: A&A **322**, 565 (1997)
Ilijić, S.: ASP Conf. Ser. **318**, 107 (2004)
Ilijić, S., Hensberge, H., Pavlovski, K., Freyhammer, L.M.: ASP Conf. Ser. **318**, 111 (2004)
Kurucz, R.L.: Kurucz CD-ROM No. 13 (1983)
Lyubimkov, L.S., Rachovskaya, T.M., Rostopchin, S.I., Tarasov, A.E.: ARep **40**, 46 (2004)
Lyubimkov, L.S., Rostopchin, S.I., Lambert, D.L.: MNRAS **351**, 745 (2004)
Meynet, G., Maeder, A.: A&A **361**, 101 (2000)
Pavlovski, K.: ASP Conf. Ser. **318**, 206 (2004)
Pavlovski, K., Hensberge, H.: A&A **439**, 309 (2005)
Przybilla, N.: A&A **443**, 293 (2005)
Simon, K.P., Sturm, E.: A&A **281**, 286 (1994)
Southworth, J., Maxted, P.F.L., Smalley, B.: MNRAS **351**, 1277 (2004)
Vrancken, M., Hensberge, H., David, M., Verschueren, W.: A&A **320**, 878 (1997)

Doppler Imaging of Rapidly-Rotating M Stars

Klaus G. Strassmeier

Received: 1 November 2005 / Accepted: 1 February 2006
© Springer Science + Business Media B.V. 2006

Abstract With the advent of 8–12m-class telescopes and powerful new spectrographs, we can now extend the Doppler-imaging technique to the cool (and faint) end of the main sequence. At a spectral type of approximately M2, stars are thought to become fully convective and cannot possess an overshoot layer between a radiative core and a convective envelope which, as in the case of the Sun and similar stars, likely harbors the dynamo. Therefore, one could expect a fundamentally different magnetic-field topology than on the Sun and thus a qualitatively different surface temperature distribution with new, hitherto unknown, magnetic activity phenomena. Unfortunately, most single M stars do not rotate sufficiently fast for Doppler imaging and one has to "use" binaries or pre-main-sequence stars in which M stars appear spun up or, in binaries, synchronized to the orbital motion.

Keywords M stars · Stellar activity · Active chromospheres · Doppler imaging

1. Towards a global understanding of the Sun

Arguably the most interesting stars from the point of view of understanding the Sun are solar analogues, or at least solar-type stars. However, slowly rotating solar analogues are still off limit for the Doppler imaging (DI) technique. DI solves the integral equation that relates the distribution of, e.g., surface temperature to the observed line profile and light curve variations and thus provides the means to reconstruct an "image" of the stellar surface with the help of numerical inversion techniques (see, e.g., the review by Rice (2002)).

One of the theoretically interesting comparisons is to see if fully-convective stars have a similar distribution of surface magnetic field as does, e.g., the Sun. The model predictions for the location of the active regions in latitude on the Sun fit well with observation. Generally, the work of Schüssler et al. (1996), Granzer et al. (2000), a.o., involves calculating the trajectory of the magnetic flux tubes in the convective envelope of cool stars from their origin at the overshoot layer between the radiative core region and the envelope up to where they erupt at the surface. For fully convective stars without a radiative core, the model fails, although there is clear proof that many of these stars are magnetically active. The detections of extreme-UV emission, flares, and periodic modulation of the light from stars like VB 8 (M7) and VB 10 (M8; $T_{\rm eff} = 2500$ K) (see, e.g. Drake et al., 1996) and even L-dwarfs suggest that these stars must harbor a magnetic field (see the review by Basri (2000)). We speculate that another type of dynamo works in these kind of stars but do not even know whether the magnetic activity behavior of such stars is indeed different from the Sun's, although it is suggested by indirect indicators as well as turbulent dynamo models (Küker and Rüdiger, 1999).

2. Doppler imaging from atomic and molecular lines

Atomic lines of cooler stars become progressively harder to observe because the thermal energy is too small to populate the higher atomic levels and also because more atoms are being bound in molecules and are missing for the atomic line formation (Fig. 1). Stars cooler than M2 could only be mapped by considering the many molecular bands, mostly TiO, OH, and CO. Spectrum synthesis of the optical TiO

K. G. Strassmeier
Astrophysical Institute Potsdam (AIP), An der Sternwarte 16, D-14482 Potsdam, Germany
website: http://www.aip.de/groups/activity

of magnetic star-disk coupling for MN Lupi that predicts a polar surface magnetic field of ≈3 kG.

The rotation rate of MN Lupi of 10.5 hours is just 3.7 times below its break-up rotation rate of 2.8 hours. It is therefore among the fastest rotators of non-degenerate stars. The simultaneous presence of high-excitation Balmer lines and periodically changing broadened photospheric absorption lines hint towards a complex evolution of the magnetic star-disk coupling and its associated angular-momentum exchange. MN Lupi seems to be an example of a CTTS that has been spun up recently, or is still being spun up, while accretion onto the stellar surface is ongoing at the same time. Clearly, MN Lupi is a very young M star and not a "normal" M dwarf and is not directly useable for solar-stellar comparisons.

4. The role of binaries

A main-sequence M dwarf in a close binary usually rotates much faster due to tidal locking than its single-star counterpart. Binary components are therefore good targets for M-star Doppler imaging but, of course, must rely on the assumption that they are representative in terms of the internal and atmospheric structure. This is highly questionable because their (external and internal) angular-momentum history must have been different. Moreover, because of their always almost identical luminosities, M-dwarf binaries come as double-lined binaries. Among the most well studied cases is YY Gem, M1V + M1V (see Ribas, this volume). Figure 4 shows part of a time series of $R = 60,000$ spectra obtained at CFHT. Both stars appear in the spectrum and their lines are rotationally broadened. Furthermore, a number of VO and TiO bands obstruct the true continuum and blend with all atomic lines in that particular wavelength region. Clearly, Doppler imaging must take into account molecules, atomic lines *and* both stellar components simultaneously. Another DI-related problem with YY Gem is that it is an eclipsing binary, its inclination thus close to 90°, and thus will produce non-unique Doppler images due to the well-known north-south mirroring. Mapping of YY Gem is ongoing at the time of writing of this paper (and continues to be difficult).

Acknowledgements This work was supported by the *Deutsche Forschungsgemeinschaft*, DFG project number STR645/1-1.

References

Basri, G.: ARA&A **38**, 485 (2000)
Berdyugina, S.V., Solanki, S.K., Lagg, A.: In Brown, A., Harper, G.M., Ayres, T.R. (eds.), 12th Cambridge Workshop on Cool Stars, Stellar Systems, and the Sun, Univ. Colorado, p. 210 (2003)
Drake, J.J., Stern, R.A., Stringfellow, G.S. et al.: ApJ **469**, 828 (1996)
Granzer, T., Schüssler, M., Caligari, P., Strassmeier, K.G.: A&A **355**, 1087 (2000)
Küker, M., Henning, T., Rüdiger, G.: ApJ **589**, 397 (2003)
Küker, M., Rüdiger, G.: A&A **346**, 922 (1999)
O'Neal, D., Neff, J.E., Saar, S.H., Cuntz, M.: AJ **128**, 1802 (2005)
Rice, J.B.: AN **323**, 220 (2002)
Savanov, I.S., Strassmeier, K.G.: A&A **444**, 931 (2005)
Schüssler, M., Caligari, P., Ferriz-Mas, A., Solanki, S.K., Stix, M.: A&A **314**, 503 (1996)
Strassmeier, K.G., Rice, J.B., Ritter, A. et al.: A&A **440**, 1105 (2005)
Wallace, L., Livingston, W., Bernath, P.F., Ram, R.S.: An atlas of the Sunspot umbral spectrum, NSO/NOAO (1999)

Disentangling of the Spectra of Binary Stars – Principles, Results and Future Development

Petr Hadrava

Received: 1 November 2005 / Accepted: 1 February 2006
© Springer Science + Business Media B.V. 2006

Abstract The method of spectra disentangling is reviewed in context with related techniques. New generalizations (disentangling with constraints) and applications (for spectroastrometry) are also described.

Keywords Spectroscopic binaries · Spectra disentangling · Line-profile variations

1. Introduction

Disentangling of spectra of multiple stellar systems is one of techniques for interpretation of spectroscopic data. The general aim of interpretation of any observational or experimental data (not only in astrophysics) is to determine some properties of the studied object by comparing the data with a theoretical model describing the behaviour of the object. One should be aware of crucial dependence of the results of this procedure on the proper choice of the model, as can be demonstrated by the following example.

In principle, the data obtained from an observation can be represented as a set $\{x_i|_{i=1}^n\}$ of points (scattered due to variability of the source or due to observational errors) in a multidimensional space of observed quantities (e.g. intensities of light in different wavelengths). The interpretation of the data thus consists in the fit of these points by a theoretical model $x = m(p)$ adjusting its free parameters p to give the best agreement with the observation (e.g. the mass of a main-sequence star corresponding best to the observed colours of a star). It is mostly performed by a least-square fit or by minimization of some other norm $\sum_i ||x_i - m(p)||$ of O–C with respect to the variation δp. However, the best fit need not be a good fit, because we cannot be sure that our model does not neglect some important feature, which may change the apparent values of p (e.g. a circumstellar envelope around the star), or that a completely different model is the correct one. Even between models explaining the observations by objects of the same kind we can choose either some more general with more free parameters, or some more restrictive one. The former is usually capable to fit the observations more precisely, but it need not be necessarily better than the latter (let us recall e.g. the principle of Ockham's razor). One has always to judge critically if restrictions or generalizations of a model are reasonable.

In view of these very general thoughts, it is desirable to develop tools for interpretation of observational data enabling to choose always a level of generality of the model appropriate for the particular case as well as to take into account information available in different kinds of data. To understand what is the role of the disentangling method and what should be done next, let us briefly summarize first the previous methods, their assumptions, advantages and shortcomings.

2. Classical methods

2.1. Radial-velocity measurements and solution of radial-velocity curve

The spectroscopy of binary stars has contributed to astrophysics first of all by determination of absolute sizes of their orbits (and thus, via Kepler's laws, also of their masses) from the Doppler shifts of component spectral lines. The wavelengths measurement is straightforward for sharp and narrow or at least symmetric and well separated spectral lines. However, for wide assymetric and blended lines, some model of the observed spectra should be taken into account.

P. Hadrava
Astronomical Institute, Ondřejov

When the radial velocities (RVs) are determined at different phases, the orbital parameters of the binary can be solved from the RV-curve, assuming that the measured wavelengths (e.g. centers of lines) correspond to the velocities of mass-centers of the component stars. This need not be exactly true in the case of line-profile variations (LPVs). Usually the standard least-square method is used to fit the observed RVs by the theoretical (Keplerian) curve. It is worth to note that the originally observed quantities, which are the intensities at the individual pixels of the observed spectrum (with more or less equal noise), are replaced for the fitting by some much less numerous auxiliary quantities, i.e. by the RVs of individual spectral lines, the observational errors of which are quite uncertain.

2.2. Methods of cross-correlation and broadening function

Several tricky methods have been developed to enable measurement of RVs for asymmetric and blended spectral lines. Widely used is the cross-correlation method, based on the cross-correlation function $F(v)$ calculated in the logarithmic wavelength scale x as

$$F(v) = \int I(x) J(x - v) dx , \quad x = c \ln \lambda , \qquad (1)$$

which characterizes the coincidence of the observed spectrum I with a template spectrum J Doppler shifted for radial velocity v (cf. e.g. Hill, 1993). Advantage of the method is that it cumulates the contributions of all lines coinciding in I and J and suppresses the noise, enabling thus measurement of even quite weak component. A disadvantage is the need of a proper template with a similar spectral type as the observed star. Another disadvantage is that F is bilinear in both I and J and its profile thus adds the line widths of I and J (cf. Rucinski, 2002).

The method of cross-correlation has been generalized by Zucker and Mazeh (1994) to a two-dimensional version in which each component of a double-lined binary has its own template.

Another method enabling to measure RV from a large region of a spectrum at once is the method of broadening function by Rucinski (1992), in which a broadening function $B(x)$ is calculated by Singular Value Decomposition (SVD) to fit the observed spectrum I in the form

$$I = J * B \qquad (2)$$

for an appropriately chosen template J. For an ideal template, B should be a shifted delta-function, for fast rotating star and non-rotating template (e.g. a model spectrum), the width of B can specify also the rotational broadening.

2.3. Decomposition of spectra, tomographic separation

While the above methods determine RVs and then orbital and basic physical parameters of the binary, there are also other methods which, for known RVs at different orbital phases, decompose the observed spectra to yield the spectra of individual components. First, it is the method of direct subtraction by Ferluga (1991), which restores both component spectra from two exposures at different phases to, which, however, suffers from cumulation of noise. Another alternative is the method of tomographic separation by Bagnuolo and Gies (1991), based on the fact that the superpositions of the component spectra at different orbital phases are equivalent to their projections (if they are placed parallelly at a non-zero separation) in different directions, like in the problem of tomographic reconstruction of the spatial distribution of optically thin absorbers or emitters. Using some of the standard numerical methods of CT, the component spectra can thus be reconstructed from a sufficient amount of binary spectra obtained at different phases.

3. Disentangling in wavelength and Fourier domain

As we saw in the previous Section, the spectra of a binary secured at different phases contain information both about the RVs (and consequently also the orbital parameters) and the component spectra. However, this information is entangled like a Gordian knot and the standard methods need solution of one of the unknowns to solve for the other. It is thus a natural task to develop a method disentangling both aspects of the problem simultaneously (cf. the scheme in Fig. 1). In this sense the notion of spectra disentangling introduced by Simon and Sturm (1994) surpasses a simple decomposition, which is its inevitable part. (Let us note that the word 'disentangling' is sometimes used as synonym for 'solution'. However, it would be better to refine the terminology by distinguishing the complexity of the solved problem.)

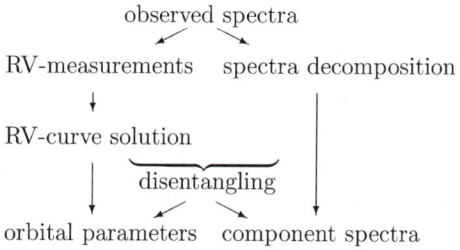

Fig. 1 Flow of information in standard methods and in disentangling: spectra decomposition gives component spectra for known RVs, RV-measurements yield orbital parameters assuming (at least qualitatively) the appearance of component spectra. Disentangling comprises solution of both standard problems

The problem of spectra disentangling can thus be formulated as simultaneous fitting of spectra $I(x, t)$ of a multiple stellar system at all times t in the form

$$I(x, t) = \sum_{j=1}^{n} I_j(x) * \delta(x - v_j(t, p)) \qquad (3)$$

by n arbitrary unknown functions $I_j(x)$, each one representing spectrum of one of the component stars (this is the decomposition part of the problem, which is linear with high dimension), and simultaneous optimizing the fit with respect to the orbital parameters p (i.e. nonlinear orbital solution). Note that not only RVs of line centers but complete spectra are fitted in this way, i.e. the available information is better exploited – the difference is like between the light curve solution vs. study of timing of light-curve minima of eclipsing binaries (cf. Hadrava, 2005). Being more sophisticated, this tool leads to more accurate results, provided (as discussed in Section 1) the underlying model given now by Equation (3) is appropriate to the observed system (and the data are sufficient to determine the solution). If the best fit leaves residuals exceeding a reasonable level of noise, it indicates, that the model neglects some important feature or it is completely wrong. The same may be true if the fit is formally good, but the resulting component spectra are not physically plausible (what is possible owing to the high freedom of the mathematical solution). In both cases, despite it may be annoying for the user, it is a warning that something is wrong either with the data (e.g. in the data processing, quite often in the rectification of spectra) or with the model. Use of a more restrictive model can lead more easily (without any warning) to results, which, however, may be wrong.

In the formulation by Simon and Sturm (1994), the spectra $I(x, t)$ and $I_j(x)$ are represented as vectors of intensities in the individual pixels and the convolution with shifted δ-functions in Equation (3) is represented as multiplication by sparse matrices with a unit off-diagonal shifted for number of pixels corresponding to the value $v_j(t, p)$. The whole system of equations for the decomposition is thus a huge (overdetermined) linear system, which can be solved by SVD method.

The disentangling developed by Hadrava (1995) is based on Fourier transform of Equation (3)

$$\tilde{I}(y, t) = \sum_{j=1}^{n} \tilde{I}_j(y) \exp(i y v_j(t, p)), \qquad (4)$$

in which the Doppler shift is given by simple multiplication with complex exponentials. The decomposition of spectra in Fourier domain is thus fitting the observed spectra $\tilde{I}(y, t)$ with a linear combination (with unknown coefficients $\tilde{I}_j(y)$) of known functions $\exp(i y v_j(t, p))$, i.e. solution of n linear equations for each Fourier mode of the component spectra separately. The representations of spectra in the wavelength and the Fourier domain are equivalent like the x- and p- representations in the quantum mechanics, and owing to Parseval theorem, also the least square fits in both domains are equivalent. The difference is, that the operator to be inverted in the spectra decomposition is reducible in the Fourier representation. The solution is thus numerically much easier and allows further generalizations.

4. Disentangling with line-profile variability

An example of possible failure of the above described disentangling in fitting observed spectra of binaries is the case of exposures taken during an eclipse. It is obvious that during a total eclipse the term in the sum in Equation (3) corresponding to the eclipsed component should be missing and hence, because the spectra are rectified with respect to the observed continuum, which is the sum of component continua, the line depths of remaining components may also be changed. During a partial eclipse, line-profiles of the partially eclipsed component may be distorted e.g. due to Schlesinger-Rossiter rotational effect, or even due to variations of limb darkening across the line-profile (cf. Hadrava and Kubát, 2003). We have thus to generalize Equation (3) into form

$$I(x, t) = \sum_{j=1}^{n} I_j(x) * \Delta_j(x, t, p), \qquad (5)$$

where $\Delta_j(x, t, p)$ is in fact the Rucinski's broadening function B able to describe both the orbital Doppler shifts as well as LPVs (e.g. due to the eclipses, ellipticity, reflection, pulsations, stellar spots etc.).

In the disentangling, Δ_j is restricted to a form described by some parameters p, which are solved simultaneously with the component spectra I_j (restricted by templates J in the Rucinski's method). Equation (5) can be solved again by Fourier transform if the transforms $\tilde{\Delta}_j(y, t, p)$ of broadening functions are calculated either analytically (as in Hadrava, 2004a) or numerically from a 'light-curve model' of line-profiles (cf. Hadrava, 2005).

The simplest example of such a generalization is the method of line-strength photometry (Hadrava, 1997), which assumes

$$\Delta_j(x, t, p) = s_j(t) \delta(x - v_j(t, p)), \qquad (6)$$

i.e. the shapes of line-profiles are constant but their strengths are modulated by line-strength factors $s_j(t)$ for each component at each exposure. They are solved explicitly by least-square fits in the author's code KOREL (cf. Hadrava, 2004c) which enables to measure the eclipse from spectra and to get correct RVs from the exposures during the eclipse, to follow line-strength changes due to other reasons (e.g. secular changes) or to disentangle telluric lines (cf. Hadrava, 2004a for details).

5. Results of disentangling

Disentangling as well as line-strength photometry have been used for many studies of different multiple stellar systems. For account of early applications we can refer to review by Holmgren (2004). Here a few selected recent works will be added and commented.

The study of Atlas in Pleiades by Zwahlen et al. (2004) proved that the parallax of Pleiades determined using Hipparchos data is wrong. It is important experience for present as well as future astrometric satellites. It also demonstrates the usefulness of spectroscopic-binary studies and disentangling in particular for measurement of distances, which already reached the extragalactic scale and pretends to be an indispensable part of basis for observational cosmology.

Another challenging possibility for application of spectra disentangling appears in combination with spectroastrometry (cf. Porter et al., 2004), which aims at measurement of components angular distance from long-slit spectra of binaries. For a non-zero projection of the distance on the direction of the slit (but smaller than the point-spread function), the spectrum at each row of the spectrogram is a superposition of the component spectra with the strength-factors s_j being different, mutually shifted, functions of the row number. Determination of these functions would be easier for component spectra disentangled from temporal changes. This application is thus promising especially for distinguishing a distant third companion of a close binary, which appears to be an important and quite common case.

Let us mention also the common use of spectra disentangling to studies of binaries with LPVs connected mostly with stellar pulsations (as a recent example see e.g. de Cat et al., 2005). In fact such LPVs violate assumption (3) and (6) of time invariance of I_j. However, it can be accepted as an approximation if LPVs are a small perturbation only to the mean line-profiles obtained by disentangling. For this purpose KOREL enables the output of residual spectra in rest-frame of any chosen component (cf. Hadrava, 2004c). A more reliable approach has to replace Equation (6) by a better model of LPVs.

6. Disentangling with constraints

From the view-point of generality of a model used for data interpretation as discussed in Section 1, the disentangling is less restrictive on the component spectra than the methods based on template spectra. On the contrary, in its standard form it is more restrictive (and slightly less in the generalization (5)) regarding the broadening function compared to the Rucinski's method and, in some sense, even compared to any RV-measurement if the assumption of Keplerian motion is imposed on the disentangling. Because each degree of freedom of the model can be useful or harmful for the solution depending on the particular case in question, it is desirable to have a tool for the data interpretation, in which the level of the model freedom could be tuned in details.

It can thus be needed to enable some restrictions for component spectra. One straightforward possibility is to put a constraint on some of the components in the form of template spectra J_j (either from another star, from model atmospheres or even from other solutions of the same system) and to disentangle the observed spectra with respect to a subset of m component spectra only (and the parameters p of any broadening Δ_j) modifying Equation (5) to the form

$$\sum_{j=1}^{m} I_j(x) * \Delta_j(x, t, p)$$
$$= I(x, t) - \sum_{j=m+1}^{n} J_j(x) * \Delta_j(x, t, p). \quad (7)$$

This may be useful e.g. for disentangling of telluric spectrum, which is basically known, but quite often its continuum is distorted due to an imperfect disentangling of lower Fourier modes of wide stellar lines.

Another type of constraints can be imposed on the orbital and other parameters of Δ-functions. Like in light- or RV-curve solution or any other data fitting, different parameters p can be either converged or fixed, if they are better determined by an independent way. However, a constraint on the parameters may also have a form of mutual dependence, which can be described by one or more conditions $F_k(p) = 0$ in the space of p. In such a case, the parameters can be substituted in a form $p = p(q)$ and the observations fitted with respect to the less dimensional space of q, or, invoking the method of Lagrange multiplicators, the fit can be performed by numerical minimization of expression

$$0 = \delta \left\{ \sum_t \int |I - \sum_j I_j * \Delta_j|^2 dx + \sum_k \lambda_k F_k^2(p) \right\}, \quad (8)$$

where sufficiently high values of coefficients λ are chosen. The later approach is predominant, if the parameter constraints are also of an observational nature with some uncertainty. In such a case we can put $F^2 \equiv (O - C)^2$ and Equation (8) is then an equation for simultaneous disentangling and solution of other type of data. A merging of disentangling (e.g. using KOREL) with solution of light- and RV-curves, astrometry etc. (e.g. using FOTEL) is thus a direction to be followed in future (cf. Wilson, 1979; Holmgren, 2004; Hadrava, 2004b, 2005).

References

de Cat, P., Briquet, M. et al.: A&A **432**, 1013 (2005)
Hadrava, P.: A&AS **114**, 393 (1995)
Hadrava, P.: A&AS **122**, 581 (1997)
Hadrava, P.: ASP Conf. Ser. **318**, 86 (2004a)
Hadrava, P.: Publ. Astron. Inst. ASCR **92**, 1 (2004b)
Hadrava, P.: Publ. Astron. Inst. ASCR **92**, 15 (2004c)
Hadrava, P.: Astrophys. Sp. Sc. **296**, 239 (2005)
Hadrava, P., Kubát, J.: ASP Conf. Ser. **288**, 149 (2003)
Hill, G.: ASP Conf. Ser. **38**, 127 (1993)
Holmgren, D.: ASP Conf. Ser. **318**, 95 (2004)
Porter, J.M., Oudmaijer, R.D. et al.: A&A **428**, 327 (2004)
Rucinski, S.: AJ **104**, 1968 (1992)
Rucinski, S.: AJ **124**, 1746 (2002)
Simon, K.P., Sturm, E.: A&A **281**, 286 (1994)
Wilson, R.E.: ApJ **234**, 1054 (1979)
Zucker, S., Mazeh, T.: ApJ **420**, 806 (1994)
Zwahlen, N., North, P. et al.: A&A **425**, L45 (2004)

Automated Analysis of Light Curves of OGLE LMC Binaries: The Period distribution

Tsevi Mazeh · Omer Tamuz · Pierre North

Received: 1 November 2005 / Accepted: 1 February 2006
© Springer Science + Business Media B.V. 2006

Abstract We present a new algorithm, the Eclipsing Binary Automated Solver (EBAS) to analyse lightcurves of eclipsing binaries. To replace human visual assessment, we introduce a new 'alarm' goodness-of-fit statistic that takes into account correlation between neighbouring residuals. We apply the new algorithm to the whole sample of 2580 binaries found in the OGLE photometric survey of the LMC and derive the photometric elements for 1931 systems. To obtain the statistical properties of the short-period binaries of the LMC we construct a well defined subsample of 938 eclipsing binaries with main-sequence B-type primaries. Correcting for observational selection effects, we derive the distributions of the fractional radii of the two components and their sum, the brightness ratios and the periods of the short-period binaries. Somewhat surprisingly, the results are consistent with a flat distribution in log P between 2 and 10 days.

Keywords LMC · Stars: Binaries

1. Introduction

The commonly used interactive way of solving an eclipsing binary lightcurve uses human guess and decisions for both the starting point and the converging iterations (e.g. Ribas et al., 2000). However, such a process is impractical when it comes to the large set of lightcurves which have become available recently (e.g. Alcock et al., 1997). The OGLE survey, for example, yielded a huge photometric dataset of the SMC (Udalski et al., 1998) and the LMC (Udalski et al., 2000), which includes a few thousand eclipsing binary lightcurves.

The SMC and LMC datasets allow for the first time a statistical analysis of the population of short-period binaries in another galaxy. A first effort in this direction was performed by North and Zahn (2003), who derived the orbital elements and stellar parameters of 153 eclipsing binaries in the SMC in order to study the statistical dependence of the eccentricity of the binaries on their separation. North and Zahn (2004) analyzed another sample of 509 lightcurves selected from the 2580 eclipsing binaries discovered in the LMC by the OGLE team (Wyrzykowski et al., 2003). However, the OGLE LMC data contain many more eclipsing binary lightcurves. An automated algorithm would have made an analysis of the whole sample possible.

It was precisely to fulfil this need that we developed EBAS (Eclipsing Binary Automated Solver), a completely automatic scheme that derives the photometric parameters of eclipsing binaries. The goal of the present study was to analyze the whole sample of 2580 lightcurves discovered by OGLE in the LMC, and to derive some statistical properties of the short-period binaries, after correcting for observational selection effects. Such a correction is essential for the derivation of the period distribution, for example, because the probability of detecting an eclipsing binary is a strong function of the orbital period. In order to apply the right correction one needs a complete homogeneous data set of lightcurves, all discovered by the same photometric survey of a well defined sample of stars. These requirements were exactly met by the OGLE data set for the LMC, enabling such analysis for a large sample of eclipsing binaries for the first time.

We present here our EBAS code and its application to the LMC eclipsing binary stars, discovered by Wyrzykowski et al. (2003) with the OGLE data. The full report of our

T. Mazeh (✉) · O. Tamuz
School of Physics and Astronomy, Tel Aviv University, Israel

P. North
Laboratoire d'astrophysique, Ecole Polytechnique Fédérale de Lausanne (EPFL), Observatoire, CH–1290 Sauverny, Switzerland

work was submitted for publication in the Monthly Notices of Royal Astronomical Society (Tamuz et al., 2005; Mazeh et al., 2005).

2. The EBAS code

EBAS is based on the EBOP subroutines (Popper and Etzel, 1981; Etzel, 1981) that generate a lightcurve for a given set of orbital elements and stellar parameters, but we rewrote the iterative code which finds the best parameters to fit the observed lightcurve, making it *fully automated*. Although the EBOP code is admittedly less accurate than the widely used Wilson-Devinney one, it has the advantage of being much simpler and faster, which makes it very well suited to our purpose.

The EBAS strategy consists of three stages. First, EBAS finds a good initial guess by a combination of grid searches, gradient descents and geometrical analysis of the lightcurve. Next, EBAS searches for the global minimum by a simulated annealing algorithm. Finally, we assess the quality of the solution using the new 'alarm' statistic that we developed for this purpose, and if necessary, perform further minimum searches.

Contrary to the usual χ^2 statistics, our 'alarm' statistics $\mathcal{A}$ is sensitive to the correlation between adjacent residuals of the measurements relative to the model. In other words, it is sensitive to the number of *consecutive* residuals with the same sign. Its definition is the following: the whole lightcurve is divided into separate sequential runs, where a 'run' is defined as a maximal series of consecutive residuals (in the folded lightcurve) with the same sign. Long runs might indicate that the residuals are not randomly distributed. Denoting by k_i the number of residuals in the $i-th$ run, we define the 'alarm' $\mathcal{A}$ as:

$$\mathcal{A} = \frac{1}{\chi^2} \sum_{i=1}^{M} \left(\frac{r_{i,1}}{\sigma_{i,1}} + \frac{r_{i,2}}{\sigma_{i,2}} + \cdots + \frac{r_{i,k_i}}{\sigma_{i,k_i}} \right)^2 - \left(1 + \frac{4}{\pi}\right), \quad (1)$$

where $r_{i,j}$ is the residual of the j-th measurement of the i-th run and $\sigma_{i,j}$ is its uncertainty. The sum is over all the measurements in a run and then over the M runs. The χ^2 is the known function:

$$\chi^2 = \sum_{i=1}^{N} \left(\frac{r_i}{\sigma_i}\right)^2, \quad (2)$$

where the sum is over all N observations. Dividing by χ^2 assures that, in contrast to χ^2 itself, $\mathcal{A}$ is not sensitive to a systematic overestimation or underestimation of the uncertainties.

It is easy to see that $\mathcal{A}$ is minimal when the residuals alternate between positive and negative values, and that long runs with large residuals increase its value. The minimal value of the summation is exactly χ^2, and therefore the minimal value of $\mathcal{A}$ is $-4/\pi$.

The EBAS algorithm was successfully tested through simulated light curves, which allowed us to verify that the elements are recovered without major biases and with reasonable accuracy, except for the ratio of radii k which is notoriously difficult to determine unless the eclipses are total. We also compared the automatic solutions of real lightcurves with 'manual' solutions obtained interactively with the EBOP code: the results are similar and even slightly better with EBAS. Yet another test was done by comparing the geometric elements (sum and ratio of radii, inclination and orbital eccentricity) obtained by EBAS with those obtained by other authors on the same sytems, but with the completely different Wilson-Devinney code. Gonzàlez et al. (2005) find the same elements as us within the errors for the 4 systems we have in common, and the results of Michalska and Pigulski (2005) are also coherent with ours for the common 85 binaries.

3. Statistical study of the LMC eclipsing binaries

The LMC OGLE-II photometric campaign (Udalski et al., 2000) provided 260 to 512 measurements in the I band for stars in 21 fields (Zebrun et al., 2001). Wyrzykowski et al. (2003) identified 2580 binaries in this database. EBAS analysed those systems and found 1931 acceptable solutions, after excluding 376 solutions with alarm too high and 269 ones with radii too large, i.e. too close to the Roche lobe radius.

Since the sample suffers from observational selection effects, we trimmed it by retaining only those systems which are clearly detectable, lie on the main sequence according to their $V - I$ index, and have a total magnitude $17 < I < 19$ and a period shorter than 10 days. The detectability parameter is defined as $D = N \frac{var(m_i)}{var(r_i)}$, where the m_i's are the values of the EBAS model at the times of observation, and the r_i's are the residuals of the measurements relative to the model. The histogram of log D rises sharply from about 2.4 to 3.0, then decreases to zero at log $D \sim 5$. The limit of detectability is thus set to $\log \mathcal{D} > 2.6$, so we are left with 938 binaries.

Figure 1 (top) shows the raw period distribution of the trimmed sample. Obviously, the short periods are overrepresented, since their probability for eclipses is larger. To correct for this, we have computed for each system the minimum inclination i_{min} for which the eclipse could still be detected according to the criterion $\log \mathcal{D} > 2.6$. The detection probability is $p = \cos i_{min}$, so we corrected the histogram by assuming that each observed binary represents $1/p$ binaries.

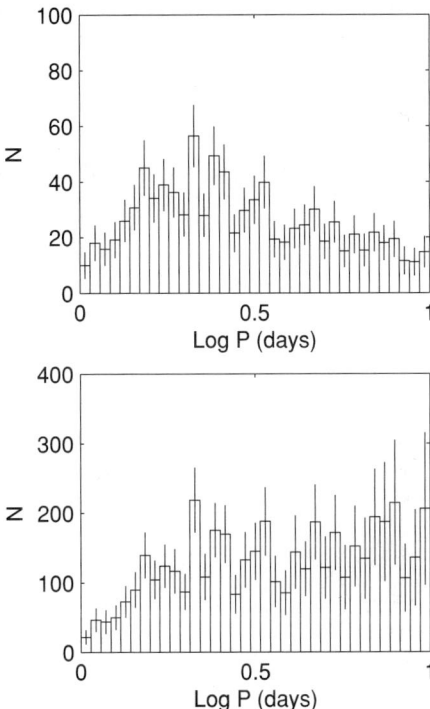

Fig. 1 *Top panel:* raw period distribution of the bias-free sample. *Bottom panel:* period distribution corrected for the detection probability

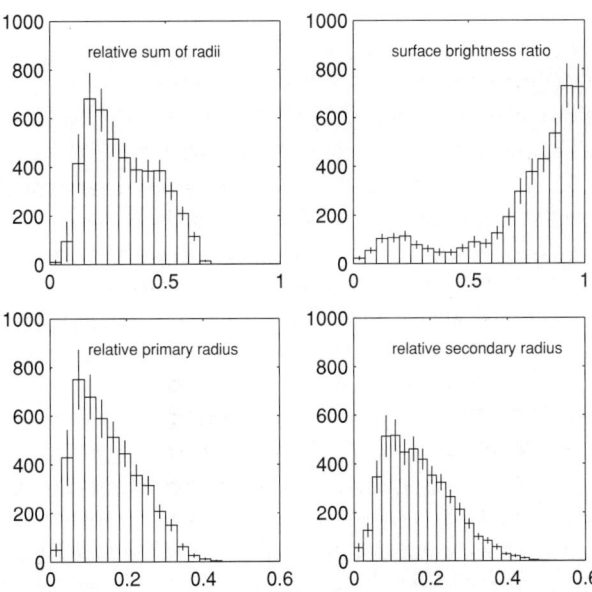

Fig. 2 Histogram of the elements, corrected for the observational selection effect

Furthermore, we took into account the fact that the detection probability depends on the phase of the primary eclipse, especially for systems whose period is an integer number of days: we therefore averaged each p over 100 different phases of the primary eclipse, and only then assigned $1/p$ binaries to each system. The result is shown on the bottom panel of Fig. 1. As far as the correction was applied properly, this panel represents the period distribution of *all* main-sequence binaries – not only the eclipsing ones – in the LMC with I-magnitude between 17 and 19, and with sum of radii smaller than 0.6.

Figure 2 shows the distributions of the primary and secondary relative radii, of their sums, and of the ratios of surface brightnesses, all corrected for the detection probability defined above. The secondary maximum in the histogram of the surface brightness ratio, at $J_s \sim 0.2$, is especially interesting since for binaries with main-sequence components, the surface brightness ratio is a monotonic function of the mass ratio. The secondary maximum is most probably due to semi-detached systems, because it almost disappears when only binaries with $\log P > 0.5$ are considered. Therefore, we suggest that the initial mass-ratio distribution of the short-period binaries with B-type primaries rises monotonically up to a mass ratio of unity.

4. Discussion

Our analysis suggests that binaries with B-type primaries in the LMC have a flat log-period distribution between 2 and 10 days. The period distribution derived here is probably not consistent with the period distribution adopted by Duquennoy and Mayor (1991): a Gaussian with $\overline{\log P} = 4.8$ and $\sigma = 2.3$, P being measured in days. Their distribution would provide within a $\log P$ interval 1.7 more systems at $\log P = 1.0$ than at $\log P = 0.3$, while the new derived distribution is probably flat.

The flat log-period distribution implies a flat log orbital-separation distribution for a given total binary mass. Such a distribution indicates that there is no preferred length scale for the formation of short-period binaries (Heacox, 1998), at least in the range between 0.05 and 0.16 AU, for a total binary mass of 5 $M_\odot$. Although this probable result refers to a very narrow orbital range, a binary formation model should account for this scale-free distribution, which might be common in short-period binaries.

References

Alcock, C., Allsman, R.A., Alves, D., Axelrod, T.S., Becker, A.C.: AJ **114**, 326 (1997)
Duquennoy, A., Mayor, M.: A&A **248**, 485 (1991)
Etzel, P.B.: In: E.B. Carling and Z. Kopal (eds.): Photometric and Spectroscopic Binary Systems, p. 111 (1981)
Gonzalez, J.F., Ostrov, P., Morrell, N., Minniti, D.: ApJ **624**, 946 (2005)
Heacox, W.D.: AJ **115**, 325 (1998)
Michalska, G., Pigulski, A.: A&A **434**, 89 (2005)
North, P., Zahn, J.-P.: A&A **405**, 677 (2003)

North, P., Zahn, J.-P.: New Astronomy Review **48**, 741 (2004)
Popper, D.M., Etzel, P.B.: AJ **86**, 102 (1981)
Ribas, I., Guinan, E.F., Fitzpatrick, E.L., DeWarf, L.E., Maloney, F.P., Maurone, P.A., Bradstreet, D.H., Giménez, Á., Pritchard, J.D.: ApJ **528**, 692 (2000)
Udalski, A., Soszyński, I., Szymański, M., Kubiak, M., Pietrzyński, G., Woźniak, P.R., Żebruń, K.: AcA **48**, 563 (1998)
Udalski, A., Szymański, M., Kubiak, M., Pietrzyński, G., Soszyński, I., Woźniak, P.R., Żebruń, K.: AcA **50**, 307 (2000)
Wyrzykowksi, Ł., Udalski, A., Kubiak, M., Szymański, M., Żebruń, K., Soszyński, I., Woźniak, P.R., Pietrzyński, G., Szewczyk, O.: AcA **53**, 1 (2003)
Żebruń, K., Soszyński, I., Woźniak, P.R., Udalski, A., Kubiak, M., Szymański, M., Pietrzyński, G., Szewczyk, O., Wyrzykowksi, Ł.: AcA **51**, 317 (2001)

Disentangling Effective Temperatures of Individual Eclipsing Binary Components by Means of Color-Index Constraining

A. Prša · T. Zwitter

Received: 30 September 2005 / Accepted: 28 November 2005
© Springer Science + Business Media B.V. 2006

Abstract Eclipsing binary stars are gratifying objects because of their unique geometrical properties upon which all important physical parameters such as masses, radii, temperatures, luminosities and distance may be obtained in absolute scale. This poses strict demand on the model to be free of systematic effects that would influence the results later used for calibrations, catalogs and evolution theory. We present an objective scheme of obtaining individual temperatures of both binary system components by means of color-index constraining, with the only requirement that the observational data-set is acquired in a standard photometric system. We show that for a modest case of two similar main-sequence components the erroneous approach of *assuming* the temperature of the primary star from the color index yields temperatures which are systematically wrong by ~ 100 K.

Keywords Binaries: Eclipsing · Stars: Fundamental parameters · Methods: Data analysis · Numerical · Techniques: Photometric

1. Introduction

Eclipsing binary stars (EBs) are renowned for their typical geometrical layout and well-understood underlying physics. Solving the inverse problem for detached EBs has become an indispensable tool for obtaining absolute dimensions of individual stars (their masses, radii, temperatures, and luminosities). The impact is very broad: EBs are used to establish color calibrations (Harmanec, 1988; Flower, 1996; Popper, 1998), evolutionary properties based on exact coevality of both components (Ribas et al., 2000; Lastennet and Valls-Gabaud, 2002) and relations between M-L-R-T quantities for different spectral types and luminosity classes (Gorda and Svechnikov, 1998). It is thus very important to model EBs as objectively as possible, otherwise it is likely that systematic errors may creep in and mislead any theoretical models built on those results. In this paper we concentrate on a particular issue of objectively determining effective temperatures of individual components in an EB system.

The common practice of determining effective temperatures of individual components is to *assume* the temperature of one star and obtain the temperature of the other star by the model. The assumed temperature is usually obtained from observed color-indices at quarter-phase by using a color calibration or by spectral analysis and comparison against synthetic spectra database. While this approach may be adequate for components having significantly different luminosities (so that the luminosity of one star is equal to the system luminosity for all practical purposes), it is very inadequate for components with similar luminosities and introduces systematical effects into model solutions. When dealing with photometric observations that are acquired in a standard photometric system (i.e. Johnson, Cousins, Strömgren, ...), effective temperatures of both components may be disentangled from two or more photometric light curves *objectively*, without any assumptions. We build this argument and demonstrate its implications on modeling in the following Sections.

2. Simulation

To evaluate the uncertainty of determining a primary star temperature a priori, we built a synthetic binary star model. Testing the method against a synthetic model may seem artificial,

A. Prša (✉) · T. Zwitter
University of Ljubljana, Dept. of Physics, Jadranska 19, SI-1000 Ljubljana, EU

Table 1 Physical parameters of the F8 V–G1 V test binary star. Linear (x) and non-linear (y) coefficients of the logarithmic limb darkening law for Johnson B and V passbands, taken from van Hamme (1993)

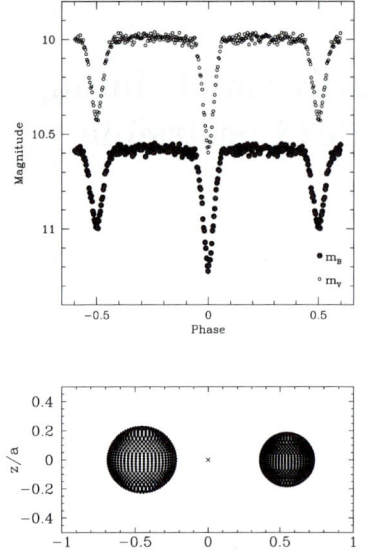

Parameter [units]	F8 V	Binary	G1 V
P_0 [days]		1.000	
a [$R_\odot$]		5.524	
$q = m_2/m_1$		0.831	
i [°]		85.000	
v_γ [km s^{-1}]		15.000	
T_{eff} [K]	6200		5860
L [$R_\odot$]	2.100		1.100
M [$R_\odot$]	1.239		1.030
R [$R_\odot$]	1.260		1.020
Ω [−]^a	5.244		5.599
$\log(g/g_0)$ [−]^b	4.33		4.43
x_B [−]	0.818		0.833
y_B [−]	0.203		0.158
x_V [−]	0.730		0.753
y_V [−]	0.264		0.242

$^a\Omega$ is potential as defined in Wilson (1979).
$^b g_0 = 1\,\text{cm}\,\text{s}^{-2}$ makes g/g_0 dimensionless.

but the obvious advantage of knowing the right solution is the only true way of both qualitative and quantitative assessment.

Our synthetic binary consists of two main-sequence F8 V–G1 V components with their most important orbital and physical parameters listed in Table 1. It is a partially eclipsing detached binary with only slight shape distortion of both components ($R_{1,\text{pole}}/R_{1,\text{point}} = 0.974$, $R_{2,\text{pole}}/R_{2,\text{point}} = 0.979$). Light curves are generated with PHOEBE (Prša and Zwitter, 2005) for Johnson B and V passbands in 300 phase points with Poissonian scatter $\sigma_{\text{LC}} = 0.015$ and quarter-phase magnitude $m_V = 10.0$. Passband transmission curves were taken from ADPS (Moro and Munari, 2000).

The simulation logic is as follows: we take Kurucz's spectral energy distribution (SED) function for both components from the spectra database compiled by Zwitter et al. (2004), doppler-shift them to the reference phase, sum their SEDs weighted by their corresponding luminosity and multiply the sum with the passband response functions of Johnson B and V filters. This enables us to compute the B–V color-index, which needs to be zero-corrected with respect to spectral type A0 (adopted temperature 9560 K). Once this is done, we have obtained a synthetic prediction of the color index. Alternatively, the same result may be obtained by using color calibrations, e.g. Flower (1996)'s corrected tables given by Prša and Zwitter (2005). Color indices may now be readily transformed to the ratio of passband fluxes $J^V_{\text{ref}}/J^B_{\text{ref}}$.

Once the light curves are built so that the color index is preserved, we follow the color-index constraining principle described by Prša and Zwitter (2005) to assess the implications of erroneous *assumed* value of T_{eff1}.

3. Results

Our analysis brings us to the following conclusions:

1. Individual temperatures without the imposed color-index constraint are fully correlated, as is well known. The reason for this is the degeneracy between the temperatures and passband luminosities L_1^i: local effects of slightly different temperatures on the *shape* of the light curve are very small, hardly noticeable (they scale roughly with the ratio of both temperatures). However, even slightly different temperatures cause an additional effect which is an order of magnitude larger: they scale both light curves, thus changing the color index. If this change is not constrained, the model relies only on secular effects, which are almost always buried in data noise and parameter degeneracy. Without color-constraining there is thus *no practical way* to pinpoint a location in the T_{eff1}-T_{eff2} cross-section that corresponds to the right solution.

2. Unless the luminosity of one component is overwhelming, the *assumed* value of the temperature is doubtful. Fig. 1 shows a range of assumed values for T_{eff1} and the color indices yielded by the model that is *not* color-constrained. Since shapes of the light curves determine the ratio of luminosities of individual stars, it has to remain approximately constant. To achieve this with changing temperatures, the model adapts passband luminosities (first order effect) and stellar radii (second order effect) accordingly. The problem now becomes immediately evident: if we are to assume a primary star temperature from the quarter-phase color index, it would be roughly $T_{\text{eff1}}^{\text{assumed}} = 6080$ K, around 120 K off just because the luminosity of the secondary star is not negligible. In

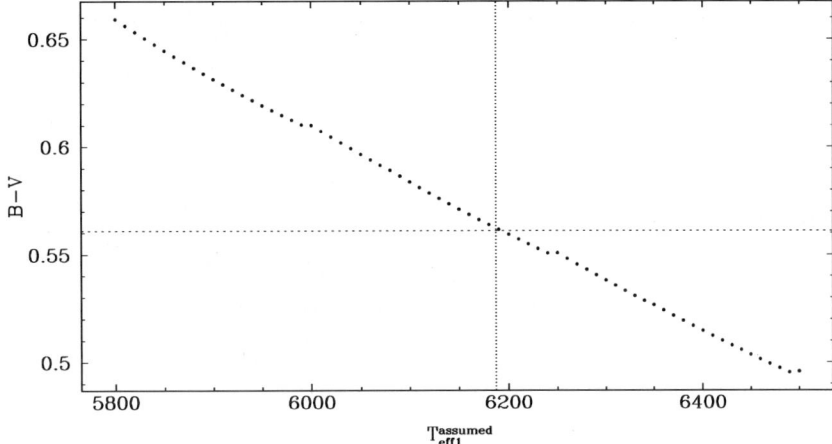

Fig. 1 Color index dependence on the assumed value of the primary star temperature. If color-constraining is not applied, all depicted points may be satisfied by the model. If we are to assume the value of T_{eff1} from the color index at quarter-phase, it would yield $T_{\mathrm{eff1}} = 6080$ K, a value which is ~ 120 K off from the correct value. If on the other hand we apply color-constraining, we immediately get a handle on which color index corresponds to the temperatures yielded by the model. Cross-hairs denote the solution obtained by color-constraining, which gives $T_{\mathrm{eff1}} = 6187 \pm 32$ K.

principle, the solution could then be *manually* re-iterated by computing the predicted color-index from the solution, comparing it to the observed color index and then modifying the assumed temperature to get closer to the right solution. This is tedious and prone to subjective considerations, it should thus be avoided.

3. Adopting the color-constraint means keeping the ratio of passband luminosities L_1^V / L_1^B accordant with the observed color index. This means that *only* L_1^V is adjusted to reproduce the V light curve and L_1^B (and other L_1^i for the remaining passbands) is then computed from this constraint. The degeneracy between the temperatures and the passband luminosities is now broken: we obtain a straight horizontal line over Fig. 1 which best represents the color-index. Thus, we obtain the primary temperature *without any assumptions*, and the secondary temperature follows directly from the model. Applying this to our test binary, we obtain individual temperatures $T_{\mathrm{eff1}} = 6\,187$ K $\pm\,32$ K and $T_{\mathrm{eff2}} = 5\,868$ K $\pm\,31$ K follows from the model. Comparing these values to true values $T_1 = 6\,200$ K and $T_2 = 5\,860$ K is very encouraging.

4. Finally, let us consider the correlation between the temperatures and the radii of the binary system components. We have mentioned earlier that this is a second order effect, but it is still very important to quantify the correlation between those two parameters. Since the radii come into the model via surface potentials Ω (defined in Wilson, 1979), we need to consider the correlation between the ratio of temperatures T_2/T_1 and surface potentials. Fig. 2 shows the convergence tracer for the T_2/T_1–Ω_2 cross-section: tracing parameter values from each starting point, iteration after iteration, all the way to the converged solution. Attractors – regions that attract most convergence traces – within these cross-sections reveal parameter correlations and

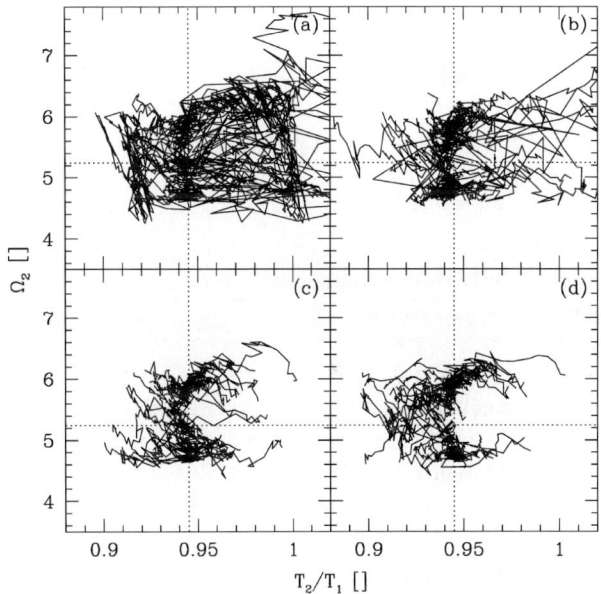

Fig. 2 Convergence tracer for the $T_{\mathrm{eff2}}/T_{\mathrm{eff1}}$-$\Omega_2$ cross-section. Crosshairs denote the correct solution. Panels (*a*) to (*d*) correspond to the solution obtained by Nelder & Mead's downhill Simplex (NMS) for 0, 1, 2 and 3 parameter kicks. See Prša & Zwitter (2005) for details on interpreting convergence tracers

degeneracy. Panels denoted with (*a*) to (*d*) represent three subsequent parameter kicks, which are introduced and thoroughly explained by Prša and Zwitter (2005).

4. Discussion

We have demonstrated how severe the implications of assuming a primary star temperature from the color index may be if extra care is not taken to submit the acquired solution to manual re-iteration. The effect is already very pronounced

for a modest case of two similar main-sequence components and it would be significantly augmented e.g. for a hot main-sequence–cool giant system, where both stars have similar luminosities; in those cases the published solution may be off by as much as ∼1000 K. By having the color-constraining method in PHOEBE, we hope to raise awareness of this issue in future solution-seeking analyses.

References

Flower, P.J.: Astrophys. J. **469**, 355–365 (1996)
Gorda, S.Y., Svechnikov, M.A.: Astronomy Reports **42**, 793–798 (1998)
Harmanec, P.: Bulletin of the Astronomical Institutes of Czechoslovakia **39**, 329–345 (1988)
Lastennet, E., Valls-Gabaud, D.: Astron. Astrophys. **396**, 551–580 (2002)
Moro, D., Munari, U.: Astron. Astrophys. Suppl. Ser. **147**, 361–628 (2000)
Popper, D.M.: Publ. Astron. Soc. Pac. **110**, 919–922 (1998)
Prša, A., Zwitter, T.: Astrophys. J. **628**, 426–438 (2005)
Ribas, I., Jordi, C., Giménez, Á.: In: NATO ASIC Proc. 544: Variable Stars as Essential Astrophysical Tools. pp. 659–670 (2000)
van Hamme, W.: Astron. J. **106**, 2096–2117 (1993)
Wilson, R.E.: Astrophys. J. **234**, 1054–1066 (1979)
Zwitter, T., Castelli, F., Munari, U.: Astron. Astrophys. **417**, 1055–1062 (2004)

A Method For Eclipsing Component Identification In Large Photometric Datasets

J. Devor · D. Charbonneau

Received: 6 October 2005 / Accepted: 3 November 2005
© Springer Science + Business Media B.V. 2006

Abstract We describe an automated method for assigning the most likely physical parameters to the components of an eclipsing binary (EB), using only its photometric light curve and combined color. In traditional methods (e.g. WD and EBOP) one attempts to optimize a multi-parameter model over many iterations, so as to minimize the chi-squared value. We suggest an alternative method, where one selects pairs of coeval stars from a set of theoretical stellar models, and compares their simulated light curves and combined colors with the observations. This approach greatly reduces the EB parameter-space over which one needs to search, and allows one to determine the components' masses, radii and absolute magnitudes, without spectroscopic data. We have implemented this method in an automated program using published theoretical isochrones and limb-darkening coefficients. Since it is easy to automate, this method lends itself to systematic analyses of datasets consisting of photometric time series of large numbers of stars, such as those produced by OGLE, MACHO, TrES, HAT, and many others surveys.

Keywords Eclipsing binaries · Techniques: photometric · Data analysis

1. Introduction

Eclipsing double-lined spectroscopic binaries provide the only method by which both the masses and radii of stars can be estimated without having to resolve spatially the binary or rely on astrophysical assumptions. Despite the large variety of models and parameter-fitting implementations (e.g. WD and EBOP), their underlying methodology is essentially the same. Photometric data provides the light curve of the EB, and spectroscopic data provide the radial velocities of its components. The depth and shape of the light curve eclipses constrain the components' brightness and fractional radii, while the radial velocity sets the length scale of the system. In order to characterize fully the components of the binary, one needs to combine all of this information. Unfortunately, only a small fraction of all binaries eclipse, and spectroscopy with sufficient resolution can be performed only for bright stars. The intersection of these two groups leaves a pitifully small number of stars.

In the past decade, there has been a dramatic growth in the number of stars with high-quality, multi-epoch, photometric data. This has been due to major advances in both CCD detectors and the implementation of image-difference analysis techniques (Crotts, 1992; Alard et al., 1998; Alard, 2000), which enables simultaneous photometric measurements of tens of thousands of stars in a single exposure. Today, there are many millions of light curves available from a variety of surveys, such as OGLE, Udalski et al. (1994), MACHO, Alcock et al. (1998), TrES, Alonso et al. (2004), and HAT, Bakos et al. (2004). But there has not been a corresponding growth in the quantity of spectroscopic data, nor is this likely to occur in the near future. Thus, the number of fully-characterized EBs has not grown significantly. In recent years there has been a growing effort to mine the wealth of available photometric data, by employing simplified EB models in the absence of spectroscopic observations (Wyithe et al., 2001, 2002; Devor, 2004, 2005).

In this paper we present a novel approach, which utilizes theoretical models of stellar properties to estimate the orbital parameters as well as the masses, radii, and absolute magnitudes of the stars, while requiring *only* a photometric light

J. Devor (✉) · D. Charbonneau
Harvard-Smithsonian Center for Astrophysics, 60 Garden St.,
Cambridge, MA 02138, USA
e-mail: jdevor@cfa.harvard.edu

curve and an estimate of the binary's combined color. This approach can be used to characterize quickly large numbers of eclipsing binaries, however it is not sufficient to improve stellar models since underlying isochrones must be assumed. We have created two implementations of this idea. The first, which we have named MECI-express, and is described in Section 2, is a "quick and dirty" program that is designed as a simple extension to the Detached Eclipsing Binary Light curve (DEBiL) fitter (Devor, 2004, 2005). The second, which we have named MECI, and is described in Section 3, is considerably more accurate, but also more computationally demanding. The source code for both MECI-express and MECI will be provided upon request.

2. Express method for eclipsing component identification (MECI-express)

The primary application of MECI-express is to identify the stellar components of a given EB. It operates after a conventional EB model-fitting program has already analyzed the given EB's light curve. In our implementation, we chose to employ DEBiL (Devor, 2004, 2005) since it is simple, fast, and fully automated. The fitted parameters are the orbital period (P), the apparent magnitudes ($mag_{1,2}$), and the fractional radii ($r_{1,2}$) of the binary components. A fractional radius is defined as the radius ($R_{1,2}$) divided by the sum of the components' semimajor axes (a). In MECI-express we iterate through a large group of MK spectral type pairings, to each of which we associate typical stellar parameters (Cox, 2000). These stellar parameters are the masses ($M_{1,2}$), the radii ($R_{1,2}$), and the absolute magnitudes ($Mag_{1,2}$) of the binary components. If the assumed values of the stellar parameters match the true values, then the stellar and fitted parameters should obey to the following equations:

$$\frac{4\pi^2 R_1^3}{G(M_1 + M_2)} = P^2 r_1^3 \qquad (1)$$

$$\frac{4\pi^2 R_2^3}{G(M_1 + M_2)} = P^2 r_2^3 \qquad (2)$$

$$Mag_1 - Mag_2 = mag_1 - mag_2 \qquad (3)$$

We also may have additional constraints from the observed out-of-eclipse combined colors of the system. For example, in the case of OGLE II targets, we have the estimated V-I color:

$$Mag_V - Mag_I = mag_V - mag_I \qquad (4)$$

We assume that the color has been corrected for reddening and that no systematic errors are present, so any inequalities would be due to an incorrect choice for the component pairing. The likelihood of each pairing is assessed by calculating the difference between the left-hand-side (stellar parameters) and right-hand-side (fitted parameters) of each equation. These differences are divided by their uncertainties, and added in quadrature. The pairing with the smallest sum is deemed the most likely pairing. For each given EB light curve, MECI-express returns the list of the top ranked (most likely) binary pairings, with their corresponding sums. MECI-express can also be used to create a contour plot of the probability distribution for all pairings. We illustrate an example of individual MECI-express components in Figs. 1a–c, which are then combined to create the result shown in Fig. 2a.

3. Method for eclipsing component identification (MECI)

MECI was developed to improve significantly upon the accuracy of MECI-express (see Table 1). This was done as follows: We replaced the use of spectral types with the more fundamental (and continuous) quantities of mass and age. Furthermore, in MECI we assume that the two binary components are coeval, thus replacing the 2-dimensional spectral type – spectral type grid, with a 3-dimensional

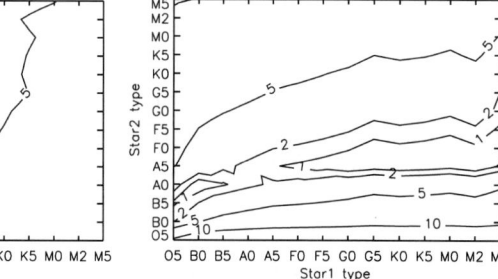
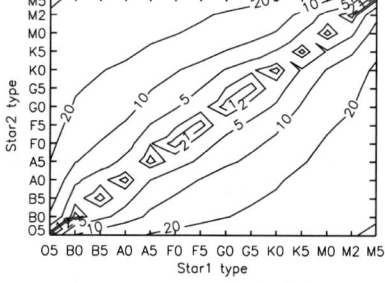

a. Constraints from (eq. 1)   b. Constraints from (eq. 2)   c. Constraints from (eq. 3)

Fig. 1 Contour plots of the absolute difference between the left-hand-side and the right-hand-side of each equation, divided by its uncertainty, as applied to the WW Camelopardalis light curve (Lacy et al., 2002). Adding these results in quadrature, produces the likelihood plot shown in Fig. 2.a

Table 1 A comparison of the results produced by MECI-express, MECI, and conventional analyses with their uncertainties (Lacy et al., 2000, 2002, 2003). The square brackets with numerical values indicate the deviation of our results from those of the conventional approach

Parameter	MECI-express		MECI			Lacy et al. (2000, 2002, 2003)		
	Mass 1 [$M_\odot$]	Mass 2 [$M_\odot$]	Mass 1 [$M_\odot$]	Mass 2 [$M_\odot$]	Age [Gyr]	Mass 1 [$M_\odot$]	Mass 2 [$M_\odot$]	Age [Gyr]
FS Mon	2.9 (A0) [77.7%]	2.0 (A5) [36.8%]	1.62 [0.6%]	1.52 [4.1%]	1.4 [0.2]	1.632 ±0.012	1.462 ±0.010	1.6 ±0.3
WW Cam	2.0 (A5) [4.2%]	2.0 (A5) [6.8%]	1.97 [2.8%]	1.89 [1.0%]	0.5 [0.0]	1.920 ±0.013	1.873 ±0.018	0.5 ±0.1
BP Vul	2.0 (A5) [15.1%]	1.6 (F0) [13.6%]	1.77 [2.1%]	1.48 [5.4%]	0.7 [0.3]	1.737 ±0.015	1.408 ±0.009	1.0 ±0.2

Fig. 2 A comparison of the MECI-express (left) and MECI (right) likelihood contour plots for WW Camelopardalis. The value of the contours are described in the body of the text. The asterisk marks the solution of Lacy et al. (2002)

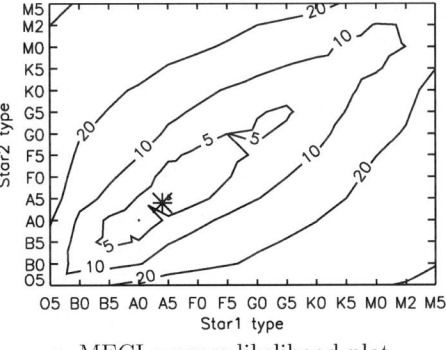
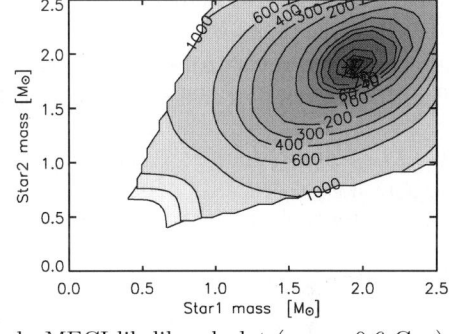

a. MECI-express likelihood plot

b. MECI likelihood plot (age = 0.6 Gyr)

mass-mass-age grid. Finally, we no longer rely on parameter fits of the components' apparent magnitudes and fractional radii directly from the light curve, which are often very uncertain, nor do we assume constant limb-darkening coefficients. Instead, we interpolate these values for the given mass-mass-age pairing, from precalculated tables [Yonsei-Yale isochrones (Kim et al., 2002); ATLAS limb-darkening coefficients (Kurucz, 1992), used when $T_{\text{eff}} \geq 10000$ K or $\log g \leq 3.5$; PHOENIX limb-darkening coefficients (Claret, 1998, 2000), used when $T_{\text{eff}} < 10000$ K and $\log g > 3.5$]. Thus, by assuming the masses ($M_{1,2}$) of the EB components and the system's age, we can look-up the radii ($R_{1,2}$), the absolute magnitudes ($\text{Mag}_{1,2}$), and the limb-darkening coefficients for the binary components. We then use these values, as well as the observationally-determined period (P) and combined magnitude out of eclipse (mag_{comb}), to calculate the apparent magnitudes ($\text{mag}_{1,2}$) and factional radii ($r_{1,2}$) of the EB components, as follows:

$$\text{mag}_1 = \text{mag}_{\text{comb}} + 2.5 \log[1 + 10^{-0.4(\text{Mag}_2 - \text{Mag}_1)}] \quad (5)$$

$$\text{mag}_2 = \text{mag}_1 + (\text{Mag}_2 - \text{Mag}_1) \quad (6)$$

$$a = [G(M_1 + M_2)(P/2\pi)^2]^{1/3}$$
$$\simeq 4.206 R_\odot (M_1/M_\odot + M_2/M_\odot)^{1/3} P_{\text{day}}^{2/3} \quad (7)$$

$$r_{1,2} = R_{1,2}/a \quad (8)$$

Besides the epochs of eclipses, which can be determined directly from the EB light curve, there are only two additional parameters required for us to simulate the light curves of the given pairing: the orbital eccentricity (e) and inclination (i). For binaries with short periods ($\lesssim 2$ days) and a secondary eclipse precisely half an orbit after the primary eclipse, it is reasonable to assume a circular orbit ($e = 0$). Otherwise, one should use the eccentricity derived by an EB model-fitting program (we use DEBiL). Finding the inclination robustly is more difficult. We employ a bracket search (Press et al., 1992), which returns the inclination that produces the best resulting fit.

To summarize, for every combination of component masses and system age of an EB, we can look-up, calculate, or fit all the parameters needed to simulate its light curve (P, limb-darkening coefficients, $\text{mag}_{1,2}$, $r_{1,2}$, epochs of eclipses, e, i), as well as its apparent combined color. We systematically iterate through many such combinations. For each one we compare the expected light curve with the observations, and calculate the reduced chi-squared value (χ_ν^2). We also compare each observed color ($O_c \pm \epsilon_c$) with its calculated value (C_c), and combine them by defining: score $\equiv (w\chi_\nu^2 + \sum_{c=1}^{N}[(O_c - C_c)/\epsilon_c]^2)/(w + N)$. Where w is the χ_ν^2 information weighting. We use $w = 1$, and assume that the smaller the score, the more likely it is that we have chosen the correct binary pairing. One can visualized this result using a series of score (M_1, M_2) contour plots, each with a constant age (e.g. Fig. 2b).

4. Conclusions

We have described a novel method for identifying an EB's components using only its photometric light curve and combined color. By utilizing theoretical isochrones and limb-darkening coefficients, this method greatly reduces the EB parameter-space over which one needs to search. This approach seeks to estimate the masses, radii and absolute magnitudes of the components, without spectroscopic data. We described two implementations of this method, MECI-express and MECI, which enable systematic analyses of datasets consisting of photometric time series of large numbers of stars, such as those produced by OGLE, MACHO, TrES, HAT, and many others. Such techniques are expected to grow in importance with the next generation surveys, such as Pan-STARRS (Kaiser et al., 2002) and LSST (Tyson, 2002).

Acknowledgements We are grateful to Guillermo Torres for many helpful conversations.

References

Alard, C., Lupton, R.H.: ApJ **503**, 325 (1998)
Alard, C.: A&AS **144**, 363 (2000)
Alcock, C., et al.: ApJ **492**, 190 (1998)
Alonso, R., et al.: ApJ **613**, L153 (2004)
Bakos, G., Noyes, R.W., Kovács, G., Stanek, K.Z., Sasselov, D.D., Domsa, I.: PASP **116**, 266 (2004)
Claret, A.: A&A **335**, 647 (1998)
Claret, A.: A&A **363**, 1081 (2000)
Cox, A.N.: Allen's astrophysical quantities, (4th ed.; New York: AIP Press; Springer) (2000)
Crotts, A.P.S.: ApJ **399**, L43 (1992)
Devor, J.: American astronomical society meeting abstracts, 205 (2004)
Devor, J.: ApJ **628**, 411 (2005)
Kaiser, N., et al.: SPIE **4836**, 154 (2002)
Kim, Y., Demarque, P., Yi, S.K., Alexander, D.R.: ApJS **143**, 499 (2002)
Kurucz, R.L.: IAU Symp. 149: The stellar populations of galaxies **149**, 225 (1992)
Lacy, C.H.S., Torres, G., Claret, A., Stefanik, R.P., Latham, D.W., Sabby, J.A.: AJ **119**, 1389 (2000)
Lacy, C.H.S., Torres, G., Claret, A., Sabby, J.A.: AJ **123**, 1013 (2002)
Lacy, C.H.S., Torres, G., Claret, A., Sabby, J.A.: AJ **126**, 1905 (2003)
Press, W.H., Teukolsky, S.A., Vetterling, W.T., Flannery, B.P.: (2nd ed.; Cambridge: University Press) (1992)
Tyson, J.A.: SPIE **4836**, 10 (2002)
Udalski, A., et al.: Acta Astronomica **44**, 165 (1994)
Wyithe, J.S.B., Wilson, R.E.: ApJ **559**, 260 (2001)
Wyithe, J.S.B., Wilson, R.E.: ApJ **571**, 293 (2002)

Hα Photometry of Two Contact Binaries

Szilárd Csizmadia · Zsolt Kővári · Péter Klagyivik

Received: 1 November 2005 / Accepted: 1 February 2006
© Springer Science + Business Media B.V. 2006

Abstract We carried out optical and Hα photometry of two contact binaries (V861 Herculis, EQ Tauri). The light curve modeling revealed stellar spots in both contact systems and strong Hα excess in the position of the observed stellar spots. A correlation was found between the $V - R$ and $R - H_\alpha$ colour indices of V861 Her.

Keywords Stars: activity of · Binaries: contact · Stars: individual: V861 Her, EQ Tau · Stars: late-type · Starspots

1 Introduction

Hα emission is a good indicator of high temperature regions located in stellar atmospheres. These regions belong to the plages which are high density and high temperature regions in the stellar atmospheres and they are connected to stellars spots via the magnetic field lines. Therefore the Hα emission is caused by stellar activity in the case of contact binaries. Our aim was to detect the time variation of Hα in some selected binaries and to investigate the relation of Hα emission to the flux in $V(RI)_C$ bands.

2 Observations and data reduction

The $BV(RI)_C$ observations were carried out by the 1 m RCC telescope of the Konkoly Observatory located at the

S. Csizmadia (✉) · Z. Kővári
Konkoly Observatory, H-1525 Budapest, P. O. Box 67, Hungary
e-mail: csizmadia@konkoly.hu

P. Klagyivik
Eötvös Loránd University, Budapest

Piszkéstető Mountain Station. As a detector an electronically cooled CCD camera was used. The Hα measurements were made photometrically throughout our Hα filter which has an effective wavelength of 656.8 nm and a full-width at half maximum of 7.7 nm. For EQ Tau we used the 1m RCC telescope while for V861 Her the 60/90/180 cm Schmidt telescope of our Observatory was applied.

All frames were bias-subtracted and flat-fielded. Dome flats were used. Instrumental magnitudes were extracted by applying the IRAF/DAOPHOT package using aperture photometry. Differential magnitudes were formed and they were transformed into standard systems. Details of transformation can be found in Csizmadia et al. (2004).

3 Results

V861 Herculis. Its light variation was discovered by Antipin (1996). V861 Her is a $V_{max} = 13.1$ magnitude contact binary star with solar-like colour index ($B - V = 0.63$). The system was studied in Csizmadia et al. (2004) where we published the results of $V(RI)_C$ photometry carried out in 2000 and 2003. The data collected on nine nights showed that the light curve of the system changed between 2000 and 2003. Moreover, we could demonstrate that its light curve varies on a time scale of a day (see Fig. 1).

The Hα measurements were carried out at only a single phase of $\varphi = 0.33$. Ten stars in the field of NGC 7790 were also observed in $VRH\alpha$. These ten stars and V861 Her had similar colour indices. The $(V - R) - (R - H_\alpha)$ two-colour diagram was constructed and this diagram revealed that V861 Her was brighter in Hα by ∼50% than the ten stars in the field of NGC 7790. (H_α stands for magnitudes measured through the Hα filter.)

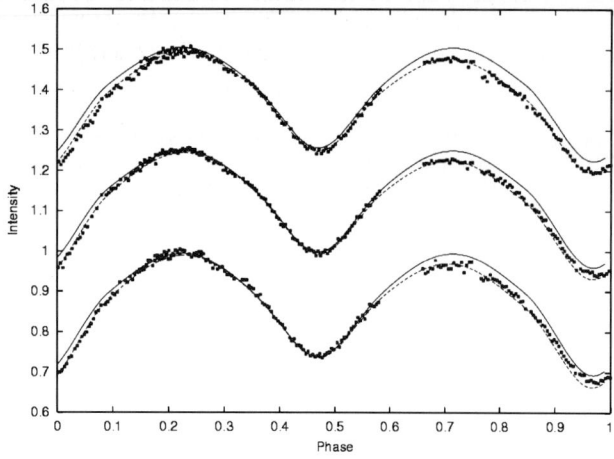
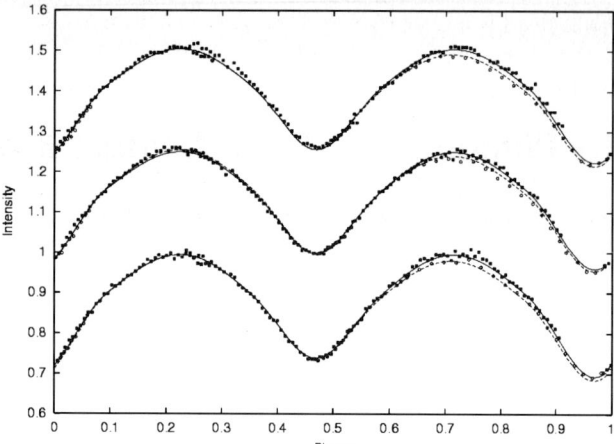

Fig. 1 $V(RI)_C$ light curves of V861 Herculis in 2000 (left) and in (2003). For sake of clarity the curves were shifted by adding a constant value to the curves. The solid line are the unspotted solution while the dashed ones represent the spotted solutions. For more explanation see Csizmadia et al. (2004).

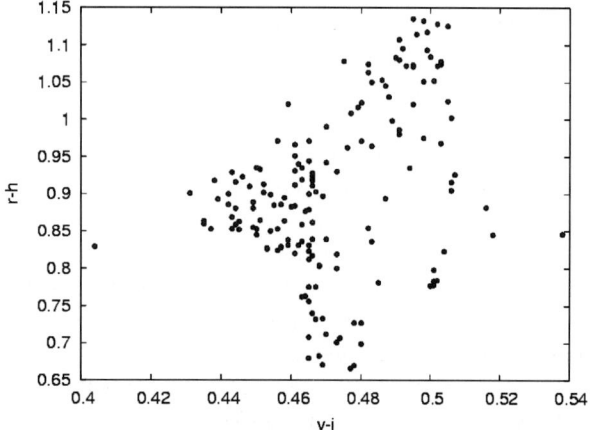

Fig. 2 $(V-I) - (R-H_\alpha)$ two-colour diagram of EQ Tau during a cycle. The data points are in the instrumetal magnitude systems and are measured to the comparison stars.

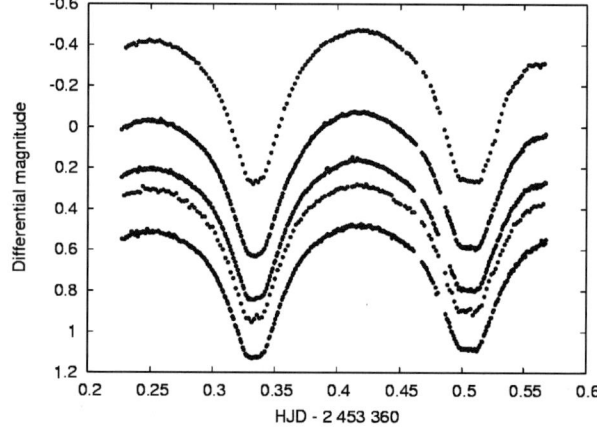

Fig. 3 From top to bottom the $BVR_C H_\alpha I_C$ light curves of EQ Tau can be seen in this diagram during a cycle.

The $H\alpha$ observations were made in August 2004 while $V(RI)_C$ observations were done in February 2004. The modeling of the $V(RI)_C$ observations revealed one spot on the secondary's surface at a longitude of $239° \pm 6°$. (Latitude of the spot was found to be $48° \pm 11°$ and its diamater was $10° \pm 1°$ and its temperature factor was 0.67.) This spot is on the central meridian at phase of $\varphi = 0.327$. We measured the $H\alpha$ emission exactly in this phase. Since we could not measure an $H\alpha$ light curve covering the whole cycle and the observations were separated by six months in time we cannot state that the $H\alpha$ emission comes from only the plage related to this spot. However, the coincidence is exactly what we expect and inspires further work on this field.

EQ Tauri. EQ Tau is a bright (V=11 mag) G1V spectral type contact eclipsing binary in the field of Pleiades but it is not a member of the cluster. We observed EQ Tau on the night 20/21 December 2004 covering a whole cycle. The light curves obtained with $V(RI)_C H\alpha$ filters are shown in Fig. 3. A preliminary fit were made applying the elements obtained by Pribulla and Vanko (2002) by adding two spots. Spot parameters are presented in Table 1.

Two spots were required to have an acceptable fit. We could not fit simultaneously the VRIHα data therefore the Hα curve was fitted separately. Both spots were found to be brighter in Hα than in other bands.

We also plotted $R - H_\alpha$ against $V - I$ (Fig. 2). Note that Hα filter has a shorter effective wavelength (656.8 nm) than that of the R filter (670 nm) therefore the correct use of this

Table 1 Spot parameters for EQ Tau on the night 20/21 Dec, 2004

Colour	Longitude	Latitude	D	TF
V,R,I	270°	0°	15	0.72
	180	0	10	0.95
H	270	0	15	1.07
	180	0	10	0.98

colour index would be $H_\alpha - R$. However, the using of $R - H_\alpha$ is more convenient. Therefore smaller $R - H_\alpha$ means that the star is relatively faint in H_α while larger $R - H_\alpha$ means that the star is relatively bright in H_α.

As one can see in Fig. 3 the $R - H_\alpha$ is smaller when the system is bluer in $V - I$. (Note that in Fig. 2 the real colour indices were not plotted, we used the instrumental differential colour indices.) This means that $H\alpha$ excess can be observed when the star is redder. The interpretation of this phenomenon is as follows. There were some plages over the spots. The spots are dark and therefore they caused redder colour index when they were close to the central meridian. The plages are bright in H_α and hence we found the increase of H_α brightness which yielded larger $R - H_\alpha$ colour index. This is why these two colour indices ($V - I$ and $R - H_\alpha$) correlate.

Acknowledgements Authors from Konkoly Observatory are grateful to the Hungarian Science Research Program (OTKA) for support under grants T-034551, T-038013, T-043504, T-048961.

References

Antipin, S.V.: IBVS 4360 (1996)
Csizmadia, Sz., Patkós, L., Moór, A., Könyves, V.: A&A **417**, 745 (2004)
Pribulla, Th., Vanko, M.: CoSka **32**, 79 (2002)

A Library of Eclipsing Binary Light-Curves: Fast Initial Parameters Estimation, and on the Uniqueness of the Inverse Problem

T. Borkovits · Sz. Csizmadia · L. Patkós

Received: 1 November 2005 / Accepted: 1 February 2006
© Springer Science + Business Media B.V. 2006

Abstract In order to improve the light curve solutions we suggest to use a new approach. A library of the Fourier-coefficients (up to 20th term) of (several millions of) light curves was constructed each of the light curves has different input parameters. A light curve solution is made in a very fast way: its discrete Fourier-transform yields not more than some simple number which are compared to the Fourier-coefficients in the library. A chi-squared test can help us to estimate very good initial parameters for the differential correction analysis. The library can also be used to study the uniqeness problem of the light curve solutions. Here we investigated the uniqueness of the spot solution of SV Cam.

Keywords Eclipsing binaries · Data analysis · Light curve analysis · Stars : individual : SV Cam

1. Introduction

The aims of this study were to investigate how accurately the light curve solutions can be determined from photometry and to apply our method to SV Cam first. The basic idea is based on the proposal of one of us (Sz. Cs.). Let us assume that we calculated and saved the Fourier-coefficients of many theoretically computed light curves. If we had the Fourier-parameters, we could compare the Fourier-coefficients of the observed light curves to the ones of the theoretically computed light curves. Since the number of the required Fourier-components is less than the number of the light curve points we could save some hard disk memory.

The light curve solution is defined by the minimum value of χ^2. From the tables one can read out the corresponding input parameters. Errors can be estimated by the χ^2 statistic. All solution is acceptable which satisfies the following relation: $\chi^2 \leq \chi^2_{\rm cr}(DF, \alpha)$, where DF is degree of freedom and α is the significance level. It is reasonable to choose $\alpha = 0.95$. There are appropriate tables from which the value $\chi^2_{\rm cr}$ can be read out. Applying this inequality one can reliably estimate the errors.

2. Initial parameter estimation, and the library

Differential correction method is a commonly used procedure to find the right parameters of an eclipsing binary stars. In close binary stars the χ^2 dependence on the parameters can be very complex and therefore the differential correction method could be failed because it generally finds the closest local minimum in the N-dimensions parameter space instead of the global minimum. The so-called 'grid method' (Maceroni et al., 1984) avoids this problem at many times. In the grid method the mass ratio is fixed at several different values and the other parameters are free ones. This method finds the correct value of the mass ratio but not in every cases. Sometimes the method has no convergence and requires a lot of CPU time (in general few ten thousands of iterations are required to reach a light curve representation). In other cases the light curves can be represented by more than one parameter set – the problem is degenerated. Such an example can be found in Niarchos et al. (1994) who found equally good spot solutions of XZ Leonis but the spot configurations were very different.

T. Borkovits (✉)
Baja Astronomical Observatory of the Bács-Kiskun County,
H-6500 Baja, Szegedi út, KT.766., Hungary

Sz. Csizmadia · L. Patkós
Konkoly Observatory of HAS, H-1525 Budapest, PF.67, Hungary

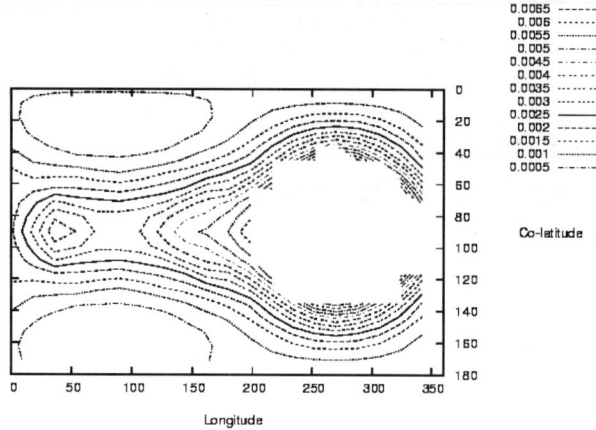
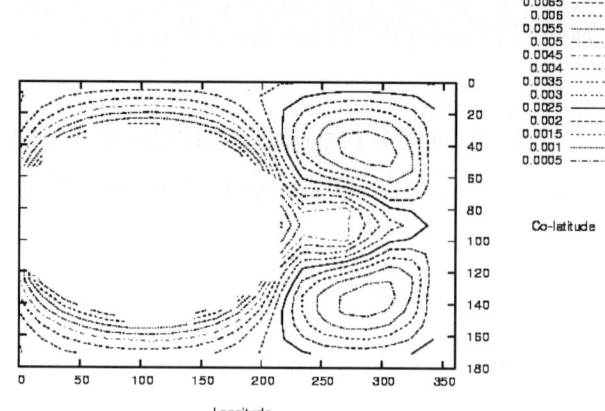

Fig. 1 Input spot parameters were as follows: *left panel:* co-latitude 9°, longitude 90°, spot's diameter 45°, temperature factor 0.85; *right panel:* co-latitude 162°, longitude 288°, spot's diameter 51°, temperature factor 0.875. The lines represent the χ^2 iso-contours at different χ^2 values. For details see the text.

Our tables were constructed carefully. First we studied how the albedo, gravity darkening and limb darkening affect the light curves. Then we decided that it is enough to calculate the tables for a few value of them (e.g. it is appropriate to compute the light curves with $x = 0.0, 0.6, 0.75$ and 0.9 limb darkening coefficients). Other free parameters were inclination, mass ratio, fill-out factor and luminosity ratio within reasonable boundaries. We used the 1998 version of the Wilson-Devinney Code (LC program) (Wilson, 1998) to calculate the light curves. In total, four million light curves of eclipsing contact binaries were calculated and their Fourier-coefficients were saved. The size of the electronic table is 3.4 GB – it is less than the capacity of a DVD.

Our first attempts have been showed that estimation of the initial parameters are very precise and fast. (Less than ten minutes we could determine the seven numbers of the initial parameter set of the test light curves and the accuracy of the resulted values was better than 5% in general.) Then the differential correction method can be applied reliably because we are searching for the final parameter values in the very narrow environment of the paramater space. Also, the uniqueness of the problem can be studied.

Now we have two tables. One of them can be used for contact binaries which will be described elsewhere. The other table was computed for SV Cam specially. This latter table has the SV Cam's physical and geometrical elements (taken from Patkós and Csizmadia, 2004) and we added a spot to it which is located on the primary. 160 000 different spot configurations (different combinations of spot's longitude, latitude, diamater and temperature) were calculated. The table consists of 160 000 lines, each line consists of the value of the input parameters and the first twenty corresponding Fourier-coefficients. A small C program generated the table so the table can be computed for other stars with different geometrical elements. Another short C program determines the χ^2 values. Then we examined how the goodness of the fit varies with the parameters.

3. Spot modeling of SV Cam

Such a database can be used to test numerically the uniqueness of the inverse problem. Here we apply the idea to test the spot determination accuracy and uniqueness of SV Camelopardalis.

No spot case. Our first attempt was as follows: we do not apply any spot and we tested what results we got. We found that in this case a small polar spot (diameter less than 20° and distance from the poles less than 10°) with large temperature factor ($TF \geq 0.9$) could also reproduce the curves. This means that the unspotted state of SV Cam cannot be determined by this way. This result also means that false polar spot can be detected from photometry. The reality of existence of photometrically determined polar spots should be very carefully checked in SV Camelopardalis! We note that this result is a natural consequence of the nearly 90° inclination of the system. In such a system the polar caps are visible under a very low angle and hence the contribution of their radiation to the integrated flux is almost negligible. Therefore the method is insensitive for the presence or absence of polar spots. As a next step we started the parameter search with spot at different locations.

Polar spots. It should be noted that polar spots are frequently observed in RS CVn systems. Jeffers et al. (2005) found polar spots in SV Cam via HST spectroscopy, too. Therefore we examined the case of polar spots. As one can see in Fig. 1a the less χ^2 values can be found at 90° longitude on both hemisphere very close to the 10° and 170° co-latitude. The fine structure of the diagram shows that the solution of the inverse problem is double-degenerated in latitude and the position of the spot can be determined by about 5° accuracy in case of a photometric accuracy of 0.005 magnitude (one sigma).

Spot at high, and medium latitudes. In Fig. 1b we illustrate the results for a northern high-latitude spot. We also investigated the case of a spot at the southern hemisphere (latitude −78°). The solution remained double-degenerated symetrically to the stellar equator. The fine structure of the diagram showed that these positions could be determined by the above mentioned accuracy in this case, too. Similar results were obtained for the case of medium latitude spots in the sense of the double-degenerated nature.

4. Conclusions

As a conclusion we can state that the absolute value of the latitude of stellar spots can be determined in case of SVCam, in spite of the fact that the inclination is close to 90°. This statement is valid for this newly proposed Fourier-decomposition method.

Although the two hemisphere is equivalent for a distant observer as we are, the spot's latitude – without its sign – can be reliable determined from photometry. Therefore we also concluded that the change in the latitude distribution of spots on the pirmary component's surface could be real effect. As we pointed out in Patkós and Csizmadia (2004) the spot's latitude was higher after 1990 than between 1972–1980 in SV Cam.

Acknowledgements This work has been supported by the OTKA Number T 034 551.

References

Jeffers, S.V., Cameron, A.C., Barnes, J.R., Aufdenberg, J.P., Hussain, G.A.J.: ApJ **621**, 425 (2005)
Maceroni, C., Milano, L., Russo, G.: A&AS **58**, 405 (1984)
Niarchos, P.G., Hoffmann, M., Duerbeck, H.W.: A&A **292**, 494 (1994)
Patkós, L., Csizmadia, Sz.: AN **325**, 424 (2004)
Wilson, R.E.: Computing the Binary Star Observables, University of Florida (1998)

On Methods for the Light Curves Extrema Determination

Zdeněk Mikulášek · Marek Wolf · Miloslav Zejda · Petra Pecharová

Received: 1 November 2005 / Accepted: 1 February 2006
© Springer Science + Business Media B.V. 2006

Abstract We present two methods for the determination of moments of extrema and their errors appropriate for the analysis of light variations of variable objects. The method I is suitable for determination of times of extrema of non-periodical variables or objects, whose light curves vary. The method II is apt for O-C analyses of objects whose light curves are more or less repeating. Both methods are displayed on the analysis of BL Cam light variations and compared with the Kwee-van Woerden method.

Keywords Methods: Data analysis · Kwee-van Woerden method · Least square method · SX Phe variables · Star: Individual: BL Cam

1. Introduction

The analysis of the differences $O-C$ between the observed timing O of an extremum of light curve and the expected (calculated) time of that extremum C, is one of the most efficient tool for deep insight in the physics of variable stars. It helps to test and improve ephemeris of variable stars, enables to reveal another body in the systems studied, it can serve as the main instrument for determination of parameters of many interesting physical processes influencing light variation.

2. Method I

Suppose that observations of a variable object obtained in the particular colour cover a time interval containing at least one extremum of its light curve. Our aim is to determine the time O when the light extremum (extrema) came into being and to estimate the uncertainty $\delta(O)$ of that determination. Let the observations consist of a set of N discrete magnitudes obtained in the same colour m_i, with individual weights (optionally) obtained at times t_i. Let us assume the continuous smooth function $F(t)$ is a good fit of observed light curve defined by observed points which can be represented as a linear combination of the set of g smooth functions, e.g. polynomials. $F(t) = \mathbf{f}(t)\,\mathbf{b}$, where the set of g coefficients $\{b_j\}$ forms a column vector $\mathbf{b}$ and the set of functions $\{f_j(t)\}$ a row vector $\mathbf{f}(t)$. The set of coefficients represented by $\mathbf{b}$ can be obtained either by the standard least square method linear regression or better by an appropriate variant of the robust regression.

Let $\mathbf{Y}$ be a column vector containing all measured magnitudes, $\mathbf{X}$ is a matrix of dimension $N \times g$, whose i,j-th element $X_{ij} = f_j(t_i)$ and $\mathbf{W}$ is the diagonal matrix, whose element $W_{ii} = w_i$, where w_i is the weight of i-th observation and $\bar{w}$ is the mean value of individual weights. The $\mathbf{b}$ is a column vector containing g coefficients of the regression:

$$\mathbf{V} = \mathbf{X}^T \mathbf{W} \mathbf{X}, \quad \mathbf{U} = \mathbf{X}^T \mathbf{W} \mathbf{Y}, \quad \mathbf{H} = \mathbf{V}^{-1}, \quad \mathbf{b} = (\mathbf{U}^T \mathbf{H})^T.$$

The time of an extremum t_{ex} of the fitted function $F(t)$ can be determined by solving the equation: $F'(t_{ex}) = 0$ by the

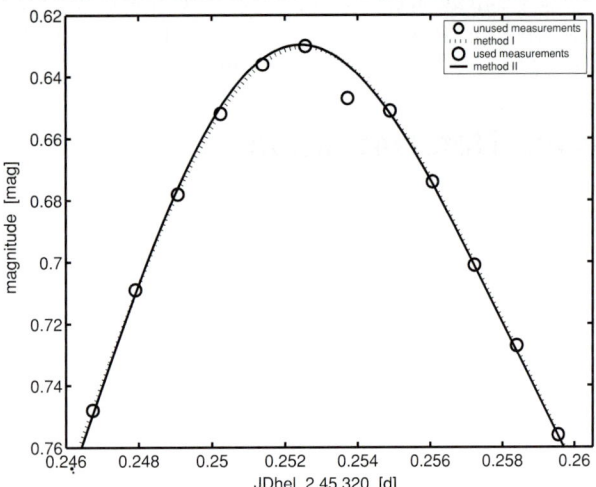

Fig. 1 The maximum of BL Cam fitted by both of outlined methods.

standard methods. The uncertainty of the time $\delta(t_{ex})$, can be estimated by the relation:

$$\delta(t_{ex}) = \frac{\delta[F'(t_{ex})]}{|F''(t_{ex})|} = \frac{\sigma\sqrt{\bar{w}\,\mathbf{f}'(t_{ex})\,\mathbf{H}\,\mathbf{f}'(t_{ex})^{\mathrm{T}}}}{|\mathbf{f}''(t_{ex})\,\mathbf{b}|},$$

where σ is the weighted standard deviation.

We recommend to transform times of observations into $\langle 0, 1 \rangle$ and to use the following set of largely trigonometric functions: $\{1;\ x;\ \sin(\pi x);\ \sin(2\pi x);\ \sin(3\pi x);\ \ldots\}$, orthonormalized by the Gram-Schmidt procedure. It allows to examine even such extrema which are close to borders of the interval covered by observations (Mikulášek and Gráf, 2005).

It is very important to choose the optimum number of free parameters g for representation of the observed light curves. Sometimes it is useful to divide the studied light curve into parts – it ordinarily leads to the radical diminution of the number of free parameters needed.

3. Method II

The majority of variable stars displays more or less periodic variations which can be described by a function of time or phase: $F(t-C) = F(\Delta t)$, where C is the expected moment of a curve extremum.

The real light curve used to be shifted both in time (O–C shift) and in the vertical direction – in magnitude q. The last shift may be due to instrumental effects, multiperiodicity, secular variations of brightness etc. The seeking time shift O–C of the time of the extreme O and the vertical shift q can be calculated by minimizing the weighted sum of squares of differences Δy_i, where using Taylor expansion we can write:

$$\Delta y_i(\Delta t_i) = y_i - F(\Delta t_i) \cong q - (O - C)\,F'(\Delta t_i).$$

The expression can be considered as the equation for a straight line intersecting the y axis at q with the slope of $-(O-C)$. The time of the observed extreme O and its uncertainty $\delta(O)$ can be written:

$$u = \overline{F'^2} - \overline{F'}^2, \quad O-C = -\frac{\overline{\Delta y F'} - \overline{\Delta y}\,\overline{\Delta F'}}{u},$$

$$q = \frac{\overline{F'^2}\,\overline{\Delta y} - \overline{F'}\,\overline{\Delta y F'}}{u},$$

$$\delta(O-C) = \sqrt{\frac{\overline{(\Delta y)^2} - q\,\overline{\Delta y} + (O-C)\,\overline{\Delta y F'}}{(N-2)\,u}},$$

$$\delta q = \delta(O-C)\sqrt{\overline{F'^2}},$$

where N is number of measurements. Quantities denoted by overlines are mean weighted values of the corresponding quantities.

As outliers are quite common in photometric data sets we strongly recommend to use a robust regression instead of the bare LSM. We used the robust regression outlined in Mikulášek et al. (2003).

4. BL Camelopardalis: Comparison of methods

BL Cam = GD 428 ($\alpha_{2000} = 03^{\mathrm{h}}47^{\mathrm{m}}19.6^{\mathrm{s}}$, $\delta_{2000} = +63°22'43''$) is a SX Phe type variable star with the primary period of 0.0391 day (Hintz et al., 1997), the secondary pulsational period is disputable. Hintz reported the constantly increasing primary pulsation period, Wolf et al. (2002) has found the damping of this tendency.

We have chosen BL Cam as an appropriate variable, where the application of both outlined general methods can be demonstrated. The O–C analysis results from own photometric material of 1059 V and R observations of 29 maxima of the star observed in 2001–5. All measurements were obtained by the CCD camera Apogee AP7 attached to the 0.65-m reflector of the Astronomical Institute of CU in Ondřejov.

We determined the mean linear ephemeris of the primary pulsational variability satisfactorily well describing the observations:

$$JD_{\mathrm{helmax}} = 2\,452\,691.87866(3) + (E - 11\,829)$$

$$0.0390979777(37)\,\mathrm{d}.$$

We found 29 well covered maxima of BL Cam and determined their times and errors by several methods including the classical Kwee – van Woerden method (1956). The resulting O–C values display a large scatter with dispersion of $0.019\,P$ caused by either quasiperiodicity of the pulsations of the star or by secondary pulsations, several times larger than typical formal uncertainties of their determination: $0.0029\,P$ – Kwee - van Woerden; $0.0026\,P$ – Method I; $0.0018\,P$ – Method II.

While the mean difference between O–C's determined by the Method I and the Method II is exactly zero, the results of the Kwee-van Woerden method are systematically shifted by $(+0.0035 \pm 0.0005)\,P$, which is due to the asymmetry of the light curve of BL Lac.

While the Method I and Kwee-van Woerden method (being nearly identical with the Method I, $g = 3$) are sensitive to changes of the form of light curve in the very close vicinity of the maximum, the Method II does not reflect them, which results in a diminishing of the observed scatter of O–C by $(11 \pm 4)\%$.

The detailed description of both methods and their comparison with other commonly used methods, as well as results for the BL Cam O–C, analysis will be published elsewhere.

Acknowledgements This work was supported by grants GA ČR No. 205/04/2063, 205/06/0217 and by the Czech-Greek project of collaboration RC-3-18 of Ministry of Education, Youth and Sport of Czech Republic.

References

Hintz, E.G., Joner, M.D., McNamara, D.H., Nelson, K.A., Moody, J.W.: PASP **109**, 15 (1997)
Kwee, K.K., van Woerden, H.: BAN **12**, 327 (1956)
Mikulášek, Z., Gráf, T.: Contr. Astron. Obs. Sk. Pleso **35**, 83 (2005)
Mikulášek, Z., Žižňovský, J., Zverko, J., Polosukhina, N.S.: Contr. Astron. Obs. Sk. Pleso **33**, 29 (2003)
Wolf, M., Crlíková, M., Bašta, M., Šarounová, L., Štěpán, J., Švéda, L., Vymětalík, O.: IBVS **5317**, 1 (2002)

LU Vel (GJ 375): A M3.5Ve Flare and Double-Lined Spectroscopic Binary

D. Montes · M. C. Gálvez · M. J. Fernández-Figueroa ·
I. Crespo-Chacón

Received: 1 November 2005 / Accepted: 1 February 2006
© Springer Science + Business Media B.V. 2006

Abstract High resolution echelle spectroscopic observations taken with the FEROS spectrograph at the 2.2 m telescope ESO confirm the binary nature of the flare M3.5V star LU Vel (GJ 375, RE J0958-462) previously reported by Christian and Mathioudakis (2002). Emission of similar intensity from both components is detected in the Balmer, Na I D_1&D_2, He I D_3, Ca II H&K, and Ca II IRT lines. We have determined precise radial velocities by cross correlation with radial velocity standard stars, which have allowed us to obtain for the first time the orbital solution of the system. The binary consists of two near-equal M3.5V components with an orbital period shorter than 2 days. We have analyzed the behaviour of the chromospheric activity indicators (variability and possible flares). In addition, we have determined its rotational velocity and kinematics.

Keywords Stars: activity · Chromospheres · Binaries: spectroscopic · Late-type · Flare · Individual (LU Vel)

1. Introduction

LU Vel (GJ 375, BD-45 5627), is a southern nearby ($d = 16$ pc) M3.5 dwarf ($V = 11.3$) with Balmer and Ca II H&K lines in emission and known to have a high level of flare activity (Doyle et al., 1990). It is an EUV source (RE J0958-462) detected by the ROSAT Wide Field Camera all-sky survey (Pounds et al., 1993). Recently, two high resolution spectra reported by Christian and Mathioudakis (2002)

D. Montes (✉) · M. C. Gálvez · M. J. Fernández-Figueroa ·
I. Crespo-Chacón
Dpto. Astrofísica, Facultad Físicas, Universidad Complutense
Madrid, Spain
e-mail: dmg@astrax.fis.ucm.es

identified it as a double-lined spectroscopic binary (SB2) with strong Balmer emission from both components (see Fig. 2), but the orbital solution has not been obtained until now. These authors also determined a rotational velocity ($v \sin i = 10 \, \text{km s}^{-1}$), but did not detect the lithium line (6708 Å) in their spectra.

2. Observations

The spectroscopic observations (high resolution echelle spectra) of this star were obtained during one observing run from 18 to 22 February 2005 using the 2.2 m telescope at the European Southern Observatory, ESO (La Silla, Chile). We have used the FEROS (Fiber-fed Extended Range Optical Spectrograph) linked to the Cassegrain focus of the 2.2 m telescope, with the CCD 2048 × 4096, 0.15 μ/pixel. This configuration gives a resolution of 48000 and a spectral range from 3500 to 9000 Å, from Ca II H&K (3933, 3968 Å) to Ca II IRT (8498, 8542, 8662 Å), in a total of 39 orders. A total of 12 spectra of LU Vel has been taken during the 5 nights of observations. Reference stars of similar spectral type and radial velocity standard stars have also been observed with the same configuration.

3. Orbital solution

Heliocentric radial velocities of both components of LU Vel have been determined by using the cross-correlation technique. The spectra of LU Vel were cross-correlated order by order, using the routine FXCOR in IRAF, against spectra of radial velocity standards of similar spectral types. The velocity is derived from the position of the cross-correlation peak. Uncertainties have been estimated from the width of

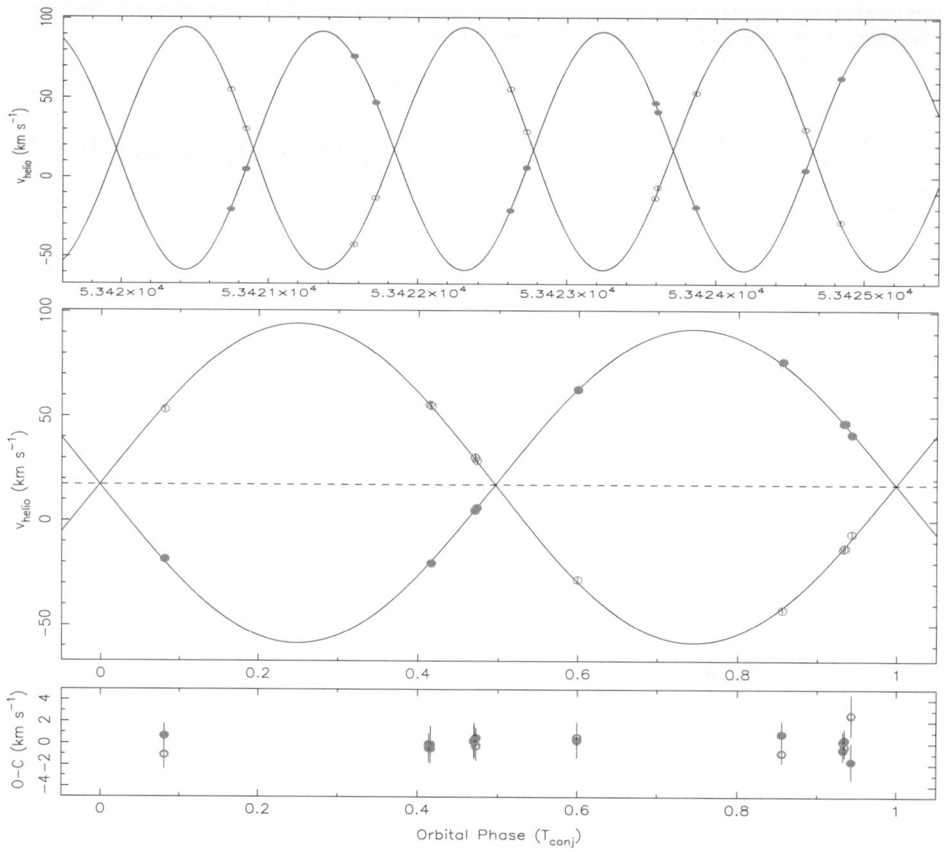

Fig. 1 Heliocentric radial velocity (determined in this work) vs. HJD (upper panel) and vs. orbital phase (lower panel). Our observations cover 3 orbital periods. Solid circles represent the primary (the component with larger chromospheric emission) and open circles the secondary. The solid line is the obtained orbital solution

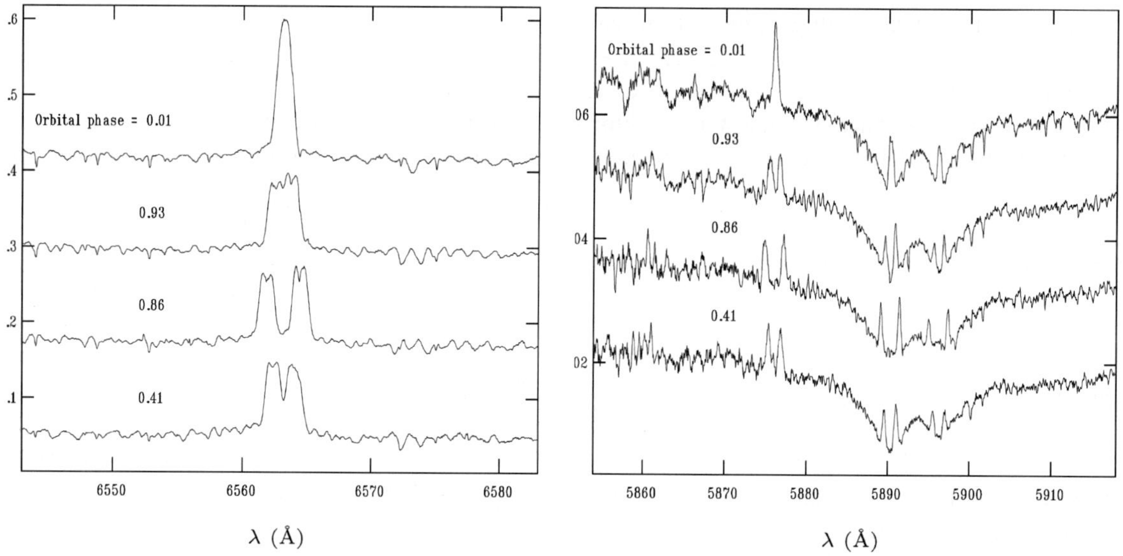

Fig. 2 Spectra of LU Vel at different orbital phases in the Hα (left) and Na I D$_1$&D$_2$; He I D$_3$ (right) line regions

the cross-correlation peak and the inter-order agreement in the derived velocities.

Using the 11 radial velocities that we have obtained, we have determined for the first time the orbital solution of LU Vel (see Fig. 1 and Table 1). We have obtained a near circular orbit ($e = 0.01$) with an orbital period of 1.8752 days. The resulting minimum masses ($M \sin^3 i$) ≈ 0.3 $M\odot$ and mass ratio are compatible with two near-equal M3.5V components.

4. Chromospheric activity

The echelle spectra analysed in this work allow us to study the behaviour of the different optical chromospheric activity

Table 1 Orbital solution of LU Vel

Element	Value	Uncertainty	Units
P	1.8752	0.0021	days
T_0 (peri)	2453418.0898	0.1110	HJD
ω	215.2045	21.5762	degrees
e	0.0113	0.0069	
K_1	75.2714	1.0829	km s^{-1}
K_2	76.4120	1.5824	km s^{-1}
γ	17.3916	0.2632	km s^{-1}
$q = M_1/M_2$	1.0152	0.0151	
$a_1 \sin i$	1.9408	0.0280	10^6 km
$a_2 \sin i$	1.9702	0.0409	10^6 km
$a \sin i$	3.9110	0.0495	10^6 km
$M_1 \sin^3 i$	0.3415	0.0137	$M_\odot$
$M_2 \sin^3 i$	0.3364	0.0165	$M_\odot$
$f(M)$	0.0828431	0.0035909	$M_\odot$

indicators from the Ca II H&K to the Ca II IRT lines, formed at different atmospheric heights (Montes et al., 2000). Strong emission from both components is observed in the Ca II H&K lines, all the Balmer lines (see Hα in Fig. 2 left), the Ca II IRT lines, as well as the Na I D$_1$&D$_2$ and He I D$_3$ lines (Fig. 2 right). The emission lines of both components have similar intensity, but the primary component have a slightly large intensity in all the orbital phases except the last night, where an enhancement of the emission of the secondary is observed. The Hα line of both components exhibits a central self-absorption (see Fig. 2 left). Although flares have been detected in this star (Doyle et al., 1990), during our observations, covering 5 nights (with 2 or 3 spectra per night), we have not found evidences of strong flares (only small variations in the emission lines has been detected). Both components of this BY Dra binary system are very active stars since even in the quiescent (or pseudo-quiescent) state they show strong emission in all the chromospheric activity indicators. High temporal resolution spectroscopic observations are needed to analyse the possible flares of both components of this binary.

Acknowledgements These observations have been funded by the Optical Infrared Coordination network (OPTICON), supported by the Research Infrastructures Programme of the European Commissions Sixth Framework Programme. This work has been supported by the Universidad Complutense de Madrid and the Ministerio de Educación y Ciencia (Spain), under the grant AYA2004-03749.

References

Christian, D.J., Mathioudakis, M.: AJ **86**, 403 (2002)
Doyle, J.G., Mathioudakis, M., Panagi, P.M., Butler, C.J.: A&AS **86**, 403 (1990)
Montes, D., Fernández-Figueroa, M.J., De Castro, E. et al.: A&AS **146**, 103 (2000)
Pounds, K.A., et al.: MNRAS **260**, 77 (1993)

Dense Spot Coverage and Polar Caps on SV Cam

S. V. Jeffers · A. Collier Cameron · J. R. Barnes ·
J. P. Aufdenberg

Received: 1 November 2005 / Accepted: 1 February 2006
© Springer Science + Business Media B.V. 2006

Abstract We have used spectrophotometric data from nine Hubble Space Telescope orbits to eclipse-map the primary component of the RS CVn binary SV Cam. From these observations and its HIPPARCOS parallax we find that the surface flux in the eclipsed low-latitude region is about 30% lower than computed from the best fitting PHOENIX model atmosphere. This flux deficit can only be accounted for if about a third of the primary's surface is covered with unresolved spots. Even when we extend the spottedness from the eclipsed region to the entire surface, there still remains an unaccounted flux deficit. This remaining flux deficit is explained by the presence of a large polar spot extending down to latitude $42 \pm 6°$.

Keywords Binaries: Eclipsing · Stars: Individual (SV Cam) · Late-type · Activity · Spots

1. Introduction

Doppler images of rapidly rotating RS CVn binary systems frequently show long-lived polar spots. Theoretical

---

*Marie Curie Intra-European Fellow

S. V. Jeffers (✉)
Laboratoire d'Astrophysique de Toulouse-Tarbes (UMR 5525), Observatoire Midi-Pyrenees, 14 ave Edouard Belin, 31000 Toulouse, France

S. V. Jeffers* · A. Collier Cameron · J. R. Barnes
School of Physics and Astronomy, University of St Andrews, North Haugh, St Andrews, KY16 9SS, Scotland

J. P. Aufdenberg
National Optical Astronomy Observatory, 950 N. Cherry Ave, Tucson, AZ 85726, U.S.A.

efforts to understand this phenomenon assume that meridional flows sweep magnetic flux from decaying active regions towards the pole (Schrijver and Title, 2001). In order to produce a polar spot, this model requires a bipolar spot emergence rate 30 times larger than in the case of the Sun, implying that the stellar surface should be peppered with many small spots. To investigate this we used the Hubble Space Telescope to observe three primary eclipses of the RS CVn close binary SV Cam. The observations spanned five days with a 30 s exposure time, a wavelength coverage of 290 nm to 570 nm, and a signal-to-noise of approximately 5000.

2. Temperature fitting

Jeffers et al. (2006) determined accurate radii of both SV Cam binary components through lightcurve fitting, which allows us to determine the relative flux contribution from the secondary star. Outside of eclipse the total flux f_{phx} is expressed as;

$$f_{ptotal} = \frac{r_{pri}^2}{d_{HIPP}^2}\left(F_{pri} + \frac{r_{sec}^2}{r_{pri}^2}F_{sec}\right) \quad (1)$$

where F_{pri} and F_{sec} are the model atmosphere fluxes for the primary and secondary stars respectively; r_{pri} and r_{sec} are the primary and secondary radii, d_{HIPP} is the HIPPARCOS distance (84.96 ± 8.5 pc). The PHOENIX spectra (Allard et al., 2000) are scaled by r^2/d^2_{HIPP} to account for flux at the Earth. We fitted this equation for each combination of primary temperature in the range T_{pri} = 4500 K to 6500 K and secondary

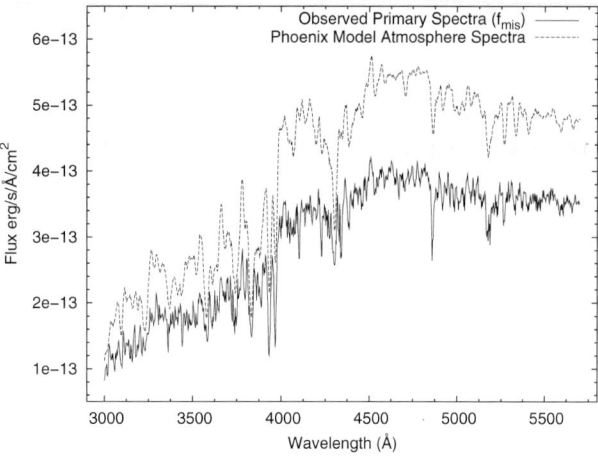

Fig. 1 Plot showing the spectrum of the primary star, f_{mis}, and the closest match PHOENIX model atmosphere (6000 K), without using a scaling factor. This plot illustrates the flux deficit due to the presence of dark star-spots on the surface of the primary

temperature $T_{\mathrm{sec}} = 2800$ K to 5500 K. The best fitting pair of temperatures was determined using χ^2 minimisation;

$$\chi^2 = \sum_{\lambda(i)=1}^{N} \left(\frac{f_{\mathrm{out}}(i) - \gamma_{\mathrm{total}} * f_{\mathrm{ptotal}}(i)}{\sigma(i)} \right)^2 \quad (2)$$

where f_{out} is an averaged spectrum outside of primary eclipse, and γ is a scaling factor to fit the shape of the spectrum rather than the absolute flux levels. It is defined as;

$$\gamma_{\mathrm{total}} = \frac{\sum f_{\mathrm{out}} f_{\mathrm{ptotal}}/\sigma^2(i)}{\sum (f_{\mathrm{ptotal}})^2/\sigma^2(i)} \quad (3)$$

3. Spot coverage

The existence of several large spots on SV Cam has been shown by Jeffers et al. (2006) and Jeffers (2005) using the eclipse mapping technique. However, these spots are insufficient to account for the flux deficit shown in Fig. 1. The total missing flux can only be accounted for if the eclipsed surface of the primary star is peppered with spots too small to be resolved with eclipse mapping techniques. The spot filling factor is given by $\alpha = 1 - \gamma$, where α is the fractional spot coverage and γ the scaling factor. This shows the fractional coverage of spots on SV Cam to be $28 \pm 8\%$.

4. SV Cam's polar spot

When the 28% flux deficit in the eclipsed flux is applied to the whole surface of the primary star, we find that in the eclipsed region there is an additional 12.5% flux deficit. However, this does not take into account that the star is a sphere rather than a

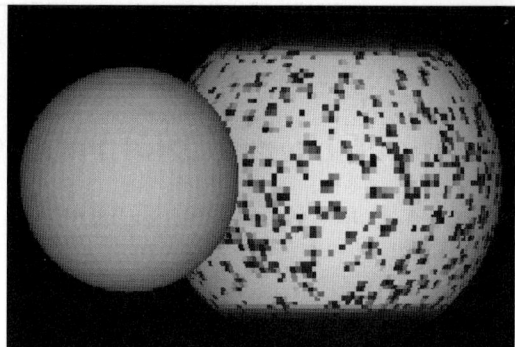

Fig. 2 A model of SV Cam showing 28% spot coverage and a 42° polar cap

disk, any limb and gravity darkening, or spherical oblateness. We use the binary eclipse-mapping tomography code DoTS (Collier Cameron, 1997), which includes these capabilities, to model the fractional decrease in stellar flux as a function of polar spot size and determine the polar spot radius to be $42 \pm 6°$. The size of the polar cap is larger than found for the G2V single star He699 (Jeffers et al., 2002), which could be a result of the presence of high latitude structure that is not included in our model.

5. Conclusions

This work provides strong evidence, independent of Doppler Imaging, that the poles of SV Cam's primary star are darkened by extensive polar caps. At lower latitudes the photosphere is peppered with small star-spots, as has been suggested by TiO band monitoring (O'Neal et al., 1998). It is important to establish that these stars are in fact peppered with small star-spots in addition to a polar cap, as a high rate of flux emergence is required for diffusive and advective models of polar-spot formation. A more detailed description of this work is given by Jeffers et al. (2005).

Acknowledgements SVJ currently acknowledges support at OMP from a personal Marie Curie Intra-European Fellowship within the 6th European Community Framework Programme.

References

Allard, F., Hauschildt, P.H., Schweitzer, A.: Spherically symmetric model atmospheres for low-mass pre-main-sequence stars with effective temperatures between 2000 and 6800 K. AJ **539**, 366–371 (2000)

Collier Cameron, A.: Eclipse mapping of binary stars. MNRAS **287**, 556–566 (1997)

Jeffers, S.V.: Spurious 'active longitudes' in parametric models of heavily spotted eclipsing binaries. MNRAS **359**, 729–733 (2005)

Jeffers, S.V., Collier Cameron, A., Barnes, J.R., Aufdenburg, J.P., Hussain, G.A.J.: Direct evidence for a polar spot on SV camelopardalis. AJ **621**, 425–431 (2005)

Jeffers, S.V., Barnes, J.R., Collier Cameron, A., Donati, J.F.: Hubble Space Telescope Observations of SV Cam: I. The Importance of Unresolved Starspot Distributions in Lightcurve Fitting. MNRAS **366**, 667–674 (2006)

Jeffers, S.V., Barnes, J.R., Collier Cameron, A.: The latitude distribution of star-spots on He 699. MNRAS **331**, 666–672 (2002)

O'Neal, D., Neff, J., Saar, S.: Measurements of starspot parameters on active stars using molecular bands in echelle spectra. AJ **507**, 919–937 (1998)

Schrijver, C.J., Title, A.M.: On the formation of polar spots in Sun-like stars. AJ **551**, 1099–1106 (2001)

Doppler Images of ζ Andromedae

Zsolt Kővári · Katalin Oláh · János Bartus ·
Klaus G. Strassmeier · Michael Weber ·
Albert Washuettl · John B. Rice · Szilárd Csizmadia

Received: 1 November 2005 / Accepted: 1 February 2006
© Springer Science + Business Media B.V. 2006

Abstract Doppler images are presented for the RS CVn-type binary ζ And. Our upgraded Doppler imaging code TempMap$_\epsilon$ takes into account the distorted geometry of the primary giant component. On the maps several low latitude spots are restored with a temperature contrast of about 1000 K. Some weak polar features are also found. Cross-correlation of the consecutive Doppler-maps suggests solar-like differential surface rotation.

Keywords Stars: Activity of · Stars: Imaging · Stars: Individual: ζ Andromedae · Stars: Late-type · Starspots

1. Stellar basics

ζ And (HD 4502) is an RS CVn-type single line spectroscopic binary with an ellipsoidal giant component. Spectral type is classified as K1 III (Strassmeier et al., 1993) with a possible but unseen F companion.

Photometric light variation is ruled by the ellipticity effect, however, long-term changes in the photometric amplitude with a period similar to the orbital one suggests spot activity (Strassmeier et al., 1989 and references therein).

Our Doppler imaging procedure (Section 2) is applied to iteratively refine some of the basic stellar parameters,

Z. Kővári (✉) · K. Oláh · S. Csizmadia
Konkoly Observatory, Budapest, Hungary

J. Bartus · K. G. Strassmeier · M. Weber · A. Washuettl
Astrophysical Institute, Potsdam, Germany

J. B. Rice
Brandon University, Canada

like inclination and the equatorial rotational velocity with a better accuracy (for the method see e.g., Weber and Strassmeier, 1998). Combining the refined parameters with the previously determined ones yielded a consistent picture for the system (see our other paper by Kővári et al. in this volume, hereafter paper I). Final adopted and derived parameters, as well as the most important astrophysical data used for Doppler imaging are summarized in Table 1.

2. Doppler imaging

For the image reconstruction we use TempMap$_\epsilon$, a new version of the code originally written by Rice et al. (1989), expanded for taking into account the distorted geometry. The elliptical distortion is approximated with a rotation ellipsoid elongated towards the secondary component. Our parameter study yielded to a distortion parameter of $\epsilon = 0.27$ ($\epsilon = 1 - (\frac{b}{a})^2$, where a and b are the long and the short axes of the ellipsoid, respectively), in good agreement with the results from photometry.

A total of 54 Ca I 6439-Å spectra covering ≈3.8 rotations were collected at NSO between Nov/96–Jan/97. From the data we formed four subsets covering full consecutive rotation cycles (for a sufficient phase coverage 'NSO-4' has an overlap of its first three spectra with 'NSO-3'). The KPNO dataset was obtained one year later during Dec/97-Jan/98. The 14 spectra span roughly one rotation, thus they are suitable only for reconstructing one single map. The typical S/N ratio of the spectra is about 250–300:1 in the continuum. Resulting Doppler maps are shown in Fig. 1.

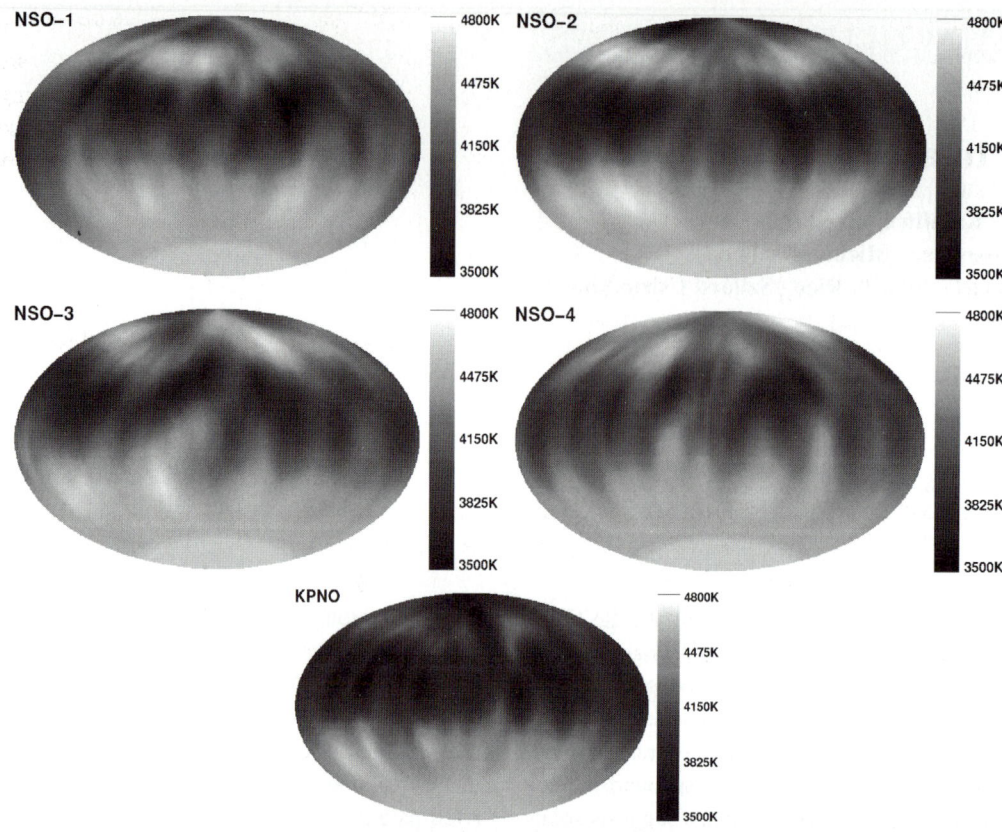

Fig. 1 Doppler images of ζ And. The four NSO maps show mainly cool, low latitude spots and some weak polar features with a maximum temperature contrast of ≈ 1000 K (in a good agreement with the photometric modelling results in paper I). Plots are presented in aitoff-projection with the zero meridian at the centre. In agreement with the NSO images, the KPNO map also shows mainly cool spots at lower latitudes

For phasing the observations we shifted the ephemeris in Fekel et al. (1999) by $\Delta \phi = 0.25$, thus, in our study, phase value $\phi = 0.0$ corresponds to the conjunction with the primary in front.

Table 1 System and stellar parameters for ζ And

Classification	K1 III + K V
Period*, ($P_{orb} = P_{rot}$) [days]	17.769426 ± 0.000040
Eccentricity, e	0.0
Inclination***, i [°]	65 ± 5
$v \sin i$*** [km s^{-1}]	41.4 ± 0.5
$a_1 \sin i$* [km]	6.14 × 10^6
$f(m)$*	0.0292
q**	0.29
M_1** [$M_\odot$]	2.7
M_2*** [$M_\odot$]	≈0.78
R_1*** [$R_\odot$]	16.0 ± 0.3
$\log g$	2.78 ± 0.5
T_{eff} [K]	4600 ± 100
Microturbulence, ξ [km s^{-1}]	1.0
Macroturbulence, $\zeta_{R,T}$ [km s^{-1}]	2.0
[Fe/H], in solar units	−0.30 ± 0.05
[Ca/H], in solar units	−0.40 ± 0.05

* Fekel et al. (1999), **Stawikowski and Glebocki (1994); *** Paper I

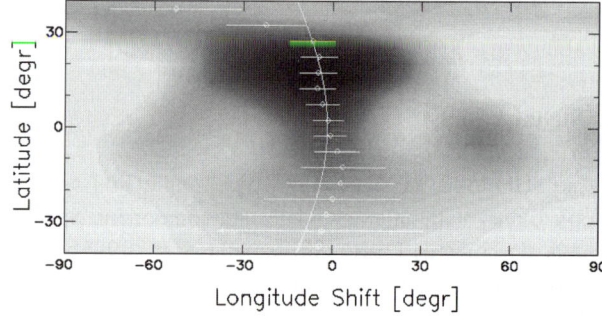

Fig. 2 Average cross-correlation map from the time-series Doppler maps, where the better the correlation, the darker the shade. For each latitude strip the maximum correlation is represented by the Gaussian peaks (dots) and the corresponding FWHMs (bars). The solid line is the best-fit solar-type differential rotation law

3. Differential rotation

From the 54 NSO 96/97 spectra we built 36 data subsets with 17 spectra in each. The first subset consists of the first 17 observations, the next subset is formed from omitting the first spectrum and adding the subsequent one to the end, etc., until the last 17 spectra are included. The result is a time series of altogether 36 Doppler-maps. We then cross correlate consecutive but contiguous, i.e. independent but neighbouring maps: we compute a cross-correlation-function (ccf) map from image #1 and #17, then one from #2 and #18 and so forth. This gives 19 ccf maps. Because the time baseline for these ccf maps varied between 18.09 and 20.46 days, we normalized the longitude shifts to the average time interval of 19.02 days, and then averaged the 19 ccf maps. We then searched for a correlation peak in each latitude strip and fit a Gaussian to it. The Gaussian peaks are then fitted with a solar-type quadratic differential rotation law of the form $\Omega(\theta) = \Omega_{eq} - \Delta\Omega \sin^2\theta$. Relevant part of the resulting ccf map is shown in Fig. 2. The continuous line in Fig. 1 is the best fit with $\Omega_{eq} = 20.164 \pm 0.122°$/day, i.e. $P_{eq} \approx 17.85$ days, which means that the equator corotates with the orbital period, while $\Delta\Omega = 1.231 \pm 0.514$ means that poles rotate a bit slower, with a period of 19.01 days. This yields a differential rotation parameter $\alpha = \Delta\Omega/\Omega_{eq}$ of 0.061 ± 0.026, which is one-third of the solar value.

Acknowledgements Authors from Konkoly Observatory are grateful to the Hungarian Science Research Program (OTKA) for support under grants T-038013, T-043504, T-048961 and T-034551. JBR thanks the NSERC-CRSNG for their support of this work.

References

Fekel, F.C., Strassmeier, K.G., Weber, M., Washüttl, A.: A&A Suppl. **137**, 369 (1999)
Rice, J.B., Wehlau, W.H., Khokhlova, V.L.: A&A **208**, 179 (1989)
Stawikowski, A., Glebocki, R.: Acta Astron. **44**, 393 (1994)
Strassmeier, K.G., Hall, D.S., Fekel, F.C., Scheck, M.: A&A Suppl. **100**, 173 (1993)
Strassmeier, K.G., Hall, D.S., Boyd, L., Genet, R.: AJ Suppl. **69**, 141 (1989)
Weber, M., Strassmeier, K.G.: A&A **330**, 1029 (1998)

Future Prospects for Ultra-High Resolution Imaging of Binary Systems at UV and X-ray Wavelengths

Margarita Karovska

Received: 1 November 2005 / Accepted: 1 February 2006
© Springer Science + Business Media B.V. 2006

Abstract UV and X-ray space-based interferometry will open unprecedented possibilities for spectral and spatial studies of a wide range of currently unresolvable interacting systems. Ultra-high angular resolution direct imaging of individual components and transport processes in interacting binary systems is essential for detailed studies and modeling of accretion and activity. Understanding the mass loss characteristics of both components, and the dynamics of the system as a function of time, will provide key inputs to evolutionary models and will revolutionize our view and understanding of the Universe.

Keywords High-angular resolution · Interacting binaries

1. Introduction

The quest for high angular resolution has been a perennial pursuit since Galileo pointed the first telescope toward the sky almost four centuries ago. Since Galileo's time telescopes with increasing apertures are being built in order to improve the diffraction-limited resolution, finally to enable high-angular resolution observations of astronomical sources.

However, in the early 20th century it became obvious that the limitations in the achievable angular resolution are not due only to the diffraction limit of the telescope but also to the degrading effects of the Earth's atmosphere. While the light-collecting power of a telescope increased as a square of its diameter, the angular resolution remained dominated by the atmospheric blurring or the seeing. In other words, even at the best observing sites in California, and later on in for

M. Karovska
Harvard-Smithsonian Center for Astrophysics

e.g. Chile and Hawaii, the effective resolution of a telescope with a diameter of a few meters remained around 0.5″, which is similar to the resolution of a telescope with a diameter of only few dozen centimeters.

With the recent development of adaptive optics some of the effects of the atmosphere can be mitigated, however, the potentially achievable resolution is still degraded by the atmosphere. With the construction of the first interferometer at Mt. Wilson by Michelson and Paese in the early twenties of the last century, a new avenue was opened for high angular resolution astronomical observations. Today there are a number of ground-based interferometers operating at wavelengths ranging from optical to radio wavelengths, producing detailed views of many astronomical objects. Although the resolution is still degraded to a degree by the atmosphere, interferometers like VLBI in the radio and KECK-I, and VLTI in the IR are able to reach a milliarcsecond-level resolution.

However, the UV and X-ray domains of the spectrum are not accessible from the ground and therefore high-angular resolution observations of powerful astrophysical processes such as accretion, with a dominant signatures at short wavelengths, are only possible from space.

Since 1990s the Hubble Space Telescope (HST) has been providing high-angular resolution views of the Universe from UV to IR wavelengths with an angular resolution of ∼0.1 arcsecond. The diffraction limit of this 2.5 m telescope is comparable to the ground-based telescopes built over 50 years ago (e.g. Mt. Wilson 100 inch telescope). However, the fact that the HST is in space and therefore free of the atmospheric seeing effects, provided unprecedented opportunities for high-resolution studies of many astronomical sources. Similarly, the X-ray astronomy was revolutionized with the launch of the Chandra X-ray Observatory in 1999. Chandra increased the resolution in the X-ray domain by over an order of magnitude when compared to ROSAT and XMM, and reached for the first time sub-arcsecond level in the X-rays. This allowed access to small spatial scales that have never before been seen in many astronomical objects. Despite this increase in resolution at UV and X-ray wavelengths, very few interacting binaries have been resolved by direct imaging either by Chandra or HST.

The next step in the quest for ultra-high resolution in astronomy will be space interferometry at UV and X-ray wavelengths. In this presentation I will discuss the prospects for ultra-high resolution imaging of interacting binaries with future UV and X-ray Interferometry Missions.

2. Interacting binary systems as astrophysical laboratories

Many sources in the Universe are powered by the potential energy released via accretion. Understanding accretion driven flows in binaries will directly affect our understanding of similar flows around YSOs, including the formation of planets in the circumstellar disk as well as the much larger scale accretion flows in active galactic nuclei (AGNs).

Compact, mass transferring binaries provide us with laboratories for testing energetic processes such as magnetically driven accretion and accretion geometries, various evolutionary scenarios, and conditions for induced stellar activity. In close binary stars the flow of material from one component into the potential well of the other is a key in determining the future evolutionary histories of each component and the system itself, and particularly the production of degenerate companions and supernovae. Our cosmological standard candles, the Type Ia supernovae, for example, may be a consequence of accretion onto a white dwarf in a close binary.

Interaction between the components in close binaries is believed to occur via Roche lobe overflow and/or wind accretion. 3-D hydrodynamic simulations show that the accretion processes in tidally interacting systems are very complex (Blondin et al., 1995). Wind accretion is even more complicated. The amount of the accreted material depends on the characteristics of both components including stellar activity and wind properties (e.g. density and velocity), the binary parameters (e.g. orbital period and separation), and the dynamics of the flow. In the case of Roche lobe overflow, the accretion may form an extended accretion disk whose turbulent magnetic dynamo drives the flow through it.

Stellar activity of the rapidly rotating donors and their impact on the accretor remains poorly constrained despite being crucial in regulating the mass transfer rate and setting the long-term evolution.

The key to further advance in accretion studies is resolving and directly imaging a wide range of interacting binaries and studying their components and mass flows.

3. The power of UV and X-ray high-angular resolution imaging

High spatial and spectral resolution in the UV and X-rays offers unprecedented opportunities for detailed studies of interacting binary systems. Currently, most of our accretion paradigms are based on time-resolved spectroscopic observations. Although spectroscopy of many interacting binaries has provided valuable information, very few systems have been spatially resolved so far.

Observations of the nearby symbiotic system Mira AB demonstrate the power of the sub-arcsecond angular resolution by having separated the components of this 0.6″ interacting binary for the first time at UV and X-ray wavelengths using modern space observatories. Symbiotics are interacting binaries showing a composite spectrum with signatures of a late-type giant and a high-temperature component, often a compact object in the form of a white dwarf (Livio, 1988). Mira AB is the nearest symbiotic system composed of an evolved mass-loosing AGB star (Mira A) and a wind-accreting white dwarf.

In 1995, HST observations of Mira AB resolved the components of the system and separated for the first time their spectra at ultraviolet and optical wavelengths (Karovska et al., 1997). The high-angular resolution HST images detected significant asymmetries in the giant's atmosphere and showed evidence for possible direct interaction between Mira A and its companion.

In 2003, Chandra resolved the components of this system for the first time at X-ray wavelengths (Karovska et al., 2005). Both components were detected, which was a great surprise, since X-rays have never before been detected from an AGB star (Fig. 1a). Furthermore, The spectra showed that the X-ray emission is due to a powerful soft X-ray outburst. A combination of Chandra and HST images of the system (obtained few months apart) identified the source of the outburst with Mira A (or in its vicinity). This outburst is possibly associated with a jet-like activity, as evidenced by the extended structures in the X-ray and UV images of Mira A.

Chandra also detected a bridge between the components showing that Mira B is accreting not only from the wind of the AGB star, but also via direct mass exchange. This was also an unexpected result, because the components are separated by ∼70 AU and it has been assumed in the past that the interaction between the components in such a system can be carried out *only* via wind accretion.

These results have important implications for our understanding of accretion processes in detached binaries and in other wind accreting systems in the Universe. They also demonstrate the power of the high-angular resolution imaging combined with spectroscopy of the individual components in the interacting systems. Increasing the resolution to *sub-milliarcsecond* level in the UV and X-rays will

High-Angular Resolution Observations

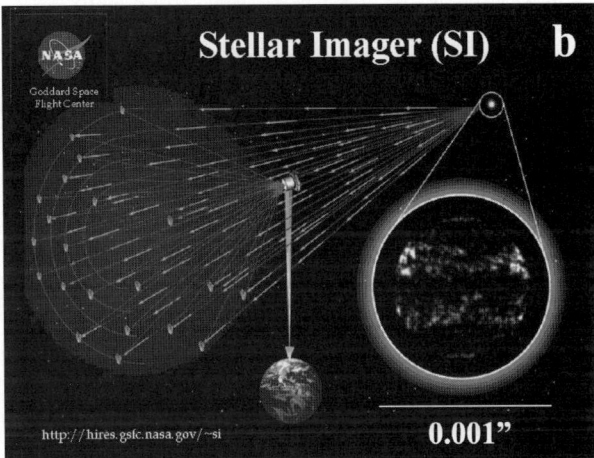

Fig. 1 The present and the future of high angular resolution imaging of interacting binaries: (a) Chandra image of Mira AB symbiotic binary (0.6″ separation) obtained with a resolution of 0.2 arcsecond. This resolution reflects the "state of the art" of the imaging capabilities of current UV and X-ray observatories. (b) Future space observatories like the SI, a 0.5 km baseline multi-mirror imaging interferometer operating in the UV/Optical, will provide a resolution of *less then* 0.1 milliarcsecond. This will be an increase of over 2 magnitudes when compared to the current resolution power of the HST and Chandra. The SI will be able to resolve a large fraction of symbiotic binaries, and many other fascinating interacting systems

revolutionize the observational astrophysics of the 21st century and provide unprecedented opportunities for studies of many interacting binaries.

4. Prospects for ultra-high resolution imaging of interacting binary systems

The next major observational advances in astronomy will require orders of magnitude leaps in angular resolution. To achieve this, large baselines, from 0.5 to many kilometers will be required. Several future missions providing ultra-high angular resolution at UV and X-ray wavelengths are currently under study. These include the Stellar Imager (SI), a UV/Optical interferometer with a sub-milliarcsecond resolution, and MAXIM – an X-ray interferometer with a microarcsecond-level resolution.

The SI is a UV-optical deep-space telescope for sub-milliarcsecond resolution imaging of astrophysical sources (Carpenter et al., 2004).[1] Its mission is to enable an understanding of solar/stellar magnetic activity and its impact on the origin and continued existence of life in the Universe, the structure and evolution of stars, the habitability of planets, and the transport of matter throughout the Universe. The current SI concept is a long-baseline (0.5 km) Fizeau interferometer with 20–30 free-flying mirrorsats (Fig. 1b).

It is designed to obtain images with sub-milliarcsecond resolution of the surfaces of solar-type and other stars and to measure their sub-surface structure via spatially-resolved asteroseismology. The SI will also record ultra-high resolution images of many other types of targets, including planet-harboring environments, AGN's, QSO's, supernovae, and numerous interacting binaries.

Sub-milliarcsecond resolution in the UV will lead to unprecedented opportunities for detailed studies of accretion phenomena in many interacting systems including symbiotics, Algol type binaries, Cataclysmic Variables (CVs) and their progenitors. The SI will be able to resolve the components of a diverse set of interacting systems and will therefore provide a unique laboratory for studying accretion processes. The binary components themselves can be studied individually at many wavelengths including Lyα, CIV, and Mg h&k lines, and the geometry of accretion can be imaged directly, giving us the first direct constraints on the accretion geometries for a range of systems.

For example, the SI will be able to separate the components of many currently unresolved symbiotic binaries. These are some of the most fascinating interacting systems because of their dramatic transformations and extremely complex circumbinary environment (Corradi et al., 2001). Symbiotics are very important because they are likely progenitors of bipolar planetary nebulae. They have also been invoked as potential progenitors of at least a fraction of Supernovae type Ia, a key cosmological distance indicators (Chugai and Yungelson, 2004). As mentioned earlier, the

---

[1] Detailed information on SI can be found on http://hires.gsfc.nasa.gov/~si/.

individual components have so far been resolved clearly in only one nearby symbiotic system – Mira AB (Karovska et al., 1997; Karovska et al., 2005). The SI will have the capability of resolving a significant fraction of currently known symbiotic systems, and will be able to image directly the individual components and the dynamical accretion flows in dozens of nearby systems.

SI will be able to study the direct impact accretion that occurs in Algols, where the mass-losing donor star feeds the primary with a stream of gas that hits the surface of the primary. The physical properties of the impact site and the associated transfer of angular momentum are crucial in determining the accretion history. For example, Algol itself (β Per) is a nearby 2 mas eclipsing binary system (at 30 pc) and its accretion environment and geometry will be easily imaged by SI.

The SI will be able to resolve few of the brightest and closest CVs. These include AE Aqr, a $\sim$10 hour system containing a K5 donor star feeding a rapidly spinning and magnetized white dwarf. At 0.1 milliarcsecond resolution, the SI will be able to image the outflow propelled by the rapidly rotating magnetosphere around the white dwarf, this providing a key to our understanding of other systems containing rapidly spinning and magnetized compact objects including accreting pulsars.

There is substantial indirect evidence that magnetic activity plays an important role in driving the formation and secular evolution of a wide range of binary systems. In some cases, tomographic imaging has identified signatures of star spots on the rapidly rotating donor stars in binaries. SI can provide a breakthrough by resolving the surface structure of such stars directly. Furthermore, with the SI, we will be able to study the presence of active coronae and stellar prominences for the first time. For example, RS CVn binaries are detached systems containing a very active sub-giant. Large star spots are inferred through light-curve modeling and the presence of emission lines and UV/X-ray emission indicates active chromospheres and coronae. Direct imaging of the stellar activity structures with SI would be invaluable in linking these active sub-giants to solar-type activity.

The SI will be able to resolve outflows and the circumbinary environments in a variety of systems, including regions of colliding winds in the photoionization wakes in X-ray binaries. However, in order to be able to resolve the components of interacting X-ray binary systems it is necessary to increase the resolution by at least an order of magnitude, to less then 10 micro-arcsecond level. X-ray interferometers such as the MicroArcsecond X-ray Imaging Mission (MAXIM) will enable resolving these and many other fascinating systems.

MAXIM is one of the concepts for a Black Hole Imager Mission that is currently being studied.[2] Its principal goal is to image the event-horizons around black holes, but with microarcsecond-level resolution it will be able to produce images of the components and the dynamic flows in many interacting binaries, and study in detail the accretion processes. The MAXIM X-ray interferometer would entail a fleet of up to 33 optics spacecraft flying in formation with a precision of 20 nano-meters, plus a detector spacecraft 500 kilometers behind the mirrors.

The UV and X-ray space interferometers are ambitious, long-term missions, composed of multiple, formation-flying spacecraft located near the Sun-Earth L2 point. There are numerous technical challenges that need to be solved before these imaging interferometers see the "first light".

A stepping stone toward the sub-milliarcsecond resolution and beyond, could be a pathfinder mission with 20–50 m baseline and 3–5 mirrorsats operating at UV wavelengths. Such a mission would allow testing these technologies on a time scale of a decade, and will provide guidance for future development of "short wavelength" ultra-high resolution interferometers. It will also provide a powerful tool for studying nearby interacting systems with a resolution exceeding that of the HST by at least one order of magnitude.

Acknowledgements We are grateful to Drs. K. Carpenter, C. Schryver, D. Steeghs and the Stellar Imager team for helpful inputs on the SI capabilities and scientific potential. Support for this work was provided by NASA through grants number GO4-5024A, GO-08298.01-99A. M. K. is a member of the Chandra Science Center, which is operated under contract NAS-839073, and is partially supported by NASA.

References

Blondin, J.M., Richards, M.T., Malinowski, M.: ApJ **445**, 939 (1995)
Carpenter, K., the SI team: SPIE 5491_28 (2004)
Chugai, N.N., Yungelson, L.R.: Astron. Lett. **30**, 65 (2004)
Corradi, R.L.M. et al.: ApJ **560**, 912 (2001)
Karovska, M., Hack, W., Raymond, J., Guinan, E.: ApJ **482**, L175 (1997)
Karovska, M, Schlegel, E., Warren, H., Raymond, J., Wood, B.E.: ApJ **623**, L137 (2005)
Livio, M.: In: Symbiotic Phenomena, Procedings of IAU Coll. No. 103 (1988)

[2] http://maxim.gsfc.nasa.gov .

Astrophys Space Sci (2006) 304:381–384
DOI 10.1007/s10509-006-9147-3

ORIGINAL ARTICLE

The Impact of CoRoT on Close Binary Research

Carla Maceroni · Ignasi Ribas

Received: 7 October 2005 / Accepted: 1 December 2005
© Springer Science + Business Media B.V. 2006

Abstract The space experiment CoRoT will provide continuous monitoring and high accuracy light curves of about sixty thousand stars. Selected binary systems will be observed in the Additional Program frame as targets of long and continuous pointed observations. Moreover, thousands of new binaries will certainly be detected and hundreds of them will have extremely accurate light curves. This will allow studies of fine effects on the light curves, monitoring of stellar activity and, in combination with ground-based observations, will provide exquisite determination of stellar parameters. Among the new discoveries of interesting systems of special value will be those of low mass binaries.

Keywords CoRoT · Close binaries · Asteroseismology

1. CoRoT in a nutshell

CoRoT (COnvection, ROtation and planetary Transits) is a french-led international "small" space mission whose launch is scheduled for October 2006. The mission[1] focuses on two parallel "core programs", asteroseismology and extra-solar planet search, both requiring high accuracy photometry and continuous monitoring.

The CoRoT payload consists of a 28cm telescope (1200 mm focal length and field of view of 3.8 degrees) which will

---

C. Maceroni (✉)
INAF – Osservatorio Astronomico di Roma, Monteporzio Catone, Italy
e-mail: maceroni@oa-roma.inaf.it

I. Ribas
CSIC – Institut d'Estudis Espacials de Catalunya, Barcelona, Spain

[1] Complete information on the mission is available at http://corot.oamp.fr.

fly in a polar inertial orbit. Its focal plane hosts two pairs of 2k × 2k CCDs, one dedicated to asteroseismology, the other to planetary transit searches, both have a field of view of $2.64° \times 1.32°$.

The defocused seismology field (PSF of 450 pixel for a G2 star) will observe a maximum of ten bright targets per run ($5.7 \leq V \leq 9.5$) with typical time sampling of 32^s. Simultaneously, the (focused) exo-planet field will observe up to 12000 fainter targets ($11.5 \leq V \leq 16$) with typical time sampling of 512^s. Color information will be available for the exo-planet targets brighter than V_{15}, thanks to a prism providing very low resolution spectra. The expected accuracy of the photometry is a few parts per million (ppm) for asteroseismology and 100 ppm for planet search.

The nominal mission duration is 2.5 years. Each year will be split in two 150^d–long "Long Runs" (LR), and two/one $20/30^d$–long "Short Runs" (SR). Pointing is restricted to the CoRoT "Eyes", i.e. two 10° radius circles centered at $\delta = 0$ and, respectively, $\alpha = 6^h50^m$ for the "Anticenter" direction and $\alpha = 18^h50^m$ for the opposite, "Center", direction. The LR fields (much smaller than the "Eyes") are chosen according to the best compromise between the two core programs, and are constrained to a short list of preselected seismology "primary targets". Short Runs can instead be anywhere inside the Eyes.

Science outside the two core programs is included in the mission as specific "Additional Programs" (APs), which can apply for both LR and SR observations. In practice, most APs will be performed in the Exo-planet field, because of the much larger number of available windows.

2. CoRoT and close binaries

The best CoRoT assets are the accurate and stable photometry and the continuous monitoring, with an estimated duty cycle

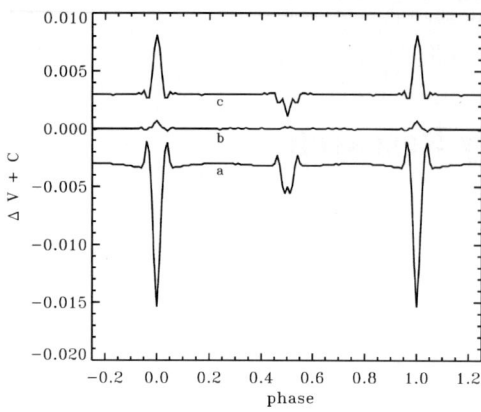
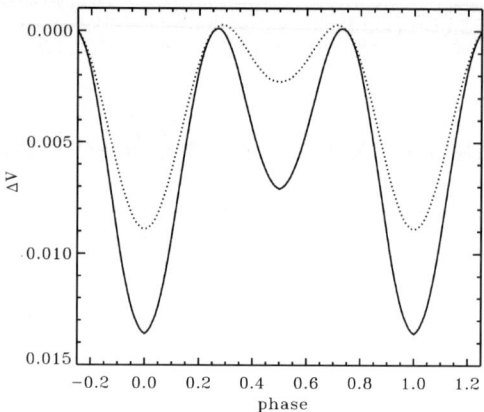

Fig. 1 Left panel: difference between synthetic light curve models with different assumptions on the limb darkening law: (a) linear ($x = 0.57$) – linear ($x = 0.67$), (b) linear ($x = 0.57$) – logarithmic, (c) logarithmic law – square root law. The parameters of the non-linear laws are taken from Van Hamme (1993) for the corresponding component temperatures and gravities. For better readability curves (a) and (c) are shifted with respect to the zero line. Right panel: the effect of gravity darkening in a system with the same parameters as V805 Aql, but with $i = 65°$. The continuous line corresponds to a gravity darkening exponent $\beta = 1.0$ (typical choice for radiative envelopes), the broken line to $\beta = 0.32$ (convective envelopes)

of 94%. These will provide close binary light curves of exquisite accuracy, excellent phase coverage and extending over a long time baseline.

The high accuracy with allow a thorough study of fine effects in the light curves by direct determination of second order effects such as limb darkening and gravity darkening. The latter can be derived for non eclipsing system as well, as it will clearly show up in the frequency domain.

The left panel of Fig.1 shows, as an example, the effect of different limb darkening laws on the light curve of a moderately close system. The synthetic light curves were computed by the 2003 version of the Wilson and Devinney code (Wilson and Devinney, 1971). For illustration purposes, we chose a model corresponding to the binary V805 Aql studied by Popper (1984); the main physical parameters of the model are: orbital period $P = 2.41^d$, eccentricity $e = 0$, inclination $i = 86°$, mass ratio $q = m_s/m_p = 0.81$ (the index p stays for primary and refers to the more massive component), fractional component radii $r_p = 0.18$ and $r_s = 0.15$, effective temperatures $T_p = 8184$ and $T_s = 7178\ K$, fractional V-band luminosity $L_s/L_p = 0.362$. The figure shows the difference in magnitude between synthetic light curves obtained with different assumptions on limb darkening (different values of the coefficient for the linear law, or different laws). The difference between linear laws of the form:

$$R_\lambda(\mu) = \frac{I_\lambda(\mu)}{I_\lambda(1)} = 1 - x_\lambda(1 - \mu), \quad (1)$$

with $\mu = \cos\theta$, is obtained by changing the value of the coefficient x_λ. Among the proposed non-linear functional forms we considered a logarithmic relation:

$$R_\lambda(\mu) = 1 - x_\lambda(1 - \mu) - y_\lambda \mu \ln\mu \quad (2)$$

and a square root one:

$$R_\lambda(\mu) = 1 - x_\lambda(1 - \mu) - y_\lambda(1 - \sqrt{\mu}) \quad (3)$$

see Van Hamme (1993) and references therein. The coefficients used to compute the differences (b) and (c) in Fig. 1 have be taken, according to the component temperatures and gravities, from the tables by Van Hamme (1993) (i.e. the differences are between two fits of the same atmosphere). For this reason the deviations are smaller than in the case of linear laws. At any rate the photometry by CoRoT, with its 10^{-3}–10^{-4} mag accuracy, should allow to distinguish at least between linear and non-linear limb darkening prescriptions. The directly determined limb darkening can then be used to constrain model atmosphere details.

The right panel of Fig.1 shows the amplitude of the ellipsoidal variation in a system differing from V805 Aql only for a lower value of the inclination (to avoid eclipses), $i = 65°$. In spite of the small deviation from spherical symmetry (the maximum difference in radius for the primary component is $< 2\%$) the effect is large enough to be easily measured by CoRoT. The determination of gravity brightening and, with the help of follow-up observations, of the related system configuration, will yield precious information on close binary tidal evolution.

The long monitoring of the same field will allow to study on a long time baseline the manifestations of stellar activity such as spots, flares, stellar activity cycles. These will provide information on rotational period and differential rotation (from spot migration) in late-type components.

A very young field, asteroseismology in close binaries, combining the information from asteroseismology with that from eclipses will certainly profit of CoRoT assets. Looking for (solar-type) oscillations in suitable eclipsing binaries

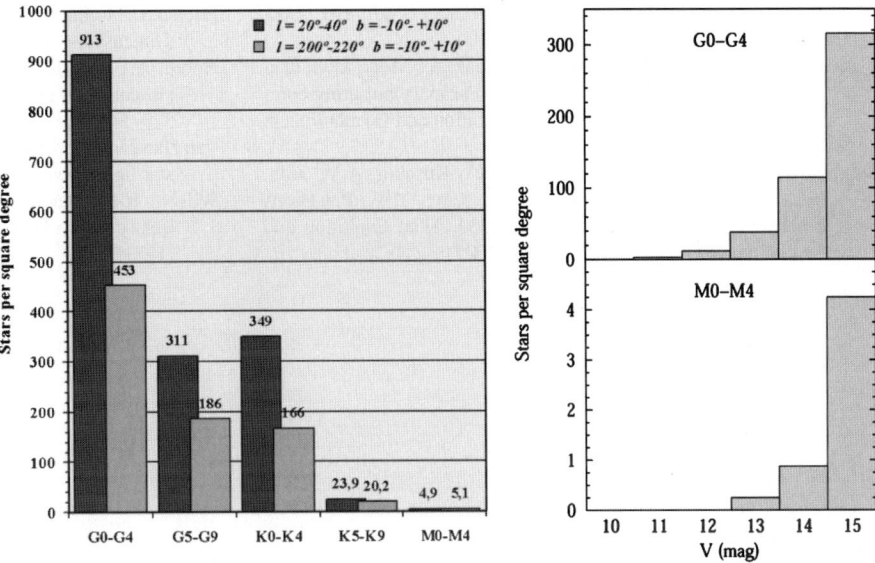

Fig. 2 Left panel: The expected number of G-M stars of luminosity class IV–V, in a 16□ field inside the CoRoT eyes. Right panel: the distribution in apparent magnitude in the winter (AC) field for G0–G4 and M0–M4 spectral type intervals

Table 1 Expected binaries per 3.4□ field

Spectral type	Expected EB number
G0-G4	7.2–31.1
G5-G9	3.2–10.5
K0-K4	2.9–11.9
K5-K9	0.35–0.8
M0-M4	0.09–0.17

would have the advantage of knowing the masses, the radii and the orbit inclination angle, and therefore the rotational velocity. While binarity might increase the complexity of analysis, the identification of even non-radial pulsation modes is possible, as recently shown by several authors (Mkrtichian et al., 2005; Gamarova et al., 2005; Escolá-Sirisi et al., 2005). In particular eclipsing systems, with their near equator-on view, favor the detection of sectorial modes (i.e modes with $\ell = |m|$, where nodal lines form lines of longitude).

The CoRoT eyes contain ~ 200 known system, but only a tiny fraction of them will actually be observed during a Long Run.[2] However, many interesting systems will certainly be discovered in the CoRoT fields. Figure 2 shows the results of a simulation with the Besançon galaxy model (Robin et al., 2003), the constraints are: spectral type G-M, luminosity class IV–V, V magnitude 5.5–15.5 (to include both sismo- and exo-fields). The results give the number of star expected in a field of $4° \times 4°$ in the CoRoT Center and Anticenter directions. The magnitude distribution of the expected stars at the spectral extremes, and for the winter (Anticenter) field, is also shown.

Photometric large surveys such as OGLE, MACHO, ASAS, STARE, Vulcan indicate that 0.5%–1% of all monitored stars turn out to eclipsing binaries. From these percentages, and the density estimate, we obtained the maximum and minimum number of eclipsing binaries for an exo-planet field (~ 3.4□). Table 1 gives the results subdivided by spectral type intervals. With a total number of five LRs and five/ten SRs some hundreds of new objects are expected only along the MS. Among those, the stars at the two extremes in mass will be especially valuable to better constrain the theoretical models. In particular, at the low mass end of the MS, the absolute elements obtainable by combining follow-up spectroscopy with the CoRoT photometry will be precise enough to clarify what appears as a serious discrepancy in the mass-radius relation between theoretical models and data from observed eclipsing binaries (see Ribas talk, these proceedings).

In summary the "small space experiment" CoRoT will provide great opportunities not only in the field of astero-seismology and extra-solar planet searches. Research on binary stars (and more generally on stellar physics) will greatly benefit of its accurate photometry and stable monitoring and many long standing questions will hopefully find a definite and clear answer.

Acknowledgements We thank the members of the CoRoT "Thematic Team" on binaries for useful discussion on the different aspects of binary research that can be addressed by the mission. CM acknowledges the support from the MIUR-Cofin 2004 "Asteroseismology".

---

[2] see the link 'Information' of the CoRoT *Thematic Team on binaries* web-page, at http://thor.ieec.uab.es/binteam/, for details about the known binaries in the "Eyes" and in the fields recently chosen for the first two Long Runs.

References

Escolà-Sirisi, E., Juan-Samsó, J., Vidal-Sáinz, J., Lampens, P., García-Melendo, E., Gómez-Forrellad, J.M., Wils, P.: Detection of a classical δ Scuti star in the new eclipsing binary system HIP 7666. Astron. Astrophys. **434**, 1063–1068 (2005)

Gamarova, A.Yu., Mkrtichian, D.E., Rodríguez, E.: Mode identification in the ecplising binary systems with primary pulsating components. ASP Conf. Ser. 333: Tidal Evolution and Oscillations in Binary Stars **333**, 258 (2005)

Mkrtichian, D.E., Rodríguez, E., Olson, E.C., Kusakin, A.V., Kim, S.-L., Lehmann, H., Gamarova, A.Yu., Kang, Y.W.: Pulsations in eclipsing binaries. ASP Conf. Ser. 333: Tidal Evolution and Oscillations in Binary Stars **333**, 197 (2005)

Popper, D.M.: Error analysis of light curves of detached eclipsing binary systems. Astron. J. **89**, 132–144 (1984)

Robin, A.C., Reylé, C., Derrière, S., Picaud, S.: A synthetic view on structure and evolution of the Milky Way. Astron. Astrophys. **409**, 523–540 (2003)

Van Hamme, W.: New limb-darkening coefficients for modeling binary star light curves. Astron. J. **106**, 2096–2117 (1993)

Wilson, R.E., Devinney, E.J.: Realization of accurate close-binary light curves: Application to MR Cygni. Astrophys. J. **166**, 605 (1971)

On the Gaia Expected Harvest on Eclipsing Binaries

P. G. Niarchos

Received: 1 November 2005 / Accepted: 1 December 2005
© Springer Science + Business Media B.V. 2006

Abstract The space experiment Gaia, the approved cornerstone 6 ESA mission, will observe up to a billion stars in our Galaxy and obtain their astrometric positions on a microarcsec level, multi-band photometry as well as spectroscopic observations. It is expected that about one million Eclipsing Binaries (EBs) (with $V \leq 16$ mag) will be discovered and the observing fashion will be quite similar to Hipparcos/Tycho mission operational mode. The combined astrometric, photometric and spectroscopic data will be used to compute the physical parameters of the observed EBs. From a study of a small sample of EBs, it is shown that the agreement between the fundamental stellar parameters, derived from ground-based and Hipparcos (Gaia-like) observations, is more than satisfactory and the Gaia data will be suitable to obtain accurate binary solutions.

Keywords Surveys · Stars: fundamental parameters · Stars: binaries: eclipsing

1. Introduction

Gaia, one of the next two 'cornerstones' of ESA's science program (2000), is designed to obtain extremely precise astrometry (in the micro-arcsec regime), multi-band photometry and medium-resolution spectroscopy for a large sample of stars. The main scientific objectives of the Gaia mission and the expected benefits for astrophysical research are given by Gilmore et al. (1998), Perryman et al. (2001) and Mignard (2005). An overview of the Gaia mission is presented by Perryman (2005), the Gaia payload and spacecraft design is detailed by Pace (2005), and the goals of Gaia spectroscopy and photometry are discussed by Munari (1999a,b; 2002) and Høg (2002). The role of Gaia photometry and spectroscopy on eclipsing binaries has been reviewed in details by Munari et al. (2001) and Zwitter (2003). The Gaia large-scale photometric survey will have significant intrinsic scientific value for the study of variable stars of nearly all types, including detached eclipsing binaries, near contact or contact binaries and pulsating stars. It is expected that about one million Eclipsing Binaries (EBs), (with $V < 16$ mag), will be discovered. Even if reliable physical parameters could be derived for only 1% of the observed EBs, this would be a great contribution to stellar astrophysics and a giant leap in comparison with what has been obtained so far from ground based observations.

2. The Gaia scientific and mission objectives

According to the new ESA Science Programme, approved by the ESA Science Programme Committee in May 2002, the Gaia spacecraft has been redesigned to be launched in mid-2011 with a Soyuz-Fregat launcher and it will be put in a Lissajous-type orbit around the Sun-Earth Lagrangian point L2. The new Gaia Technical Baseline at satellite level is summarized in detail by Pace (2005). The objectives of the Gaia mission are to build a catalogue of one billion stars with accurate positions, parallaxes, proper motions, magnitudes and radial velocities. The catalogue will be complete up to $V = 20$ mag. The overall mission contains three parts: **(1)** extremely precise astrometry (in the micro-arcsec regime), with the measurement of accurate stellar positions, parallaxes and proper motions; **(2)** photometry, with measurements in 15 (11 medium and 4 broadband) filters; **(3)** radial velocity measurements up to $V = 17$ mag.

P. G. Niarchos
Department of Astrophysics, Astronomy and Mechanics, National and Kapodistrian University of Athens, GR 157 84 Zografou, Athens, Greece

2.1. Astrometry with Gaia

The main objective of the Gaia mission is to perform global or wide-field astrometry instead of local or narrow field astrometry (Pace, 2005). The expected accuracy for the astrometry mission is 10 μ as in positional and parallax accuracy and 10 μ as yr^{-1} for proper motion accuracy, for stars of $V = 15$ mag. The final catalogue will be built from the best estimate of all star positions and of the instrument parameters computed from the total amount of data collected during the mission. The final accuracy is only reached at the end of the mission, i.e., after 5-year of operation, and after a complete processing of all the data.

2.2. Photometry of eclipsing binaries with Gaia

The photometric information obtained by the Gaia satellite will be far superior to that of Hipparcos. The photometric accuracy for objects brighter than $V \sim 18$ will be better than 0.01 mag. The photometry mission will be multi-colour and multi-epoch. Magnitudes will be obtained in 4 wide spectral bands for all stars up to $V = 20$ mag, complemented with measurements in 11 narrow bands for stars up to $V = 17$ mag. The system features four broad wavelength bands (the Broad Band Photometer, BBP) and eleven bands of medium to narrow width (the Medium Band Photometer, MBP). The bands span the complete spectral range from 280 to 1000 nm.

It is expected that about 1 million eclipsing binaries (with $V < 16$ mag), will be discovered. In the five-year mission lifetime the number of photometric observations per target and for each one of the 4 broad bands will be about 100 and about 150 for each of the 11 intermediate bands, and the observing fashion will be quite similar to the Hipparcos/Tycho mission operating mode. In many cases the observations obtained by Gaia will be good enough to determine system parameters at 1–2% accuracy level. The large amount of data requires an automatic process in all stages of reduction and interpretation. The photometric data analysis is carried out on the ground.

2.3. Spectroscopy of eclipsing binaries with Gaia

Eclipsing binary stars represent the most astrophysically relevant type of binary stars that Gaia will observe spectroscopically during its mission. The spectrograph aboard Gaia will have a square primary mirror with a surface area of 0.5×0.5 m^2 and a field of view of $1.65° \times 1.6°$. There is no slit on the spectrograph and the stars drift across the focal plane at a rate of 60″/s. This gives a fixed exposure time of 99 seconds for each transit of the star across the focal plane. Because of a precessional motion of the spin axis of the satellite the whole celestial sphere will be covered, and each position on the sky will be observed on average 100-times during a 5-year mission.

Gaia will record spectra covering the 8490–8750 Å wavelength range, at 0.75 Å/pix dispersion as currently baselined in the Concept and Technology Study Report (ESA SP-2000-4). The Gaia 8490-8750 Å interval is centered on the near-IR Ca II triplet and head of the Paschen series. It also includes numerous N I, Fe I, Ti I, Cr I, Si I, Ni I, S II, Mg I and Mn I lines (Munari, 2002). A review of Gaia spectroscopy has been presented by Munari (1999a, 2002), while the performance of the spectrograph and the expected radial velocity accuracy as a function of magnitude and spectral type was discussed by Zwitter (2003). The Gaia spectral targets (approaching 10^8 in current estimates) will require fully automatic data treatment and analysis. The development of reliable classification and analysis procedures is one of the major tasks facing the scientific community before the launch of Gaia.

3. Accuracy of the fundamental parameters

It is expected that the number of Eclipsing Binaries discovered by Gaia will reach hundreds of thousands. Combined astrometric, photometric and spectroscopic observations of these systems will be reduced and analyzed by suitable codes so that reliable orbital and physical parameters are determined. Some crucial questions regarding the reliability of the derived stellar parameters from Gaia observations are: How can these observations be compared with the state-of-the-art ground-based observations? Can Gaia observations permit us to derive fundamental parameters as good as or better than previous work? What is the accuracy to which Eclipsing Binaries can be investigated using Gaia data alone?

The accuracy of the computed elements depends strongly on both the quality and number of photometric and spectroscopic observations and the methods of light curve and spectral analysis. For Gaia observations, it was shown by Munari et al. (2001), Zwitter et al. (2003) and Marrese et al. (2004), that stellar parameters could be determined (in many cases) at about 2% accuracy level. Given that Gaia observations will be superior to those of Hipparcos/Tycho mission, the results obtained by using Hipparcos observations should present a lower limit to the Gaia expected accuracy.

Niarchos and Manimanis (2003) made a first evaluation of the Gaia performance on the photometry of contact and near-contact eclipsing binaries by analysing the light curves of selected systems using ground-based and Hipparcos (Gaia-like) observations. In the present study the sample is enriched and includes systems of all the basic spectral types from A to K. The criteria of selection were: (1) systems observed by Hipparchos/Tycho mission in H_p filter (very close to V filter); (2) high quality ground-based V observations

Table 1 Fundamental parameters expressed in solar units

System	M_1	M_2	R_1	R_2	L_1	L_2	Sp. type
RZ Dra (1)	1.40(4)	0.62(3)	1.62(1)	1.12(1)	10.1(4)	1.01(6)	A5 + K2
RZ Dra (2)	1.42(5)	0.59(4)	1.55(1)	1.11(1)	9.60(47)	0.86(5)	
% difference	1.4	4.8	4.3	0.9	5	15	
V1010 Oph (1)	1.87(12)	0.90(5)	2.08(6)	1.48(8)	10.7(7)	3.47(35)	A5 + F6
V1010 Oph (2)	1.88(14)	0.89(7)	2.08(1)	1.46(1)	11.7(8)	3.07(8)	
% difference	0.5	1.1	0	1.4	9.3	12	
XZ Leo (1)	1.83(5)	0.64(2)	1.71(2)	1.07(3)	7.29(29)	2.56(16)	A5
XZ Leo (2)	1.82(13)	0.63(6)	1.71(2)	1.07(3)	7.26(29)	2.69(19)	
% difference	0.5	1.6	0.2	0.2	0.4	5.1	
ϵ CrA (1)	1.75(4)	0.21(2)	2.20(3)	0.80(1)	11.1(1)	1.08(1)	F2
ϵ CrA (2)	1.69(7)	0.22(5)	2.12(12)	0.80(4)	10.3(20)	1.07(14)	
% difference	3.4	4.8	3.6	1.2	7.2	0.9	
V566 Oph (1)	1.40(3)	0.33(1)	1.47(1)	0.79(1)	4.57(4)	1.26(1)	F4 + F8
V566 Oph (2)	1.45(11)	0.35(4)	1.518(3)	0.819(4)	4.76(24)	1.26(2)	
% difference	3.5	6.1	3.5	3.7	4.5	0.8	
YY CrB (1)	1.41(9)	0.34(2)	1.40(7)	0.77(10)	2.46(26)	0.81(20)	F8
YY CrB (2)	1.37(16)	0.33(4)	1.36(7)	0.74(10)	2.33(25)	0.74(20)	
% difference	2.8	2.9	2.9	3.9	5.3	8.6	
V839 Oph (1)	1.62(4)	0.49(1)	1.53(6)	0.93(8)	3.12(26)	1.32(23)	G0
V839 Oph (2)	1.61(17)	0.49(5)	1.52(6)	0.92(8)	3.10(26)	1.22(21)	
% difference	0.6	0.8	0.7	1.1	0.6	7.6	
AB And (1)	1.01(2)	0.56(1)	1.04(4)	0.80(5)	0.87(7)	0.46(5)	G5
AB And (2)	1.00(9)	0.55(5)	1.06(4)	0.83(5)	0.91(7)	0.45(5)	
% difference	1.0	1.8	1.9	3.7	4.6	2.2	
XY Leo (1)	0.82(4)	0.41(2)	0.87(1)	0.64(2)	0.38(2)	0.14(1)	K0
XY Leo (2)	0.83(21)	0.41(10)	0.86(1)	0.62(2)	0.37(2)	0.17(1)	
% difference	1.2	0.7	1.1	3.1	2.6	21.4	
AH Vir (1)	1.45(13)	0.44(4)	1.36(1)	0.77(1)	2.45(8)	0.71(2)	K0
AH Vir (2)	1.33(20)	0.40(6)	1.44(1)	0.88(1)	2.76(9)	0.91(3)	
% difference	8.3	9.1	5.9	14.3	12.7	28.2	

(1): Ground-based observations
(2): Gaia (expected) observations

for the same systems; (3) accurate spectroscopic mass-ratios for these systems derived by modern techniques. The systems selected are: the near contact systems V100 Oph (Shaw et al., 1990) and RZ Dra (Kreiner et al., 1994); and the contact systems: XZ Leo (Rucinski and Lu, 1999), ϵ Cra (Goecking and Duerbeck, 1993), V566 Oph (Van Hamme and Wilson, 1985), YY CrB (Rucinski et al., 2000), V839 Oph (Rucinski and Lu, 1999), AB And (Hrivnak, 1988), XY Leo (Hrivnak et al., 1984), AH Vir (Lu and Rucinski, 1993). The Wilson-Devinney code was used for the light curve analysis and the results are given in Table 1. From an inspection of Table 1 it is obvious that, for the near-contact systems in particular, the derived absolute elements from Hipparcos observations differ from those of ground-based observations within the limits of the combined errors.

For the contact systems, although these differences are larger than in the case of near-contact systems, they are mostly within the limits of the combined errors. For systems with variable light curves (like AH Vir) or with poor phase coverage, the differences become larger. But, since Gaia photometry will have excellent precision and much larger number of photometric bands than Hipparcos, it is expected that the resulting stellar parameters will be comparable to those obtained from high quality ground-based observations.

4. Summary and conclusions

Gaia will provide light curves for millions of faint eclipsing binaries and also their absolute luminosities and temperatures (from parallaxes and colours). These data, combined with the radial velocity measurements obtained from spectroscopic observations, will allow us to determine the fundamental parameters (radii, masses, luminosities, temperatures) of hundreds of thousands eclipsing systems. Although the expected accuracy will be moderate, the large amount of data will allow us to look for large deviations from the 'normal' mass-radius-luminosity relations.

An analysis of photometric observations for a small sample of eclipsing systems, obtained by ground-based and (Gaia-like) Hipparcos photometry, shows that the fundamental parameters of these systems can be determined at $\sim 2\%$ accuracy level. This will definitely have an immense impact on theories of stellar structure and evolution.

Added note: The spacecraft and payload configuration was recently re-optimised by the industrial teams in their Phase B2/C/D proposal in response to the mission requirements document issued by the ESA project team in 2005. As a result, from early 2006, the final Gaia payload looks somewhat different from the previous design, although all functionality is preserved. It is not yet clear by how much the Gaia science on EBs has been compromised.

References

ESA: GAIA Consept and Technology Study Report, ESA-SCI, 4 (2000)
Gilmore, G., Perryman, M., Lindegren, L. et al.: Proc. SPIE Conf. **3350**, 541 (1998)
Goecking, K.-D., Duerbeck, H.: A&A **278**, 463 (1993)
Høg, E.: In: Bienayme, O., Turon, C. (eds.) GAIA: A European Project. EAS Pub. Ser. **2**, 27 (2002)
Hrivnak, B.J.: ApJ **335**, 319 (1988)
Hrivnak, B.J., Milone, E.F., Hill, G., Fisher, W.A.: ApJ **285**, 683 (1984)
Kreiner, J.M., Pajdosz, G., Tremko, J., Zola, S.: A&A **285**, 459 (1994)
Lu, W.-X., Rucinski, S.M.: AJ **106**, 361 (1993)
Marrese, P.M., Munari, U., Siviero, A. et al.: A&A **413**, 635 (2004)
Mignard, F.: In the three-dimensional universe with GAIA, 4–7 October 2004, Paris, ESA SP-576, 5 (2005)
Munari, U.: In: Proceedings of the ESA Leiden Workshop on GAIA, 23–27 Nov. 1998. Baltic Astronomy **8**, 73 (1999a)
Munari, U.: In: Proceedings of the ESA Leiden Workshop on GAIA, 23–27 Nov. 1998. Baltic Astronomy **8**, 123 (1999b)
Munari, U., Tomov, T., Zwitter, T. et al.: A&A **378**, 477 (2001)
Munari, U.: In: Bienayme, O., Turon, C. (eds.) GAIA: A European Project, EAS Pub. Ser. **2**, 39 (2002)
Niarchos, P., Manimanis, V.: A&A **405**, 263 (2003)
Pace, O.: In the three-dimensional universe with GAIA, 4–7 October 2004, Paris, ESA SP-576, 23 (2005)
Perryman, M., De Boer, K.S., Gilmore, G. et al.: A&A **369**, 339 (2001)
Perryman, M.: In the three-dimensional universe with GAIA, 4–7 October 2004, Paris, ESA SP-576, 15 (2005)
Rucinski, S.M., Lu, W.-X.: AJ **118**, 2451 (1999)
Rucinski, S.M., Lu, W.-X., Mochnacki, S.W.: AJ **120**, 1133 (2000)
Shaw, J.S., Guinan, E.F., Garasi, C.J.: BAAS **22**, 1296 (1990)
Van Hamme, W., Wilson, R.E.: A&A **152**, 25 (1985)
Zwitter, T.: In: Munari, U. (ed.) GAIA Spectroscopy, Science and Technology. ASP Conf. Ser. **298**, 329 (2003)
Zwitter, T., Munari, U., Marrese, P.M. et al.: A&A **404**, 333 (2003)

The *Kepler Mission*: Astrophysics and Eclipsing Binaries

D. Koch · W. Borucki · G. Basri · T. Brown ·
D. Caldwell · J. Christensen-Dalsgaard · W. Cochran ·
E. Dunham · T. N. Gautier · J. Geary · R. Gilliland ·
J. Jenkins · Y. Kondo · D. Latham · J. Lissauer ·
D. Monet

Received: 1 November 2005 / Accepted: 1 February 2006
© Springer Science + Business Media B.V. 2006

Abstract The *Kepler Mission* is a photometric space mission that will continuously observe a single 100 square degree field of view (FOV) of the sky of more than 100,000 stars in the Cygnus-Lyra region for four or more years with a precision of 14 parts per million (ppm) for a 6.5 hour integration including shot noise for a twelfth magnitude star. The primary goal of the mission is to detect Earth-size planets in the habitable zone of solar-like stars. In the process, many eclipsing binaries (EB) will also be detected. Prior to launch, the stellar characteristics will have been determined for all the stars in the FOV with $K < 14.5$. As part of the verification process, stars with transits (about 5%) will need to have follow-up radial velocity observations performed to determine the component masses and thereby separate grazing eclipses caused by stellar companions from transits caused by planets. The result will be a rich database on EBs. The community will have access to the archive for uses such as for EB modeling of the high-precision light curves. A guest observer program is also planned for objects not already on the target list.

Keywords Extra-solar planets · Planet detection · Eclipsing binaries · Precision photometry

Introduction

The *Kepler Mission* is the tenth principal investigator-led mission selected in the NASA Discovery program series. The primary objective is to detect Earth-size and smaller planets in the habitable zone of solar-like stars. The mission concept is given in detail by Borucki et al. (2005). In summary, *Kepler* is a photometric space-based mission designed specifically for finding habitable planets, that is, those between about one-half to ten Earth masses and in the habitable zone (HZ) of solar-like stars, that is, near 1 AU. The HZ is taken to be the distance from a star where liquid water can exist on the

D. Koch (✉) · W. Borucki · J. Lissauer
NASA Ames Research Center, Moffett Field, CA 94035

G. Basri
Univ. California-Berkeley, Berkeley, CA 94720

T. Brown
High Altitude Observatory, Boulder, CO 80307

D. Caldwell · J. Jenkins
SETI Institute, Mountain View, CA 94043

J. Christensen-Dalsgaard
Aarhus University, Denmark

W. Cochran
Univ. Texas at Austin, Austin, TX 78712

E. Dunham
Lowell Observatory, Flagstaff, AZ 86001

T. N. Gautier
Jet Propulsion Laboratory, Pasadena, CA 91109

J. Geary · D. Latham
Smithsonian Astrophysical Observatory, Cambridge, MA 02138

R. Gilliland
Space Telescope Science Institute, Baltimore, MD 21218

Y. Kondo
NASA Goddard Space Flight Center, Greenbelt, MD 20771

D. Monet
US Naval Observatory, Flagstaff, AZ 86002

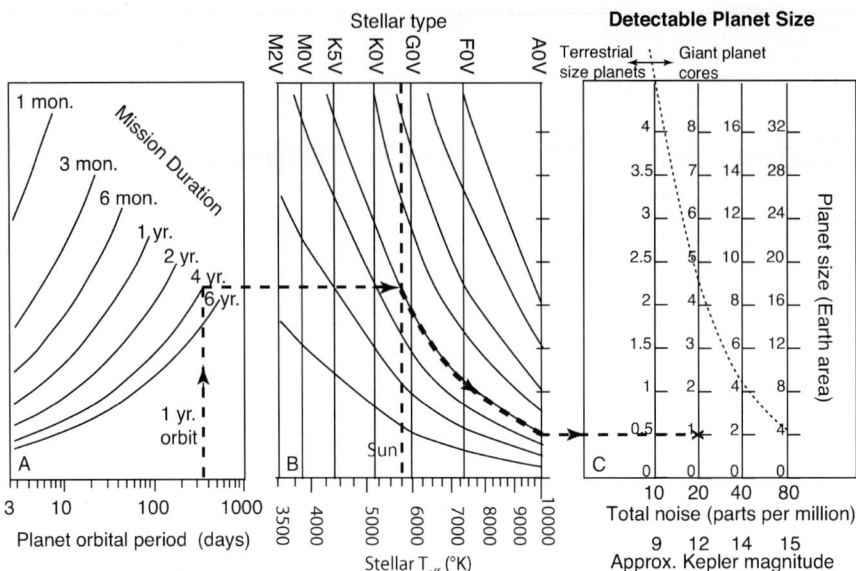

Fig. 1 *Detectable planet size*. The dashed line illustrates, that for a one-year planetary orbit, a four-year mission, a solar like star, G2V (T_{eff} = 5780°K), and a total noise of 20 ppm (including 10 ppm of stellar variability), that a one-Earth-size planet is detectable at 8σ. For other cases, select an orbital period in plot A and draw a vertical line up to the curve for the mission duration. Then draw a horizontal line to the desired stellar type or stellar effective temperature in plot B. The intersection of the horizontal line with the stellar type defines the "noise/size" curve to use or path to follow in between curves. Follow along the curve to the right edge of plot B. Draw a horizontal line to plot C to the appropriate total noise scale and read the value for the planet size. The dotted line in plot C is the approximate boundary between terrestrial-size planets and giant-planet cores

surface of the planet (Kasting et al., 1993). Although *Kepler* is optimized for the above case, its envelope for detection will clearly span many possibilities, especially the case for planets in shorter period orbits and for sizes significantly less than Earth. Figure 1 shows the detectable size dependence on 1) the planet's orbit, 2) the mission duration, 3) the stellar type and 4) total noise (1σ) defined as CDPP. The combined differential photometric precision (CDPP) includes shot noise, instrument measurement noise and an assumed stellar variability of 10 ppm in the time domain of a transit – a few hours to about a half of a day. The figure is based on equation (1) (Koch et al., 2004) for grazing transits that are only half as long as a central transit, detection of at least four transits, and a combined signal to noise ratio of 8σ for all the transits from a given planet detection. The figure should be applicable to any mission with continuous data and the appropriate CDPP.

Kepler will continuously and simultaneously observe over 100,000 main-sequence stars, using a one-meter Schmidt telescope with a FOV of greater than 100 square degrees and an array of 42 CCDs with 95 million pixels. The *Kepler* photometer (Fig. 2) will be the ninth largest Schmidt ever built and the largest ever launched into space. The spacecraft will be launched into an Earth-trailing heliocentric orbit. This allows for continuous viewing of the selected FOV throughout the orbital year. In the process of performing the search for extra-solar terrestrial-size planets, the mission will produce a photometric database of unprecedented precision, duration and number of stars. In addition, *Kepler* has the capability to perform photometric observations of other astrophysical objects that are within its FOV as part of a guest observing program.

The photometric capabilities

Kepler will continuously observe the same FOV for four to six years, except for less than one day every three months when the spacecraft needs to be rotated 90° to keep the Sun on the solar arrays and the radiator used to cool the CCDs pointed to deep space. Data will have fifteen-minute time resolution, except that for 512 objects a one-minute time resolution is available. This will be used primarily to obtain finer time resolution of high signal to noise ratio transits (a Jupiter-size transit will have an SNR = 800) and for measurement of p-mode oscillations in stars brighter than $R = 12$. The combined noise from the instrument, stray light, background stars, etc. is expected to be 6.6 ppm for a 6.5 h integration. The combination of observational noise and photon shot noise for $R = 12$ is expected to be 16 ppm in 6.5 h. Assuming a stellar variability in the time domain of a transit to be 10 ppm, yields a CDPP = 20 ppm. A broad bandpass from 430 to 890 nm (50%) was chosen to minimize the shot noise. The bandpass was also chosen to avoid the variability in Ca II H&K in the blue and fringing in the CCDs in the red. The point spread function FWHM is about 12 arc sec (one pixel is 3.98 arc sec) and was chosen to minimize the effects of pointing jitter

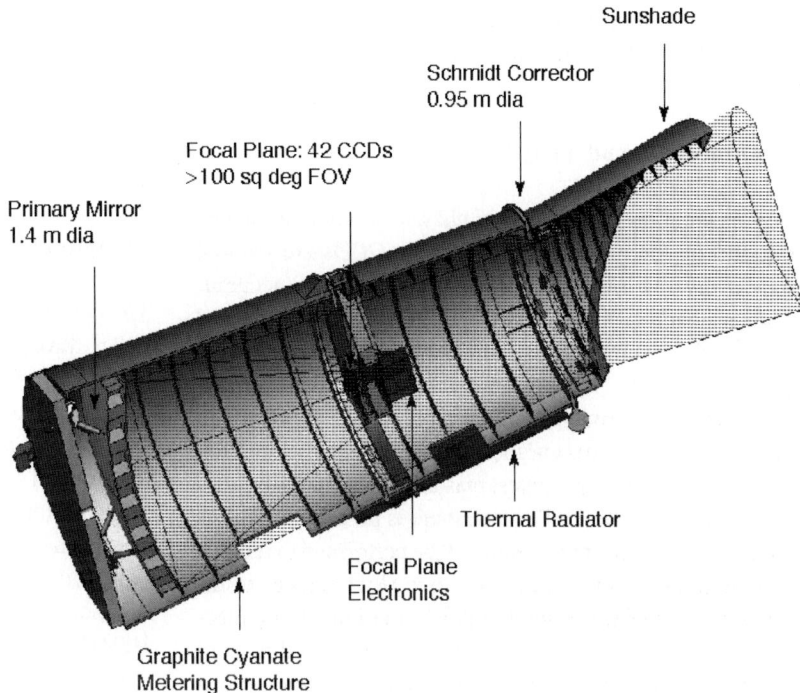

Fig. 2 Cross sectional view of the *Kepler* photometer

and sensitivity to spatial variations in the CCDs (Koch et al., 2000).

Field of view

Unlike almost all other astrophysics missions, the *Kepler Mission* will point to a single FOV for the entire mission. The objective was to select a single region with the richest field and to obtain the longest possible continuous exposure. The FOV choice was constrained to be above 55° ecliptic latitude, so as not to be blocked by the Sun at anytime during its orbit. The net result was a field slightly off the galactic plane centered at RA $19^h\ 22^m\ 40^s$ and declination $44°30'$. The positions of the individual CCDs on the sky are shown in Fig. 3. The layout is four-fold symmetric, which permits the 90° rotation while maintaining the same FOV on active silicon. Except for the central module, the orientation of CCD rows and columns is the same for all four rotations.

Stellar classification program

Given the scientific goal of finding planets around solar-like stars, we want to preferentially select late dwarfs of spectral types F, G, K and M. Since there does not exist a catalog that has luminosity classes to the necessary magnitude limits, the *Kepler* project is producing a catalog for the field of view. The stellar classification program (SCP) is making new multi-band photometric observations using the SDSS filter bandpasses plus a few additional filters to aid in estimating log(g). These data are analyzed using model fitting to derive, effective temperature, log(g) and radius, metallicity, extinction and reddening. The catalog has an observing limit to 18th magnitude and a classification limit of $K = 14.5$. The catalog will be federated with other catalogs, such as 2MASS

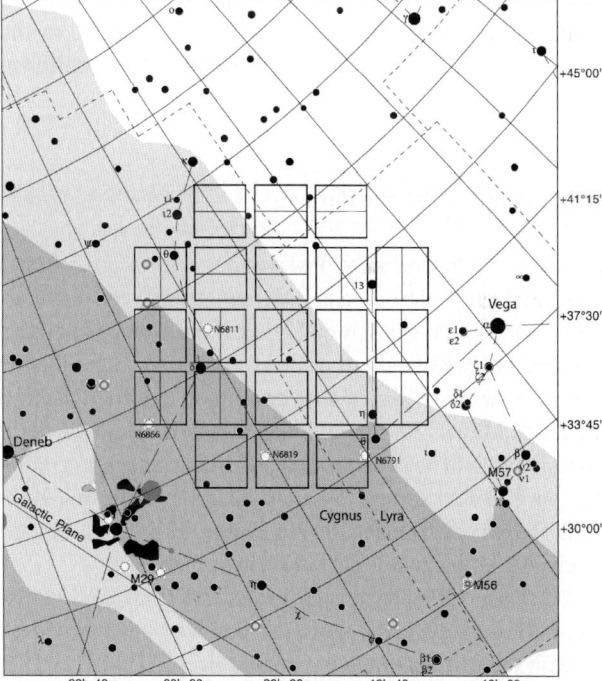

Fig. 3 Layout of the CCD modules within the FOV in the Cygnus-Lyra region

and USNO-B, to provide a cross reference. The catalog will be available to the community prior to launch.

Data processing and analysis

The data are processed in several phases on-board the spacecraft and on the ground. On-board, the CCDs are readout about every five seconds. All 95 megapixels are accumulated on board in memory for fifteen minutes. At the end of each fifteen-minute integration only the pixels of interest associated with each preselected star are stored and later transmitted to the ground. Data are requantized and Huffman encoded (compressed). On the ground, the data processing will remove cosmic-ray spikes, bias and the smear produced during readout of the CCD, since there is no shutter. Differential ensemble renormalization will be performed to remove common-mode signal variations due to sky background effects such as zodiacal light and gain variations in the electronics. The flux time series will be analyzed for threshold crossing events. These events will then be further scrutinized to determine which are consistent with transits and those that are simply false positives, such as grazing eclipsing binaries. Follow-up observations will be performed to further eliminate false positives, such as background eclipsing binaries.

Other results from the data

Follow-up radial-velocity observations with a precision of about 300 m/s will be used to differentiate between planetary transits and grazing eclipsing binaries. High-precision radial velocity observations will provide masses for giant planets and even for short period terrestrial mass planets (Rivera et al., 2005; Mayor, 2005). For binary cases, the high precision photometry will be capable of detecting extreme mass ratios, e.g., an A-dwarf orbited by a late M-dwarf with $M_2/M_1 \sim 0.02$. We expect to detect 1000 to 1500 eclipsing binaries and produce high precision light curves for them.

Other astrophysical results we expect from the data include: Measurement of p-mode oscillations, which provides a window to the stellar interior, namely, the stellar mass and age (Brown and Gilliland, 1994); Measurement of stellar activity yielding star spot cycles, white light flaring data and in particular the frequency of Maunder minimums of solar analogs, which has implications for terrestrial climatic changes and paleoclimatology; Measurement of stellar rotation rates providing an extensive data set for correlation with other measured characteristics especially those already provided from the *Kepler* stellar classification program; Measurement of cataclysmic variables providing data on pre-outburst activity, which is typically too low to have been previously seen;

and Measurements of active galactic nuclei variability at a much lower level, which has implications on "engine" size in BL Lac, quasars and blazers. Thus, *Kepler* will provide a wealth of photometric results that can be used to address a broad range of astrophysical questions.

Community access and participation

The community can participate in the mission in two ways, either through accessing the data archive or by proposing additional objects to observe.

All mission data will become available to the community through a data analysis program. Data will be accessible through the Multi-mission Archive at Space Telescope (MAST) for at least ten years after the end of the mission. The primary content of this archive will be the calibrated light curves for all objects observed during the mission life time, approximately 170,000 for at least the first year and 100,000 for the remainder of the mission. Data will be in the form of flux time series (light curves) for each object. Original pixel data will also be available to permit investigations such as astrometry or for those who prefer to conduct there own photometric processing. A data release policy has been developed to release mission data at the earliest time that allows for data calibration and validation and ensures against false-positive planetary detections. The first three-month data set will be released approximately one year after commissioning and then supplements will be released annually. Publicly released *Kepler* observations will be freely available to all interested parties for data mining from the MAST.

There will also be a guest observer (GO) program, whereby scientists can propose to observe objects that are not already being observed as part of the *Kepler* planet search. Particular objects of interest to the community need to be specified, since only the data for the pixels of interest for the selected objects will be transmitted, processed and archived. One can assume that *Kepler* will already be observing most late-type-dwarf stars down to about $R = 15$ and for M-dwarfs down to perhaps $R = 16$. All proposed objects must be in the *Kepler* FOV. The FOV will not be moved. A tool for finding whether an object is on or off of a CCD can be found on the *Kepler* web site at http://Kepler.NASA.gov/sci/GODAP.html. Objects can be brighter or fainter than the nominal dynamic range, which was chosen strictly for terrestrial planet detection reasons. The duration of the observation can be from three months to the length of mission based on the scientific justification. Up to 3,000 objects are set aside for the GO program at fifteen-minute cadence and an additional 25 can be assigned at one-minute cadence. All GO objects will be processed with the standard pipeline and then placed in the archive.

Summary

The *Kepler Mission* will observe more than 100,000 dwarf stars continuously for four or more years with a precision of 17 ppm (20 ppm when including 10 ppm for stellar variability) in 6.5 h for an $R = 12$ G2V star. The mission design is based on reliably detecting one-Earth radius planets in the habitable zone of solar-like stars. The mission will identify all eclipsing binaries in the FOV, with the expectation of detecting extreme mass ratio cases. The scientific community will have access to the entire mission data set with full photometric precision. The community can propose to observe additional objects within the field of view that are not part of the planetary detection program. Launch is planned for 2008.

References

Borucki, W.J., Koch, D., Basri, G., Brown, T., Caldwell, D., DeVore, E., Dunham, E., Gautier, T, Geary, J., Gilliland, R., Gould, A., Howell, S., Jenkins, J., Latham, D.: Kepler mission: design, expected science results, opportunities to participate. In: Livio, M. (ed.) A Decade of Extrasolar Planets around Normal Stars, Cambridge: Cambridge University Press, in preparation (2005)

Brown, T.M., Gilliland, R.L.: Asterioseismology. ARAA **32**, 37–82 (1994)

Kasting, J.F., Whitmire, D.P., Reynolds, R.T.: Icarus **101**, 108 (1993)

Koch, D.G., Borucki, W., Dunham, E., Jenkins, J., Webster, L., Witteborn, F.: CCD photometry tests for a mission to detect earth-size planets in the extended solar neighborhood. SPIE Conference **4013**, UV, Optical and IR Space Telescopes and Instruments, Munich, Germany, March (2000)

Koch, D.: A model for estimating the number of stars for which terrestrial planets can be detected using transits. IAU Symp 213 Bioastronomy 2002, Life Among the Stars, Great Barrier Reef Aust, Astron. Soc Pacific July (2002)

Koch, D., Borucki, W., Dunham, E., Geary, J., Gilliland, R., Jenkins, J., Latham, D., Bachtell, E., Berry, D., Deininger, W., Duren, R., Gautier, T. N., Gillis, L., Mayer, D., Miller, C., Shafer, D., Sobeck, C., Stewart, C., Weiss, M.: Overview and status of the Kepler Mission. SPIE Conf **5487**, Optical, Infrared, and Millimeter Space Telescopes, Glasgow, Scotland (2004)

Mayor, M.: Planetary statistics. In: Livio, M. (ed.) A Decade of Extrasolar Planets around Normal Stars, Cambridge: Cambridge University Press, in preparation (2005)

Rivera, E.J., Lissauer, J.J., Butler, R.P., Marcy, G.W., Vogt, S.S., Fischer, D.A., Brown, T.M., Laughlin, G., Henry, G.W.: A $\sim$7.5 earth-mass planet orbiting the nearby star, GJ 876. Astrophys. J. (2005) in press

STELLA: Two New Robotic Telescopes for Binary-Star Research

Klaus G. Strassmeier · The AIP STELLA team*

Received: 1 November 2005 / Accepted: 1 February 2006
© Springer Science + Business Media B.V. 2006

Abstract Several new robotic telescopes had or will see first light in 2005/2006 and are designed for either wide-field imaging, high-precision photometry or even for high-resolution echelle spectroscopy. These telescopes are in the 1–2 m class and therefore will focus on very specific tasks. Here, I present an update of the robotic STELLA facility currently under construction in Tenerife and emphasize its science capabilities for binary-star research. Among the many science applications of STELLA is the monitoring of magnetic activity in single and binary stars and their relation to age, rotation rate, metallicity and binarity per se.

Keywords Robotic telescopes · STELLA · Stellar activity · Active binaries

1. Introduction

STELLA is a robotic observatory with two 1.2 m fully automatic telescopes (STELLA-I and STELLA-II) located at the Teide Observatory of the IAC in Tenerife, Spain. Not only the telescopes are automatic but also the entire observatory, no human presence is needed for observing. STELLA-I supports a wide-field, multi-band CCD imaging photometer and an adaptive optics testbed while STELLA-II supports a high-resolution, fiber-fed and bench-mounted echelle spectrograph. Many references and articles related to robotic telescopes can be found in the proceedings of the Third Potsdam Thinkshop on "Robotic Astronomy" (eds. Strassmeier and Hessman (2004)) and we refer the reader to this source. The current paper updates the presentation by Strassmeier et al. (2004).

The central scientific objective is to understand the structure and dynamics of stellar surface activity and its impact on stellar evolution. The magnetic fields that permeate stellar atmospheres affects the transport of energy and momentum and give rise to surface features similar to the spots and plages we observe on the Sun. These features change with time in dependence of the cyclic behavior of magnetic activity. Reconstructing the spatial distribution of these features from high-resolution spectra is still a great challenge (see Strassmeier, this volume) and requires simultaneous photometric and spectroscopic observations. Together with a systematic determination of rotational periods in open clusters, time-series Doppler imaging is the core science program for STELLA.

2. An observatory for STELLAr activity

2.1. The telescopes

Both telescopes were build for AIP by Teleskoptechnik-Halfmann near Augsburg, Germany. STELLA-I is a 1.2 m f/8 Cassegrain system with a ZEISS Zerodur mirror, an alt-az mount, and two Nasmyth foci (see Fig. 1). One of the two foci has an image derotator and will be equipped with the Wide-Field STELLA Imaging Photometer (WIFSIP) based on a large-format CCD. The second focus currently feeds the STELLA Echelle Spectrograph (SES), but that will be moved to STELLA-II once it arrives at the end of 2005. The second focus of STELLA-I is reserved for testing a low-cost adaptive optics for later implementation of a NIR instrument.

*The AIP STELLA team members are T. Granzer, M. Weber, M. Woche, M. I. Andersen, J. Bartus, S.-M. Bauer, F. Dionies, T. Fechner, H. Korhonen, J. Paschke, E. Popow, A. Ritter, A. Schwope, A. Staude, A. Washuettl.

K. G. Strassmeier (✉)
Astrophysical Institute Potsdam (AIP), An der Sternwarte 16, D-14482 Potsdam, Germany (www.aip.de/stella)

Fig. 1 The STELLA-I telescope in its home on Tenerife (Nov. 2004)

The second telescope is scheduled for arrival in December 2005 and is principally a mechanical/electrical copy of STELLA-I. Because it is designed for just fiber feeding the SES through a modified Newton focus it uses comparably simple and effective optics. A spherical f/1.95 primary mirror qualified for space and produced by former ZEISS Jena allows a field-of-view of 2' with a three-lens focal extender (to f/10) and a corrector with one aspheric surface.

2.2. A self-focusing acquisition- and guiding unit for both STELLA telescopes

The acquisition- and guiding unit (AG unit) of a robotic telescope is naturally more complex than for a manual telescope. In the case for the SES it must assure that a star is positioned in the 80–160 μm (1.7–3.4″) diaphragm with sub-arc-second precision and then kept at this position for an hour. Our AG-unit is based on a grey beam splitter that separates $\approx$4% of the starlight out of the beam, reflects it back to overlap with itself and then is re-imaged through the beam splitter onto a single guiding CCD. The guiding CCD is a 2184 × 1072 6.8-μm pixel uncooled Kodak detector. This design allows to control the telescope focus with the same read-out cycle than for the guiding signal. STELLA-I will be equipped with a similar system but designed to work fully off axis outside the science field of view.

2.3. The STELLA control system (SCS)

The SCS is the central node of the control system (Fig. 2). Its duty is the delivery of commands in a correct timing sequence to the different subsystems. Error recovery is also part of the SCS. For ease of use, the SCS also provides access to variables of general interest like airmass or position of the actual target. The SCS sends commands to all other subsystems except the data archive. All of these devices react to the commands with replies. Replies are always sent through the identical channel as the command received. Errors occurring in the devices must be sent back over a dedicated error channel, regardless of their occurrence within a command cycle, e.g., moving of the telescope fails or within a passive phase, e.g., power-loss in the CCD readout-system, bad weather, a.s.o.. Further details are given in Granzer (2004) and Strassmeier et al. (2004).

3. Instrumentation

3.1. The STELLA Echelle Spectrograph (SES)

The STELLA Echelle Spectrograph (SES) is a white-pupil spectrograph with a fixed wavelength format of 390–860 nm (Fig. 3). The instrument is located in a separated room on

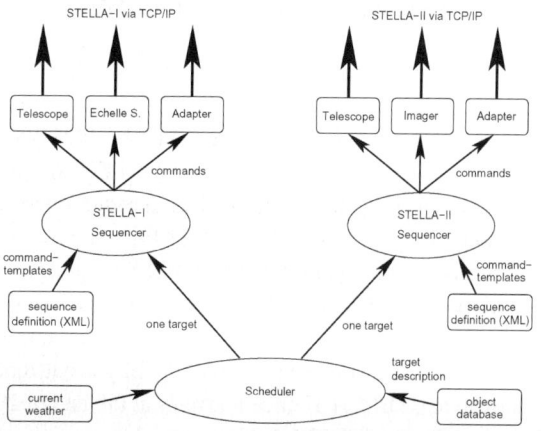 

Fig. 2 Left: Schematic description of the STELLA Control System (SCS). Right: The operator and archive consoles at the Robotics Center at AIP in Potsdam

Fig. 3 The SES in its thermal chamber during installation (June 2005)

a stabilized optical bench and is fed with up to two optical fibres, each 15 m long. A pair of either 50 μm or 100 μm core-diameter fibres will enable resolutions of either 50,000 or 25,000, corresponding to entrance apertures of 1.7" and 3.4", respectively. Its heart is a 31 lines per mm R2 grating from RGL. Two off-axis parabolic collimators, one folding mirror, and two prisms as cross disperser transport the light into the f/2.7 katadioptric camera with a 20 cm corrector and a 40 cm spherical mirror. Beam diameter is 130 mm. The CCD is a E2V42–40 2048 × 2048 13.5 μm pixel chip. Its quantum efficiency is 90% at 650nm, and 65% at 400 nm and 800 nm, respectively. Together with a second-generation CUO controller the read-out-noise is 3–4 electrons rms. A closed cycle cooler keeps the detector cooled to −130 degree Celsius. Because the original Marconi-E2V chip had a bias read out in the central detector area that were slightly above

Fig. 4 Design and construction of WIFSIP. Top left: overall design. Middle: the 24-hole filter wheel in the AIP mechanics workshop. Top right: Design detail with the CCD camera (left, in blue), filter wheel and shutter (middle), pick-off mirror for the off-axis guider and the schematic position of the guider CCD, and the telescope flange (right). Bottom left: field of view and CCD size. Bottom middle: The 4k × 4k CCD in the hands of Mike Lesser. Bottom right: the Kodak guiding CCD and the science dewar and camera head

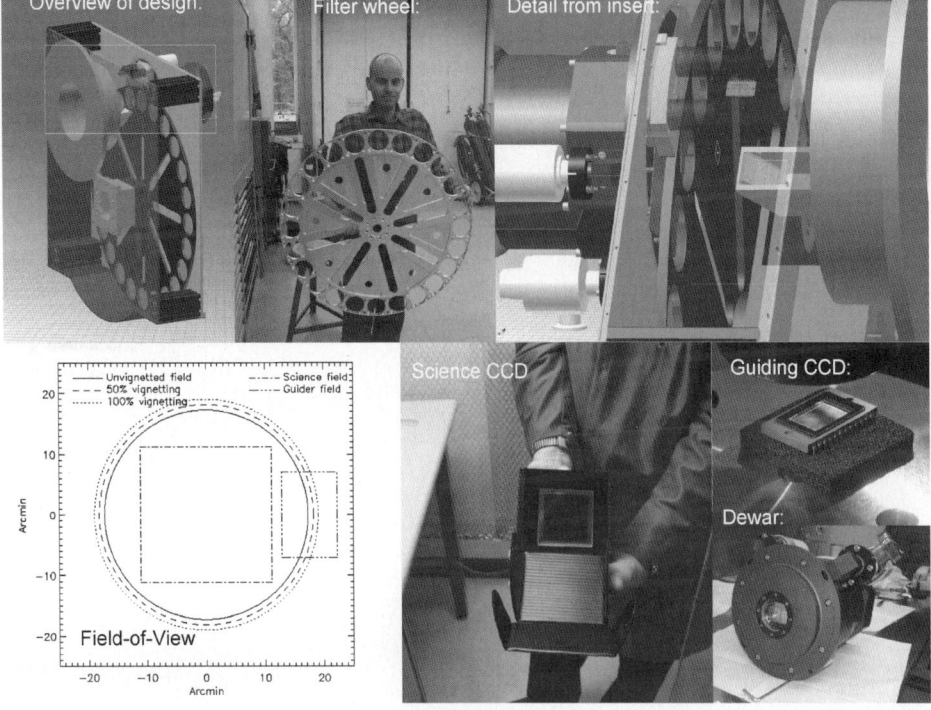

specifications, E2V kindly delivered a replacement without additional costs. This chip is now built into a copy of the first camera and is being used as a spare.

The magnitude limit is determined by the brightness-S/N product but also by the tracking of the telescope and the guiding capability of the AG unit. We estimate the practical limit to be around $V = 12$ mag with a S/N of 50:1 in 1.5 hours with a guiding limit near $I = 14$ mag.

3.2. The Wide-Field STELLA Imaging Photometer (WIFSIP)

WIFSIP is a single-channel optical CCD photometer with off-axis guiding. The science CCD is a monolithic 4096 × 4096 15-μm pixel back-illuminated CCD from Semiconductor Technology Associates and was thinned and AR coated in M. Lesser's Steward-Imaging Technology Lab. The field-of-view with a 3-lens corrector is $22' \times 22'$ with a scale of 0.3"/pixel. The photometer is equipped with 90 mm sets of Strömgren-$uvby$, narrow and wide Hα and Hβ, Johnson-Bessell UBVRI, and Sloan $ugriz$ filters. For time-series work, an external photometric precision of below 1 milli-magnitude is envisioned with Strömgren y and 3–5" defocusing for stars between 8.5–12 mag. Limiting magnitude will be around 19 mag in V with a guiding limit near 17 mag and an expected σ of 0.01 mag. An astrometric imaging mode is foreseen to allow the combination of many subimages into a very wide-field imager. Figure 4 shows some design and construction images of WIFSIP.

3.3. Automatic data reduction

The automatic data-reduction pipeline for SES is based on the NOAO Image Reduction and Analysis Facility (IRAF). It consists of a number of IRAF-CL scripts which are invoked by a master script. The reduction process is split into two parts. First, the bad-pixel correction, bias subtraction, scattered-light subtraction, cosmic-ray correction, flat fielding and aperture extraction are done at Teide Observatory on a local computer. The relatively small one-dimensional spectra (800kB) are automatically transferred to the AIP archive, where the second step continues with Thorium-Argon emission-line identification, wavelength calibration, radial-velocity measurement and normalization.

The WIFSIP pipeline is basically split into three modes: imaging, large-field photometry, and single-target photometry. In its current version it employs the GaBoDS image reduction package (Erben et al., 2005) for astrometrically calibrated imaging and the Aahrus University's MOMF package for photometry. Basic frame reduction is queued from a data base containing updated sky flats, biases, and darks, and is done automatically at the telescope and then transferred with lossy compression to the AIP. The original raw data (32MB per image) will be shipped regularly on DAT or similar and are then again reduced in Potsdam. In any case, a user will get pre-reduced data, but will additionally have the option to queue for re-reduction from the archive.

Acknowledgements This work is supported by the State of Brandenburg and the Federal Government of Germany.

References

Erben, T., Schirmer, M., Dietrich, J.P.: AN **326**, 432 (2005)
Granzer, T.: AN **325**, 513 (2004)
Strassmeier, K.G., Hessman, F.V.: AN **325**, 455 (2004)
Strassmeier, K.G., Granzer, T., Weber, M. et al.: AN **325**, 527 (2004)

Getting SMARTS on Novae: Highlights of the Early Evolution of Nova V475 Sct

Guy S. Stringfellow · Frederich M. Walter

Received: 30 September 2005 / Accepted: 7 October 2005
© Springer Science + Business Media B.V. 2006

Abstract There is a disconcerting global trend of retiring telescopes of modest aperture, supplanting them instead with fewer expensive telescopes of quite large aperture. As a consequence, the available time and feasibility of following transient objects in astrophysics is diminishing. We show the utility of having a suite of small to moderate aperture telescopes capable of conducting imaging and spectroscopic observations in a service queue mode. The example we provide is the high-cadence early observations of the classical nova V475 Scuti (2003) carried out with the SMARTS suite of telescopes located at CTIO.

Keywords Novae · Photometry · Spectroscopy · Light curves · Line profiles

1 Introduction

The study of transient eruptive stars, such as novae, are plagued by the lack of readily available facilities that can provide the multi-wavelength, high-cadence temporal coverage required to properly define and extract crucial physical properties that provide insight into the physical processes operating during, and the mechanisms responsible for, their transient behavior. Whether it be immediate activation of observations upon discovery, or long-term temporal photometric or spectroscopic coverage, access to telescopes of modest aperture equipped with imaging and spectroscopic instruments are evaporating. These smaller telescopes are being supplanted by fewer telescopes of very large aperture which are highly over subscribed and often too powerful to carry out studies of relatively bright transient objects, and for which less time is generally available. In this paper we show the rich data that can be acquired on transient objects provided that telescopes with the proper instruments, and the means for scheduling and conducting the observations can be performed in service queue mode, are available. Specifically, we present a sampling of photometry and spectroscopic results for the classical Nova V475 (2003) Scuti during the first many months of its outburst (Stringfellow and Walter, 2004).

2 Photometric light curves

We have an ongoing program studying the outbursts of classical novae using the suite of telescopes known as SMARTS.[1] The 1.3 m telescope (the southern hemisphere 2MASS telescope) has a dedicated optical and near-infrared imaging camera (ANDICAM) providing wavelength coverage in the UBVRIJHK bands. Nova Scuti 2003 was discovered by H. Nishimura from films obtained on the nights of August 28.58 and 29.436, and reported by Nakano and Sato (2003) on August 30. Our photometric observations of Nova Scuti 2003, subsequently designated V475 Sct, were immediately triggered and our first observation was obtained on August 31 (JD 2452882.5). Our complete set of UBVRI photometry covering the first 68 days are shown

G. S. Stringfellow
Center for Astrophysics & Space Astronomy, University of Colorado, 389 UCB, Boulder CO 80309-0389, USA
e-mail: Guy.Stringfellow@colorado.edu

F. M. Walter
Department of Physics and Astronomy, Stony Brook University, Stony Brook NY 11794-3800, USA
e-mail: fwalter@astro.sunysb.edu

---

[1] SMARTS, the Small and Medium Aperture Research Telescope Facility, is a consortium of universities and research institutions that operate the small telescopes at Cerro Tololo under contract with AURA.

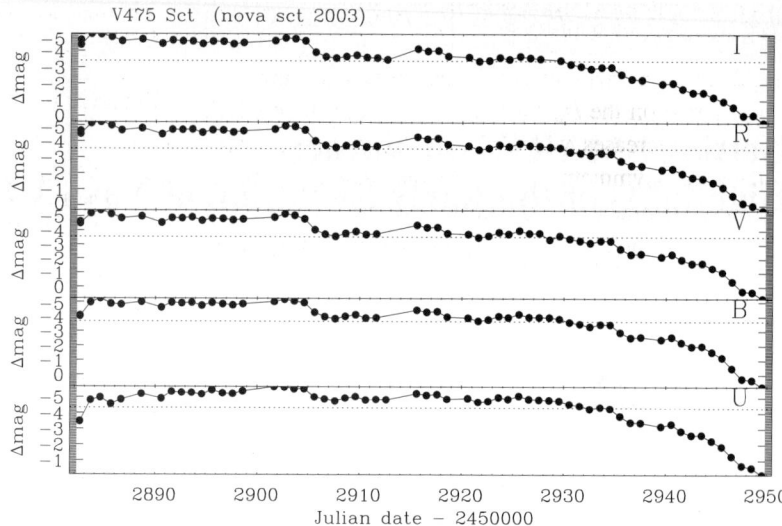

Fig. 1 UBVRI light curves for the first 68 days in the life of V475 Sct.

in Fig. 1, measured differentially with respect to field stars. Maximum light was reached on September 2 (mJD 2884.5) with $V \sim 8.36$, $B \sim 9.14$, and $B-V \sim 0.78$. Defining maximum light can be challenging for many novae, and V475 Sct is a prime example of this. For the next 17 days V475 Sct hovered around (but below) maximum light with modest variability. However, on September 19 the nova once again reached an equivalent maximum to that of Sept. 2, though the U-band was actually brighter. During this peak plateau the nova was producing a massive optically thick wind. Instabilities in the wind and the mass loss rate prior to discrete shell formation causes variations in the light curve. This behavior differs markedly from that exhibited by very fast novae, which narrowly defines a peak light from which rapid decline quickly ensues. In very fast novae the concept of peak light is less ambiguous, though often imprecisely measured or observationally missed entirely. In order to proceed with our analysis of V475 Sct we will take as peak light the first instance achieved on September 2.

There are many ways to determine the reddening, and consequently the distance and intrinsic bolometric luminosity (or magnitude) of novae; we provide one example calculation for illustration. Observations of novae at maximum light typically display an early F-supergiant (absorption) spectrum, which has an intrinsic color of $B-V \sim 0.23$. Assuming a standard reddening relation, this implies maximum of $m_V = 6.61$. The light curves are complex with apparent modulations, though discussion is deferred to Stringfellow and Walter (2006) (in preparation).

Downes and Duerbeck (2000) derive rate of decline relations using all classes of novae, and separate relations for very fast versus slower novae. Taking September 2 as maximum light, we obtain a rate of decline of $t_2 = 46 \pm 2$ days and $t_3 = 53 \pm 2$ days, placing V475 Sct into the moderately fast speed class (Warner, 1995). We then obtain absolute magnitudes in the V band of $M_V (t_2) = -7.08$ and $M_V (t_3) = -7.61$ using the relations for all speed classes. However, we can do better by considering which subclass V475 Sct fits into, based on the general behavior of the light curves alone. Downes and Duerbeck note the clear separation at $\log t_3 = 1.5$ in the (M_V, t_3) domain between novae with Super-Eddington luminosities (very fast ONeMg novae with large white dwarf masses; rapid declines with smooth light curves) and those near the Eddington limit (slower Fe II novae with smaller white dwarf masses; slower declines with irregularities). Duerbeck (1981) has created a simple classification scheme based on light curves, and V475 Sct fits in the B-class where decline of the light curve involves irregularities and fluctuations with plateaus and double maxima possible. When using relations derived for the slower subclasses (BCD, excluding A types) we obtain $M_V (t_2) = -7.00$ and $M_V (t_3) = -7.15$ – self-consistent, in excellent agreement. The values derived by Chochol et al. (2005) are consistent with those derived above. From these values the distance modulus in V is ~ 13.7, which implies a distance of ~ 5.5 kpc to V475 Sct.

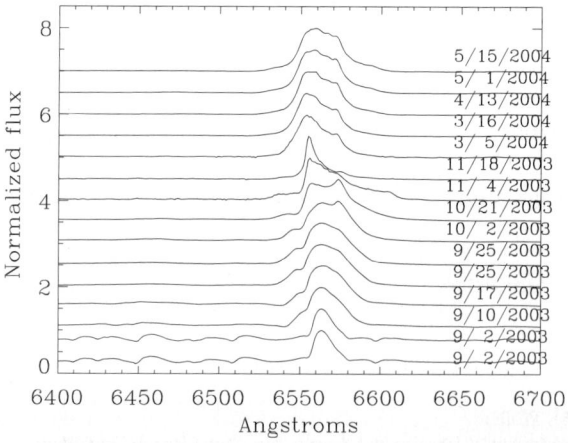

Fig. 2 Evolution of the H_α profile. The equivalent width continuously increases early on during formation of the wind and ejecta.

3 Spectroscopic development of the ejecta

Figure 2 shows a small subset of the spectra we have obtained centered on the H_α line, where the equivalent width continuously increases with time, and goes through various stages of asymmetry related to the underlying burning and establishment of the shell. High-cadence spectral data enable interpretation and modeling such complicated behavior.

Acknowledgements This work has been supported by NASA grant NAG5–11963 (GSS), and NSF grant AST-0307454 (FMW). Stony Brook's participation in SMARTS is made possible by support from the offices of the Provost and the Vice President for Research.

References

Chochol, D., Katysheva, N.A., Pribulla, T., Schmidtobreick, L., Shugarov, S. Yu., Skoda, P., Slechta, M., Vittone, A.A., Volkov, I.M.: Contributions of the Astronomical Observatory Skalnaté Pleso **35**(2), 107 (2005)
Downes, R.A., Duerbeck, H.W.: AJ **120**, 2007 (2000)
Duerbeck, H.W.: PASP **93**, 165 (1981)
Nakano, S., Sato, H.: IAUC 8190 (2003)
Stringfellow, G.S., Walter, F.M.: Bull. Am. Astron. Soc. **36**, 776 (2004)
Warner, B.: Cataclysmic Variable Stars. Cambridge: Cambridge Univ. Press (1995)